International Conference: Actual Issues of Mechanical Engineering (AIME 2020)

IOP Conference Series: Materials Science and Engineering
Volume 1111

Online
27-29 October 2020

Editors:

Vladimir A. Zhukov
Pavel A. Garibin
Vakhit Kh. Akaev
Artyom A. Butsanets

ISBN: 978-1-7138-4602-4
ISSN: 1757-8981

Printed from e-media with permission by:

Curran Associates, Inc.
57 Morehouse Lane
Red Hook, NY 12571

proceedings.com

Some format issues inherent in the e-media version may also appear in this print version.

Institute of Physics
Dirac House, Temple Back
Bristol BS1 6BE UK

Phone: 44 1 17 929 7481
Fax: 44 1 17 920 0979

techtracking@iop.org

Additional copies of this publication are available from:

Curran Associates, Inc.
57 Morehouse Lane
Red Hook, NY 12571 USA
Phone: 845-758-0400
Fax: 845-758-2633
Email: curran@proceedings.com
Web: www.proceedings.com

Determination of stripping zones and axle spacing between the cylinders in two-cylinder stripping reaper (header) 211
V A Sakharov, A A Kuvshinov

Problem solution of optimal pathfinding for the movement of vehicles over rough mountainous areas 219
K Zhigalov, D K-S Bataev, E Klochkova, O A Svirbutovich, G A Ivashchenko

Planning model of a group UAV flight in three-dimensional space 229
N A Maksimov, K Zhigalov, A V Gorban, I V Ignatev

Economic efficiency of increasing the reliability of electric transport in urban transport systems 237
M K Chernyakov, K Ch Akberov, B V Malozyomov, V Yu Listkov, I Yu Kharlamova, E N Saricheva, I A Shuraev

Classification of heavy high-viscosity oils for improving the efficiency of reservoir development 243
I Sh Mingulov, V V Mukhametshin, Sh G Mingulov, L S Kuleshova

Techniques for optimization of gas extraction from production wells annulus 249
Sh G Mingulov, I Sh Mingulov

Developing a drilling fluid compensator for increasing the technological condition of drilling equipment 254
T N Minnivaleev, A N Minnivaleev

Methods for increasing the efficiency of hydrochloric-acid treatments of wells 260
V V Mukhametshin, L S Kuleshova

Use of criticality principle in information protection systems at machine-building enterprises 265
K Yu Zhigalov, R R Turluyev, O V Dudareva

The effect of the grain size on the cycle life of spring steel 50 CrMnV 270
V G Pachurin, V V Galkin, A A Filippov, G V Pachurin, M V Mukhina, Zh V Smirnova

Rationale for choosing the sound-absorbing materials for the operator's cabins of the rail-grinding machine 275
A A Nazaretov, I A Yaitskov, N M Yaitskova

Liable Bluetooth tracking technology for enhancement of location-based services 282
I M Daudov, Zh L Gavrilova, V A Kudashkin

Influence of plasma arc power on the melting of the envelope layer containing NiCrBSi alloy and the surface coating quality during plasma heating 288
Van Chie Nguyen, A E Balanovskiy, S K Kargapol'tsev, V E Gozbenko, Yu I Karlina, M G Shtayger, A I Karlina

Modelling the movement of log trucks with a hydro-mechanical transmission 295
E A Pitukhin, V I Skrypnik

Use of hydraulic drive of sucker-rod pump 120-6-24 in oil and gas industry as an alternative to import substitution of small-sized pumps 301
I A Pogrebnaya, S V Mikhaylova

Improvement of heat transfer of liquid-oil heat exchangers for internal combustion engines 307
Iu S Prikhodko, D A Akhmetshin, Ya S Bolkov, A N Shishkov

Modeling the sunflower seeds separation process in air-sieve grain-cleaning machines 314
I E Priporov, V S Kurasov, E E Samurganov, A B Shepelev

Optimization of the sunflower seeds extrusion process in obtaining sunflower cake 320
I E Priporov, A B Shepelev

Developing optoelectronic devices based on AFV properties to control the physical and chemical parameters of oil products 326
N Rakhimov, V Sh Mukhametshin, K T Tyncherov

Development of digital models and information systems for machine-building enterprises........................... 330
K Yu Zhigalov, Kh A Daudov, A V Gorban

Modern equipment for core sampling while drilling wells in the fields of Tomsk region 334
N N Saveleva, S N Shed

Comparative study of state of charge estimation algorithms for Lithium-Ion Battery 339
N Shamarova, P Komarnicki, C Wenge

The development of microwave installations for separating the fluff from skins of rabbits in the agricultural enterprises of low power 345
E Shamin, G Novikova, O Mihkailova, M Prosviryakova, N Kirillov, V Storchevoj, S Zaitsev

Positioning automation when assembling panel structures 352
M V Kovalevich, E S Shemonaeva

Correlation analysis to determine the relationship between the geometric dimensions and the electromagnetic moment of a switched reluctance motor... 359
A V Shevkunova, A S Shapshal, T Z Talakhadze, S A Shapshal

Methods for determining endurance limits for welded joints of marine stationary platforms......................... 367
I V Starokon

Adjustment of mud pump parameters by the frequency conversion method.. 374
R I Suleimanov, M Ya Khabibullin, A Yu Davydov, L M Zaripova

Selection of optimal operating modes for a block cluster pumping station... 381
R I Suleimanov, M Ya Khabibullin, L Z Zainagalina, L M Zaripova

Mathematical modeling of the sunflower cake grinding process 386
A B Shepelev, I E Priporov

Features of information visualization in coal quarrying... 393
K Yu Zhigalov, R R Turluyev, T Ya Druzhinina

Models and classifications of secure resource management methods in distributed information communication networks.. 397
R R Turluyev, M Yu Ivanov, G M Beregova

Increasing the efficiency of the wireline in the drilling rig lifting complex 405
R I Suleimanov, M Ya Khabibullin, L M Zaripova, A Yu Davydov

Application of the least square method for calculating the optimal well profile 412
F K Usmanova, L Z Usmanova

Transmission type parallel to ship with hybrid power plant.. 418
R Yu Dobretsov, S B Dobretsova, A I Vasiliev

Design and technological support for the assembly of long thin-walled buildings 424
A S Yamnikov, E N Rodionova, I A Matveev

Automation security in a machine-building plant.. 430
K Yu Zhigalov, E R Guzueva, S I Polovneva

Technology for prediction of salt deposition in oil production 436
L E Lenchenkova, L Z Zainagalina, N D Bulchaev, Yu A Kotenev, F Z Bulyukova, A R Safiullina

Selection of the drilling fluid density as specified with the expected deformation of the cross-section of the well.. 441
L Z Zainagalina, F Z Bulyukova

Author Index

PREFACE

International Conference "Actual Issues of Mechanical Engineering–2020" was successfully held at Admiral Makarov State University of Maritime and Inland Shipping (Saint-Petersburg, Russia) on 27-29 October, 2020. Because of the travel restrictions and to avoid public gathering, it was decided to hold the conference online. This gave participants from across Russia an opportunity to present their reports despite the lockdown. The geography of AIME–2020 includes the researchers from Moscow, Saint Petersburg, Ufa, Irkutsk, Nizhniy Novgorod, Rostov, Bryansk and other Russian cities.

AIME–2020 became a scientific forum to discuss various issues of mechanical engineering, such as: automotive and locomotive engineering, farming machinery, water transport, special purpose civil and military vehicles. Researchers and practitioners presented their papers on construction equipment, reaction engines, power and heat generators, machinery manufacturing automation, and machinery lifecycle management. After careful reviewing, 80 papers were accepted for publication.

Due to the complicated epidemiological situation and the travel restrictions applied by the Russian Government, the conference was held remotely. The work of the sections was organized in a video conference mode, where the most relevant reports selected by the organizing committee of the conference were presented using the Cisco Webex platform (https://www.webex.com/downloads.html). The traditional 15-minute time limit on speeches was preserved. The attendance for virtual participation without the report and without the opportunity to participate in discussion with the voice was free of charge. However, all participants could take their part in the discussion via moderators.

We express our sincere gratitude to the Organizing Committee of AIME–2020 as well as to all the participants, reviewers and editors for their significant contribution in preparing the conference proceedings. We hope that this volume will be of great interest for specialists in the field of mechanical engineering and non-specialist readers.

With the best regards
On behalf of the editorial board,
Artyom A. Butsanets

IOP Conf. Series: Materials Science and Engineering 1111 (2020) 011001 doi:10.1088/1757-899X/1111/1/011001

EDITORIAL BOARD

PROGRAMME COMMITTEE

ORGANIZING COMMITTEE:

IOP Conf. Series: Materials Science and Engineering 1111 (2020) 011002 doi:10.1088/1757-899X/1111/1/011002

Peer review declaration

All papers published in this volume of *IOP Conference Series: Materials Science and Engineering* have been peer reviewed through processes administered by the Editors. Reviews were conducted by expert referees to the professional and scientific standards expected of a proceedings journal published by IOP Publishing.

- **Type of peer review:** Single-blind peer review
- **Conference submission management system:** submissions were received and handled via e-mail
- **Number of submissions received:** 159
- **Number of submissions sent for review:** 119
- **Number of submissions accepted:** 80
- **Acceptance Rate (Number of Submissions Accepted / Number of Submissions Received X 100):** 50%
- **Average number of reviews per paper:** 2
- **Total number of reviewers involved:** 28
- **Any additional info on review process:** all the papers were reviewed
  - by the member of program committee;
  - by the external referee (chosen by the editorial board among the authors or staff of organizing institutions).

  The applied review criteria are described below.

  - Does the contribution fit the scope?
  - Does the contribution feature original research?
  - Are the results and conclusions adequate?
  - Does the presentation style meet academic standards?
  - Is the referencing appropriate?

  For each question reviewer had to choose one reply from the list. There were 3 variants:

  - No
  - Not sure
  - Yes

All the reviewers were supplied with the link to the questionnaire made with the Survey Monkey web kit. The questionnaire had two text fields that could be used to submit short communications (up to 500 words) to the editors or authors. All individual responses are immediately become accessible to the member of the editorial team. After this, the editor has to make a decision about the papers based on the reviewers' feedback. Then the editor relays the decision and remarks to the authors.

Reviewers submit

- **Contact person for queries:** Artyom Butsanets, secretary of the editorial board, head of the department of intellectual property and technical information, Admiral Makarov State University of Maritime and Inland Shipping, Saint Petersburg, Russia, ButsanetsAA@gumrf.ru

Development of the performance control algorithm of the blower motors of electric locomotives for various operating modes

A R Akhmetshin[1], K V Suslov[2], N P Astashkov[3], V A Olentsevich[3], M G Shtayger[4] and A I Karlina[2]

[1] Kazan State Power Engineering University, Kazan, Russia
[2] Irkutsk National Research Technical University, 83, Lermontov str., Irkutsk, 664074, Russia
[3] Irkutsk State Transport University, 15, Chernyshevsky str., Irkutsk, 664074, Russia
[4] MC Mechel Steel, 1, Krasnoarmeysky Street, Moscow, 125167, Russia

E-mail: souslov@ex.istu.edu

Abstract. The widespread use of electric motors in various electric drive systems and the automatic regulation of their performance are largely determined by the reliability of these machines, the failures of which cause significant material damage and are caused by significant yard time at the depot repair positions. Improving their reliability is a major scientific and technological challenge. When electric locomotives operate on the East Siberian Railway, severe climatic conditions have a direct impact on the reliability of electrical equipment, so the development of an algorithm to control the performance of blower motors for various operating modes is an urgent task. The main parameters of the entire electric locomotive energy system, which largely determine the safety of trains and the amount of throughput capacity of the sections, are the reliability of cooling machines and the algorithm of their operation in different operating modes, taking into account weather and climatic conditions. The immediate accounting and analysis of the causes of the failure of asynchronous electric motors with a short-circuited rotor in the blower drive, the use of which is justified by the simplicity of the design, was the basis for the development of the algorithm proposed in this article. The electric locomotive cooling control algorithm is focused on eliminating or minimizing snow, dust insulation of traction electrical equipment in a given range of regulation. Elimination of the unauthorized shutdown of cooling machines by locomotive crews while the train runs in the slow-down mode, as well as the elimination of the start processes of asynchronous electric motors of the blower motor actuator, will improve the reliability of the entire electric locomotive in general.

1. Introduction

The increase in the throughput and operating capacity of the transport infrastructure, caused by the increase in the volume of transport by rail, leads to physical and functional obsolescence of fixed assets, which requires their strengthening and constant control in strict operating conditions. In order to ensure the safe and uninterrupted organization of transport infrastructure facilities, special attention should be paid to the work of electrical machines, the quality of which can be presented as a set of properties that determine their suitability for normal operating conditions. Due to the fact that reliability is the most important both technical and economic indicator, characterizing the quality of operation of any electrical machine, it allows determining its ability to work smoothly during a given period of time without changing the technical characteristics under certain operating conditions [1-4].

With a wide range of electrical vehicles in various electric drive systems and automatic regulation, production is largely determined by the reliability of these machines. The failures of electrical machines cause significant material damage, so increasing their reliability is the most important scientific and technical problem [5].

The algorithm of operation and reliability of the blower motors used directly affect the parameters of operation of the entire electric locomotive energy system and largely predetermine the safety of trains. The substantiation of the use of asynchronous electric motors with a short-circuited rotor in the blower drive is the simplicity of the design, which affects not only in reducing the time of its repair, but also the yard time of the electric locomotive in general on the positions of depot service. The long-term mode of blowers' operation determines the high requirements for the reliability of the considered electric motors in question.

When electric locomotives operate on the East Siberian Railway, severe climatic conditions have a direct impact on the reliability of electrical equipment, so the development of an algorithm to control the performance of blower motors for various operating modes is an urgent task.

Abroad, when choosing the most profitable offer for the supply of traction rolling stock, the concept of the maintenance of the locomotive at all stages of its life cycle is increasingly used. The Japanese firm Hitachi considers the maintenance system of new locomotives to be the key to guaranteeing a high level of reliability. However, as the world experience shows, now, on average, up to 40% of the time is spent on troubleshooting, rather than on routine maintenance [6, 7].

2. Analysis of the reliability of traction electrical and auxiliary machines

The distribution of failures of assemblies and equipment of AC locomotives based on depot statistics shows that auxiliary machines account for up to 15%, and up to 30% of the total damage to traction electrical machines. It should be noted that the cost of repairing the traction electrical machines compared to other electric locomotive equipment is the most expensive. In addition, the complex process of repair, which involves rolling out the wheel-engine unit, is reflected in the time of the electric locomotive's yard time in the repair positions, bringing huge losses to the company, both temporary and financial. The reliability of the electric part of the traction electrical machines is emphasized due to the algorithm of cooling machines in operation.

The number and characteristics of failures of traction electrical machines are divided into liability categories for the companies that repair and operate them. Failures, the responsibility of which is attributed to the Russian Railways OAO, are preconditioned by the following operational violations:

- overheating of insulation due to the speed below the estimated limiting ascents with full-weight trains;
- spinning and slipping of wheel pairs, in which there may be overexertion on the traction electrical machines collector and the electric arcing;
- disabling the blower motors by locomotive crews in the slow-down mode to save power for the drive of auxiliary machines.

During the blowers' operation, excessive pressure is created in the body of the electric locomotive, which prevents dust or snow from getting on the traction electrical equipment of the electric locomotive. This is taken into account in the proposed algorithm of cooling machines. The cooling air enters the traction engine from the side of the collector, through the vent, passes through the ventilation channels in the armature and through the air gap between the armature and the main poles of the traction electrical motor. The air exits from the traction engine from the opposite collector, through a special casing, up under the body of the electric locomotive.

When the electric locomotive runs in the slow-down mode with the disconnected blowers there is a swirl of dust, snow, and a loose adjoining of the covers of the lower and upper viewing hatches, the presence of a technological hole in the frame of the traction engine and air gap lead to their ingress to the main and additional poles, compensatory winding, collector and the brushing gear of the traction electrical machine.

It should be specially emphasized that moisture and its properties largely predetermine the life of the insulation of traction electrical equipment, as its ingress into the pores reduces not only the resistance and dielectric strength, but also leads to the emergence of leaks, partial discharges and other phenomena, the impact of which increase the probability of a breakdown. The impregnation does not always protect the material from damping, but only reduces the rate of moisture absorption [8, 9].

The presence of moisture in isolation leads to a sharp decrease in leakage resistance, which leads to an increase in dielectric losses. As a result, the heat breakdown voltage is reduced and the insulation is further heated, which leads to an acceleration of the rate of thermal aging [10]. Strong insulation damping can affect the dielectric permeability of the moisturized layers. If it can be unevenly moisturized, this circumstance can distort the electric field in insulation and reduce the disruptive voltage [11]. The use of energy-intensive calorific furnaces allows removing moisture from insulation by drying. However, this process is not only energy-intensive, but also requires a lot of yard time as part of the depot repair.

Under the influence of solid particles of dust penetrating the machine, there is an abrasive wear of the insulation of windings. When following the electric locomotive in the slow-down mode with the disabled blowers, the withdrawal of lubricants from motor-armature bearings is possible. This process causes the deterioration of insulation parameters due to the grease ingress in the frame of the traction engine. In addition, due to the small amount of lubricating material in motor-armature bearings, the armature shaft may be jammed. The thermal current relays and overload relays used do not fully protect the blower motors, as evidenced by the failure statistics of these electric motors. Based on the data obtained on the failures of the auxiliary machines of the locomotives of the East Siberian Railway, it has been established that the electrical damage account for up to 55% of the total damage to the electric motors under consideration.

3. Development of the automatic control system for the auxiliary machines of a locomotive

As part of the presented work, the authors examined the haul distance of the circulation of electric locomotives attributed to the operating locomotive depot Nizhneudinsk (Figure 1) with a total length of 2,582 km.

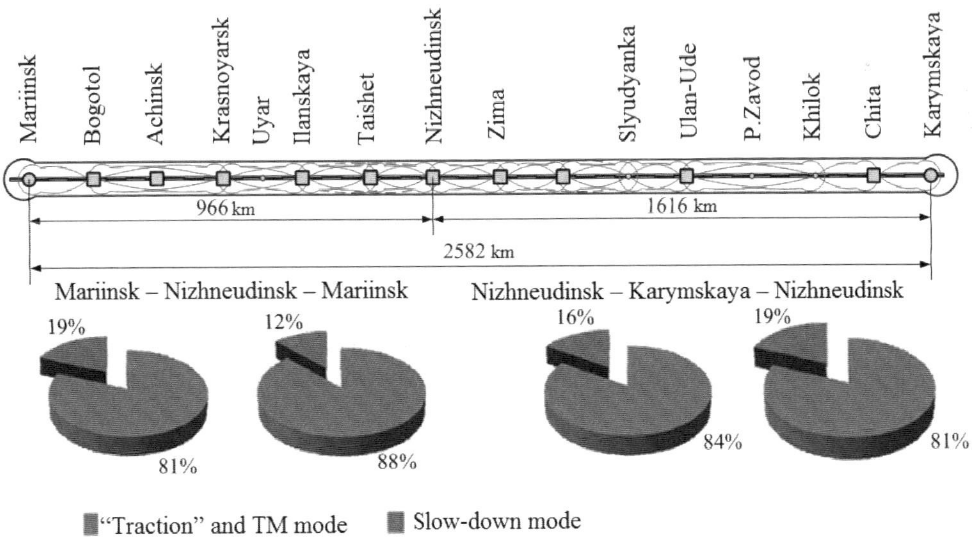

Figure 1. Haul distances of electrical locomotives attributed to the operating locomotive depot Nizhneudinsk

According to the train driving mode charts and decryption of the registration files of the parameters of the automatic brake control system, the order of the train running on the Mariinsk-Karymskaya section in both directions in different operation modes of the electric locomotive was analyzed.

On the Nizhneudinsk-Mariinsk section, when running with down trains, the electric locomotive runs for an average of 16 hours 25 minutes and is in the slow-down mode for 3 hours 39 minutes (12%). On the Mariinsk-Nizhneudinsk section, when running with up trains, the train runs for 16 hours and is in the slow-down mode for 3 hours 40 minutes (19%).

On the Nizhneudinsk-Karymskaya section, when running with up trains, the electric locomotive runs for an average of 39 hours 50 minutes and is in the slow-down mode for 6 hours 30 minutes (16%). On the Karymskaya – Nizhneudinsk section, when running with down trains, the train runs for 31 hours 10 minutes and is in the slow-down mode for 5 hours 50 minutes (19%).

The data of the analysis allow concluding that it is advisable to control the performance of auxiliary cooling machines for traction electrical equipment in the slow-down mode.

The development of an automatic control system (ACS) for auxiliary machines of an electric locomotive with adaptation to operating conditions is aimed at eliminating the ingress of snow and dust on the insulation of traction electrical equipment. The operation of an asynchronous electric motor in a controlled electric drive on an electric locomotive is characterized by significant features, which determine the specific requirements imposed on them [12]. These features are associated with varying values of the RMS voltage and current consumption of the motor. Due to this, mathematical models of electromagnetic, electromechanical, and energy processes in steady-state and transient operating modes of motors are specific.

To technically implement the proposed microprocessor-based ACS and create a mathematical model that takes into account not only environmental parameters, but also the operating modes of the electric locomotive, it is necessary to take into account:

- limiting the minimum amount of cooling air during the slow-down mode and stops at intermediate stations to create excess pressure in the body of the electric locomotive;
- heating temperatures of the insulation of traction electrical equipment of an electric locomotive when implementing the algorithm for controlling the performance of blower motors.

The basic design principles of the developed control system for cooling machines should take into account the parameters of their operation as part of an electric drive for various operating modes, taking into account the positive aspects of the developed and operated systems, taking into account their shortcomings.

It is necessary to use a systematic approach to analyze the operation of an asynchronous motor in interaction with other elements of the power system:

a) circuit implementation of the power part of the semiconductor converter;

b) an algorithm for controlling traction electrical equipment cooling machines, adaptive to operating conditions;

c) datasheet specifications of the electric motor, taking into account the design parameters of the blower wheel to estimate the output value of the system.

The elements of an electric drive should be implemented in a mathematical model, taking into account individual characteristics to implement an effective system approach when designing the system being developed. The correctness of the calculation of the parameters of the individual links of the system will make it possible to assess the adequacy of the mathematical model of the system to real physical processes. It should be noted that a change in one of the parameters of the system will affect the parameters of the output signal, which must be adequately changed depending on the steady-state value of the insulation temperature of traction electrical equipment.

Guided by the task set within the framework of the article, when developing an algorithm, it is necessary to take into account:

- providing soft start of blower motors at a reduced frequency;
- control of the amount of current in traction electrical equipment using a current sensor;

- duplicated control of the temperature mode of traction electrical equipment through the use of temperature sensors with reasonable settings for switching cooling machines from one frequency to another.

National and international experience in operating systems for controlling the performance of cooling machines for electric locomotives made it possible to determine the disadvantages of their use – low energy indicators of frequency converters PChF and SAUV. It should be noted that a threefold decrease in the performance of blower motors is reflected in the reliability of the insulating materials of the traction electrical equipment of the electric locomotive, due to the ingress of snow and dust on its surface. In addition, the frequent switching of cooling machines from one frequency to another is due to the insufficient amount of cooling air for unloaded operation.

The boundary of creating excessive pressure in the body of an electric locomotive to eliminate the ingress of snow and dust on the insulation of traction electrical equipment begins with a reduced rotation frequency of the cooling machine shaft of at least 25 Hz.

Guided by the well-known analytical dependencies, it follows that the air consumption for electric motors with a blower-type load is proportional to the power consumption. Productivity Q, head H and load moment M on the shaft of the cooling machine depend on the speed of rotation of the impeller n, formulas 1, 2.

$$\frac{Q}{Q_n} = \frac{n}{n_n} \tag{1}$$

$$\frac{H}{H_n} = \frac{M}{M_n} = \frac{n^2}{n_n^2} \tag{2}$$

In formulas 1, 2, the index "n" refers to the parameters of the nominal operating mode.

The calculation of the useful power P on the shaft for different modes is determined by the following relationship, formula 3.

$$\frac{P}{P_n} = \frac{n^3}{n_n^3} \tag{3}$$

It should be noted that the steady-state value of the temperature of traction electrical equipment in order to ensure its normal thermal mode, taking into account its current load and the temperature of the cooling air, is fundamental in regulating the performance of cooling machines.

The dependences of the required air supply on its temperature obtained in the works [13, 14] indicate the possibility of using a control system for cooling machines of an electric locomotive, since at 0 °C the supply of cooling air can be reduced by 13-38%, and at a temperature of -40 °C – by 28-60%, taking into account the parameters of the cooled equipment.

The required supply of cooling air, depending on the change in current load, in the reference literature is determined from the conditions of thermal equilibrium in the steady state, formula 4.

$$k \cdot I^2 \cdot r = \rho \cdot c \cdot \tau_\infty \cdot Q^n \tag{4}$$

where k is a coefficient that takes into account heat losses; I is the current strength in traction electrical equipment, A; r is the active resistance depending on temperature, Ohm; ρ is the air density, kg/m^3; c is the heat capacity of ambient air; τ_∞ is the steady-state value of the excess of the machine winding temperature over the ambient temperature, °C; n is the coefficient for power equipment.

Required air supply when the current strength is different from the nominal, formula 5.

$$Q = Q_n \cdot \left(\frac{I}{I_n}\right)^{\frac{2}{n}} \tag{5}$$

The values of the coefficient n for the main power electrical equipment are shown in Table 1.

Table 1. Numerical values of the coefficient n

	Electrical equipment		
	of the traction electrical machine, reversible converter, smoothing reactor	Traction transformer	Power resistors
n	0.8	0.6	0.72

The current load of electrical equipment of AC electric locomotives is unstable and depends on the following random factors: train weight; section profile; conditions of wheel-rail adhesion; the nature of the train environment. The ability to regulate the performance of blowers can provide a static analysis of the current loads of electric locomotives and the calculation of the total operating time in the slow-down mode.

To justify the control algorithm in the slow-down mode and stops at intermediate stations, it is necessary to take into account the permissible temperature rise of the insulation of conductive parts.

The presented well-known formulas 4-5 do not take into account the modes of operation of an electric locomotive when slowing down and stopping at intermediate stations, which implies the introduction of an additional coefficient and a minimum amount of air, which will prevent snow from entering the traction electrical machine due to the shutdown of the blower motors, taking into account the magnitude of the steady-state value of the temperature rise of the traction winding machines, formulas 6, 7.

$$k_{\text{slo}} + k \cdot I^2 \cdot r = \rho \cdot c \cdot \tau_\infty \cdot Q^n \quad (6)$$

$$Q = Q_{\text{slo}} + Q_n \cdot \left(\frac{I}{I_n}\right)^{\frac{2}{n}} \quad (7)$$

On the basis of the above mentioned calculations, using Matlab computer mathematics, a block diagram of the blower motors control was developed (Figure 2).

At the instant when the current in the traction electrical equipment became equal to 0A, the control algorithm of the blower motors was determined by the parameters of the technological process of the routine diagram. The proposed algorithm is possible when the train follows a favorable profile.

Based on the obtained dependences, it can be concluded that the steady-state value of the temperature regime for the presented control algorithm tends to the value of the environment. This allows drawing a conclusion about the adequacy of the mathematical model to real physical processes [14-15].

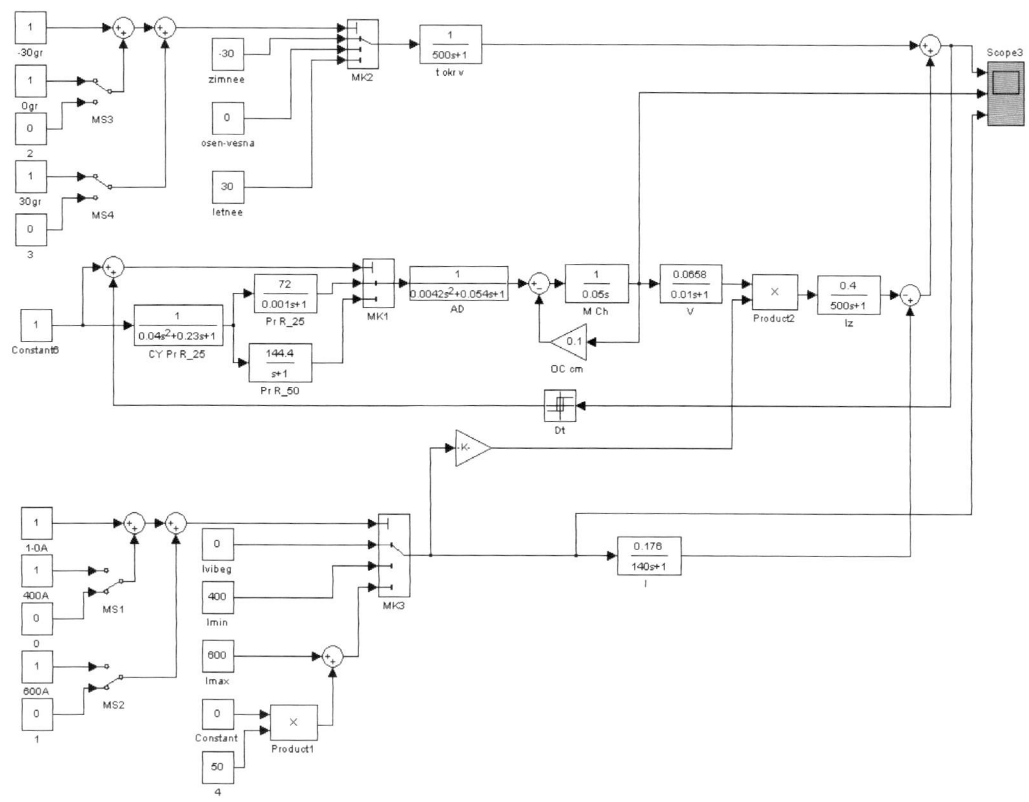

Figure 2. Block diagram of the control of blower motors of an AC electric locomotive

4. Conclusion

Substantiated minimum performance limit of blower motors will create sufficient overpressure in the locomotive body. The use of the proposed control algorithm for the cooling machines of the electric locomotive will allow minimizing the ingress of snow, dust on the insulation of traction electrical equipment within the entire control range. With that, the practical application of the developed automatic control system will allow eliminating the unauthorized shutdown of electric locomotive cooling machines by locomotive crews when the train is in the slow-down mode, and will also eliminate starting processes of asynchronous electric motors of the electric locomotive blower motors' drive. The developed technical solution will reduce the number of failures of traction electrical equipment and the yard time of electric locomotives for unscheduled types of repair.

References

[1] Ivanov V N, Konovalenko D V, Olentsevich D A, Sidorov V V and Lytkina E M 2008 Reliability of electric traction vehicles *Scientific problems of transport of Siberia and the Far East* **1** 196-198

[2] Olentsevich V A, Belogolov Yu I and Grigoryeva N N 2020 Analysis of reliability and sustainability of organizational and technical systems of railway transportation process *IOP Conf. Ser.: Mater. Sci. Eng.* **832** 012061

[3] Suslov K, Gerasimov D and Solodusha S 2015 Smart grid: Algorithms for control of active-adaptive network components *IEEE Eindhoven PowerTech* 7232462

[4] Upyr' R Yu, Goncharova N Yu and Eremenko M N 2020 Heavy-haul traffic is a reserve for increasing carrying capacity *Railway transport* **7** 10-13

[5] Kotelenets N F and Kuznetsov N L 1988 *Testing and reliability of electrical machines* (Moscow: Vysshaya shkola) p 232
[6] 2009 Maintenance of rolling stock in European countries *World Railways* **4** 50-52
[7] Osyaev A T and Nikiforov V A 2012 About the system of locomotive service abroad *VNIIZHT Scientific J.* **5** 56-62
[8] Khudonogov A M, Khudonogov I A, Dulskiy E Y, Ivanov P Y, Lobytsin I O and Khamnaeva A A 2020 Reliability analysis of power equipment of traction rolling stock within the eastern region *IOP Conf. Ser.: Mater. Sci. Eng.* **760** 012018
[9] Suslov K, Piskunova V, Gerasimov D, Ukolova E, Akhmetshin A, Lombardi P and Komarnicki P 2019 Development of the methodological basis of the simulation modelling of the multi-energy systems *E3S Web of Conf.* **124** 01049
[10] Dulskiy E, Khudonogov A, Khudonogov I, Astrakhantsev L, Ivanov P and Tuigunova A 2020 Modeling of the oscillating mode of the ir-energy supply in the technology of restoration of insulating fingers of electric motors of locomotives *Advances in Intelligent Systems and Computing* **982** 546-555
[11] 2007 *Rules for the arrangement of the traction power supply system of the railways of the Russian Federation TsE-462* (Moscow: Transport) p 450
[12] Marchenko V S, Sergeev A A, Ivanchenko V T et al 2003 *Electric equipment for diesel electric locomotives: A reference book* (Moscow: ICC "Akademkniga") p 248
[13] Akhtziger V V 1974 Study of the effects of the operating conditions of electric locomotives on the reliability of the insulation of the armature winding of traction engines. Study of electrical equipment performance and issues of the strength of electric rolling stock, in: *Scientific works of OMIIT* (Omsk) pp 72-74
[14] 2000 Trends in the manufacture, maintenance and repair of rolling stock, in: *Railroads of the world* **12** 28-34
[15] Kuzmich V D, Rudnev V S and Frenkel' S Ya 2005 *The theory of locomotive traction* (Moscow: Marshrut) p 446

Threaded lock conic connection of drill pipes and method to increase its carrying ability and resource of work

D S Zhovtikhanov, M I Korabelnikov and N A Aksenova

Industrial University of Tyumen, 38, Volodarskogo str., Tyumen, 625000, Russia

E-mail: nnsavelieva@yandex.ru

Abstract. This article presents the rationale and relevance of the use of an improved threaded keyhole conical connection of drill pipes for the oil and gas industry. The proposed threaded joint for drill pipes is characterized by an increased service life and a high rate of bearing capacity of the threaded joint in the conditions of drilling wells with complex spatial parameters, including when drilling sidetracks, horizontal and branched trunks of deep oil and gas wells.

1. Introduction

Threaded locking conical joints of drill pipes must fulfill a number of important functions, which are aimed at quick and reliable connection (screwing) of pipes together, the ability at high values of axial tensile and compressive loads to transmit the calculated torque to the drill pipe and rock cutting tool, provision of reliable sealing of pipes at high pressure flushing fluid. However, the practical use of drill pipes in well construction has shown that threaded tool joints, which accept complex alternating bending and tensile loads, remain the most vulnerable unit in the drill string assembly. During repeated screwing and unscrewing operations, the thread profile wears out and deforms, tightness and mechanical strength are lost, and the risk of thread breakage and an accident increases. In this regard, the task of increasing the bearing capacity indicators, reducing the accident rate and the service life of a threaded locking joint under complex cyclic loads is urgent and should be considered, taking into account the developed innovative technical and technological solutions.

The increase in drilling depths, the complication of the well profile and the length of the wellbore, including those with a horizontal end, has led to an increase in requirements and the need for continuous improvement of threaded pipe joints. Along with changes in the drilling technology, Russian State standards that regulate the requirements for threaded joints of drill pipes have changed: GOST 5286- (53) 75 [1]; GOST 28487-90 [2]; GOST R 50864-96 [3]. GOST 34438.2-18 [4]. Since 2020, GOST 28487-2018 [5] "Threaded joints with a locking thread of drill string elements has been enacted." General technical requirements".

At present, according to GOST 28487-90 [2] "Tapered keyhole thread for drill string elements" and according to GOST R 50864-96 [3], most of the domestic drill pipes are equipped with locking threaded joints, the outer diameter of which extends to locks with a diameter from 35 to 279 mm. The main technical requirements of GOST R 50864-96 standard coincide with the requirements of API 7 standard (American Petroleum Institute, USA). According to the above GOSTs, the tapers of threaded locking joints are regulated with a ratio of 1: 4 and 1: 6; and also the pitch of the threads of the triangular profile: - 5 or 6 threads over a length of 25.4 mm (inch).

When developing an improved type of threaded connection of drill pipes, the authors of the article carried out a patent analysis of the solutions closest in design, analyzed their advantages and disadvantages. Thus, to increase the strength of the threaded connection, the thread is made with run, ending near the resistant ends of the elements for drill pipes [6]. In the threaded connection of drill pipes according to patent RU No. 2354799 [7], two ratchet half-couplings are made for transmitting axial force at the ends of the nipple lock and the coupling, which complicates the threaded connection and does not increase its service life.

The threaded connection [8] is characterized in that the threads of the nipple and the coupling are made with a taper of 1: 8 and a pitch of 8.467 ± 0.05 mm, and the profile of the thread of the threads of the nipple and the coupling has an angle of inclination of the reference face of 20-30 ° to the normal to the axial line of the thread and the angle of inclination of the embedded face 35-44 ° to the normal of the axial line of the thread, while the hollow of the thread profile of the thread of the nipple and the coupling is made in the form of an ellipse arc showed an increase in torque resistance in relation to the standard connection of drill pipes by 81%.

The threaded TMK UP EXD drill joint, although it has several advantages over the first-generation double-joint, and can withstand higher torques than the TMK UP TDS, does not fully contribute to the longer service life of the joints due to the weakening and relaxation of stresses on the supporting surfaces during operation, and also complicates the repair of threaded parts in case of critical wear [9].

2. Results and discussion

Drilling of directional wells, including those with horizontal completion, often tack and tighten the drill string in the wellbore. As a result, drill pipes experience additional torque compared to the moment that occurs when screwing up the drill during assembly of the string on the surface of the well. The additional make-up torque has an increased effect on the threaded joint and can lead to exceeding the yield strength of the coupling material or nipple and their destruction. To prevent such accidents with a drilling tool, it is necessary to have a margin of safety of the threaded joint by the value of the expected increase in torque when the drill string is stuck in the well.

The increase in the bearing capacity of a threaded joint is achieved by the fact that the full-sized threaded part of the nipple of a typical locking joint in accordance with GOST 28487-2018 [5] and [8] is increased by a threadless conical extension. The linear size of the extension cord is in the range from 0.55 to 0.65 of the length of its threaded part, and in the coupling part of the locking joint, after the end of the threaded type cutting, an internal conical bore (surface) is provided congruent with the conical extension of the nipple part.

The abutting supporting surfaces of the nipple and coupling parts are structurally made with the possibility of their phased interaction when making up. So, at the stage of screwing the threaded connection, 30 ° -45 ° to the stop position (limiting fixed position of the nipple and coupling elements at a regulated torque) and the conical surface of the extension of the nipple part is introduced into the tensile radial contact with the conical surface of the coupling part. Then, in the next step, with further rotation of the nipple part relative to the coupling part, a tight grip (interference fit) of the threaded parts of the nipple and coupling is created.

And at the final stage of the screwing-up screw driving cycle, with the creation of a torque calculated for each size of the screwed threaded connection, the end surfaces of the nipple and coupling are closed, providing an interference fit (tension) at their ends. Figure 1 (on the left) shows the threaded tool conical connection of drill pipes in the screwed-up state, (on the right) shows fragment II of the parameters of the conical threaded connection, where: S - the thread pitch; h_{pr} - thread profile height; σ - the width of the protrusion of the thread; r1 - the radius of conjugation of the vertices of the thread profile; r2 - the radius of curvature of the thread depression.

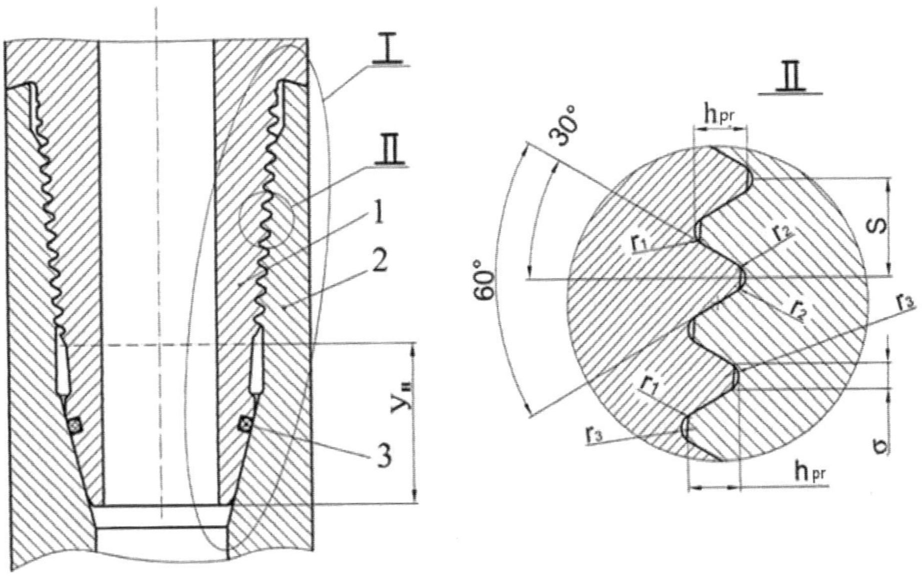

1- nipple part of the lock connection;
2- coupling part of the lock connection;
3- o-ring.

Figure 1. Threaded locked conical drill pipe joint.

In the developed threaded connection, the end parts on the nipple and the coupling are made with negative taper. The value of the closing angle parameter (A) is 2-3 times higher than the angle of friction of steel on steel. This complex technical solution is aimed at balancing the axial load along the threads in the direction of its normalization and eliminating the effect of uneven loading, proved by Professor N.E. Zhukovsky in 1902 At the same time, the influence of the concentration of internal stresses arising under bending loads in the threaded grooves of the nipple part used in known solutions, as well as the technical requirements of the standard version, is minimized. To increase the sealing ability of a threaded conical connection, an option is provided for equipping a conical extension with a groove, with a ring placed in it.

Figure 2 visualizes the forces (force reactions) acting in the elements of a threaded conical connection after screwing up with a regulated torque, where: R is the reaction of the forces of the threaded connection section; F1 - reaction of the forces of the end face of the coupling part; F2 - reaction of forces on the conical section of the coupling part; P1; P2; T2; K1 - components of the reaction of forces, laid out along the axes.

The diagrams of the torques arising from the friction forces of the nipple part of the lock in the coupling part at the final stage of screwing the threaded connection are shown in Figure 3. Where: φ - the central angle of the beginning of the radial contact of the nipple cone is extended with the conical bore of the coupling part; φ1 - the central angle of the beginning of the radial contact of the threaded nipple with the internal conical thread of the coupling; φ2 - the central angle of the beginning of the end contact of the nipple with the end face of the coupling; M (F2) - the maximum value of the braking (torque) moment from the reaction of the coupling part from the contact of the conical extension; M (R) - the maximum value of the braking (torque) moment from the reaction of the threaded section of the coupling part from the contact with the thread of the nipple; M (cr) - the maximum value of the torque from the reaction of the coupling part, at the moment of termination of the relative rotation of the nipple and the coupling (position "stop").

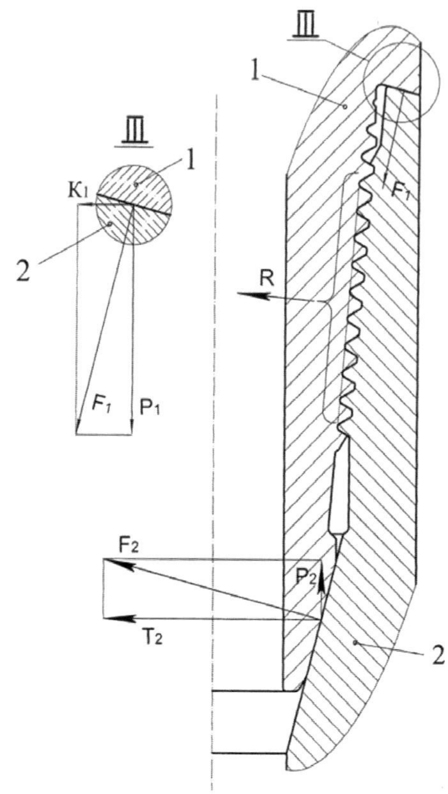

Figure 2. Scheme of a threaded connection, with visualization of forces (reaction of forces) acting in the elements of a threaded conical connection after screwing up with a regulated torque

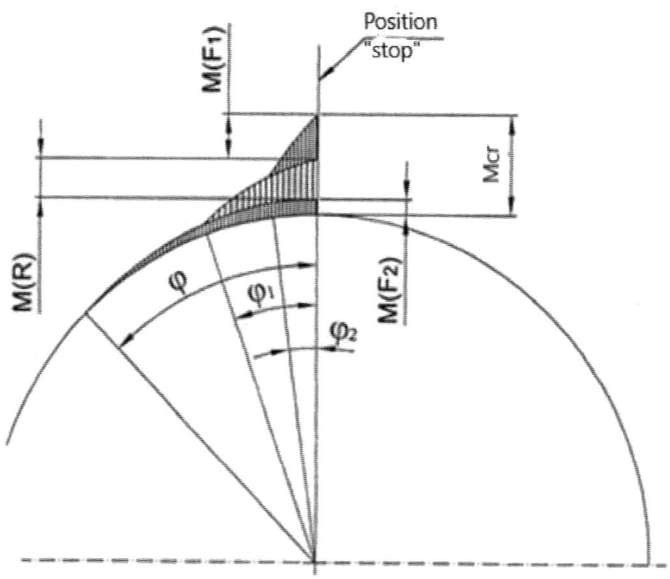

Figure 3. Diagrams of torques from the friction forces of the nipple part of the lock in the coupling part at the final stage of screwing the threaded connection

Figure 4 shows (for visualization) the diagrams of the axial load distribution for the threaded part of the coupling after screwing with the nipple, where: 4 - load distribution curve in the threaded connection of the coupling from reaction P2; 5 - load distribution curve in the threaded connection of the coupling from the reaction P1; 6 - the final curve of the load distribution in the threaded connection of the coupling from the reaction of forces P1 and P2.

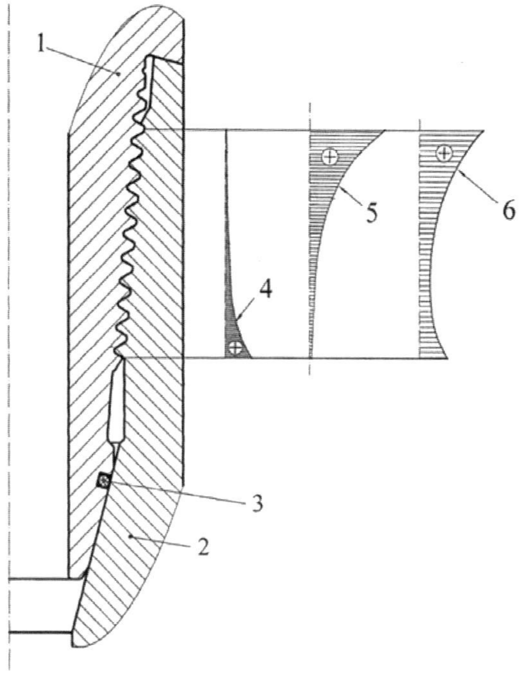

Figure 4. Diagrams of axial load distribution diagrams for the threaded part of the coupling after screwing with a nipple

When using the proposed threaded connection, preliminary, in preparation for screwing, it is necessary to clean the foreign and internal particles of the inner and outer surfaces of the nipple and coupling parts from the particles of water and steam and apply a grease. Screwing is done with a mechanical key after combining the movable element of the nipple part of the pipe with the fixed-mounted coupling part of the pipe at the wellhead. The angular speed (ω) for make-up is taken to be ω = (3-6) s-1, with the creation of a regulated torque (M_{cr}) set for each size of the lock connection.

Moreover, in sector β (the angular parameter of the generatrix of the conical surface of the nipple threaded part) - 30 ° -45 ° (0.5 ... 0.8 mm before the thrust ends of the nipple and the coupling close) to the stop position - (maximum fixed position the elements of the nipple and the coupling with a regulated torque) are first sequentially introduced into the tense (radial) contact conical surfaces of the extension of the nipple and coupling parts.

This solution is aimed, first of all, at counteracting the influence of negative alternating bending moments during the operation of threaded joints in the wellbore with intense curvature; as well as to equalize the axial load on the turns of the threaded joint assembly. With further rotation of the nipple part relative to the coupling part, a tight grip (tightness) of the threaded parts of the nipple and the coupling is created, and at the final stage of the screwing cycle of the threaded connection, providing the rated torque for each size of the threaded connection, the end surfaces of the nipple and coupling with an end interference are closed (voltage).

IOP Conf. Series: Materials Science and Engineering 1111 (2021) 012002 doi:10.1088/1757-899X/1111/1/012002

3. Conclusion

The use of the proposed threaded connection of drill pipes allows transmitting increased torque to the drill pipe and the layout of the bottom of the drill tool (BHA) in conditions of alternating bending loads in deviated wells. In addition, this threaded connection is characterized by increased tightness and increased resource of trouble-free operation.

References

[1] 1975 GOST 5286-(53) 75 *Locks for drill pipes* (Publishing house of standards)
[2] GOST 28487-90 *Conical locking thread for drill string elements. Profile, dimensions. Profile.*
[3] GOST R 50864-96 *Conical locking thread for drill string elements. Profile, dimensions, technical requirement*
[4] GOST 34438.2-2018 *Drill pipes and other elements of drill strings in the oil and gas industry. Key parameters and control of threaded connections. Basic technical requirements*
[5] GOST 28487-2018 *Threaded connections with a locking thread of the elements of the drill string. General technical requirements*
[6] 1984 Description to patent SU No. 1131481 F 16L 15/00 (France)
[7] Patent RU No. 2354799 *Device for connecting drill pipes* (18 April 2008)
[8] Patent RU 2508491 *Threaded connection of drill pipes* Bulletin of inventions No. 6 (2014)
[9] 2017 *Operation manual for drill pipes with welded locks TMK UP EXD. REP From 02-015-2015,* 2nd edition (LLC TMK Premium Service)
[10] GOST 27834-95 *Welded tool joints for drill pipes. Technical conditions*

Analysis of automatic control structure of transfer pipeline stations on reclamation channels

M S Ali[1], D S Beglyarov[1], E E Nazarkin[1], Yu V Korchevskaya[2] and I A Trotsenko[2]

[1] Russian State Agrarian University - Moscow Agricultural Academy named after K.A. Timiryazev, 49, Timiryazevskaya Str., Moscow, 127550, Russia
[2] Omsk State Agrarian University named after P.A. Stolypin, 1; Institutskaya square, Omsk, 644008, Russia

e-mail: ia.trotsenko@omgau.org

Abstract. Hydro-reclamation system operation management comes down to the management of pumping stations and dams. Transfer pumping stations are one of the most complex, critical and energy-rich local management objects on the channels of irrigation and drainage systems. These stations are complex aggregates of hydromechanical and electrical equipment, whose optimization can be rather significant, and are important elements determining the quality of a hydro-reclamation system as a whole. The main function of the automatic control system for transfer pumping stations is providing solutions on the optimal way to compensate for mismatches arising in the system "channel-transfer pumping station" and finding the best way to compensate for these mismatches with the available means. The first problem is hydrodynamic and is solved by studying a mathematical model of the pool, coupled with a transfer pumping station. The solution of the second problem is associated with regarding the technical characteristics and condition of the equipment of the transfer pumping station and the permissible operating modes of the channel. The article discusses one of the methods for determining the mismatch in the system "channel-transfer pumping station", which contributes to building economical and efficient control systems for transfer pumping stations.

1. Introduction

In the Russian Federation, the legislative foundation is laid in national programs for the development of land reclamation and water management. The spheres of water consumption in the agro-industrial production of Russia include water supply to settlements, agricultural enterprises, irrigation and drainage of lands, water use for fish farming and energy. The main areas of activity in Russia are water management construction, implementation of scientific and technical information in the field of both water management and land reclamation, and solution of the problem of high-quality customer service with minimal costs [1,2].

Hydro-reclamation system operation management comes down to the management of pumping stations and dams. The specificity of these structures as management objects is that they are rather complex, energy-saturated local objects, but at the same time they are also elements of a transport system [3, 4].

Currently, most of the pumping stations on the reclamation canals are ineffective. When operating in this mode, water leaks in the system may occur due to a mismatch between water supply and water consumption, as well as an increase in energy consumption.

The paper discusses the issue of increasing the efficiency of the "channel-transfer pumping station" system. Improving the efficiency of transfer pumping stations is an important task since they are the largest consumers of electricity in the world [5, 6]

The aim of the work is to select a mathematical model of the dynamic system "channel-transfer pumping station" and the algorithm for the operation of the automatic control system of a transfer pumping station. The methodological basis of the work is mathematical modeling of unsteady water movement in the canal section of the irrigation system and the measurement theory.

2. Research Methods

A pumping station with a supply channel (Figure 1), at the beginning of which there is a retaining structure, is used as a channel-transfer pumping station. With a planned water distribution, the main function of the transfer pumping station in such a system is to raise all the water entering through the canal. The water level in the canal can vary within certain limits. If no additional restrictions, for example, the requirement to stabilize the water level in a given zone are imposed on the system, then the lower limiting water level in the supply channel of the normal water level is determined by the value of the normal flow depth at a known supply of the pump station and matching it with the water flow in the channel. The upper limiting level in the area of small flows of the transfer pumping station is determined by the value of the nondepositing velocity of the water flow, and in the area of large flows is determined by the location of the emergency weir in the supply channel. The main function of the transfer pumping station is fulfilled by the automatic control system. The change in the transfer pumping station supply occurs when the water level in the controlled section of the channel reaches one of the limit values. Since the control section is usually located in the backwater area, there is no unambiguous relationship between the ordinate of the free water surface Z and the flow rate Q. At the same time, to compensate for the mismatch that occurs in the "channel-transfer pumping station" system, it is important to know not so much the water flow in the channel as the value of the mismatch ΔQ itself. Thus, the problem can be formulated as follows: determine the mismatch ΔQ between the water flow rate in the channel and the pumping station supply using water level sensors installed in the zone of hydraulic influence of the pump station as a source of measuring information.

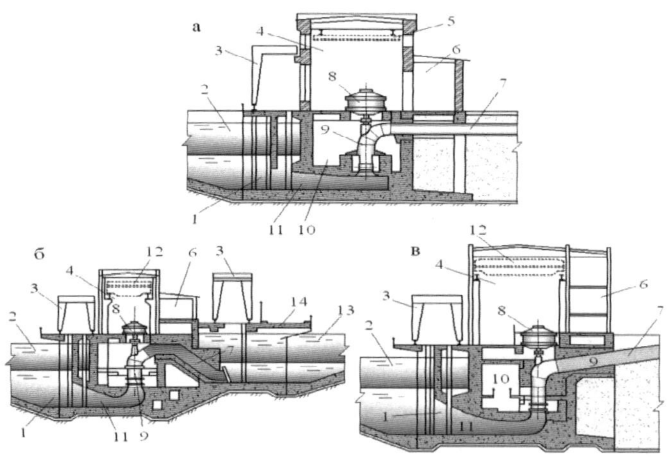

Figure 1. Building arrangements of pumping station equipped with axial-flow pump: a, b, c are the stations built on dense soils (sediment up to 5 cm), soft soils (sediment more than 5 cm), rocky foundation (with a large fluctuation of the water level in the water source), respectively; 1 is water source; 2 is water intake part of the arrangement; 3, 5 are gantry and bridge cranes; 4 is machine room; 6 is office; 7 is cable cellar; 8 is force main; 9 is butterfly valve; 10 is electric motor; 11, 14 are centrifugal and variable pumps; 12 is curved suction pipe; 13 is transformer substation

The problem was solved based on the assumption that there is a connection between the sought value ΔQ, the ordinate of the free surface of the water in the controlled section of the channel Z, and the rate of its change $\partial x / \partial t$ in some interval. To determine the mismatch ΔQ in the "channel-transfer pumping station" system by the known values of Z, a mathematical model of this dynamic system was compiled [7-10].

3. Results and discussion

Generally, water movement in the supply channel is unsteady due to the mismatch between the water flow in the channel and the supply of the transfer pumping station, which exists almost always. We assume that its characteristics have no discontinuities and that the water surface is horizontal in any cross section. Thus, it is possible to consider the movement of water in the channel in a one-dimensional approximation.

The continuity equation and the dynamic equation for slowly varying motion (Saint-Venant's equation) presented in differential form are as follows [11]:

$$B + = q \quad (1)$$

$$\frac{\partial Q}{\partial t} + \frac{\partial VQ}{\partial x} = - gw \left(\frac{\partial Z}{\partial x} + \frac{Q|Q|}{K^2} \right), \quad (2)$$

where x is cross-sectional coordinate, kg; Z is ordinate of free water surface, m; t is time, s; Q is water consumption, m3/s; w is cross-sectional area of flow, m2 ; B is free flow surface width, m; V = Q/w is average flow velocity, m/s; q is inflow per unit length of the channel directed to the normal of axis x, m3/s; K is wC - transport capacity of the section, C is Chezy coefficient, R is hydraulic radius, m; g is gravitational acceleration, m/s2.

Let us also denote the flow depth:

$$h (x, t) = Z (x, t) - Z0(x), \quad (3)$$

where Z0(x) is channel bottom ordinate.

We assume that w (x, h), B (x, h) = , q (x, h, t), K (x, h), (x) are the given functions.

The sought-for functions are the ordinate of the free water surface Z (x, t), the rate of its change and consumption Q (x, t).

The equations are hyperbolic partial differential equations; the system of equations on the plane (x, t) has two characteristic directions [5]:

$$= V \pm C ,$$

where C = is small disturbance propagation speed, m/s.

These equations can be reduced to the following form:

$$\frac{\partial Q}{\partial t} - + (V \pm C)\frac{\partial Q}{dx} - B (V \pm C)[\frac{\partial Z}{\partial t} + (V \pm C)\frac{\partial Q}{dx}] = F1,2 \ , \quad (4)$$

where

$$F_{1,2} = V^2 \left(- B\frac{dZ_0}{dx} + \frac{\partial w}{dx} | h = const \right) - gw \frac{Q|Q|}{K^2} - g (V \pm C). \quad (5)$$

Calculation of unsteady water flow in a channel is reduced to solving equation (4) with initial and boundary conditions, as well as matching condition in places of concentrated inflow, a sharp change in cross-section, or at hydraulic structures.

The initial conditions are given in the following form:

$$Z (x, 0) = , Q (x, 0) = (x) . \quad (6)$$

The boundary condition is set by one of the following dependencies:

$$Z = Z (t), \quad Q = Q (t) , Q = Q (Z) \quad (7)$$

While only flow regimes |V|< C are considered, then only one boundary condition is specified in each section. For points along the length of the channel, in which there are discontinuities in the

IOP Conf. Series: Materials Science and Engineering 1111 (2021) 012003 doi:10.1088/1757-899X/1111/1/012003

geometry, concentrated inflow and similar factors, the conjugation conditions apart from the boundary ones for two adjacent sections are noted.

Pairing conditions nth and (n + 1)th sectors can be presented as follows:

$$Zm\ (xm, t) = \ (xm, t) \qquad (8)$$

$$Qm\ (xm, t) = Qm{+}1\ (xm, t) - Qm{+}1/2(t) - \Omega_{m+1/2}\frac{\partial Z}{\partial t}, \qquad (9)$$

where Qm+1 is сосредоточенный приток, given function of t; $\Omega_{m+1/2}\frac{\partial Z}{\partial t}$ is inflow stipulated by additional capacity at the border of the sites.

If it is necessary to take into account the influence of local resistances, the pairing conditions are as follows:

$$Qm\ (xm, t) = \ (xm, t) \qquad (10)$$

$$Zm\ (xm, t) = \ (xm, t) + \ Qm\ Qm\ , \qquad (11)$$

where = (t) is given function of t; .

Equations (4)...(11) specifically describe the problem of unsteady water flow in a channel.

Having approximated relations (4) ... (11) using an implicit scheme, it is possible to obtain a determined system of algebraic equations for the unknown variables (n = 0, 1, 2,....N).

To solve the problem of determining the mismatch between the flow rate of water in the channel and the supply of the transfer pumping station, the dependence Z = Z () is of particular interest.

The pumping in most cases changes discretely, by alteration of the composition of the constituent units. Consequently, the mismatch ΔQ in the general case is determined with an excess using the family of curves Z = Z ().

Taking into account the above, the determination of the mismatches that have arisen in the "channel-transfer pumping station" system is carried out using an algorithm, the block diagram of which is shown in Figure 2.

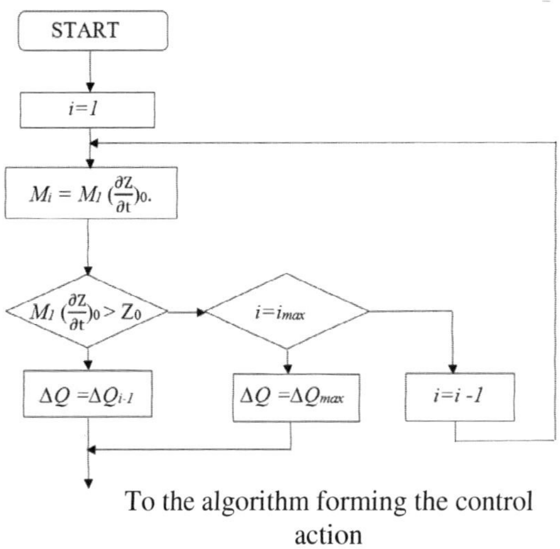

Figure 2. Block diagram of the algorithm for determining the mismatch in the "channel-transfer pumping station" system

4. Conclusion

1. The presented method and algorithm for determining the mismatch in the "channel-transfer pumping station" system allows synthesizing an effective control system for the pumping station,

which operates on the basis of data received from the water level sensor installed in the zone of hydraulic influence of the transfer pumping station.

2. The considered concept of a self-regulating irrigation system and its methodology enables the following:

- to dramatically reduce water losses from canals due to better coordination of water supply and water consumption,
- to increase the economic efficiency of systems by improving design technology based on mathematical modeling.

Acknowledgments [if any]

The research was carried out using the equipment of the Center for collective use of the Omsk State Agrarian University "Agricultural and technological research".

References

[1] Tusupbekov Z, Ryapolova N and Nadtochiy V 2019 Assessment of Changes in Water Use Indicators in Agricultural Sector in the South of Western Siberia T*he First Technological Order: Prospects for the Development and Modernization of the industrial complex of the Russian Agro-Industrial Sector: Advances in Social Science, Education and Humanities Research* ed Z Karssen vol. 393 pp. 190-193

[2] Ushakova I, Korchevskaya Y, Trotsenko I and Kondrateva T 2019 Issues of Drinking Water Supply and Possible Solutions on the Example of Gorkovsrky District in Omsk Region T*he First Technological Order: Prospects for the Development and Modernization of the industrial complex of the Russian Agro-Industrial Sector: Advances in Social Science,* ed Z Karssen vol. 393 pp 36-41

[3] Ali M S, Beglyarov D S and Chebaevsky V F 2015 *Pumps and pumping stations: textbook.* (M .: publishing house RGAU-Moscow Agricultural Academy)

[4] Borodiy O F, Trotsenko I A and Korchevskaya Yu V 2018 Problems of restoration of the reclamation complex (on the example of the Omsk region). *Bulletin of the Omsk State Agrarian University* **4 (32)** 12-17

[5] Ali M S and Nazarkin E E 2019 Methodology for calculating a pumping unit with a raised suction pipe *Natural and technical sciences* **12** 397-400

[6] Kabanets S O, Vdovina K A, Stepanova L V and Korchevskaya Yu V 2018 Complex solutions for the modernization of the water supply system on the example of a pumping station of the second rise. C*ollection of the I regional (correspondence) scientific and practical conference of young scientists and students dedicated to the 100th anniversary of the Omsk State Agrarian University: Problems of environmental protection and rational use of natural resources.* (Omsk: publishing house Omsk GAU)

[7] Ali M S and Vishnevsky K P 1990 Mathematical modeling of processes in pressure systems with automatic control of pumping stations *Melioration and water management* **2** 12

[8] Zats L I and Tsysar V G 1979 Algorithm for determining the mismatch in the "channel-pumping station" system. *Collection of scientific works of V.O. "Soyuzvodproekt" No. 51 "Improving the design of irrigation and drainage systems",* pp. 164-170.

[9] Karambirov S N 2012 *New approaches in modeling and optimization of pipeline systems. Basics, concepts, methods (*LAP Lambert Academic Publishing)

[10] Leznov B S 2013 Frequency-controlled electric drive of pumping units (Moscow: Mashinostroenie)

[11] Shterenlikht D V 2004 *Hydraulics: textbook for universities.* (Moscow: Kolos)

IOP Conf. Series: Materials Science and Engineering 1111 (2021) 012004 doi:10.1088/1757-899X/1111/1/012004

Structure of the surface layer and the microhardness of high-carbon instrumental steel after laser treatment

N A Astafyeva[1], S K Kargapol'tsev[2], V E Gozbenko[2,3], Yu I Karlina[1], M G Shtayger[4] and A I Karlina[1]

[1] Irkutsk National Research Technical University, 83, Lermontov Street, Irkutsk, 664074, Russia
[2] Irkutsk State Transport University, 15, Chernyshevsky St., Irkutsk, 664074, Russia
[3] Angarsk State Technical University, 60, Chaykovskogo Street, Angarsk, 665835, Russia
[4] MC Mechel Steel, 1, Krasnoarmeysky Street, Moscow, 125167, Russia

E-mail: ksk@irgups.ru

Abstract. The article presents the results of the study of the effect of laser treatment on the structure and microhardness of the surface of high-carbon steel samples. The treatment is carried out in order to obtain surface layers, parts with high properties for industries of various purposes. The samples from high-carbon steel U10 A after preliminary thermal treatment by modes of 780–800 °C heating for 2 minutes with cooling in 10 percent water solution of sodium chloride with tempering at 270–320 °C and cooled in the air and without the thermal treatment are treated with the Nd:YAG laser with the capacity up to 1 kW. Metallographic studies and microhardness properties illustrate the influence of laser surface processing on the increase in mechanical properties. It is shown that the laser thermal treatment of steel allows obtaining a surface microhardness of about HV 800-1000 MPa and depends on the original state of the metal and the previous thermal treatment.

1. Introduction

To improve the surface properties of metals, it is possible to use several surface hardening methods [1-4]. These may include changes in the chemical composition of the surface, such as alloying, or changes in the microstructure of the surface, such as quenching [5-9]. To enhance the durability of most machine parts that work in wear conditions [1, 5, 6], cavitation, cyclical loads, corrosion at cryogenic or high temperatures [2, 8] and working surfaces of the production instrument, including the deformed instrument, surface thermal hardening found its application [10-14]. The hardening of the surface is a process of thermal treatment in the localized heating zone [15-18] which uses concentrated sources of energy, such as laser [1, 2, 19-22], electronic beam [3], plasma arc [4,5], welding arc [5, 7]. The application of surface hardening on steel parts and combining it with a rapid tempering method, which is called martensite quenching, allow increasing fatigue durability and limit the deformation of steel parts [9-12].

The laser beam is able to quickly raise the surface temperature to a temperature above the austenization temperature (As_3) [1-4, 7, 8]. This intensive heating process converts the existing surface microstructure into austenite, and then, through rapid cooling, creates a very dispersive martensite microstructure with high hardness [19-22]. To get a homogeneous solidity profile on details with complex geometry, one needs to adjust the parameters of laser-quenched processes, such as laser power and scanning speed [1, 2, 9-14]. This uniform hardness profile is recommended for uniform distribution of mechanical loads. This layer is formed as a uniform martensite layer on the surface of quenched parts. This homogeneous hardened layer prevents the concentration of stresses on a solid surface layer of parts and limits the spread of cracks during periodic loads, which improves resistance to fatigue of the parts [10–12].

In the case of slip joints, a homogeneous hardness profile on the teeth geometry can effectively reduce the wear and tear caused by attrition [13]. To achieve this uniform hardness profile, it is important to calibrate such parameters as laser power, scanning speed and rotation speed. The work [14] proposed an experimental approach, backed up by dispersion analysis (ANOVA), to develop a model to optimize laser processing parameters. Despite the fact that the technology of laser treatment of the surface is constantly evolving and many works consider these issues [1, 2, 9-22], however, there is also interest in further works on the influence of the complex of laser irradiation parameters on the change of surface morphology, structural changes and characteristics of the material, the evolution of microstructure over time, as well as combinations of laser treatment with other types of treatment.

The relevance of this work is substantiated by the need to increase the service life of machine parts and working surfaces of the production tool. The issues of technologies of hardening and durability enhancement, as well as their introduction into production are of particular interest to various industries.

2. Materials and methods of study

The samples of high-carbon instrumental steel U10 A GOST 1435-99 were made to study the influence of laser treatment parameters on the characteristics of the hardened layer, namely its thickness, microstructure and hardness. The chemical composition of the material is presented in Table 1.

Table 1. Chemical composition of the instrumental steel U10A GOST 1435-99.

Element t	C	Si	Mn	S	P
Content (%)	0.95 – 1.09	0.17 – 0.33	0.17 – 0.28	≦0.018	≦0.025

Laser treatment was carried out for samples without preliminary thermal treatment and samples that were quenched by 780 – 800 °C heating for 2 minutes with cooling in 10 percent water solution of sodium chloride with tempering of 270 – 320 °C and cooling in the air. The surface of the samples was hardened with the "Kvant-15 MSZ" laser setup in an inert argon gas environment. Equipment is included in the setup (Figure 1).

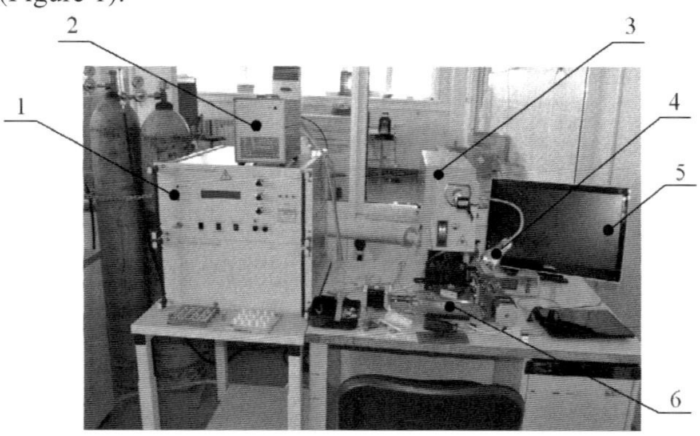

Figure 1. Setup for laser treatment of metals "Kvant-15 MSZ": 1 – pulse power source of solid-state laser P138; 2 – control unit of the coordinate table and treatment process; 3 – solid-state laser with optical system; 4 – television surveillance system; 5 – video control device; 6 – coordinate table.

The samples were treated with a pulse Nd:YAG laser. Radiation wavelength is 1,064 microns. Frequency of radiation pulses is up to 1-50 Hz. Pulse duration, adjustable within 0.1-6.0 ms. Spot size in the treatment area, 0.3-1.3 mm. The average capacity is 1,658 88W. The most important parameter that is characteristic of the thermal influences of radiation on the material, is the density of radiation power of the laser q_l (V/m^2), which is associated with other parameters of radiation by the following dependence: q_l= Rl/ S= E_l/ τ; S, where the Rl is the full radiation power, W; S is the area of radiation material, m^2; τ is the duration of exposure to radiation, s; E_l is the radiation energy, J.

To conduct the experiment, the selection of permanent and variable factors of the experiment is based on the analysis of literature data on laser thermal treatment of metals, as well as on the main purpose of the study – the search for optimal laser hardening modes that allow achieving the highest hardness of the metal surface. Taking into account the technical capabilities of the "Kvant-15 MSZ" laser setup in the inert argon gas environment and the requirements for maximum performance at the maximum possible width of the laser impact zone (LIZ) the following factors were accepted as permanent: the maximum diameter of the laser beam spot is 1.3 mm, the maximum duration of the pulse is 5 ms.

The energy density of radiation that determines the effect and result of laser impact on the metal surface, in this case depends only on the energy of the pulse of radiation, which, in turn, is the function of voltage of pumping capacitors of the setup "Kvant-15 MSZ" in the environment of the inert gas argon. This voltage is accepted as the main variable factor. The power of the radiation was regulated by changes in the voltage of the pumping capacitors. The first series of experiments was conducted when the laser impact spot was located without overlap when the voltage of the capacitors within the range of 450...508V changed, as well as with a heating spot overlap factor of 0.50 and 0.75. The required thermal treatment modes to harden the surface of the selected samples were determined by the calculation methods [9, 10] and presented in table 2.

Processing was carried out on five, experimentally selected modes (see table 1), varying the energy of radiation by the voltage of pumping capacitors while other parameters are constant: frequency f – 3.7 Hz, pulse duration τ – 1.63 ms, focus – 1.6 mm. After laser thermal treatment, microslices were prepared, the microstructure of the samples was examined on the MET-3 microscope in a bright field with an increase of 50...1000 times, microhardness was determined by the results of measurements of the PMT-3 device.

Table 2. Parameters of the laser treatment mode

Pass mode	I	II	III	IV	V
Pumping capacitors voltage, U, V	452	483	520	531	580

3. Results and discussion

The microstructure of samples unhardened by volume thermal treatment after laser treatment (mode 1) is presented in Figure 1. In the photo of the microstructure (Figure 1) there is a non-etchable white surface layer (1, a), its depth is on average 630 cm. It is adjacent to the transitional layer and the main metal, the structural components of which, coarsely plastic perlite and cementite particles, are located in the form of a distinct mesh (Figure 1, b). On average, the microhardness of the non-etchable white layer is 946 HV that of the main metal is 264 HV, the transition zone between the main metal and the white (non-etchable layer) is on average 657 HV.

In general, it should be noted that the quenching zone from the solid phase consists of two areas: at the top, this is an area with a homogeneous structure and at the bottom, this is an area with a heterogeneous structure. In the area with a homogeneous structure, a fine-dispersed martensite with a microhardness of 720...850 HV is formed. In place of perlite grains, martensite is formed (microhardness is 557-644 HV) with a small amount of residual austenite. In the area with a heterogeneous structure, as it has already been described in the works, as long as the depth increases, there is an increase in the structure heterogeneity: first the martensite-troostite is formed, then the martensite and troostite mesh that goes into the troostite-ferrite, and at the boundary with the original structure - to the ferritic structure. Figure 2 shows the microstructure of the main metal and the surface treated layer on mode II. With an increase in the energy of the laser radiation, the thickness of the hardened layer decreased (to 15 microns), much of the material evaporated. Further energy increase causes local heating to high temperatures exceeding the melting point of the material, which leads to a change in the geometry of the sample plane, the material evaporates intensely.

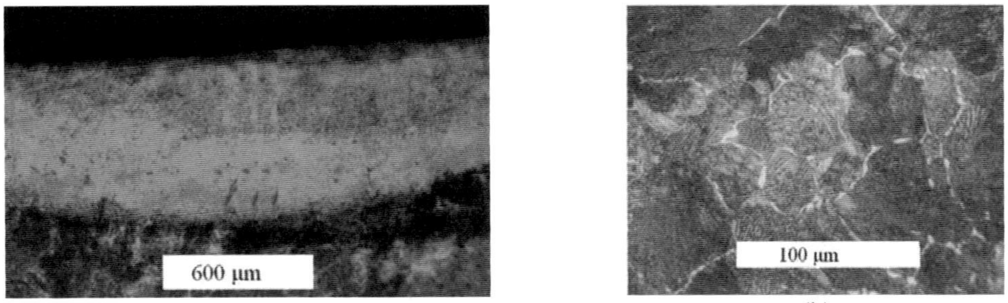

(a) (b)

Figure 2. Microstructure of the main metal and the surface treated layer (mode I).

The photos of microslices of samples that were given a preliminary volume thermal treatment (quenching+tempering) followed by surface hardening by pulse laser are shown in Figure 3.

Figure 3. Microstructure of the main metal and the surface treated layer (mode II x200).

For all processing modes in steel after annealing or normalization, the heat-affected zone is very small, there is a sharp change in mechanical properties in the surface layers of the metal [1, 2]. It is known [9-22] that during laser hardening of steels that have previously undergone volume quenching and tempering, after laser hardening, a tempering zone is revealed at the boundary with the original structure. The entire hardening zone from the solid phase is an area of homogeneous martensite with a microhardness of 700... 850 HV, which is higher than the hardness of the martensite obtained by the oven hardening.

There is a transition zone on the border with the original structure. In this case, this tempering area with a troostite structure, this has a microhardness of 300-400 HV. In our experiments, Mode III provides maximum depth of hardening without changing the geometry of the product surface. With laser hardening of steels that were previously quenched and tempered, after laser hardening there is an increase in the microhardness values by 25-30%. In modes IV and V there are significant changes in geometry, the surface of the metal is melted. At the same time, the melting area of the heat-affected zone has a needle-shaped martensite structure, which is transformed into the structure of the hardened layer formed as a result of phase transformations (Figure 3).

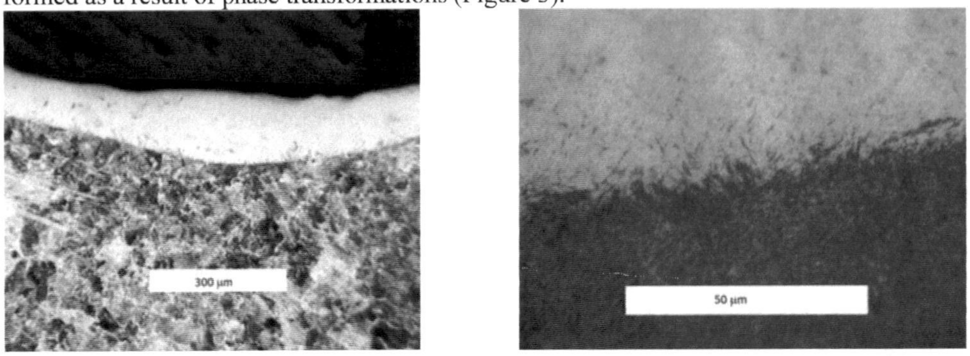

Figure 4. Microstructure of the main metal and the surface-treated layer (mode IV, x100)

At the same time [1, 2, 9, 19-22] it is necessary to notice that a feature of laser processing with the fusion of eutectoid and hypereutectoid carbon steels is the known fact of the presence of residual austenite in the melted zone, the amount of which, according to various authors, can be up to 39% for U8 steel and up to 45% in U10 steel. Meanwhile, it is important to clarify that in the melting zone of these steels, there is no non-dissolved cementite. It is obvious that the structure of the formed martensite is overly saturated with carbon, as well as the phase of residual austenite. This fact indicates a growth in microhardness with an increase in the carbon content in the structure of martensite in the formed surface-hardened layer.

However, the presence of residual austenite in the melted zone reduces microhardness compared to hardening without melting. Based on the results of the first series of experiments, histograms (Figure 4) of the dependence of the hardness (HV) of the zones of laser impact on the voltage (U) of the pump capacitors (radiation pulse energy) were obtained in order to determine the optimal level of this parameter. We can see that each hardening mode has the maximum depth of the hardened layer. For example, mode I has a maximum depth of 110 microns, mode II -160 microns, mode III - 180 microns, mode IV -260 microns, mode V - 240 microns. The decrease in the latter case is due to intense evaporation and splashing in the zone of melted metal. Comparative histograms are based on the results of comparison of microhardness values at different modes of voltage of pumping capacitors. At the same time, the comparison was conducted for all at the same distance from the surface (according to the depth of the hardened layer).

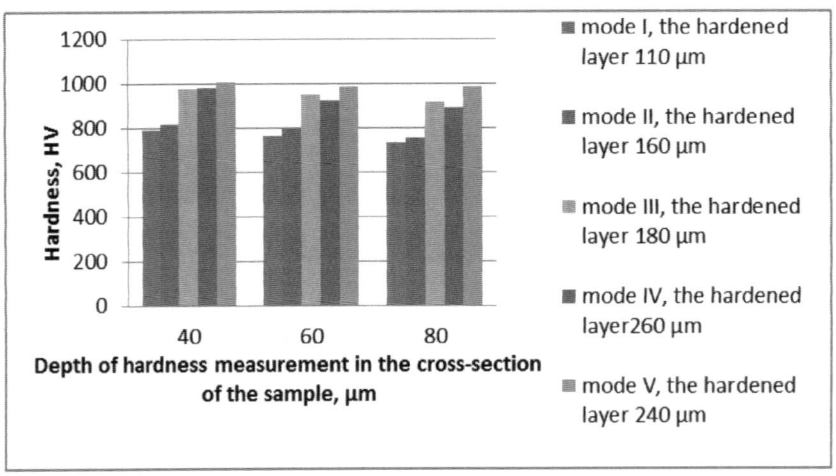

Figure 5. Diagram of the dependence of hardness values on hardening modes

The analysis showed that as the energy of laser radiation increases in the beginning (at U 452-520 V) the values of microhardness generally increase. Later on, the growth of laser radiation energy at first within the range (U 520-580 B) does not significantly increase the microhardness value and even shows some decrease (the reasons we have considered above), but it increases the overall depth of the surface hardened layer. At the same time, if we take the values of microhardness by the depth of the layer, all the modes are characterized by a reduction of microhardness from the surface to the depth of the main metal. For technological purposes, when optimizing the hardening process, it is important to establish the pulse spots overlap coefficient.

Traditionally, the ratio of maximum depth and uniform distribution of hardness is considered optimal. In our case it can be seen that, depending on the laser pulse energy, which we control by the voltage of the pumping capacitors, there are structural features in depth, which are clearly manifested in modes with surface melting. The analysis of the results obtained showed that with the investigated overlap coefficients, the depth of the zones is practically equal to and is 150-180 microns, and the boundaries of these zones are quite clear (the transition zone in depth does not exceed 10-15 microns). Thus, when using the "Kvant-15 MSZ" laser setup in the inert argon gas environment in the conditions of real production to obtain a hardened layer that is sufficiently deep and structurally uniform in width on tool carbon steel (after preliminary thermal treatment by quenching and tempering), the following mode of laser thermal treatment can be recommended: the diameter of the laser exposure zone D = 1.6 mm; the charge voltage of the pumping capacitors U = 520 V; the overlap coefficient Kp = 0.5, the pulse duration $\tau = 3.7$ ms; with a frequency f = 1.63Hz.

4. Conclusion

1. High efficiency of laser thermal treatment has been confirmed. The resulting hardened layer at a depth of 110-260 microns has a homogeneous structure and properties. The structure of the hardened layer is a freshly quenched martensite with microhardness HV = 1000 with clear boundaries in depth and transition zones of insignificant length (less than 10 microns). Average hardness HV – 880 (HRC 63-64).
2. The obtained depth of the hardened layer (150-260 microns) is sufficient for 3-5 regrinds of a worn-out tool blade (with gentle sharpening modes). During the regrinds within certain limits, their width will change slightly.
3. The research results allow recommending the use of the developed mode of laser heat treatment to harden the cutting edges of the tool.

5. Acknowledgments

This paper was prepared using the equipment provided by the Joint Instrumentation Centre “Baikal Center of Nanotechnologies”.

References

[1] Shina H J et al 2008 Microstructural and Hardness Investigation of Hot-Work Tool Steels by Laser Surface Treatment *J. of Materials Processing Technology* **201(1-3)** 342-347. DOI: 10.1016/j.jmatprotec.2007.11.232
[2] Obergfell K, Schulze V and Vohringer O 2003 Classification of Microstructural Changes in Laser Hardened Steel Surfaces *IOP Conf. Ser.: Mater. Sci. Eng.* **355(1-2)** 348-356. DOI: 10.1016/S0921-5093(03)00099-6
[3] Zenker R and Buchwalder A 2013 Electron beam surface hardening *ASM Int.* **14** 462-475
[4] Konstantinova M V, Balanovskiy A E, Gozbenko V E et al 2019 Application of plasma surface quenching to reduce rail side wear *IOP Conf. Ser.: Mater. Sci. Eng.* **560** 012146
[5] Balanovskiy A E et al 2019 Surface hardening of structural steel by cathode spot of welding arc *IOP Conf. Ser.: Mater. Sci. Eng.* **560** 012138
[6] Balanovskii A E and Van Huy V 2018 Estimation of Wear Resistance of Plasma-Carburized Steel Surface in Conditions of Abrasive Wear *J. of Friction and Wear* **39(4)** 311-318

[7] Balanovskii A E 2018 Study of the Attachment Stage of a Welding Arc Discharge of Direct-Current Straight Polarity on Aluminum Surface *High Temperature* **56(4)** 486-495
[8] Guseva E A, Kargapoltsev S K, Balanovskiy A E et al 2019 Comparative evaluation of corrosion resistance of wheel and rail steels in various media *IOP Conf. Ser.: Mater. Sci. Eng.* **560** 012181
[9] Kovalenko V 1998 Ways to intensify laser hardening technology *CIRP Annals – Manufacturing Technology* **47** 133-136
[10] Morozov V P, Misyurov A I, Grigoryants A G et al 1987 Analysis of the conditions of crack formation in the overlay laser welding of the Ni-Sg-V-Si system powders *Welding industry* **5** 32-34
[11] Misyurov A N, Grigoryants A G, Shiganov I N et al 2002 *Recovery of parts from heat-resistant alloys by laser surfacing. Welding at the turn of the century* (Moscow: Bauman MSTU Publ.) p 81
[12] Grum J and Sturm R 1997 Laser surface melt-hardening of gray and nodular irons *Applied Surface Science* **109/110** 128-132
[13] Grum J 2007 Comparison of different techniques of laser surface hardening *J. of Achievement in Materials and Manufacturing Engineering* **24(1)** 17-25
[14] Hirogaki T, Nakagawa II, Hayamizu M, ECita Y and Kakino Y 2001 ln-situ heat treatment system for die steels using YAG laser with a machining center *J. of the Int. Societies for Precision Engineering and Nanotechnology* **25** 212-217
[15] Hua M, SeDao, Shao T M and Tarn H Y 2007 Surface transformation of DF-2 steel alter continuous mode laser irradiation *J. of Materials Processing Technology* **192-193** 89-96
[16] Hwang J H, Lee Y S, Kim D Y and Youn J G 2002 Laser surface hardening of gray cast iron used for piston ring *J. of Materials Engineering and Performance* **II** 294-300
[17] Kaula R, Cianesh P, Tiwari P, Nandedkar R V and Natli A K 2005 Characterization of dry sliding wear resistance of laser surface hardened Ln8 steel *J. of Materials Processing Technology* **167** 83-90
[18] Safonov A N 1990 Microstructure of nickel-chrome-borium-silicon compounds melted with laser radiation *Welding production* **11** 34-36
[19] Lusquinos F, Conde J C, Bonss S, Riveiro A, Quintero F, Comesaña R and Pou J 2007 Theoretical and experimental analysis of high power diode laser (HPDL) hardening of AISI 1045 Steel *Applied Surface Science* **254** 948-954
[20] Meijer J and Sprang I V 1991 Optimisation of laser beam transformation hardening by one single parameter *Annals of the CIRP* **40(I)** 183-186
[21] Miokovic T, Schulze V, Vohringer O and Lohe D 2006 Prediction of phase transformations during laser surface hardening of AISI 4140 including the effects of inhomogeneous austenite formation *Materials Science and Engineering A* **435-436** 547-555
[22] Obergfell K, Schulze V and Vohnnger O 2003 Classification of microstructural changes in laser hardened steel surfaces *Materials Science and Engineering* **355** 348-356

Increase of the throughput and processing capacity of the railway line mountain pass section by strengthening the devices of the system of traction power supply

N P Astashkov[1], V A Olentsevich[1], A R Akhmetshin[2], K V Suslov[3], M G Shtayger[4] and A I Karlina[3]

[1] Irkutsk State Transport University, 15, Chernyshevsky St., Irkutsk, 664074, Russia
[2] Kazan State Power Engineering University, Kazan, Russia
[3] Irkutsk National Research Technical University, 83, Lermontov Street, Irkutsk, 664074, Russia
[4] MC Mechel Steel, 1, Krasnoarmeysky Street, Moscow, 125167, Russia

E-mail: olencevich_va@irgups.ru, karlinat@mail.ru

Abstract. According to the forecast target indicators values of the JSC "Russian Railways" integrated development project until 2023, it is necessary to increase the East Siberian railway capacity of the mountain-pass section of the Bolshoy Lug – Slyudyanka-II up to 137 pairs of trains per day. The freight trains number should be 107 pairs per day, a raft of pushing locomotives in the odd direction up to 19 units. The section transportation capacity will increase to 116.7 million tons per year. According to the specified operation parameters, with the growth of cargo turnover on the Irkutsk – Slyudyanka section by 2023, the available capacity shortage will be 14 pairs of trains. In freight traffic, if all provided reserve lines of the the regulatory schedule are used, the capacity deficit will be 11 pairs of various categories trains. The article offers a solution of the existing section power supply devices strengthening problem until the inter-train interval of 8 minutes is reached, for this purpose, the technical parameters of the section substations are considered, and the performance indicators that can provide reliable and cost-effective working conditions are determined. The operation conditions of a traction power network (TPN) of the Bolshoy Lug – Slyudyanka-II section under the existing traffic schedule and under the prospective train schedule are compared. Traction calculations have a special role, since they serve as the basis for drawing up optimal train schedules, determining the railway lines capacity, effectively placing separate points, traction substations, fuel depots, determining the locomotives electricity and fuel consumption rate, solving other practical problems. Traction operation calculation of the electrified mountain pass section of the Bolshoy Lug – Slyudyanka-II railway line was performed using the KORTES software package. The power supply parameters of a given section were used as input data: the separate points number and location, the current speed limits for trains on this section, the longitudinal profile of the railway track, the categories and types of trains that pass through the section, the trains weights and the electric rolling stock types used in the calculation.

1. Introduction

The planned and implemented measures of the JSC "Russian Railways" within the modernization program framework, complete renovation of railway transport infrastructure facilities, introduction of new operation organization ways of the traction power network for railway lines with a constantly increasing traffic volume, impose new requirements for the work of the industry structural divisions. In order to optimally address technical equipment issues and improve the work technology, it is necessary to ensure such a ratio between the volume of work performed and the available fixed assets of the railway transport infrastructure facilities, their technical means, so that the value of total operating costs tends to a minimum value, which in turn will reduce the transportation cost and reach a higher level of this transport type attractiveness.

The planned volume of the company's investment development program for 2020 amounted to 657.4 billion rubles, the investment program structure includes seven main sections, the main of which are ensuring the transportation process safety and increasing the throughput and transport infrastructure facilities processing capacity. The optimal solution of this problem can be achieved only if the actual operating conditions of railway transport infrastructure objects are taken into account, i.e. if all technical devices are loaded unevenly. The unevenness amount in the train traffic organization on a section is affected by a number of factors: the concentrated passenger trains passage, including suburban ones; the duration of transport infrastructure devices repair; the process of organizing train traffic on adjacent sections, etc. [1, 2, 18].

Russian President Vladimir Putin instructed to include in the comprehensive plan for the main infrastructure measures of modernization and expansion in the development of the Baikal-Amur mainline and the TRANS-Siberian railway, these railway lines capacity should be increased by 1.5 times by 2024. The most important measure to ensure a high level of unified standards for the freight trains weights on these sections is to organize their pushing on calculated and difficult ascents.

The history of Russian railway transport development shows that at all times the increase in traffic volumes was realized by pushing trains, this technology was used as the most effective way to increase the train mass on the transit section, which was assuring minimal operating costs level and it was creasing the carrying capacity of railway lines.

For the same purpose, train pushing is also used in Europe, Asia and the Americas, where the use of this technology allows changing the amount of locomotive power per ton of train weight, depending on the complexity of the mountain-pass sections profile of the railway lines. As a result, it is possible to avoid not only changes in the mass of the train at technical stations, but also to increase the traction force use efficiency of the head locomotive throughout the railway direction.

2. The problem of increasing the bandwidth of the railway line mountain pass section

The East Siberian Railway (hereinafter referred to as the ESR) is one of the main and largest transport enterprises in Eastern Siberia today. The operational length of the analyzed transport company is more than 4% of the entire railway transport system network (hereinafter RTSN). Over the past 10 years, freight turnover on the main line has increased by 60%, which was made possible only thanks to a large-scale program of railway infrastructure modernization. Cargo turnover since the beginning of 2020 amounted to 160 billion tariff tons-kilometers.

The most difficult section in the operational work field of the ESR, in all years, was and remains the mountain-pass section of the railway line between the Bolshoy Lug and the Slyudyanka-II railway stations, which is 83 kilometers long. The track profile is characterized by a prolonged rise in the area Bolshoy Lug of Podkamenaya, which greatly complicates its operational and technical specifications, generates a large amount additional train operation, because the freight traffic organization feature on the analyzed railway line section is the procedure of pushing by locomotives more than 80% of freight trains on the even directions from the Bolshoy Lug train station, and also pushing of the freight trains in the odd direction from the Slyudyanka-I to the Bolshoy Lug stations. The standard value of pushing locomotives for freight trains on the section is 37 units. The speed on the section does not exceed 60 km/h. The difficult terrain does not allow speeding up the trains' passage along the section [2-4, 6].

The average unified length of freight trains that follow the same even and odd numbered sections of the road under consideration is 71 conventional railway wagon. The freight trains weight standards data is presented in table 1.

Table 1. Average freight trains weight standards on the considered section of the same railway line.

Locomotive	Freight trains weight standards, t			
	Unified (for transit trains)		Power of the locomotive (for railway line trains)	
	Odd direction	Even direction	Odd direction	Even direction
VL85	6000	6000	6300	6300
VL80t	2000	4200	2000	4200
VL80t	4200	4400	4200	4400
2ES5K	4200	4400	4200	4400
2VL80v	7500	7500	7500	7500
2*2ES5	7500	7500	7500	7500

Five years ago, 75 freight trains pairs passed through the section per day, the capacity factor was approximately 0.8. Today, this value has reached 96 pairs, and the coefficient is close to one. However, the site is operating almost at the limit of its capabilities. The qualitative composition of the railway wagon traffic on the section fully depends on the functional purpose and the cargo turnover size, so the freight trains transit traffic without processing mainly pass through the precinct stations, while 20-30% are transit trains with processing and local section trains. Figure 1 shows a diagram of the train traffic organization on the Bolshoy Lug-Slyudyanka-II railway section, Figure 1.

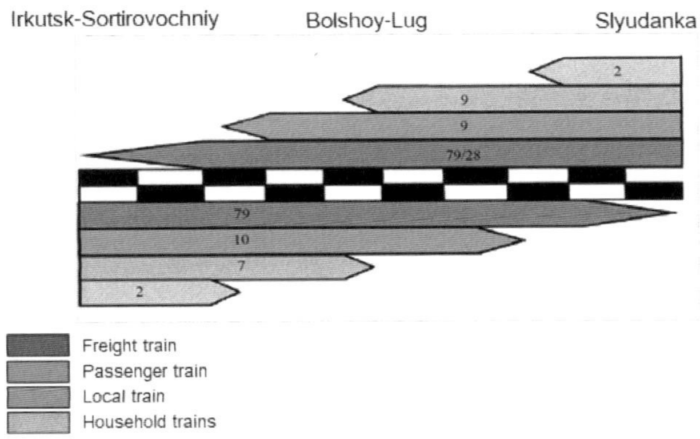

Figure 1. Average daily train traffic diagram.

The site is consist of five traction substations. Table 2 shows data about the installed transformers types at substations and the power of three-phase short circuits.

The forecast growth value in the volume of traffic to the Eastern Russian Federation ports puts forward among the most important tasks the further increase in the capacity of this section, the planned value is aimed at 137 pairs of trains per day. The number of freight trains should be 107 pairs per day, a pushing locomotives raft in the odd direction should be 19 units. The section transportation capacity will increase to 116.7 million tons per year [2, 4, 6].

Table 2. Parameters of substations of the Bolshoy Lug-Slyudyanka II section.

Substations	Short-circuit power, MVA	Type of transformer	Number
Rassoha	1584	TDTNJ-40000/110-71U1	1
Podkamenaya	1663	TDTNJ-40000/110-71U1	1
Andrianovskaya	1235	TDTNJ-40000/110-71U1	1
Angasolka	1125	TDTNJ-40000/110-71U1	1
Slyudyanka	1570	TDTNJ-40000/110-71U1	1

3. A proposal to increase the throughput and processing capacity of the railway line mountain-pass section by strengthening the power network system devices

The purpose of mathematical calculations of the power network system for railway lines and sections is to establish such parameters for the functioning of this system that would ensure the reliability and cost-effective conditions for the operation of an electrified railway section [7, 8]. In this case, traction calculations play a special role, since they serve as the basis for drawing up optimal train schedules, determining the railway lines capacity and effective placement of separate points, traction substations and fuel depots, determining the rate of electricity and fuel consumption by locomotives and solving other practical problems [11, 12].

Traction calculation of the operation of the electrified mountain pass section of the Bolshoy Lug-Slyudyanka-II railway line was performed using the KORTES software package. The given section power network parameters were used as input data: the number and location of separate points, the current speed limits for trains on this section, the railway track longitudinal profile, the categories and trains types passed through the section, the trains weight and electric rolling stock types used in the calculation [5, 6, 13].

In order to further calculate the traction power network load modes when passing trains, it is necessary to make a traction calculation for the basic freight trains weights. The traction calculation results are: the total travel time; the travel time under current; the active and full electricity consumption; the specific consumption of active and full electricity for each inter-substation zone [9-10, 14-15].

The traction calculations results for even and odd directions of the train movement on the considered section for different weight norms are presented in Table 3.

Table 3. Traction calculation results for even and odd directions of train movement.

Name of the stage	Length, km	Locomotive serie	Train weight, t	Total running time, min.	Running time under the current min.	Consumption of active energy, kW*h	Total power consumption, kW*h
Weight 6300 tons							
Even direction							
Bolshoy Lug-Podkamennaya	25.8			31	26.0	7762.1	9481.6
Podkamennaya - Glubokaya	12.4			14	13.3	3794.1	4641.9
Glubokaya – Andrianovskaya	11.9	ВЛ85	6300	14	11.4	2155.8	2645.4
Andrianovskaya – Angasolka	11.9			15	2.6	609.9	769.9
Angasolka – Slyudanka-II	21.4			26	0.0	401.2	515.8
Odd direction							
Bolshoy Lug-Podkamennaya	25.8			31	0.6	607.7	774.7
Podkamennaya - Glubokaya	12.4			14	0.2	269.9	345.7
Glubokaya – Andrianovskaya	11.9	ВЛ85	6300	14	2.9	690.5	859.0
Andrianovskaya – Angasolka	11.9			15	13.0	4443.1	5426.9
Angasolka – Slyudanka-II	21.4			26	22.2	7792.7	9513.9
Weight 4400 tons							
Even direction							
Bolshoy Lug-	25.8	2ЭС5К	4400	31	31.0	4508.3	5161.3

Podkamennaya							
Podkamennaya - Glubokaya	12.4			14	13.9	2288.4	2611.7
Glubokaya – Andrianovskaya	11.9			14	12.6	1288.4	1479.6
Andrianovskaya – Angasolka	11.9			15	3.1	323.4	386.4
Angasolka – Slyudanka-II	21.4			26	0.0	200.3	257.6
Weight 4200 tons							
Even direction							
Bolshoy Lug-Podkamennaya	25.8			31	33.2	4721.6	5986.3
Podkamennaya - Glubokaya	12.4			14	14.1	2373.0	2995.6
Glubokaya – Andrianovskaya	11.9	1.5 ВЛ80р	4200	14	12.7	1272.9	1585.9
Andrianovskaya – Angasolka	11.9			15	1.4	290.2	363.1
Angasolka – Slyudanka-II	21.4			26	0.0	200.3	257.6
Odd direction							
Bolshoy Lug-Podkamennaya	25.8			31	0.5	284.2	361.8
Podkamennaya - Glubokaya	12.4			14	1.1	201.2	246.6
Glubokaya – Andrianovskaya	11.9	2ЭС5К	4200	14	1.2	267.5	323.8
Andrianovskaya – Angasolka	11.9			15	14.0	2535.5	2897.9
Angasolka – Slyudanka-II	21.4			26	25.7	4461.1	5103.2
Weight 2000 tons							
Odd direction							
Bolshoy Lug-Podkamennaya	25.8			31	3.1	294.7	372.9
Podkamennaya - Glubokaya	12.4			14	2.4	158.4	196.9
Glubokaya – Andrianovskaya	11.9	ВЛ80р	2000	14	2.8	170.4	212.6
Andrianovskaya – Angasolka	11.9			15	12.6	1339.8	1684.7
Angasolka – Slyudanka-II	21.4			26	22.6	2468.5	3106.3
Weight 7500 tons							
Even direction							
Bolshoy Lug-Podkamennaya	25.8			31	31.8	7622.7	8731.1
Podkamennaya - Glubokaya	12.4			14	13.9	3811.8	4352.7
Glubokaya – Andrianovskaya	11.9	2ЭС5К	7500	14	12.8	2239.5	2570.3
Andrianovskaya – Angasolka	11.9			15	5.8	630.7	753.5
Angasolka – Slyudanka-II	21.4			26	0.0	333.8	429.4

IOP Conf. Series: Materials Science and Engineering 1111 (2021) 012005 doi:10.1088/1757-899X/1111/1/012005

Odd direction							
Bolshoy Lug-Podkamennaya	25.8			31	0.3	518.9	663.2
Podkamennaya - Glubokaya	12.4			14	0.0	247.9	318.9
Glubokaya – Andrianovskaya	11.9	2ЭС5К	7500	14	3.5	854.2	1010.9
Andrianovskaya – Angasolka	11.9			15	13.0	4595.7	5230.9
Angasolka – Slyudanka-II	21.4			26	22.4	8012.0	9123.9

Graphical images of the performed calculations are shown in figures 2-5.

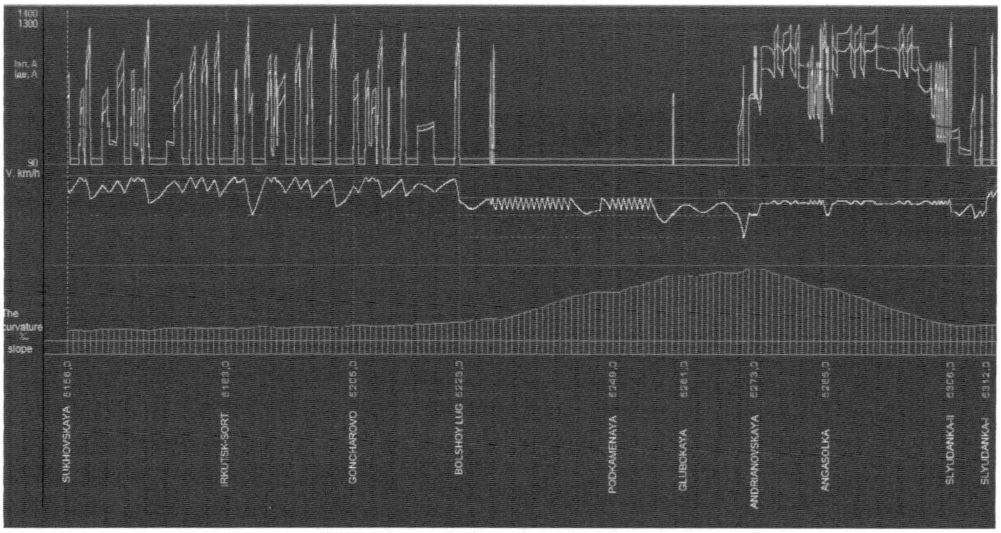

(a) 6300 tons weight trains of the even direction

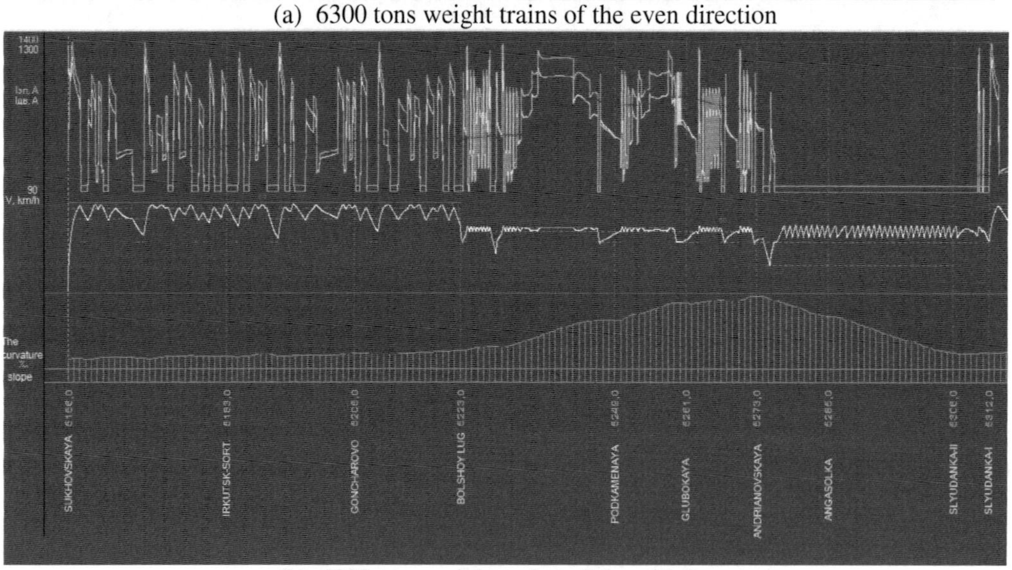

(b) 6300 tons weight trains of the odd direction

Figure 2. Graphic representation of the traction calculation result for a 6300-tons train of even (a) and odd directions (b).

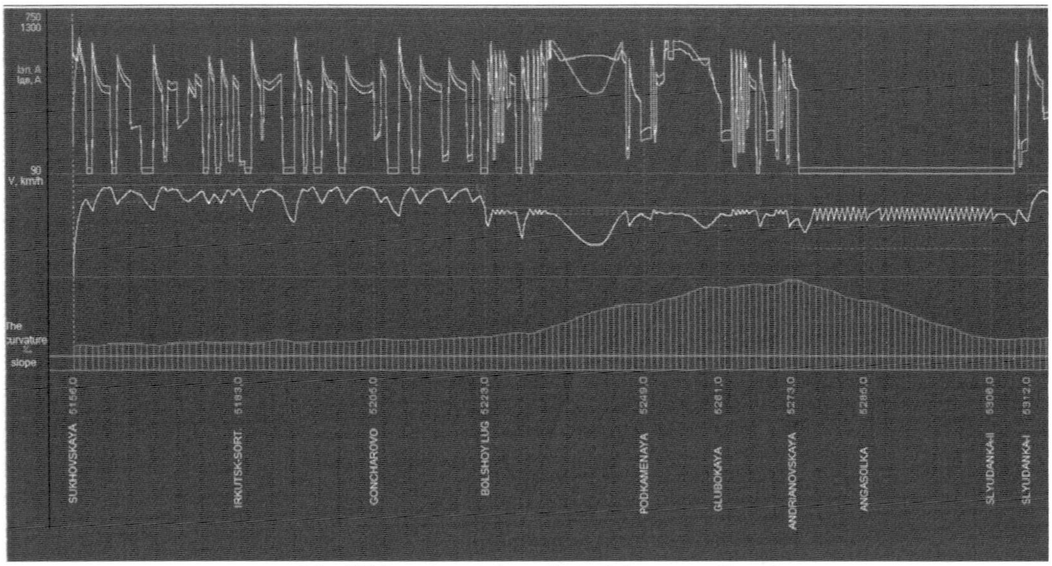

(a) 4200 tons weight trains of the even direction

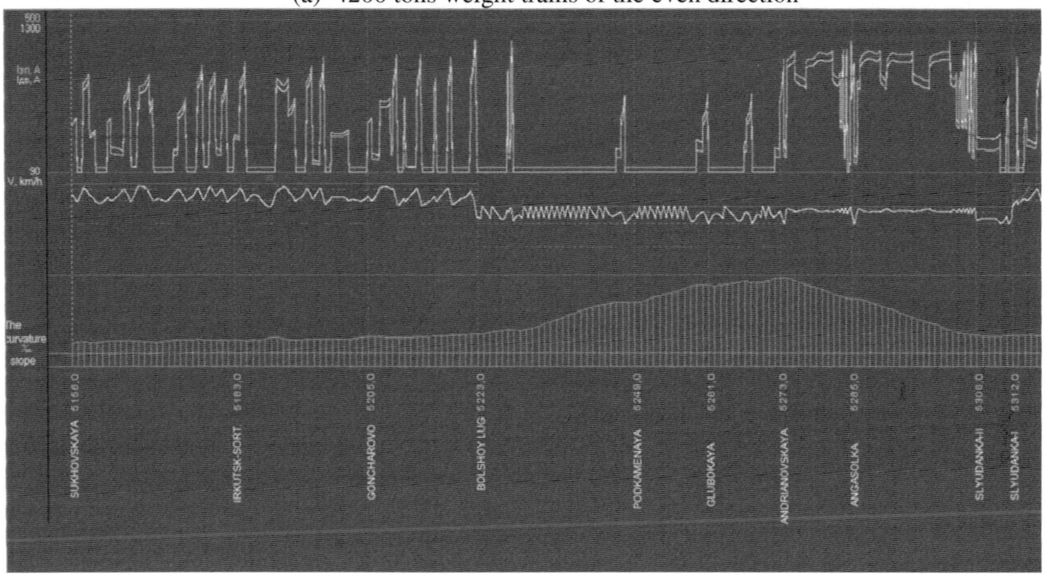

(b) 2000 tons weight trains of the odd direction

Figure 3. Graphic representation of the traction calculation result for a 4200-tons train of even (a) and for a 2000-tons train odd directions (b).

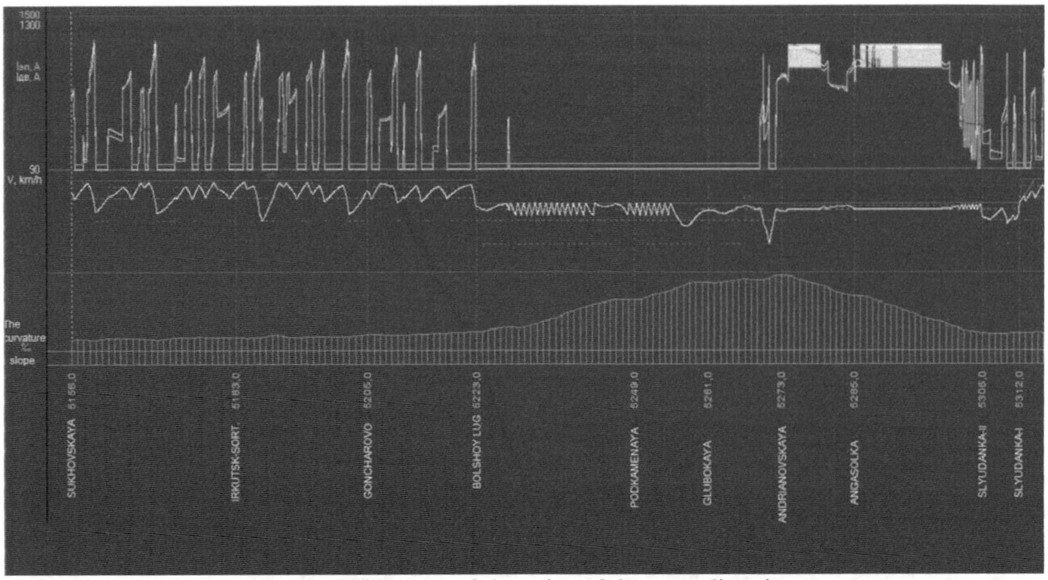

(a) 7500 tons weight trains of the even direction

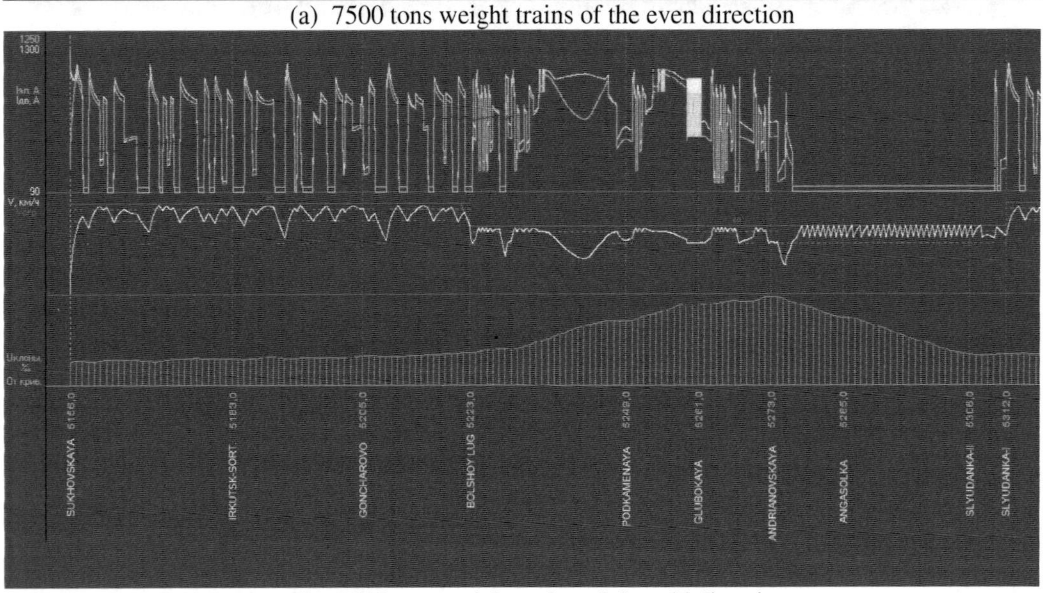

(b) 7500 tons weight trains of the odd direction

Figure 4. Graphic representation of the traction calculation result for a 7500-tons train of even (a) and odd directions (b).

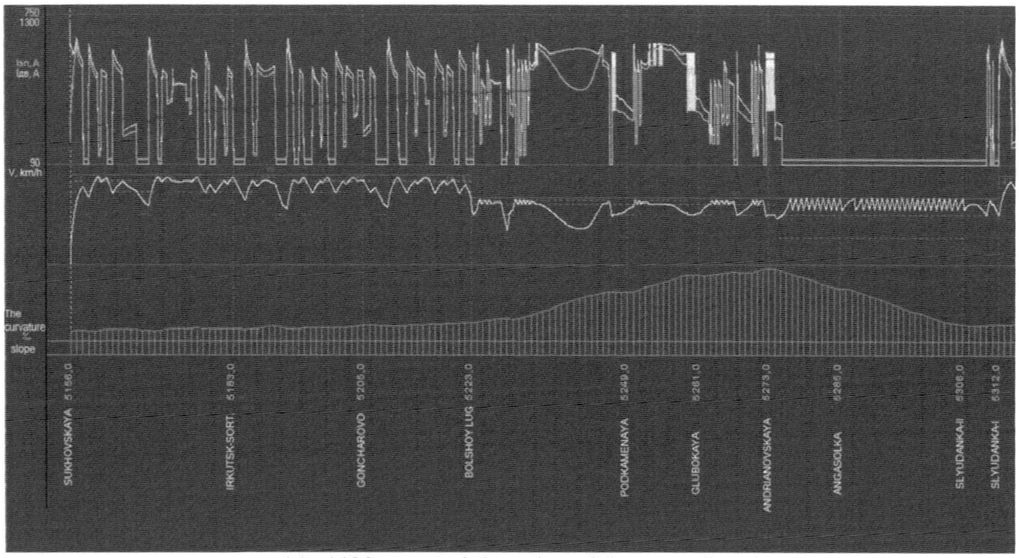

(a) 4400 tons weight trains of the even direction

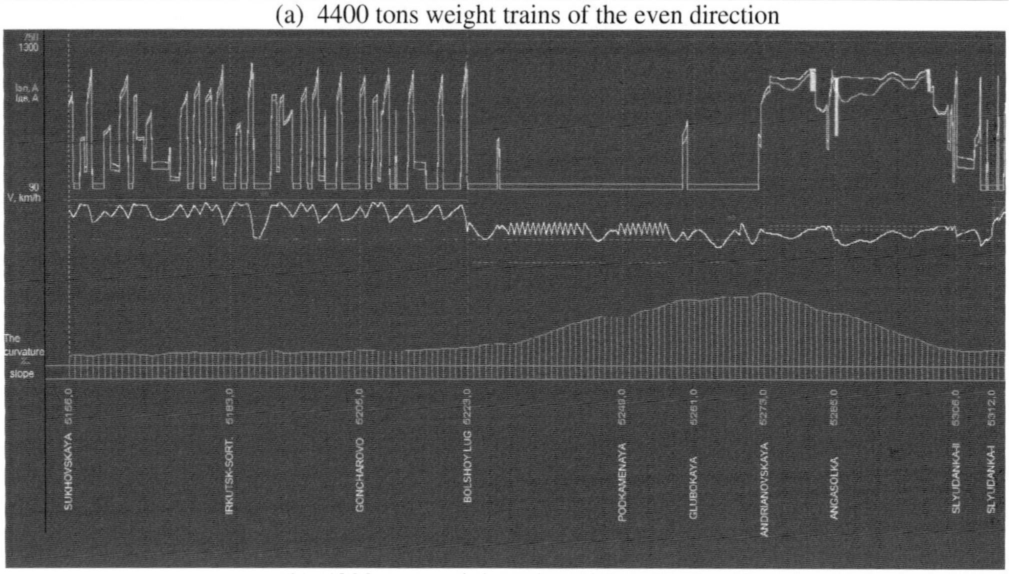

(b) 4200 tons weight trains of the odd direction

Figure 5. Graphic representation of the traction calculation result for a 4400-tons train of even (a) and for a 4200-tons train odd directions (b).

The operation parameters of the traction power network systems (TPNS) of the Bolshoy Lug – Slyudyanka-II section under the existing traffic schedule are shown in Figure 6.

The calculations results of the minimum allowable inter-train interval under the existing power scheme for the existing train traffic value are presented in Table 4.

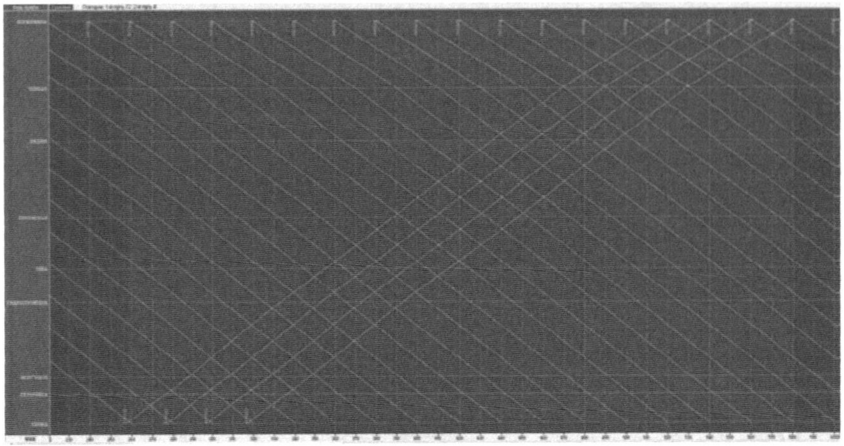

Figure 6. Parameters of the section's TPNS operation under the existing train schedule.

Table 4. Calculations results of the minimum allowable inter-train interval for the existing train traffic value.

Inter-station section name	Interval value, min, limited			
	Step-down transformers power value	Voltage in the contact network	Heating the contact network wires	Resulting value
Bolshoy Lug-Podkamennaya	7	6	6	12
Podkamennaya - Glubokaya	8	10	6	8
Glubokaya – Andrianovskaya	8	16	6	14
Andrianovskaya – Angasolka	7	11	6	16
Angasolka – Slyudanka-II	6	13	6	13

The calculations results of the minimum allowable inter-train interval for prospective traffic value showed that there are several limiting stages on the considered mountain-pass section Bolshoy Lug – Slyudyanka-II:

- Bolshoy Lug-Podkamennaya, inter-train interval of 12 minutes;
- Glubokaya-Andrianovskaya, inter-train interval of 14 minutes;
- Andrianovskaya-Angasolka, inter-train interval of 16 minutes;
- Angasolka-Slyudyanka-II, inter-train interval of 13 minutes.

Figure 7 shows the determination of the parameters of the section's TPNS operation in the perspective train schedule on the Bolshoy Lug – Slyudyanka-II section.

The calculations results of the minimum allowable inter-train interval under the existing power network scheme for the prospective traffic values, taking into account the measures implementation to strengthen the traction component, see table 5.

Taking into account the obtained calculations results, we can conclude that to ensure the passage of trains of the specified traffic value with a given inter-train interval of 8 minutes, are needed measures to increase the voltage level on the limiting stages [4, 15-17].

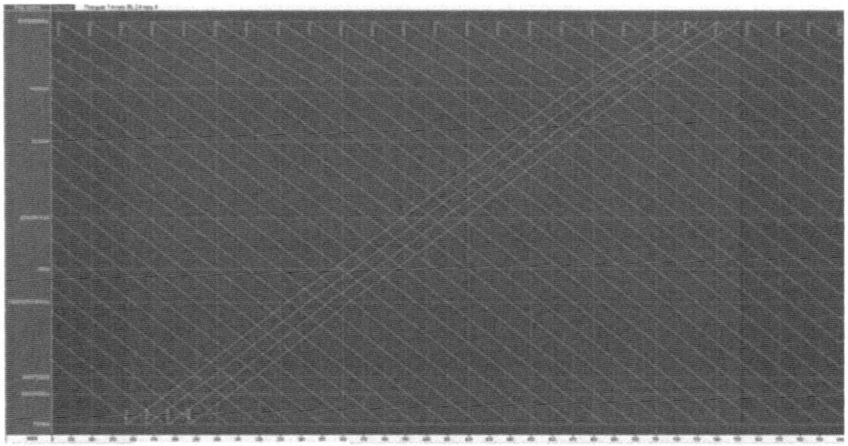

Figure 7. Parameters of the section's TPNS operation in the perspective train schedule.

Table 5. Calculations results of the minimum allowable inter-train interval for prospective traffic values.

Inter-station section name	Interval value, min, limited			
	Step-down transformers power value	Voltage in the contact network	Heating the contact network wires	Resulting value
Bolshoy Lug-Podkamennaya	7	6	6	8
Podkamennaya - Glubokaya	7	8	8	8
Glubokaya – Andrianovskaya	8	9	8	9
Andrianovskaya – Angasolka	7	8	8	8
Angasolka – Slyudanka-II	7	8	7	8

4. Conclusion

With the existing scheme with a given inter-train interval of 8 minutes, based on the performed traction calculations, it is determined that the inter-train interval can be reduced to 8 minutes by strengthening the traction power network systems of the mountain-pass section. Based on the obtained calculations, on the section under consideration was organized train schedule and its indicators were determined. The number of even-numbered freight trains can be increased to 144, which are 35 trains more than with the existing power network technology. In the odd direction, the increase will be 26 trains and will reach the level of 105 trains per day. Accordingly, the number of pushing locomotives will increase from 28 to 37 units. Strengthening the TPNS will increase the section speed by 7.8 km / hour, and the locomotive's performance by 536 tons-km-gross. The locomotive's turnover will be reduced by 1.02 hours.

The required financial resources value to enhance traction substations will amount to 66 million rubles. Annual operating costs for new equipment maintenance and service – 108 million rubles/year. This project payback period is 5 years.

References

[1] 2013 *Development strategy of the Russian Railways Holding for the period up to 2030* (Moscow: JSC Russian Railways)
[2] *Official website of Russian Railways:* http://rzd.ru

[3] Kokurin I M and Timchenko V S 2013 Methods for determining "Bottlenecks" limiting the throughput of railway lines *Bull. of the Petersburg University of Railways* **1(34)** 15-21
[4] Olentsevich V A, Belogolov Yu I and Kramynina G N 2019 Set of organizational, technical and reconstructive measures aimed at improvement of section performance indicators based on the study of systemic relations and regularities of functioning of railway transport system *IOP Conf. Ser.: Mater. Sci. Eng.* **832** 012038
[5] 2007 *Rules for the arrangement of the traction power supply system of the railways of the Russian Federation TsE-462* (Moscow: Transport) p 450
[6] 2018 *Technological process on the organization of pushing traffic on the section Bolshoy Lug - Slyudyanka-I.* 21 August 2018, No. VSZD-800 / r (Irkutsk) p 29
[7] Alekseeva T L, Ryabchenok N L and Astrakhantsev L A 2018 Electric power system of railways and electric traction of trains. State and development prospects *Transport infrastructure of the Siberian region* **2** 234-239
[8] Ryabchenok N A, Alekseeva T L, Astrakhantsev L A and Martusov A L 2019 Energy efficiency of electric traction of trains *Modern technologies. System analysis. Modeling* **1(61)** 144-156
[9] Cherepanov A V, Tikhomirov V A and Kutsy A P 2017 Influence of voltage unbalance on energy efficiency *Modern technologies. System analysis. Modeling* **4(56)** 207-217
[10] Kryukov A V and Bezridny E S 2020 Modeling of modes of traction networks when passing heavy trains *Modern technologies and scientific and technical progress* **1(7)** 207-208
[11] Osipov S I and Osipov S S 2000 *Basics of train traction. Textbook for students of technical schools and colleges of railway transport* (Moskow: UMK Ministry of Railways of Russia) p 592
[12] Rosenfeld V E, Isaev V P, Sidorov N N and Ozerov M I 1995 *The theory of electric traction* (Moskow: Transport) p 294
[13] 1985 *Rules for traction calculations for train operation* (Moscow: Transport) p 287
[14] Bondarenko O A 2016 Issues of development of heavy traffic of freight trains *Science and education for transport* **1** 91-93
[15] Tanaino Yu A 2015 Investigation of the circulation parameters of heavy and connected trains on the West Siberian railway *Scientific problems of transport in Siberia and the Far East* **4** 9-11
[16] Akhmetshin A, Mendeleev D and Marin G 2020 Improvement of Electricity Quality Indicators in Electric Networks with Voltage of 0.4-10 kV *Int. Russian Automation Conf. (RusAutoCon)* (Sochi, Russia) DOI: 10.1109/RusAutoCon49822.2020.9208158
[17] Marin G, Mendeleev D and Akhmetshin A 2020 Modeling of asynchronous motor operation modes for the correct selection of voltage regulation devices *E3S Web of Conf.* **178** 01015. DOI: 10.1051/e3sconf/202017801015
[18] Fedotov A, Misbakhov R, Chernova N, Bakhteev K and Akhmetshin A 2019 Calculation of electrical loads for decentralized power supply systems and the choice of power gas piston installations which use electrochemical energy storage *E3S Web of Conf.* **124** 05068. DOI: 10.1051/e3sconf/201912405068

Problems and prospects of using mobile devices when performing operational monitoring of water supply structures

M A Bandurin[1], I F Yurchenko[2], I P Bandurina[1] and A P Bandurin[3]

[1]Kuban State Agrarian University named after I. T. Trubilin, 13, Kalinina st., Krasnodar, 350044, Russia
[2]All-Russia Research Institute of Hydraulic Engineering and Reclamation named after A. N. Kostyakov, 44, Bolshaya Akademicheskaya st., Moscow, 127550, Russia
[3]Novocherkassk Engineering and Reclamation Institute named after A. Kortunov, Don State Agrarian University, Pushkinskaya st. 111, Novocherkassk, 346400, Russia,

E-mail: bandurin.m@edu.kubsau.ru

Abstract. This article presents the results of work on a mobile device for conducting operational monitoring of the technical condition of water supply structures, designed to determine various parameters of defects and damages, as well as calculate the projected remaining life of their elements. The modern development of mobile devices, network computer technologies, cellular communication and mathematical methods opens up new prospects in the automation of operational monitoring of water supply structures. The technical part of the mobile device for conducting operational monitoring includes various technical schemes of frames for each water supply structure separately. It can be used for conducting operational monitoring of the technical condition and determining the remaining life of the protective coating of water supply channels and soils under them, identifying dangerous defects and damages, as well as evaluating and predicting its technical condition and further suitability for operation. The technical result achieved by this mobile device is to identify the influence of defects and damages, both of the protective coating itself and the condition of the soil located under it, on the formation of decompressions and subsidence at an early stage of their formation. The mobile device is developed using the Data Mining database management system and Microsoft Access. The use of a mobile device makes it possible to improve the quality of operational monitoring of water supply channels due to the fact that the survey is carried out over the entire cross-section of the water supply channel for defects, damages, decompression and subsidence of the soil under the structure by non-destructive testing methods. And the presence of a vehicle that transports the complex allows you to speed up significantly the survey of water supply channels.

1. Introduction

The modern development of mobile devices, network computer technologies, cellular communication and mathematical methods opens up new prospects in automation of operational monitoring of water supply structures [1]. Until now, Russia has not had an operational, reliable, cheap universal digital technology for effective operational monitoring of water supply structures. The proposed mobile device for carrying out operational monitoring of the technical condition of water supply structures is designed to determine various parameters of defects and damages, as well as calculate the projected remaining life of elements of hydraulic structures [2]. Your mobile device allows you to assess the geometric parameters of each defect, an operational residual life assessment (prediction of the total number of cycles of freezing and thawing as the past operating period, and remaining until the loss of bearing capacity of reinforced concrete elements of water conveyance structures) and the complex influence of several factors on the reliability of structures [3], the most characteristic ones of which are abrasion, leaching, wear in areas with different hydraulic characteristics.

Here are a number of new features of modern mobile devices [4–6]: collecting and transmitting data from sensors without a computer using mobile communication media that has access to the global Internet;

– collecting data on the state of water supply structures using web cameras to analyze differences between the current image and previously taken images [7];

– placement of data analysis and management programs on a computer network server outside of water supply facilities;

– data analysis using Data Mining and Microsoft Access (DBMS) methods [8], a method of group accounting of arguments that allows you to identify behavior patterns and create models of property change dynamics;

– creating virtual models of water supply structures, organizing continuous monitoring of measurement results and modeling results, using models to predict the state of structures.

The results of visual observations and field studies were used as initial data when creating a complex of operational monitoring of the technical condition of water supply structures [9]:

– results of visual inspection of water supply structures with identification of characteristic damage to individual elements;

– geometric parameters of damage obtained using non-destructive monitoring devices [10].

2. Materials and methods

As a result of numerical experiments there were identified zones of hydraulic structures that may contain the same types of characteristic damage, which makes it possible to streamline the process of laying GPR profiles, determining the points where it is necessary to measure the strength of concrete during field surveys.

To perform the necessary calculations of the technical condition, the mobile device has an information and reference database on hydraulic structures operated on the territory of Russia [11].

The technical part of the mobile device of operational monitoring includes various technical schemes of frames of each hydraulic structure separately and can be used for operational monitoring of the technical condition and determining the remaining life of the protective coating of water supply channels and soils under them, identifying dangerous defects and damages, as well as evaluating and predicting its technical condition and further suitability to operation.

3. Results

The technical result achieved by the mobile device is the detection of defects and damages, both of the protective coating itself and the condition of the ground base located under it, for the formation of decompressions and subsidence at an early stage of their formation [12]. This technical result is achieved by the fact that the mobile device for conducting operational monitoring contains a mover for free movement along the bottom of the water supply channel with an operator with a processing module located in it, and the frame consists of rods and consists of three parts connected by hinges [13].

Figure 1 shows the technical part of a mobile device for conducting operational monitoring in an axonometric projection, which includes a frame 1 consisting of rods that repeat the shape of the water supply channel and consists of three parts connected by hinges 2 with rubber wheels 3, a processing module that includes the GLONASS system 4, a motion sensor 5, and antenna blocks 6 located along the perimeter of the frame 1 [14].

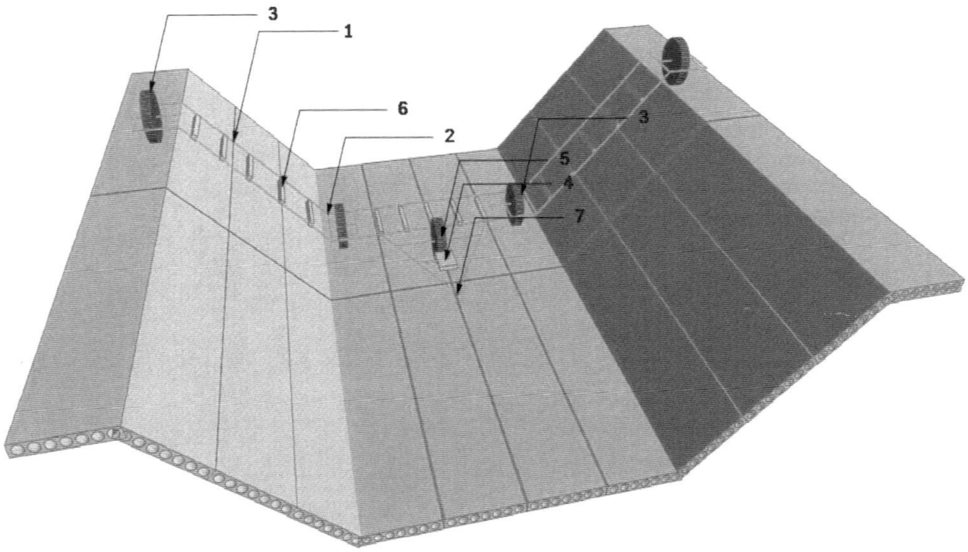

Figure 1. Mobile device for performing operational monitoring and determining the remaining service life for channels

For operational monitoring, the technical part of the mobile device is placed in the water supply channel in the absence of water, after which power is supplied to the processing module 4, which confirms the readiness of all systems for operation. With the help of the mover 7, the frame 1 moves along the water supply channel and data from the motion sensor 5 and antenna blocks 6 are sent to the processing module 4, where the data received from the antenna blocks is detailed and the location of defects and damages is positioned using the GLONASS system [15]. Using classical solutions of the theory of elasticity, according to the known angular displacement can be calculated from the spatial displacement. Moreover, the recalculation program can be located on a remote server of a computer network and serve any number of sensors located even on several objects [16].

The use of a mobile device makes it possible to improve the quality of operational monitoring of water supply channels due to the fact that the survey is carried out over the entire cross-sectional area of the water supply channel (the entire area of the structure is covered) for defects, damage, decompression and subsidence of the soil under the structure by non-destructive testing methods, and the presence of a vehicle that transports the complex can significantly speed up the survey of water supply channels [17].

The technical part of the mobile device can be used to conduct operational monitoring and determine the remaining life of the protective coating of water supply pipelines, storm drains and soils under them, as well as to conduct operational monitoring of reinforced concrete pipelines located in the ground, identify dangerous defects and damage, assess and predict their technical condition and further suitability for operation.

Before starting work at the entrance to the water supply pipeline, the operator attaches the stop 6 with fixing screws to the walls of the pipeline 7, and then, using a telescopic rail 5, moves the frame 3 with rubber wheels 1 inside the pipeline 7, located in the ground 8 (Fig. 2). Data from the motion sensor 4 and antenna blocks 2 are sent to the processing unit, where data is processed, the technical condition of the elements is evaluated for defects and damage, as well as the surrounding soil 8 to the formation of decompression and subsidence, a detailed decoding of the data obtained from the antenna blocks and positioning of the location of defects and damage using the GLONASS navigation system [18].

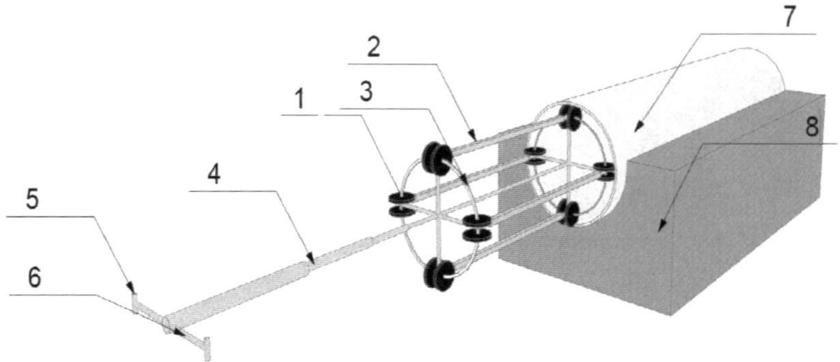

Figure 2. Mobile devices for operational monitoring and determination of the remaining life of the robot capacity of rainwater drainage structures

Software of the mobile device is based on natural research of a technical condition of hydraulic structures, data obtained by testing the water resistance of concrete and data of concrete abrasion, as well as information obtained during the testing of structural properties of concrete samples to cyclic freezing and thawing.

A generalization and analysis of the results of studies of the state of long-term water supply structures in order to determine their residual resource of their working capacity is performed. Reinforced concrete water supply structures are most often destroyed by leaching. The relatively high filtration coefficient, high porosity, and flow pressure determined the filtration of water through concrete. To calculate the time of safe lime leaching during water filtration through the concrete lining there were studied the filtration conditions, the thickness of the structure and the filtration coefficient of the concrete lining were determined [19].

The filtration coefficient of concrete of monolithic linings of water supply channels is determined. Water resistance of concrete is estimated by reducing the strength under water saturation and is characterized by a softening coefficient that expresses the ratio of the strength of water-saturated concrete to the strength of dry concrete. But such studies were not conducted, since the reduction in the strength of concrete is insignificant and will not affect the load-bearing capacity of the structure.

A generalized analysis of the results of studies of the state of long-term water supply structures is performed to determine their residual resource of resistance to cyclic freezing and thawing. On the basis of local empirical dependences of the loss of strength characteristics of structural structures on the cycles of freezing and thawing, the mathematical support of a mobile device was formed in the format of a calculation model of the structure as a whole. This allows calculating the behaviour of defects and damage to water supply structures under the further influence of cyclic freezing and thawing [20].

The degree of accident risk of elements of water supply structures (small – $k \leq 0.15$; obtained using non-destructive testing devices) was also evaluated, moderate – k=0.16...0.30; large – k=0.31...0.50; emergency – $k>0.51$).

Information about the state of the water supply system is evaluated by a variety of criteria. If the quantitative value of each local criterion falls within a certain range of values, we can make a conclusion about the state of the structure. Taking into account the multiplicity of criteria, it is advisable to "collapse" them into a single integral criterion, for which close ranges of values are preferred, which is achieved by normalizing the latter [21].

A priority direction of future improvement of mobile device software may be the creation of a virtual model of a hydraulic structure, which will increase the efficiency of monitoring and forecasting the state of hydraulic structures. Each update of information about the state of the structure is accompanied by a comparison with the forecast on the model, if the mismatch is higher than a certain value, the forecast on the model is recalculated, if the mismatch is growing at a rate higher than a certain set, then the model needs to be refined.

4. Conclusion

Modern trends in the modernization of the system of operation of the reclamation water management complex are associated with automation and digitalization of monitoring the technical condition of hydraulic structures. The development of mobile devices for monitoring water supply structures, performed as a part of these studies, solves the actual problem of technical operation - the formation of a high-tech and effective system of monitoring, accounting and evaluating the performance of GTS resources and is very timely.

To determine the residual life of the protective coating of water supply channels and soils under them, identify dangerous defects and damages, as well as assess and predict its technical condition and further suitability of operation, technical devices differentiated by types of hydraulic structures are proposed as a part of a mobile device: open channels, tubular hydraulic structures.

As initial data for the operation of the mobile device software environment, we used the results of the conducted research, which include:

– visual inspections of water supply structures with identification of characteristic damage to individual elements;

– quantitative parameters of damage to elements of water supply structures obtained using non-destructive testing devices (width, depth and length of damage, etc.);

– quantitative assessment of damage to elements of water supply structures.

The software package of the mobile device is developed using the Data Mining and Microsoft Access database management systems.

References

[1] Olgarenko V I, Olgarenko G V and Olgarenko I V 2018 A method of integral efficiency evaluation of water use on irrigation systems *Int. Multidisciplinary Sci. GeoConf. SGEM* **18(3.1)** 3–9

[2] Ovchinnikov A S, Bocharnikov V S, Skorobogatchenko D A, Borisenko I B, Chernyavsky A N, Abezin V G, Ryadnov A I, Shaprov M N, Kuznetsov N G, Nekhoroshev D A, Sedov A V, Grigorov S M, Fomin S D and Ol'garenko V I 2018 The optimum geometrical form modeling of the "striegel" type harrow *ARPN J. of Engineering and Appl. Sciences* **13(23)** 9138-9144

[3] Kireicheva L V and Zakharova O A 2002 The effect of cyclic irrigation with wastewater on the properties of gray forest soils *Eurasian Soil Sci.* **35(9)** 990-995

[4] Kuznetsov E V, Khadzhidi A E, Poltorak Y A and Kuznetsova M 2019 Operator model to control process of obtaining vermicompost *EurAsian J. of BioSciences* **13(1)** 315-321

[5] Kireicheva L V and Karpenko N P 2015 Evaluation of the efficiency of irrigation in a zonal soil sequence *Eurasian Soil Sci.* **48(5)** 524-532

[6] Bandurin M A, Yurchenko I F and Bandurina I P 2019 Computer technology to assess the capacity reserve of the irrigation facilities of the agro-industrial complex *Int. Multi-Conf. on Industrial Engineering and Modern Technologies (FarEastCon)* pp 1-5. doi: 10.1109/FarEastCon.2019.8933970

[7] Kuznetsov E V, Khadzhidi A E, Kilidi K I and Kurtnezirov A N 2018 Management of agro-resource potential for agricultural landscape stability increase *Plant Archives* **18(2)** 2151-2158

[8] Abdrazakov F K, Orlova S S, Pankova T A, Mirkina E N and Mikheeva O V 2018 Risk assessment and the prediction of breakthrough wave during a dam accident *J. of Int. Research* **8(1)** 154-161

[9] Abdrazakov F K, Pankova T A, Zatinatsky S V, Orlova S S and Trushin Yu E 2017 Increasing efficiency of water resources use in forage crops irrigation *Int. J. of Advanced Biotechnology and Research* **8** 283-293

[10] Degtyareva O G, Degtyarev G V, Togo I A, Terleev V V, Nikonorov A O and Volkova Yu V 2016 Analysis of stress-strain state rainfall runoff control system – buttress dam *Procedia Engineering* **165** 1619-1628

[11] Kireicheva L V and Khokhlova O B 2000 Elemental composition of different fractions from the

sapropel organic matter *Eurasian Soil Sci.* **33(9)** 947-949

[12] Degtyareva O G, Degtyarev G V, Lavrov N L and Aliev D U 2018 Constructive-technological decisions in regulating the flow of atmospheric precipitation *Magazine of Civil Engineering* **82(6)** 32-48

[13] Bandurin M A, Volosukhin V A, Vanzha V V, Mikheev A V and Volosukhin Y V 2018 Finite-element simulation of possible natural disasters on landfall dams with changes in climate and seismic conditions taken into account *J. Phys.: Conf. Ser.* **1015(3)** 032011

[14] Degtyarev G V, Belokur K A and Sokolova I V 2018 Modeling of the building by numerical methods at assessment of the technical condition of structures *Mater. Sci. Forum* **931** 141-147

[15] Gaydzhurov P P, Kravchenko G M and Savelieva N A 2014 Finite element modeling of elastic plastic bending of steel beams with use of rod end elements *Construction mechanics and design of structures* **2** 17-22

[16] Chesnokov B P, Abdrazakov F K, Naumova O V, Krivoschapov D S and Strelnikov V A 2017 The use of ionizing radiation for the tungsten preparation *J. of Industrial Pollution Control* 1-12

[17] Bandurin M A, Yurchenko I F, Volosukhin V A, Vanzha V V and Volosukhin Ya V 2018 Ecological and economic efficiency of diagnostics of technical condition of water supply facilities of irrigation systems *J. Ecology and Industry of Russia* **22(7)** 66-71

[18] Chesnokov B P, Naumova O V, Strelnikov V A, Abdrazakov F K and Tronin B A 2016 Polyethylene production from granules using high voltage *Int. J. of Appl. Engineering Research* **11** 2140-2144

[19] Fen N, Kozlov D B and Rumyantsev I S 2016 Hydraulic studies of stepped spillways of various design *Power Technology and Engineering* **49(5)** 337-344

[20] Abdrazakov F K, Ryzhko N F, Ryzhko S N, Horin S A and Botov S V 2018 Electricity consumption decrease at pump stations during watering by multi-support sprinkling units *J. of Fundamental and Appl. Sci.* **10** 1464-1481

[21] Degtyarev V G and Kozhenko N V 2019 Consolidation processes in clay soils, taking into account the multistory building construction staging *IOP Conf. Ser.: Mater. Sci. Eng.* **698(2)** 022016

IOP Conf. Series: Materials Science and Engineering 1111 (2021) 012007 doi:10.1088/1757-899X/1111/1/012007

Development of a methodology for automated control of the reliability indicators of the designed roads in the framework of the program-targeted approach

A P Bazhanov and R S Shamanov

Penza State University of Architecture and Construction, 28, Germana Titova ave., Penza, 440028, Russia

E-mail: bajan_p@mail.ru

Abstract. At present, the reliability of domestic roads in terms of their main indicators is significantly inferior to the reference indicators of the reliability of roads in advanced world countries. In this regard, in the field of modern road construction in the country, the task arises of strengthening the requirements for the reliability indicators of domestic roads and the transition from passive methods of ensuring these indicators to active methods of managing the reliability indicators of the country's roads at the stages of their design and stages of construction, reconstruction, overhaul, repair and maintenance. The work proposes a new program-targeted approach based on a functional model for managing the reliability indicators of roads, based on the implementation of procedures for forecasting, planning, rationing, accounting, control, analysis, regulation, and feedback, which are the initial components of the process of managing these reliability indicators of the designed roads. The main advantages of the proposed functional program-target approach to the management of the reliability indicators of roads are stated, which consist of its consistency and fairly complete development of the mathematical apparatus used in it.

1. Introduction

At present, the practical reliability of domestic roads [1] in terms of their failure-free, durability, maintainability, and preservation [2] is significantly inferior to the reference indicators of the reliability of roads in advanced world countries [3]. Therefore, the current situation in the field of modern road construction in the country raises the need to strengthen the requirements for indicators of the reliability of domestic roads and the transition from passive methods of managing these indicators to active methods [4]. This allows corrective actions to be made in the numerical values of these indicators at the design stages [5] and stages of construction, reconstruction, overhaul, repair, and maintenance of the country's roads [6].

Under the management of indicators of the reliability of roads in the framework of the program - target approach is understood as the establishment, provision, and maintenance of their required levels carried out through systematic and targeted control of the impact on these indicators of weather and climatic factors and operating conditions of roads [7].

To manage the reliability indicators of roads within the framework of the program-target approach, it is necessary to have appropriate mathematical models that reflect the causal and formal relationships of these reliability indicators with the main transport and operational conditions for the operation of

IOP Conf. Series: Materials Science and Engineering 1111 (2021) 012007 doi:10.1088/1757-899X/1111/1/012007

roads. That affects these indicators of their reliability in general, as well as their separate structural layers and elements [8].

2. Materials and research methods

Based on this point of view, one of the most noteworthy provisions of the solution to the problem outlined above should be considered an approach based on a functional model for managing the reliability of roads, based on the implementation of procedures for forecasting, planning, rationing, accounting, control, analysis, regulation and reverse connections. These are the initial constituent parts of any process of managing various objects [9].

The main feature of the proposed at the conceptual level of the functional scheme for managing the reliability indicators of roads [10] is that this model is focused on the management of specific indicators of the reliability of roads and constant monitoring of the scientific and technical level of design of roads according to their specific reliability indicators [11] regarding the reliability indicators of some design solutions for the reliability indicators of roads in the world's leading countries [12].

Taking into account the priority of the functional model for managing the reliability indicators of roads in comparison with other analytical models of automated regulation and management of various objects, in the material presented below, we will consider in detail the content of only the first three of the eight above-mentioned procedures for forecasting, planning and standardizing these indicators. We provide for the possibility of further presentation in new publications on the content of accounting, control, analysis, regulation, and feedback procedures. They are the initial components of the management of reliability indicators for new road design options.

The main advantage of the functionality of the program-targeted approach to the management of roads reliability indicators lies in its consistency and fairly complete development of the mathematical apparatus used in it. In addition, this approach fits well into the methodological concepts of the developed automated control systems for design and industrial enterprises, as well as computer-aided design system (CAD) of various technical objects [13]. The indicated important advantages of the approach proposed below make it very effective and promising in relation to the management of the reliability indicators of domestic roads in order to ensure the reliability indicators of these roads during their construction [14], reconstruction, and operation [15], exceeding the reliability indicators of roads of the world's leading countries.

3. Results of the study

Predicting the reliability indicators of roads means identifying the probabilistic tendencies of their behavior in the environment of disturbing factors (changes in the surface temperature, soil moisture of the road structure, characteristics of the inhomogeneity of the structure of road materials, etc.) [16] at a certain lead interval from the start to the end of the design of a given road option. It also includes obtaining probabilistic estimates (point and interval) of these trends. In a rather general setting, the prediction of the reliability indicators of roads can be carried out using the methods of regression analysis and the theory of random processes [17] in the following sequence.

For the projected categories of roads, a representative retrospective of reliability indicators of the same type of roads is established. For the vast majority of motor roads of the same type, which are designed within the framework of traditional principles, the values of their reliability indicators change quite monotonously from development to development. At the same time, the changes in the average values of these indicators over time (regression curves) are close to linear (Figure 1).

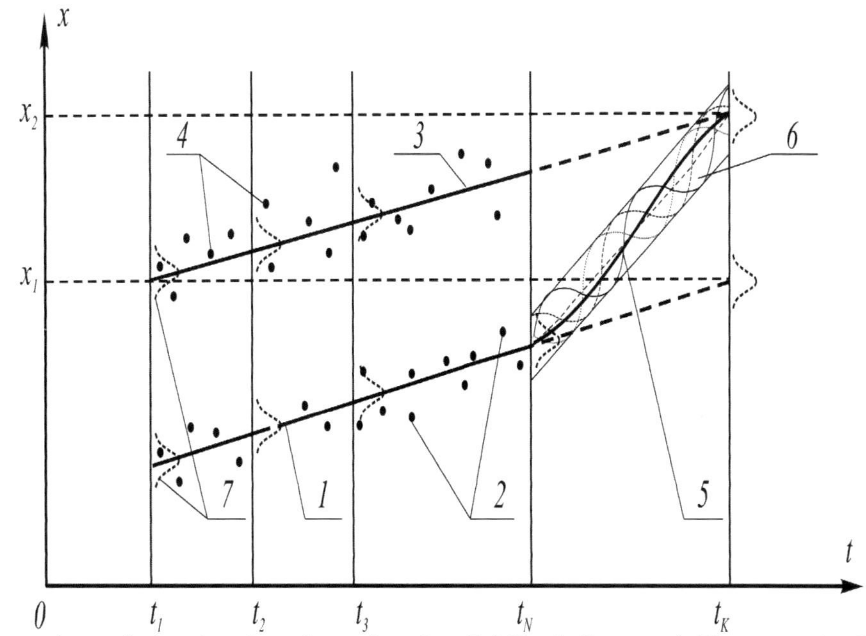

Figure 1. Forecasting and planning the values of roads reliability indicators. 1. The regression line of the road reliability indicator obtained from retrospective test points of 2 roads of this category. 2. Line of regression of the road reliability indicator – the type representatives, obtained from retrospective test points 4 of the road for the advanced world model. 3. Line of regression of the program for the development of the indicator of the reliability of the road - the type representative in section t_k, t_N. 4. The zone of possible trajectories of the development program of the 5th indicator of the reliability of the road - the type representative. 5. Densities of distribution of probabilities of values of reliability indicators for roads - type representatives of the projected road and road of the advanced world standard.

Let for a given category of similar projects of roads a sufficiently representative retrospective be established, in which the values of the reliability indicators of roads - analogs at moments t_1, t_2, ..., t_N (Figure 1). Here t_1 is the moment of the beginning of the presentation of the retrospective of reliability indicators roads of the same type, t_N - the moment of the end of this retrospective (the beginning of the design stage). Taking into account the received retrospective information, a certain correlation field (x, t) of the presence of this information is established. Further, according to the available statistical data of retrospective information, the method of least squares establishes its regression dependence of form $x = at + b$, $x = at^2 + bt + c$ etc. The adequacy of the presented mathematical model of retrospective information is checked by F - Fisher's criterion [18]. The resulting dependence can be identified with an estimate of the mathematical expectation of a certain random discrete process that characterizes the statistical behavior of the reliability indicator x of a road in its specific design version. By the behavior of the correlation function of the random process $x(t)$ (by the value of the decay of the correlation function), the correlation interval is found.

For a stationary random process, you can always indicate such a time moment t_0. in case of exceeding the time, that is, when the time moment $t > t_0$ occurs, the value of a certain indicator of the reliability of the road x at two adjacent points on the curve (see Figure 2) can be considered practically uncorrelated if, for example, correlation function $r(t)$ decreases to a value of 0.05 (Figure 2).

The value of the time moment t_0 is called in this case the correlation interval. In practice [19], the correlation interval is usually taken as half the width of the base of a rectangle of unit height, the area of which is equal to the area under the curve of the modulus of the normalized correlation function $r(t)$, i.e.

$$t_0 = \frac{1}{2}\int_{-\infty}^{\infty} |r(t)| dt \quad (1)$$

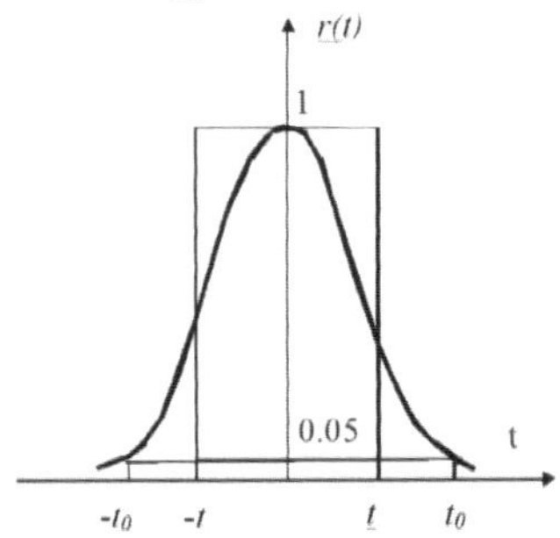

Figure 2. Substantiation of the correlation interval.

By the value of the found correlation interval, the action (extrapolation interval) of this random process is established, and, consequently, the correct use of the obtained correlation equation $x(t)$ on the lead interval t_N, t_k (Figure 1).

Thus, the forecasting of the reliability indicator x is carried out as a statistical extrapolation of the tendency of its change to the moment t_k (Figure 1) of the completion of the development of the next version of the road of a certain category.

In a similar way, the forecast of the basic (reference) version of the road is carried out, which is taken as an advanced world model. As a result of the implementation of these procedures, point and interval estimates of the predicted values of the reliability indicator of the designed road are obtained. In the case when the retrospective information is not representative, the control of its regression dependence is in the form:

$$x = \varphi(z_1, z_2, ..., z_N), \quad (2)$$

where $(z_1, z_2, ..., z_N)$ - tactical and technical characteristics of a road of a certain category.

The specific form of dependence (2) in a rather general case is determined in the form of regression polynomials, for example, of this form:

$$x = b_0 + \sum_{i=1}^{N} b_i z_i + \sum_{ij}^{N-1} b_{ij} z_{ij} + \sum_{i=1}^{N-1} b_{ii} z_i^2 + ..., \quad (3)$$

where $b_i = \frac{dx}{dz_i|_{\bar{z}_i}}$, $b_{ij} = \frac{d^2x}{dz_i dz_j|_{\bar{z}_i \bar{z}_j}}$, $b_{ii} = \frac{d^2x}{dz_i^2|_{\bar{z}_i}}$ are the coefficients of equation (3).

For fundamentally new options for roads, the predicted values of their reliability indicators (point and interval) are found by expert methods.

Prediction of reliability indicators using the described procedures is carried out at the stage of development of pre-design documentation [20]. At subsequent design stages, the results of the existing

IOP Conf. Series: Materials Science and Engineering 1111 (2021) 012007 doi:10.1088/1757-899X/1111/1/012007

forecast are constantly updated as new information becomes available on similar projects and on the basis of the results of the accumulated information on a fundamentally new design option.

With incomplete (not representative) information, in some cases, it is possible to use data on roads-analogs to obtain representative retrospective information; however, in this case, the specific main transport and operational conditions of their operation should be considered.

Planning refers to the development of models of the future state of roads based on predictive information. Since the objects of their control are indicators of reliability, then a plan for its practical implementation is developed for a certain category of roads in relation to a specific indicator. That is, the procedure for planning the values of the reliability indicators of roads is carried out on the basis of information obtained from the results of their predictive estimates (point and interval) for the newly designed and basic versions of roads. Let curve 1 (Figure 1) describe the behavior of the x index for new development, and curve 3 for the reference one (some advanced world sample). Then, if the task is set by the end of development (that is, by the moment of time t_k, Figure 1) to reach the level of the advanced world standard for this indicator, it is necessary to develop a certain plan (development program) in the form of curve 5 (Figure 1).

In accordance with it for the time period t_N, t_k the output of indicator x will be ensured at the end of the time period t_k of the end of development at the level of the advanced world standard. Consequently, in the development process, it is necessary to implement such a design solution (or a set of solutions.) Along the projected road, so that by time t_k we go to point x_2 (Figure 1). For this purpose, in order to practically realize the value of the indicator of level x_2, it is necessary to consider a certain set of possible plans in the form of probabilistic network models (probabilistic graphs with backtracks [21]). In this case, the edges of the directed graphs characterizing the dominant processes of the development program for the reliability indicator x_2 should be represented in the form of some growth models. Their parameters are set according to similar design options with the subsequent forcing these parameters to execute the development program in order to reach the value of indicator x to level x_2 (Figure 1). An indicator that in the process of developing the next modification of a road of a certain category is the movement of the reliability indicator x along a certain "tube" of possible development programs (Figure 1), the quantitative measure of N.S. Streletsky [22].

The planning function of reliability index 1 is carried out at the stage of development of pre-design documentation with the subsequent refinement of plans at the stages of development of an engineering project and working documentation as new information becomes available and the development of options for design and technological solutions for a projected road of the corresponding category.

The rationing of the reliability indicators of roads is understood as the determination of such areas of their values, at which the efficiency of the operation of roads is maximum at a given cost of the main types of resources. Or the costs of the main types of resources are minimal at a certain specified efficiency of the operation of roads. When rationing indicators of the reliability of highways, two main tasks are solved. The solution to the first problem involves the establishment of optimal values of the reliability indicators of the road as a whole. When solving the second problem, the reliability indicators are distributed (normalized) for all composite structural layers and elements of the road. At the same time, the issue of rationing the reliability indicators of structural layers and elements, which are typical design solutions for roads, is of great importance.

One of the possible formulations for solving the problem is reduced to the following.

Let $M_{ijk}^{(zt)}$ be the number of operations of k - types performed in the process of construction of a j - road. In this case, the concept of a road is understood as its generalized concept. This can be a structural layer or an element that is part of the road structure of a road. Let the construction of a road be carried out according to i - technology and be intended for operation in z - natural - climatic zone and t - calendar period. And $R_{ijk}^{(zt)}\left(\overline{B_{j_1}, B_{j_P}}, N_j, T_j\right)$, $q_{ijk}^{(zt)}\left(\overline{B_{j_1}, B_{j_P}}\right)$, $W_{ijk}^{(zt)}\left(\overline{B_{j_1}, B_{j_P}}\right)$ are respectively specified costs associated with the maintenance of road construction equipment (B_{j1}), carrying out its repair and preventive work (B_{jp}). Its annual operation is in the amount of T_j hours and the volume of the fulfillment of the shifting task for the construction of a highway by N_j percent, including labor costs for its construction, and some useful effect associated with the implementation of the construction of a road, with parameters $\overline{B_{j_1}, B_{j_P}}$ measured in conditional units.

The task is to determine such a number of roads, the values of their design parameters (indicators of functional purpose), as well as the required amount of their construction in a certain time period so that the total reduced costs for the design, construction, and operation of roads would be minimal.

In relation to the formulated conditions, the mathematical model of the problem is presented in the form of a functional:

$$\sum_{z=1}^{\eta}\sum_{t=1}^{s}\sum_{j=1}^{m}\sum_{i=1}^{n}\sum_{k=1}^{q} R_{ijk}^{(zt)}\left(\overline{B_{j_1} B_{j_P}}, N_j, T_j\right) M_{ijk}^{(zt)} \tag{4}$$

under the following restrictions:

$$0 < B_{j_1} \le B_{j_1}^{\lim},\ 0 < B_{j_2} \le B_{j_2}^{\lim},\ \ldots,\ 0 < B_{j_P} \le B_{j_P}^{\lim}; \tag{5}$$

$$\sum_{j=1}^{m}\sum_{i=1}^{n} M_{ijk}^{(zt)} = \varphi_k^{(zt)}; \tag{6}$$

$$\sum_{i=1}^{n}\sum_{k=1}^{q} \frac{M_{ijk}^{(zt)}}{W_{ijk}^{(zt)}\left(\overline{B_{j_1} B_{j_P}}\right)} \le Q_j^{(zt)}; \tag{7}$$

$$\frac{1}{\sum_{k=1}^{q} \varphi_k^{(zt)}} \sum_{j=1}^{m}\sum_{i=1}^{n}\sum_{k=1}^{q} q_{ijk}^{(zt)}\left(\overline{B_j, B_{,P}}\right) \cdot M_{ijk}^{(zt)} \le q_{\lim}^{(zt)}; \tag{8}$$

$$\sum_{z=1}^{n} \max \sum_{i=1}^{n}\sum_{k=1}^{q} \frac{M_{ijk}^{(zt)}}{W_{ijk}^{(zt)}\left(\overline{B_{j_1}, B_{j_P}}\right)} = N_j; \tag{9}$$

$$\frac{1}{N}\sum_{z=1}^{n}\sum_{t=1}^{s} \tau^{(zt)} \sum_{z=1}^{n}\sum_{t=1}^{s} \frac{M_{ijk}^{(zt)}}{W_{ijk}^{(zt)}\left(\overline{B_{j_1}, B_{j_P}}\right)} = T \tag{10}$$

where $\tau^{(zt)}$ - total operating time in hours t - period z - zone;

$$M_{ijk}^{(zt)} \ge 0,\ N_j \ge 0,\ T_j \ge 0. \tag{11}$$

The obtained ratios make it possible to ensure the prevention of exceeding the maximum permissible limits by the values of the indicators of failure-free and durability of roads (5); compliance with the specified deadlines for the implementation of all technological operations of their construction (6). It provides the prevention of exceeding the required number of roads and a certain number of their design solutions (7); the impossibility of exceeding the average labor costs for performing one typical operation by a motor road or certain (preset) values (8); calculation of construction volumes during the year with known values of $M_{ijk}^{(zt)}$ (9) and (10).

The mathematical model for solving the problem under consideration in the form of relations (4) - (11) is, in general, rather complicated, nonlinear, and multi-extreme. However, as applied to specific roads for which the necessary initial data are available, the problem in this formulation is solved by decomposing it, i.e. by dividing them into a number of simpler problems and solving them in the framework of an iterative computational procedure.

4. Conclusions

1. As a result of the research, a method of automated control of the reliability indicators of the designed roads was developed within the framework of the program-targeted approach. This approach is based on the implementation of a functional model for managing the reliability indicators of roads, based on the implementation of procedures for forecasting, planning, rationing, accounting, control, analysis, regulation, and feedback, which are the initial components of any process for managing various objects.

2. The proposed functional model for managing the reliability indicators of roads is focused on managing their specific indicators and constantly monitoring the scientific and technical level of the design of roads for specific indicators of their reliability relative to the reliability indicators of some design solutions for the reliability indicators of roads in the world's leading countries.

3. The expediency of the creation of an automated performance management system reliability projected roads has been substantiated. Such a structure of an automated control system fits into the generally accepted methodology of automated control systems to CAD of new equipment and technologies for its manufacture created in our country, both at the level of individual enterprises and at the industry level.

References

[1] Zolotar I A, Nekrasov V K and Konovalov S V 1977 *Improving the reliability of roads* 183 p
[2] 1990 GOST 27.002-89 *Reliability in technology. Basic concepts. Terms and Definitions* (Moscow: Publishing house of standards) 37 p
[3] Ansell J I and Phillips M J 1994 *Practical Methods for Reliability Data Analysis* (New York: Oxford University Press)
[4] Arzhanukhina S P, Sukhov A A and Kochetkov A V 2011 Regulatory and methodological support for the development of innovative activities in the road sector *Innovations* **7** 82-85
[5] Fedotov G A and Pospelov P I 2010 *Research and design of roads.* In 2 books. Book 2 (Moscow: Higher School) 519 p
[6] 2017 Federal Law "On the roads and road activities in the Russian Federation and on Amending Certain Legislative Acts of the Russian Federation N 257-FZ of November 8, 2007" with the changes of "The Code" on February 7, 2017. 88 p
[7] Alexandrovsky N M, Egorov S V and Kuzin R E 1973 *Adaptive control systems for complex technological processes* 347 p
[8] Bazhanov A P 2018 Structural diagrams and basic design relationships for the calculation of road safety indicators *Advances in Engineering Research (AER)* **157** 81-84
[9] Bazhanov A P and Saxonova E S 2012 Mathematical models and procedures for automated management of the main transport and operational indicators of roads *Roads and bridges* **27/1** 11-18
[10] Bazhanov A P and Shamanov R S 2019 The Concept of Functional Road Condition Management *Advances in Engineering Research* **188** 15-19
[11] Stolyarov V V 1994 *Design of roads taking into account the theory of risk:* in 2 parts. Part 1, Part 2 (Saratov: Saratov State Technical University) 232 p
[12] Magdi M E 2015 Zumrawi Investigating Causes of Pavement Deterioration in Khartoum State, Sudan *Int. J. of Civil, Environmental, Structural, Construction and Architectural Engineering* **9(11)** 1458-1463
[13] Golovitsina M V 2016 *Methodology of automation of works of technical preparation of production*, 2nd ed (National Open University "INTUIT") 209 p

[14] Podolsky V P, Pospelov P I, Glagoliev A V and Smirnov A V 2012 *Technology, and organization of road construction. Road surfaces: a textbook for students* (Institutions of higher prof. education ed. Podolsky V.P. Publishing Center "Academy") 304 p
[15] Vasiliev A P 2010 *Operation of roads*, vol 1(2) (Moscow: Publishing Center "Academy") 320 p
[16] Kokodeeva N E 1997 Analysis of the operational state of non-rigid road pavements *Problems of transport construction and transport: Mater. of the Int. Sci. and Tech. Conf.* (Saratov: SSTU) iss 2 pp 9-15
[17] Gmurman V E 2004 *Probability theory and mathematical statistics* (Moscow: Higher School) 479 p
[18] Fry T 1934 *Probability theory for engineers* (Moscow: Gostekhizdat) 383 p
[19] Filipsky Yu K 1978 *Random processes for radio engineering purposes* (Kiev: Head publishing house of the publishing association) 112 p
[20] Bazhanov A P 2019 Development of a methodology for research and assessment of the reliability of roads at the stages of their design *Transport structures* **6(2)** 1-13
[21] Mironosetsky N B 1976 *Modeling the processes of creating and releasing new products* (Novosibirsk: Nauka)
[22] Streletsky N S 1947 *Fundamentals of statistical accounting for the safety factor of structures* (Stroyizdat) 94 p

Determination of the theoretical trajectory of a tractor at a turn

A N Belyaev[1], V I Orobinsky[1], V P Shatsky[1], T V Trishina[1], V V Sheredekin[1], I A Vysotskaya [2], A I Zavrazhnov[3] and N V Mikheyev[3]

[1] Voronezh State Agricultural University named after Emperor Peter the Great, 1, Michurina ave., Voronezh, 394087, Russia

[2] Military Educational and Scientific Center of the Air Force N. E. Zhukovsky and Y. A. Gagarin Air Force Academy (Voronezh) the Ministry of Defense of the Russian Federation, 54 A, The Old Bolsheviks ave., Voronezh, 394064, Russia

[3] Michurinsk State Agricultural University, 101, International ave., Michurinsk, 393760, Russia

E-mail: aifkm@agroeng.vsau.ru

Abstract. The most complete idea of the stability of the curvilinear motion of a tractor is given by the nature of the change in the trajectory of motion of its kinematic center and the coordinates of the turn: the abscissa and ordinate of the headland, showing both the transverse and longitudinal deviations of the machine-tractor unit, including those determining its productivity and the required quality of agricultural operations. According to the operating conditions, the methods of motion of the unit are chosen, both on the main array of the field and on the headland. The paper presents explicit equations for the analytical calculation of the theoretical coordinates of the points of the kinematic center trajectory of the tractor when it performs a circular loopless rotation, which allowed determining the theoretical trajectory of a complete cycle of circular loopless rotation of a tractor depending on the following parameters. These parameters are as follows: design - base, distances between the axes of the pivots, maximum angles of rotation of the internal wheels; operational - forward speed of a tractor, the angular speed of the steered wheels in the transverse plane. The author made the attempt to analyze and compare the most universal ways of turn: the front steered wheels and the front and rear steered wheels by turning them in different directions relative to the core.

1. Introduction

In order to reduce economic costs, to achieve high productivity of the machine and tractor unit (MTU) and the required quality of agricultural operations, it is necessary to ensure such a mode of operation that would most fully comply with the technical characteristics and agrotechnical requirements and operating conditions, for which the following aspects are taken into account:

1. The types of agricultural machines, which are determined by agrotechnical requirements
2. The number of machines in MTU and the speed mode of its operation, which satisfies the permissible limits of agricultural requirements, depending on the technical and operational properties of the machines.
3. The modes of motion of the unit, both on the main array of the field and on the headland based on the operating conditions

Thus, the technology for the production of MTU should provide the compliance with the kinematic characteristics of its motion and agrotechnical conditions of the work.

2. Problem statement

MTU in the field conditions during technological operations for sowing agricultural crops, processing soil, plants, etc. undergoes a path of considerable length, alternating productive work (working strokes) with unproductive work (idle runs and turns). The constantly and almost cyclically repeating nature of changes in working strokes, idle runs and turns exactly determines the way the unit moves across the field at each of its sections. At the same time, for example, when working on medium-sized fields in the aisles of row crops, an MTU spends up to a third of the shift time on turns.

It is necessary to take into account that the deterioration in the quality of work, the increase in dynamic loads on both a tractor and an agricultural machine, the increase in the level of psychomotor costs of a driver, the decrease in the technical and economic indicators of the performance of agrotechnical operations, partial and in some cases complete, destruction of cultivated plants directly depends on the wrong way of driving the machine on the headland. In addition, on the headland, the MTU propellers compact and spray the soil more significantly than in the main section of the field thereby reducing its crop yields.

3. Materials and methods

It is recommended to study MTU curvilinear motion by the nature of its behavior during a circular loopless turn, since most of the used turning methods are based on its combinations.

The curvilinear motion of MTU, carried out mainly on the headlands, is the most complex element of its kinematics, due to the fact that individual points of the unit move at different speeds and describe trajectories that are significantly different from each other. As a rule, in practice, a turn is a gradual transition from an infinitely large radius to a minimum one (during the transition from motion along a straight section to motion along a curve of constant radius – "entering a turn") [1]; then motion along a section of constant curvature ("steady turn") [2, 3] and, finally, the transition from the minimum - to an infinitely large radius (when passing from a section of constant radius to straight motion – "the exit from the turn"). However, in most studies, it is accepted that turns are made with the minimum permissible constant radius.

The most difficult ones are the sections both in terms of dynamics and kinematics – "corner entry" and "steady turn" [1-6], on which the operational properties of the unit are noticeably deteriorated: due to large abruptly varying loads and intense lateral slip, significant deviation from the required trajectory of motion, the dynamic loading of the tractor, agricultural machine and soil increases.

The studies showed that the necessary and fairly complete assessment of the kinematic characteristics of the curvilinear motion of MTU can be carried out by the nature of the change in the trajectory outlined on the surface of one of its characteristic points (center of gravity, middle of the bridge, etc.), taken as the kinematic center. Since the experimental studies to assess the influence of the design and operational parameters of the tractor and MTU based on it on the change in kinematic characteristics during curvilinear motion are a very laborious and rather expensive process, they require a large amount of experiments and the determination of many parameters, including in the field conditions of real operation, then, it is obvious that a calculated assessment is more acceptable - the analytical description of the process of cornering by methods of mathematical modeling [2, 3, 7-13].

Therefore, we set the task of the determination of the theoretical trajectories of the kinematic center of the unit for various methods of wheel control, their comparison and evaluation when making a circular loopless turn, based on the condition of wheel rolling without slipping and the coincidence of the instantaneous centers of rotation of all tractor wheels, that is, correct turn. In order to solve this problem, we chose two of the most universal kinematic methods of turn as the object of research. In these methods the direction of motion was changed by turning either only by the front (turning method I), or only front and rear, asynchronously - in different directions (turning method II) by controlled wheels relative to tractor core [1]. The design diagram (Figure 1) shows the method of turning II. Turning method I was obtained when the rear axle wheels were in neutral position.

The point on the longitudinal axis of a tractor corresponding to the intersection of the perpendicular to it, drawn from the instantaneous center of rotation O_T (point E in Figure 1), is taken as the kinematic center. In case of turning by the front wheels, the kinematic center of a tractor corresponds to the middle of its rear axle (coincides with point D).

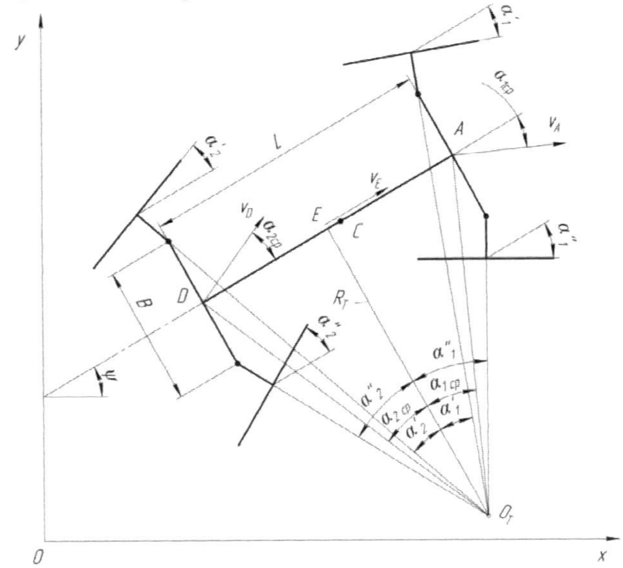

Figure 1. Design diagram of a wheeled tractor with all steering wheels

In the work [1], we obtained in a parametric form the equations to determine the current coordinates *x(t)* and *y(t)* of the theoretical curve of the trajectory of the kinematic center of a vehicle at the entrance to the turn, corresponding to points *D* - for the method of turning I or *E* - for the method of turning II, according to the scheme in Figure 1 [14-17] .

$$x(t) = v\int_0^t \sin\left[v\int_0^{t_1} \frac{\sin(\omega_1\tau + \omega_2\tau)}{\left[\frac{B}{2}(tg\omega_1\tau + tg\omega_2\tau) + L\right]\cdot\cos\omega_1\tau\cos\omega_2\tau} d\tau \right] dt_1; \quad (1)$$

$$y(t) = v\int_0^t \cos\left[v\int_0^{t_1} \frac{\sin(\omega_1\tau + \omega_2\tau)}{\left[\frac{B}{2}(tg\omega_1\tau + tg\omega_2\tau) + L\right]\cdot\cos\omega_1\tau\cos\omega_2\tau} d\tau \right] dt_1 \quad (2)$$

where L – tractor base, *m*; B – distance between the axles pivots of the steering linkage of the tractor, *m*; v – forward speed of the tractor kinematic center, $m \cdot s^{-1}$; ω_1 and ω_2 – the angular speeds of rotation of the front and rear steerable wheels of a tractor in the transverse plane relative to its core, s^{-1}; t, τ – time of motion of a tractor at the turn, *s*.

The authors also performed the assessment of the influence of some design and operational indicators of a tractor on the parameters of the curve of this trajectory.

However, it is known [14, 17] that during the construction and analysis of the theoretical dependence of a curvilinear trajectory, it is more convenient to use its explicit analytical specification.

After a nonlinear approximation of the data obtained using the parametric equations of coordinates (1) and (2) of the curve of the trajectory of the entrance to the turn [1, 14], we obtained an explicit form of the function described in general form by the formula [18-21].

$$y(x) = p x^{q}, \tag{3}$$

where p and q – quite definite constants, $0 < q < 1$.

Thus, at the turn section, the trajectory of motion with a high degree of accuracy is approximated by an explicit function (3), since the comparison of the curves of the turn entry trajectories calculated from the parametric setting of functions (1, 2) and by formula (3) showed that they practically coincide, since the maximum error in the approximation was 2.10^{-3} m [1].

At the end of the "turn entry" section, when the steered wheels reach the maximum angles at a certain time $t = t_{max}$, the tractor switches to a section of steady motion along a trajectory with a constant radius of curvature R_T.

Since the curve of the turn entry trajectory (3), must smoothly and continuously pass into a circle of constant radius R_T at the end of the section - in its upper part, then in order to determine the theoretical trajectory in the section of steady motion, using the conditions of its smoothness and continuity, we carried out gluing of functions of the initial section of the turn (turn entry) and the circle (steady turn), having previously determined the coordinates of its center O_T and received a piecewise-smooth function of the form y = f (x) [18-21]:

$$f(x) = \begin{cases} p(v)x^{q}, \quad x \le x_Q, \\ \left(R_T^2 - \left(x - x_Q - \dfrac{R_T p(v) q \cdot x_Q^{\,q-1}}{\left(1+\left(p(v) q \cdot x_Q^{\,q-1}\right)^2\right)^{1/2}} \right)^2 \right)^{1/2} + p(v)\cdot x_Q^{\,q} - \dfrac{R_T}{\left(1+\left(p(v) q \cdot x_Q^{\,q-1}\right)^2\right)^{1/2}}, \quad x > x_Q. \end{cases} \tag{4}$$

Thus, matching the circle of radius R_T with the curve of the turn entry section at the top point, we obtained the current coordinates of the theoretical MTU trajectory, corresponding to the selected kinematic center of a tractor, for two stages of turn – "turn entry" and "steady turn" [1-3].

For the section "exit from the turn" we will consider the trajectory of the "mirror" trajectory of the entrance into the turn, since when leaving the turn, a transition occurs from a more dynamically loaded state to a less loaded one.

The curves of motion trajectories according to (4) by turning method I $\boldsymbol{\omega_1}$=0.157 $\boldsymbol{c^{-1}}$, ω_2=0 c^{-1}, t=4 c, a_{1av}=28.84° (the maximum average angle of rotation of the front axle wheels relative to the core in the transverse plane - corresponds to the maximum internal turn with respect to the center of rotation of the wheel of the front axle α_1''=36°), = 0° (the maximum average angle of rotation of the wheels of the rear axle relative to the core in the transverse plane - corresponds to the maximum angle of rotation of the internal turn with respect to the center of rotation of the wheel of the rear axle α_2''= 0°), B = 2.2 m, L = 2.5 m, R_{T1} = 4.54 m [1]) and turning method II (ω_1=0.157 c^{-1}, ω_2=0.157 c^{-1}, t=4 c, a_{1av}=23.9° corresponds to a_{1av}= 36°), a_{2av} = 23.9° (corresponds to α_2''=36°), $\boldsymbol{B}$=2.2 m, L=2.5 m, $\boldsymbol{R_{T2}}$=2.82 [1]) at speeds v = 1.5 m*s^{-1} are presented in Figure 2.

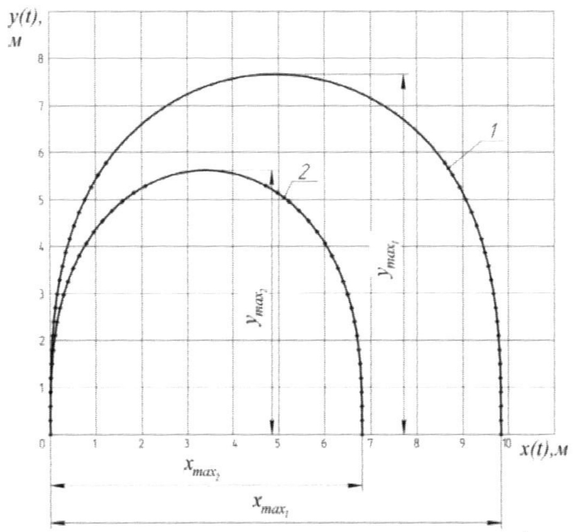

Figure 2. Trajectories of turn at v=1.5 m·s^{-1}: 1 – method I: ω_1 =0.157 s^{-1}, ω_2 =0 s^{-1}; 2 – method II: ω_2 =0.157 s^{-1}, ω_2=0.157 s^{-1}

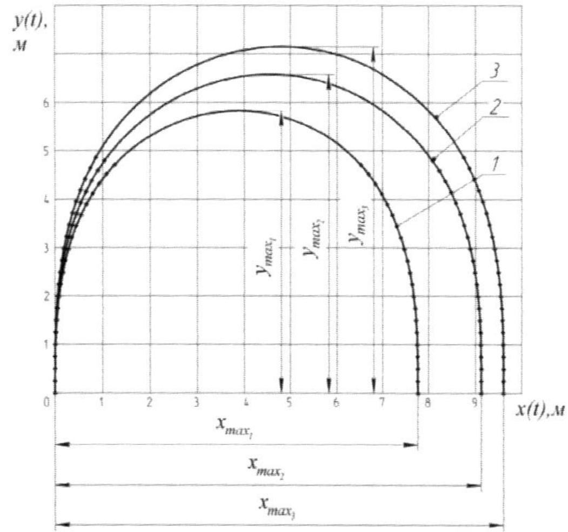

Figure 3. Trajectories of turn by method I at v=1.25 m·s^{-1}: 1 – L=2.0 m, B=1.8 m; 2 – L=2.5 m, B=1.8 m; 3 – L=2.5 m, B=2.2 m

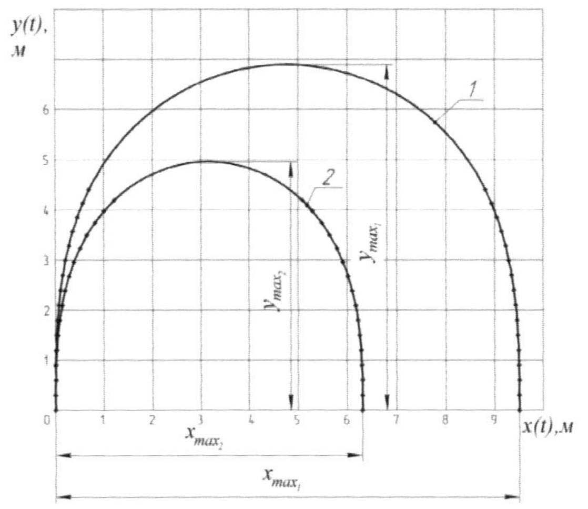

Figure 4. Trajectories of turn at v=1,5 m·s^{-1}: 1 – method I: ω_1 = 0.21 s^{-1}, ω_2 = 0 s^{-1}; 2 – method II: ω_2 = 0.21 s^{-1}, ω_2 = 0.21 s^{-1}

At the same time, it was revealed that with the turning method I, the maximum abscissa (width) of the headland was x_{max1}=9.81 m, the maximum ordinate (height) of the headland y_{max1}=7.67 m, the turning time t_{max1} = 13.74 s and the length of the turning trajectory S_{max1} = 20.62 m, and the method turn II - x_{max2} = 6.78 m, y_{max2} = 5.59 m, t_{max2} = 9.86 s and S_{max2} = 14.8 m (Figure 2). The application of method I lead to the increase in the above mentioned parameters compared to method II by 3.03 m (30.9%), 2.08 m (27.12%), 3.9 s (28.24%) and 5.82 m (28.24%) respectively.

It is obvious that from the point of view of ensuring better maneuverability the turning method II is the most rational one, in which the wheels of the front and rear axles are steered (Figure 1). However, due to the fact that at the same time, because of significant (more than 40%) decrease in the instantaneous turning radius in comparison with method I (from RT2 = 2.82 m to RT1 = 4.54 m) at the same angles of rotation of the corresponding wheels of the axles and the consequent increase of the centrifugal force of inertia acting on the machine almost 1.6 times of the lateral force, the directional stability decreases and the dynamic loads on MTU and the soil increase. The motion along an arc of trajectory of greater curve and a smaller radius also leads to the increase in the moment turning a tractor around the vertical axis, thereby increasing the risk, not only of loss of directional stability, but also of the development of the rear axle skidding.

The lengths of the sections of the steady turn for turning methods I and II were S_{y1}=8.62 *m* and S_{y2}=2.79 *m* respectively and the time of their passage - t_{y1}=5.74 *s* and t_{y2}=1.86 *s*. However, with equal length and time of passage of transitional sections of the turn S_v = 12 m and t_v = 8 s using method II the share of the total length of the trajectories of the stages "turn entry" and "turn exit" in its total length S_{max2} is 1.4 times and, consequently, the fraction of the transit time, where both kinematic and dynamic factors change more intensively than in the section of a steady turn, is 1.4 times greater. Moreover, curvilinear motion with a variable radius of curvature of the trajectory makes up a significant part, more than a half of the entire trajectory of the unit at a turn: method I - 58.1%, method II - 81.1%.

Figure 3 shows the trajectories of turning the kinematic center of a tractor using the most common method I, also constructed from the results of calculations using formula (4). It was determined that with a change in the tractor base (Figure 1) from L = 2.0 m (Figure 3, curve 1) to L = 2.5 m (Figure 3, curve 2) with the same following initial parameters: B = 1.8 m; v = 1.25 m*s^{-1}; ω1 = 0.157 s^{-1}; ω2 = 0 s^{-1}; t = 4 s; α_1''=36°; α_2''=0°, the width of the headland increases by 14.5% (x_{max1} = 7.79 m and x_{max2} = 9.11 m), and its height - by 12.6% (y_{max1}=5.76 *m* and y_{max2}=6.59 *m*), the minimum turning radius increases by 15.8% (R_{T1} = 3.653 m and R_{T2} = 4.341 m), the maximum average steering angle of the front axle wheels increases by 4.1% (a_{11av}= 28.7° and a_{12av}= 29.94°), the total length of the trajectory and the time to complete the entire maneuver increases by 11.1% (S_{max1} = 16.21 m and S_{max2} = 18.22 m; t_{P1} = 12.97 s and t_{P2} = 14.6 s), the length of the section of a steady turn and the time of its passage increases by 24.5% (S_{y1} = 6.21 m and S_{y2} = 8.22 m; t_{y1} = 4.97 s and t_{y2} = 6.56 s).

With the total length of the trajectories of the transitional turning sections S_v = 10 m and the time of their passage t_v = 8s at L=2.5 m, the share of S_v and t_v in its total length S_{max2} is 54.9%, which is 1.12 times less than at L=2.0 m. when we have 61.7% of S_v share in S_{max1}.

With the increase in the distance between the axles of the tractor pivots (Figure 1) from B=1.8 m (Figure 3, curve 2) to B=2.2 m (Figure 3, curve 3) and the same other initial data, according to the work (4), we have: the increase in x_{max3} by 5.3%, y_{max3} - by 7.96%, S_{max3} and $t_{П3}$ - by 6.83%, S_{u3} and t_{y3} - by 14.0%, R_{T3} - by 4.4% and a slight decrease by 2.0%. The share of S_v and t_v in its total length S_{max3} at B=2.2 m is 51.13%, which is only 1.074 times less than at B=1.8 m.

Thus, both with the decrease in the longitudinal base and the track width of a tractor with all other equal conditions the turning radius of a machine, the area of the headland, the length of the trajectory of the circular loopless turn and the time for its implementation decrease. However, at the same time, the most unpleasant tractor vibrations as longitudinal and transverse, transverse-angular and longitudinal-angular ones increase.

The increase in the forward speed of a tractor, for example, with the turning method I (Figure 1), from v = 1.25 m*s^{-1} (Figure 3, curve 3) to v = 1.5 m*s^{-1} (Figure 2, curve 1), with all other equal conditions leads to the increase in the width by 2% and in the height by 5.65% of the headland and the length of the trajectory by 5.2%. However, at the same time, the time for making a turn is reduced by 13.5% and is 15.64 s at v=1.25 m*s^{-1}, and at v = 1.5 m*s^{-1} - 13.75 s.

At the same time, the reduction in the time of entering the turn from t=4 s (Figure 2) to t=3s (Figure 4), which corresponds to the rotation method I ω_1=0.21 s^{-1}, ω_2=0 $s^{-1}$and when method II – ω_1=0.21 s^{-1}, ω_2=0.21 s^{-1}, with the remaining parameters unchanged, reduces the width by 3.2% and the height of the headland by 11.7%, the length of the trajectory - by 7.9% in the first case (Figure 2, curve 1 and Figure 4, curve 1), and in the second (Figure 2, curve 2 and Figure 4, curve 2) - by 7.2%, 11.1% and 9.5% respectively.

However, the increase in the speed of a tractor, including during curvilinear motion and, especially, at the transitional stages of turning and with the decrease in the time of their passage, contributes to the increase in the centrifugal force of inertia, uneven resistance to motion, is accompanied by the increase in variable and alternating dynamic loads, which cause the occurrence of increased oscillatory processes in the system "supporting surface-tractor-operator". Moreover it degrades its smoothness and stability. Therefore, in order to achieve their minimum values, it is necessary to give preference to scientifically grounded rational modes of MTU operation, for example, using the formulas (1), (2) and (4).

4. Conclusion

Therefore, in order to achieve high productivity of MTU and the required quality of agricultural operations, it is necessary:

- to choose one or another rational way of its motion on the headland, corresponding to a specific technological process,

- to operate automatically the control system for turning the rear wheels and identifying the patterns of changing the angles of rotation of the wheels of front and rear axles [9-11], based on the conditions for the increase in the stability of motion, the absence of skidding, reducing the dynamic load and providing, if necessary, the possibility of returning to the entrance to the row spacing with offset by the working width of technological equipment.

The determined analytically valid trajectory of MTU kinematic center for all the stages of circular loopless rotation by nonlinear approximation of the experimental function of an explicit form (4) with a minimum error of no more than 2.2% allows estimating the kinematic characteristics of MTU for various ways of rotation, depending on the main design and operational parameters and also studying it motion along a given trajectory.

References

[1] Belyaev A N, Kozlov V G, Vysotskaya I A and Trishina T V 2019 Computation of Vehicle Motion Path upon Entering Turn *Int. J. of Engineering and Advanced Technology* **9(1)** 4527-4531

[2] Skrypnikov A V, Kozlov V G, Belyaev A N and Chernyshova E V 2018 Theoretical Foundations of the Method of Designing a Clothoid Track with Approximation of Succession of Points *Perspectives on the use of New Information and Communication Technology (ICT) in the Modern Economy. Part of the Advances in Intelligent Systems and Computing book series (AISC)* **726** 654-667

[3] Troyanovskaya I P and Voinash S A 2018 Model for stationary turn of an arbitrary vehicle *IOP Conf. Ser.: Mater. Sci. Eng.* **450**

[4] Pozin B M, Troyanovskaya I P and Yusupov A A 2015 Optimal power distributing between wheels of the mobile vehicle under different soil conditions *Procedia Engineering* **129** 713-717

[5] Troyanovskaya I P, Ulanov A G, Zhakov A and Voinash S A 2019 Friction forces at the wheel's contact with the ground in a turning vehicle *Tribology in Industry* **41(2)** 166-171

[6] Troyanovskaya I P and Pozin B M 2015 Forces of friction at the wheel-to-ground contact in a turning vehicle *Procedia Engineering* **129** 156-160

[7] Field R V and Hurtago E J 2003 *Modeling of dynamic forces of a tractor in the MATLAB-simulink program environment* (New York: Society of Automotive Engineers) 112 p

[8] Itoh H, Oida A and Yamazaki M 1999 Numerical simulation of a 4WD–4WS tractor turning in a rice field *J. of Terramechanics* **36(2)** 91-115

[9] Karogal I 2009 Independent Torque Distribution Strategies for Vehicle Stability Control *SAE Technical Papers*

[10] Mammar S and Baghdassarian V B 2000 Two-degree-of-freedom formulation of vehicle handling improvement by active steering *Proc. of the American Control Conference* **1** 105-109

[11] Mokhiamar O and Abe M 2002 Active wheel steering and yaw moment control combination to maximize stability as well as vehicle responsiveness during quick lane change for active vehicle handling safety *J. of Automobile Engineering* **216(2)** 115-124

[12] Segel L 1956-1957 Theoretical prediction and experimental substantiation of the response of the automobile to steering control *Proc. of the IMechE Automobile Division* 310-330

[13] Winkler Ch 1998 Simplified Analysis of the Steady-State Turning of Complex Vehicle *Vehicle System Dynamics* **29** 141–180

[14] Dreizler R M and Lüdde C S 2011 *Theoretical Mechanics: Theoretical Physics* (Berlin) 776 p

[15] Jazar R N 2008 *Vehicle Dynamics: Theory and Application* (Springer, Germany)

[16] Pacejka H B 2002 *Tyre and Vehicle Dynamic* (Butterworth, Heinemann)

[17] Tyn Myint-U 2007 *Lokenath Debnath, Linear Partial Differential Equations for Scientists and Engineers, 4th ed.* (Birkhäuser Verlag AG, Basel-BostonBerlin) 778 p

[18] Aliev A V, Mishchenkova O V and Lipanov A M 2016 *Mathematical modeling and numerical methods in chemical physics and mechanics* pp 1-534

[19] Avellar J, Duarte L G S and da Mota L A C P 2012 PSsolver: A maple implementation to solve first order ordinary differential equations with liouvillian solutions *Computer Physics Communications* **183(10)** 2313-2314

[20] Pirôpo M, Moraes F and Santos F A N 2019 Surfing on curved surfaces The maple package surf *Computer Physics Communications* 107002

[21] Siddique M and Mitchell K 2010 Mathematical concepts and maple animations *Paper presented at the AIP Conference Proceedings* **1281** 706-708

IOP Conf. Series: Materials Science and Engineering 1111 (2021) 012009 doi:10.1088/1757-899X/1111/1/012009

Numerical Simulation of NOx Reduction in a SCR System

A S Blinov, N S Malastowski and L L Myagkov

Bauman Moscow State Technical University, 5, 2-ya Baumanskaya str., Moscow, 105005, Russia

E-mail: as_blinov@bmstu.ru

Abstract. The stringent limits of emission standards require advanced emission control technologies to be used in modern on and off highway diesel engines. They include both in-cylinder and aftertreatment measures where the latter now have become almost mandatory. Selective catalytic reduction aftertreatment systems are widely used for nitrogen oxide (NO_X) conversion in exhaust gases into harmless N_2. To reduce time and costs, at the design stage of SCR systems numerical modelling is applied. Mathematical models and methods providing high prediction accuracy with and acceptable level of computational efforts are required. In this work an approach for complete simulation of SCR systems based on the coupling of commercial CFD software with developed multichannel 1D catalyst model is presented. The first one is used to carefully describe processes occurring upstream in the catalytic converter, particularly, during urea water solution injection and flow mixing. As a result, the distributions of flow parameters at catalyst inlet are derived. They are subsequently imported as boundary conditions into a developed multichannel catalyst model that allows one to take them into account when calculating NOx conversion efficiency. Based on the proposed approach a SCR system was simulated. The effect of non-uniform distributions of NH_3 concentration and the gas flow velocity at the catalyst inlet on its performance was investigated. It has been shown that they have a great impact on NO_X conversion and should be taken into account during the catalyst modelling.

1. Introduction

To comply with tight ecological standards diesel engines, it is necessary to incorporate different emission control technologies [1 - 7]. Selective catalytic reduction systems are a widely used techique for NO_X reduction in exhaust gases [8, 9]. They convert NO_X to N_2 and H_2O through catalytic chemical reactions with NH_3 (6). The latter is supplied to a SCR system in the form of urea-water-solution (UWS) injected in the exhaust pipe before the catalytic converter. Due to high exhaust gas temperature the UWS evaporates (1) and decomposes to NH_3 and CO_2 through chemical reactions of thermolysis (2) and hydrolysis (3) [10]:

$$CO(NH_2)_2 \quad (solution) \rightarrow CO(NH_2)_2 + 6.9H_2O; \tag{1}$$

$$CO(NH_2)_2 \rightarrow NH_3 + HNCO; \tag{2}$$

$$HNCO + H_2O \rightarrow NH_3 + CO_2. \tag{3}$$

Many physical and chemical processes occur in SCR systems: the exhaust gas movement, the UWS injection and its interaction with the flow and the walls of exhaust pipes, the heat-transfer processes, the UWS evaporation and thermal decomposition, chemical reactions taking place in the gas flow, on the catalyst surfaces, etc. [11]. They should be properly optimized to provide high NO_X conversion efficiency for meeting emission standards requirements [12].

Numerical modeling has become an effective tool to solve such kind of optimization tasks [13]. At different design stages, models of various complexity are applied; that provides reduction of time and computational efforts [11].

At early design stage the simplified mathematical models are used [14]. They allow one in a short time to evaluate different system layouts and to find out the most promising ones. These kind of models are often implemented in thermodynamic engine simulation software, e.g. AVL BOOST, GT-POWER, ExACT, GASDYN etc. They allow one to perform the coupled modeling and optimization of an engine and aftertreatment systems. The results of numerical simulations of preliminary size dimensions and the layout of a SCR system, the catalyst characteristics, the UWS injection strategy that provide the required conversion efficiency are defined.

These simplified models can't properly describe gas flow dynamics in exhaust pipes, the UWS injection and decomposition processes, the mixing of gas flow chemical compounds that determine distributions of flow parameters at catalyst inlet and directly affect the conversion efficiency [15]. For this purpose, CFD models and related software, e.g. ANSYS FLUENT, CONVERGE, AVL FIRE, STAR-CCM+ etc., are used [16]. On the other hand, the application of such complex models for complete simulation of SCR systems requires high computational efforts. This is due to different timescales for UWS spray dynamics (ms) and catalyst transients (min) [17].

To reduce the time and computational efforts, the simulation can be divided into two sequential steps: modeling processes before and in the catalyst. For the first one, CFD models are used; for the second one – different catalyst models [18]. In this study, such integrated approach based on the coupling of ANSYS Fluent with the developed multichannel 1D catalyst model is presented.

2. Materials and methods

A schematic representation of the developed integrated approach for complete simulation of SCR systems is shown in Figure 1.

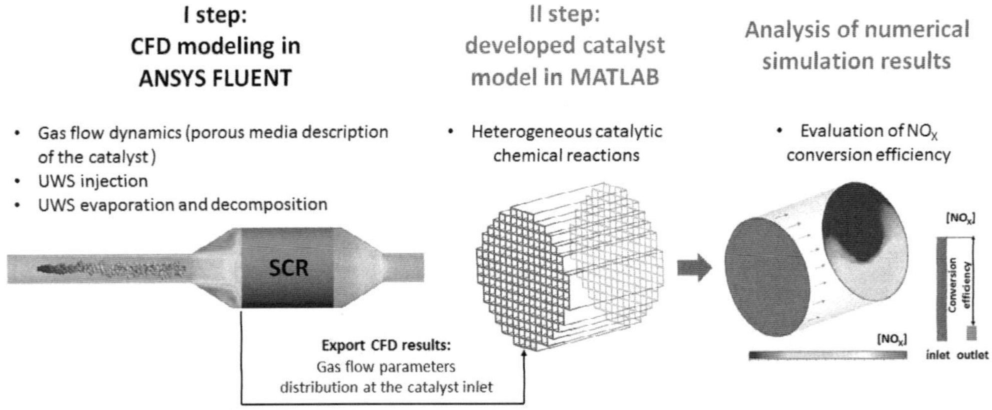

Figure 1. Integrated approach for complete simulation of a SCR system.

During the 1-st step of simulation, the gas flow dynamics throughout the SCR system, the UWS injection, evaporation and decomposition are modeled in ANSYS Fluent. The purpose of the simulation is to obtain data of flow parameters distributions at the catalyst inlet: temperature, velocity, chemical composition, etc.

The gas flow is described by Reynolds-averaged Navier-Stokes (RANS) equations coupled with the Realizable k-ε turbulence model and Enhanced Wall Treatment model [19]. For the catalyst, the brick porous medium approximation is adopted. The permeability resistance of the flow is governed by Darcy's law where linear and inertial loss terms are determined based on experimental data [20]. The

UWS injection is modelled using the Lagrangian Discrete Phase Model [19]. The primary spray breakup is not simulated but is specified by Rosin-Rammler droplet size distribution and droplet speeds derived from experimental data [21]. The secondary droplet breakup is calculated using the Taylor Analogy Breakup model [22]. The evaporation of UWS droplets is described by the convection-diffusion controlled model for multicomponent particles [23]. Two-way coupling of mass, momentum and energy is considered between the gas flow and droplets. The decomposition process of evaporated urea is modelled through single step volumetric chemical reactions of thermolysis and hydrolysis (2, 3). The formation of intermediate reaction by-products is neglected. The reaction rates are calculated according to the law of mass action. The kinetic parameters and thermal effect of the reactions are taken from the literature [23].

In the next simulation step, physicochemical processes occurring in a SCR catalyst are simulated using the developed mathematical model implemented in MATLAB. It is based on the multichannel approach whereby a 1D single channel model is applied for each of them. It allows one to take into account gas flow parameters nonuniformity at the catalyst inlet. A detailed description of the applied single channel catalyst model can be found in the previous author's work [24]. It simulates four dominant catalytic surface reactions: ammonia adsorption (4), desorption (5) and oxidation (6), NO_X reduction (7):

$$NH_3(g) \rightarrow NH_3(s); \tag{4}$$

$$NH_3(s) \rightarrow NH_3(g); \tag{5}$$

$$4NH_3(s) + 3O_2(g) \rightarrow 6H_2O(g) + 2N_2(g)\ ; \tag{6}$$

$$4NH_3(s) + 4NO(g) + O_2(g) \rightarrow 6H_2O(g) + 4N_2(g). \tag{7}$$

Reaction rates are calculated according to equations (8 - 11). Kinetic parameters of chemical reactions (pre-exponential constants and activation energies) are defined based on the experimental data:

$$R_{ads} = k_{ads} C_{NH3}(1 - \theta)\ ; \tag{8}$$

$$R_{des} = k_{des} exp(-\frac{E_{a,des}(1-\alpha\theta)}{RT_w})\theta; \tag{9}$$

$$R_{oxid} = k_{oxid} exp(-\frac{E_{a,oxid}}{RT_w})\theta; \tag{10}$$

$$R_{red} = k_{red} exp(-\frac{E_{a,red}}{RT_w}) C_{NO}\theta; \tag{11}$$

$$\theta = \frac{C_{NH3\ (s)}}{\Omega}. \tag{12}$$

where $k_i, E_{a,i}$ – pre-exponential constant and activation energy of the i-th reaction, respectively; R – universal gas constant; T_w – catalyst wall temperature; θ - NH_3 fraction loading onto the catalyst surface; Ω – number of reaction-sites per volume of washcoat.

Species equations of NH_3 and NO in the flowing gas and NH_3 adsorbed on the catalyst surface are written as:

$$\frac{\partial C_{NH3}}{\partial z} = \frac{s}{A_g u}(-R_{ads} + R_{des}); \tag{13}$$

$$\frac{\partial C_{NO}}{\partial z} = \frac{s}{A_g u}(-R_{red}); \tag{14}$$

$$\frac{\partial \theta}{\partial t} = (R_{ads} - R_{des} - R_{red} - R_{oxid}). \tag{15}$$

where s - number of reaction-sites per channel length; z - coordinate along the channel centerline.

Gas and catalyst wall temperatures are calculated based on the energy equations:

$$\frac{\partial T_g}{\partial z} = -\frac{hP}{\rho_g A_g C_{p,g} u}\left(T_g - T_w\right); \quad (16)$$

$$\frac{\partial T_w}{\partial t} = -\frac{hP}{\rho_w A_w C_{p,w}}\left(T_w - T_g\right) \quad (17)$$

where T_g – gas temperature; h – convective heat transfer coefficient; P – cross section perimeter; ρ – density, A – cross section area; C_p – specific heat capacity; t – time.

Boundary conditions at each channel entrance are determined based on CFD results imported from ANSYS Fluent in the way shown in Figure 2. The procedure consists of the following steps: the import of gas flow parameters, the distribution at the catalyst inlet, the interpolation of imported data at channels corners, the averaging of values at channel corners.

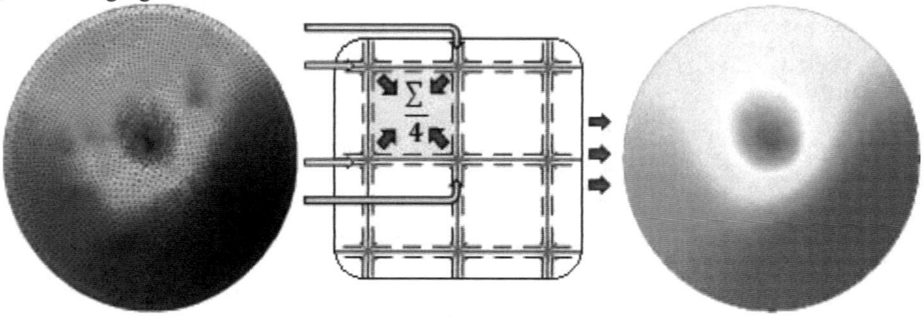

Figure 2. Estimation of boundary conditions at channels inlet: a – import of CFD results, b – interpolation and the averaging procedure, c – obtained boundary conditions.

The upwind difference scheme is used to approximate space derivatives of the equations (13, 14, 16). This leads to a system of time-dependent ordinary differential equations that are solved using standard MatLab ode23s algorithm [25].

As a result of numerical simulations, the concentrations of NO_X along the catalyst channels and at the catalyst outlet are derived. They are further used to calculate the NO_X conversion efficiency.

3. Results and discussion

Using the described integrated approach, a simple SCR system was investigated. It represents an insulated exhaust pipe equipped with a SCR catalyst block and UWS injector according to Figure 3.

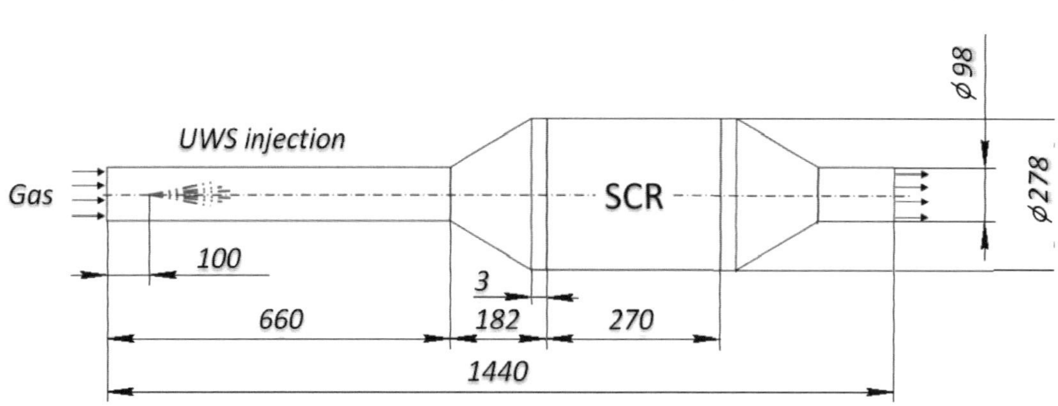

Figure 3. Computational domain.

At the exhaust pipe inlet, the hot air with an average velocity of 13.2 m/s and a temperature of 261 °C was supplied. It contained 205 ppm of NO, 18% of O_2, 3% of H_2O [26].

The SCR catalyst block represented the cylinder honeycomb-type substrate with flow-through square channels. It had an outer diameter of 278 mm, cell density of 300 cpsi and a wall thickness of 0.2 mm. The substrate was made of synthetic cordierite with the following properties: density of 2100 kg/m^3, specific heat capacity of 1000 J/(kg·K), thermal conductivity of 2 W/(m·K). The catalyst material was Cu-zeolite. Kinetic parameters of catalytic reaction rates (8 - 11) used in the simulations were taken from the experimental work [27] and are summarized in Table 1.

Table 1. Kinetic parameters of catalytic reaction rates [27].

Reaction	Pre-exponential constant, k_i	Activation energy, $E_{a,i}$, [J/mol]
Ammonia adsorption	0.25	0
Ammonia desorption	$1.4 \cdot 10^6$	$1.01 \cdot 10^5$
Ammonia oxidation	$2.2 \cdot 10^8$	$1.4 \cdot 10^4$
NO reduction	$1.5 \cdot 10^8$	$7.9 \cdot 10^4$

In ANSYS Fluent, the catalyst substrate was defined as a porous medium in terms of porosity equal to 0.7, inertial and viscous resistance coefficients were 20.414 m^{-1}, 3.846 × 10^7 m^{-2} in the direction of flow and 20414 m^{-1}, 3.846 × 10^{10} m^{-2} – шn the directions perpendicular to the flow. The resistance coefficients were calculated based on experimental data [20] to reproduce actual pressure losses across the monolith.

At the pipe centerline, the urea water solution (32.5% urea by weight) was injected along the gas flow. The spray is described as a solid cone pattern with an injection cone angle of 40°. The initial droplet size distribution was specified using the Rosin–Rammler function with $n = 2$, $\bar{d} = 20$ taken from the experimental work [28]. The initial droplet velocity was 10 m/s. The UWS mass flow rate was 45 mg/s. It was calculated based on the assumption of equal molar concentrations of NH_3 and NO with complete urea decomposition.

Results of the numerical simulation of gas flow dynamics throughout the SCR system and the UWS injection performed in ANSYS Fluent are shown in Figure 4.

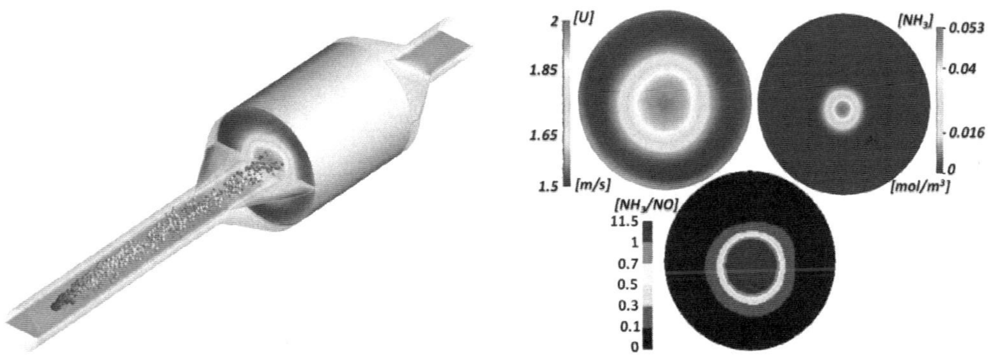

a b

Figure 4. Numerical simulation of a SCR system in ANSYS Fluent: a – velocity profile and UWS injection; b – distributions of velocity, NH_3 molar concentration and NH_3 to NO ratio at the catalyst inlet.

Figure 4a shows that not all of the injected UWS had evaporated and converted to ammonia before the catalyst inlet. The reason is the insufficient length of the mixing channel for the specified gas flow and injection conditions. As a result, the average concentration of produced NH_3 (0.0021 mol/m^3) at catalyst inlet was less than that of NO (0.0045 mol/m^3). The large diffuser cone angle and axial position of the injector led to flow maldistribution at catalyst inlet that was especially pronounced for gas velocity and NH_3 concentration (Fig. 4b). The concentration distributions of other species (NO, O_2, etc.) as well as the gas temperature profile were nearly uniform.

The simulation of catalytic reactions was carried out using the developed mathematical model implemented in MATLAB. The obtained CFD results at the catalyst inlet were used as boundary conditions. The gas flow velocity and ammonia concentration profiles were imported to the model according to the described procedure shown in Figure 2. The other flow parameters were considered to be uniformly distributed and were set equal to their average values. As a result of numerical simulations, the change in the concentration of NO along the catalyst was derived (Fig. 5).

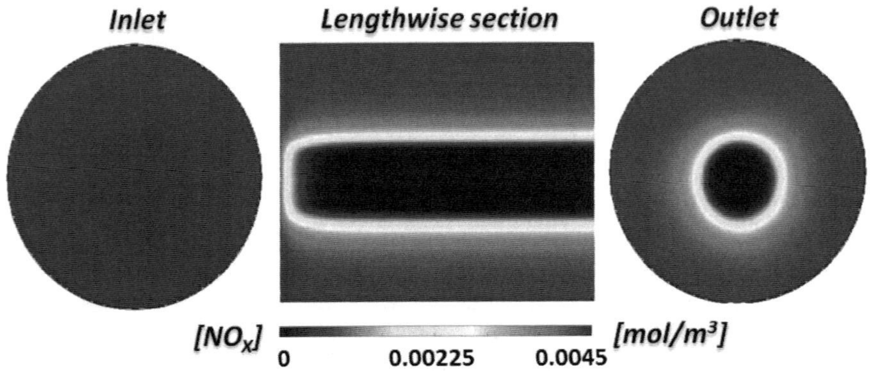

Figure 5. Contours of the NO concentration within the SCR catalyst.

Due to the maldistribution in the NH_3 concentration at the catalyst inlet (Fig. 4b), NOx conversion dominantly occurred in the central region of the converter (Fig. 5). The concentration of available NH_3 decreased in radial direction that led to reduction of NOx conversion efficiency to near zero values on the periphery of the block. The overall catalyst performance was analyzed by the average conversion of NOx along the catalyst length (Fig. 6). To assess the effects of the incomplete UWS conversion and flow maldistribution at the catalyst inlet on SCR performance, additional catalyst simulation with the uniform flow distribution and equimolar concentrations of NH_3 and NO was performed. The results were compared with the previous case (Fig. 6).

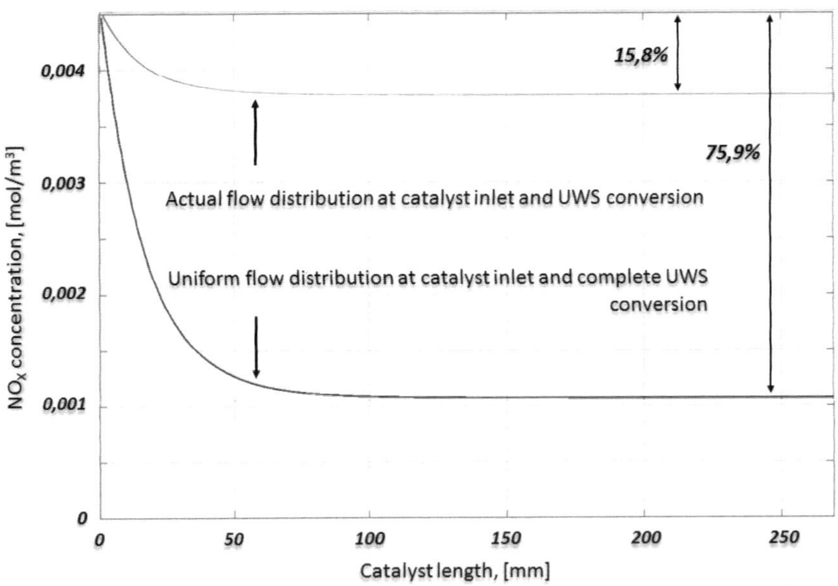

Figure 6. The change in the average NO concentration along the SCR catalyst under the actual and uniform flow distribution at the catalyst inlet.

The comparison of results showed a significant reduction in the NOx conversion efficiency under non-uniform conditions. The overall SCR performance for this case was 15,8% that almost 5 times less than for uniform conditions (75,9%). To increase the flow uniformity at the catalyst inlet, static mixers are commonly installed in real constructions of SCR systems [29]. They improve the flow mixing by creating a turbulent flow as well as the UWS evaporation breaking up the spray into smaller droplets.

Thus, processes occurring upstream of the catalyst have a great impact on its efficiency and should be carefully modeled. For this purpose the suggested integrated approach can be applied.

4. Conclusion

The integrated approach for the complete simulation of SCR systems based on the coupling of ANSYS Fluent with the in-house multichannel 1D catalyst model was developed. The simulations of the urea-SCR system using the suggested approach were performed. The results of CFD numerical simulations indicated the incomplete UWS conversion to NH_3 as well as the presence of flow maldistribution in NH_3 concentrations and gas velocity at the catalyst inlet. The effect of these factors on the performance of the SCR catalyst was analyzed by comparing NO_X conversion efficiency with that at uniform flow distributions and equimolar concentrations of NH_3 and NO at the catalyst inlet. For that purpose, the developed multichannel 1D catalyst model implemented in MATLAB was used. The comparison showed that results differed almost 5 times as compared to the lower value for non-uniform conditions. Thus, the suggested integrated approach has confirmed the ability to significantly improve the prediction accuracy with the acceptable level of computational efforts.

References

[1] Joshi A 2020 *SAE Tech. Pap.* (SAE International) 2479–2507
[2] Sharp C, Webb C C, Neely G, Carter M, Yoon S and Henry C 2017 *SAE Int. J. Engines* **10** 1697-1712
[3] Zhao J, Grekhov L, Ma X and Denisov A 2020 *Appl. Therm. Eng.* **179** 115699
[4] Zhao J, Grekhov L and Yue P 2020 *Int. J. Automot. Technol.* **21** 649
[5] Kuznetsov A, Kharitonov S and Ryzhov V 2018 *J. Eng. Gas Turbines Power* **140(12)**
[6] Kamaltdinov V G, Markov V A, Lysov I O, Zherdev A A and Furman V V 2019 *Energies* **12** 2643
[7] Markov V A, Devyanin S N and Kamaltdinov V G 2020 *Lect. Notes Mech. Eng.* 845–853
[8] Koebel M, Elsener M and Kleemann M 2000 *Catal. Today* **59** 335
[9] Johnson T V 2015 *SAE Int. J. Engines* **8** 1152
[10] Birkhold F, Meingast U, Wassermann P and Deutschmann O 2007 *Appl. Catal. B Environ.* **70** 119
[11] Faltsi R and Mutyal J 2012 *SAE 2006 World Congr. Exhib.* 40
[12] Bielaczyc P and Woodburn J 2019 *Emiss. Control Sci. Technol.* **5** 86
[13] Echtle H, Schöffel S, Wenninger G, Fischer S, Lauer T and Möltner L 2010 *MTZ Worldw.* **71** 54
[14] Chi J N and Dacosta H F M 2005 *SAE Tech. Pap.* (SAE International)
[15] Mauviot G, Berr F, Raux S, Perretti F, Malbec L M and Millet C N 2009 *Oil Gas Sci. Technol.* **64** 285
[16] Praveena V and Martin L J 2018 *SAE Tech. Pap.* (SAE International)
[17] Nova I and Tronconi E 2014 *Urea-SCR technology for deNOx after treatment of diesel exhausts* vol 5 (New York: Springer)
[18] Wurzenberger J C and Wanker R 2005 *SAE Tech. Pap.* (SAE International)
[19] ANSYS Inc. 2013 *ANSYS Fluent Theory Guide* 15317
[20] Dammalapati S, Aghalayam P and Kaisare N 2019 *Ind. Eng. Chem. Res.* **58** 20247
[21] Kim J Y, Ryu S H and Ha J S 2004 *Proc. 2004 Fall Tech. Conf. ASME Intern. Combust. Engine Div.* (American Society of Mechanical Engineers) 165–170
[22] O'Rourke P J and Amsden A A 1987 *SAE Tech. Pap.* (SAE International)

[23] Shirodkar V S 2016 *Front. Heat Mass Transf.* **7**
[24] Blinov A, Malastowski N and Myagkov L 2019 *E3S Web Conf.* 06013
[25] Shampine L F and Reichelt M W 1997 *SIAM J. Sci. Comput.* **18** 1
[26] Nadareishvili G G 2020 *Int. J. Emerg. Trends Eng. Res.* **8** 119
[27] Na H, Reed D, Annaswamy A, Laing P M and Kolmanovsky I 2011 *SAE 2011 World Congr. Exhib.*
[28] Spiteri A, Eggenschwiler P D, Liao Y, Wigley G, Michalow-Mauke K A, Elsener M, Kröcher O and Boulouchos K 2015 *Fuel* **161** 269
[29] Mehdi G, Zhou S, Zhu Y, Shah A H and Chand K 2019 *Processes* **7**

Modern methods of modifying the frictional state of the wheel-rail system

D Ya Antipin, V I Vorobiev, O A Shishkina and D A Bondarenko

Bryansk State Technical University, 7, Boulevard 50 let Oktyabrya, Bryansk, 241035, Russia

E-mail: unit@tu-bryansk.ru

Abstract. The paper analyzes and provides a systematization of possible methods for increasing the coefficient of adhesion of locomotive wheels to rails. Ensuring a rational value of the friction coefficient at the point of contact between the wheel and the rail makes it possible to reduce the energy consumption for traction, significantly reduce the wear of the wheel and rail rolling surface when passing the curved sections of the railway track, and increase the smoothness of the carriage and its safety during braking. Two fundamentally different methods of improving the locomotive traction qualities have been established - by means of structural improvements of the carriage and a direct effect on the physicochemical properties of the wheel and rail contacting surfaces. Practical methods of increasing the value of the friction coefficient in the contact between the wheel and the rail which include the use of sand, materials of natural origin, abrasive magnetic powder, braking of moving wheels, mechanical cleaning of rails, chemical cleaning of rails, electric spark cleaning of rails, plasma cleaning of rails, cleaning with high-energy sources, cleaning with high-energy sources, transmission of electric current and magnetic flux in the contact zone, the use of various modifiers of friction surfaces, are considered in detail. Their main advantages and disadvantages are noted. It is established that when wheels friction is on rails, mechanical, electrical, vibration, thermal, chemical, and magnetic processes occur simultaneously. In view of this, the practical implementation of a multivariate analysis of the entire complex of these phenomena is in most cases impossible. In this connection, it is suggested to be limited to a small set of factors that could most fully characterize the tribological system under study and determine the basic requirements for the experimental and theoretical studies.

1. Introduction

The instability of the values of the coefficient of adhesion of the locomotive wheels to the rails leads to increased energy consumption and, as a consequence, to a decrease in the locomotive functional efficiency. In practice, two fundamentally different methods to improve the traction properties of a locomotive are used. The first one is to increase the efficiency of using the adhesion coefficient by means of structural improvements of the carriage, traction drive and anti-skid systems. The second method is influencing the physicochemical properties of the tire and rail contacting surfaces. I.V. Kragelsky [1], proceeding from the concept of a discrete contact structure, proposed to calculate the friction forces F by summing up the resistances arising from molecular and mechanical interaction at individual contact areas:

$$F = \tau_{mol} \cdot A_{mol} + \tau_{mech} \cdot A_{mech}, \quad (1)$$

where τ_{mol}, τ_{mech} are specific molecular and mechanical interactions or friction forces;
A_{mol}, A_{mech} - actual molecular and mechanical contact area.

Therefore, the overall coefficient of friction f is the sum of the molecular and mechanical components:

$$f = f_{mech} + f_{mol} \cdot \quad (2)$$

The necessary measures to increase and stabilize the adhesion coefficient of wheels to rails can be conditionally divided into two groups. Belonging to a certain group is determined by the method of affecting the molecular f_{mol} and mechanical f_{mech} components of the adhesion coefficient (Fig. 1).

Let us consider in more detail the given methods.

2. Methods of solution

2.1. The use of sand

Stable adhesion of the wheels of traction rolling stock is not always realized, which requires a large consumption of sand, as the most effective and widespread method of increasing adhesion at present [2–9]. Due to the use of sand, the coefficient of adhesion of the wheels to the rails sometimes increases up to 70%. But the use of sand leads to intensive wear of wheels and rails, contaminates the upper structure of the track. Wheels "squeal" appears and wavy rail wear with short vertical irregularities develops. The low accuracy of sand feeding into the friction contact zone leads to the ingress of sand on the side surface of the rail and, consequently, to an increase in the wear rate of the wheel rails.

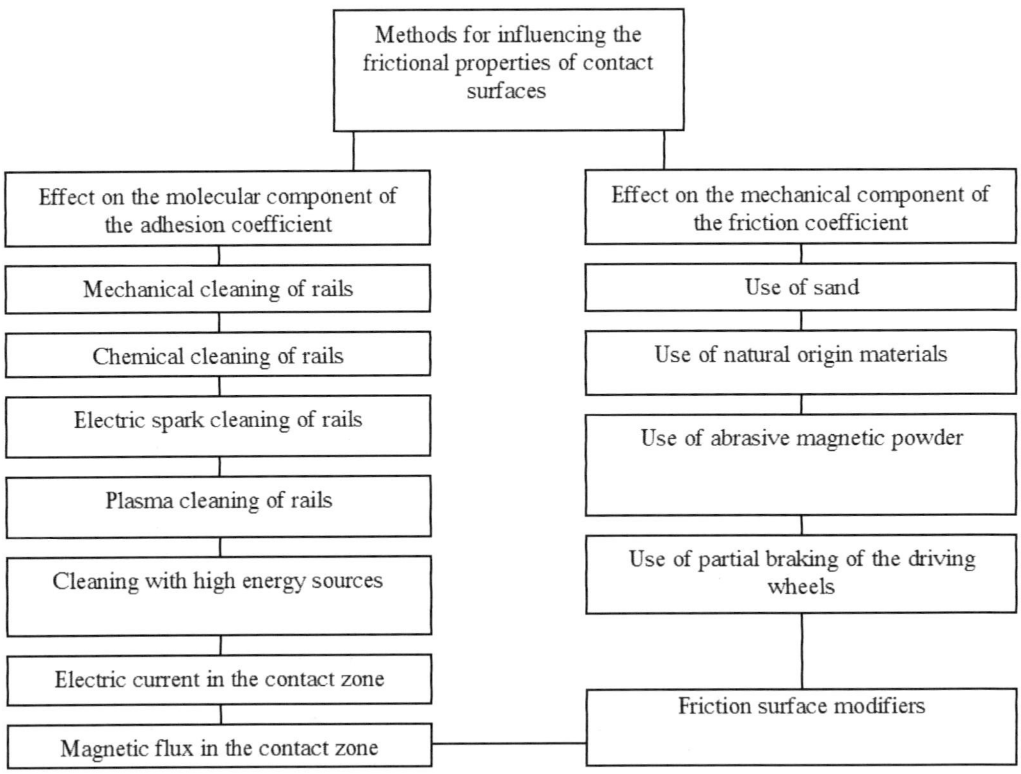

Figure 1. Methods of influencing the frictional properties of contact surfaces.

The results of experiments in which sand was spread on the tire surface using special prepared briquettes, consisting of sand and some binders are presented. However, there were difficulties in obtaining a uniform layer on the rolling surface, low reliability at high speeds.

2.2. Use of natural origin materials

In [10] it is noted that when using marble chips, potassium carbonate, blast-furnace slag, carborundum, etc. as an abrasive, no difference was found in the use of materials. Professor I.P. Isaev pointed out that the value of the adhesion coefficient depends on the hardness of the powder grains, but there is no single point of view on friction processes in the presence of solid particles in this zone. In [8], experiments on the supply of a jet of non-freezing suspension, consisting of sand, ethylene glycol, water and a binder, to the contact. The positive effect was seen on wet and oily rails. However, this abrasive delivery system proved to be complex and unreliable.

2.3. Abrasive magnetic powder use

Testing of the system [11, 12] that feeds, doses, distributes and retains abrasive magnetic powder on the rolling surface (grinding waste was carried out on a full-scale stand manufactured at the Lugansk Machine-Building Institute and on the test ring of the Lyudinovo diesel locomotive plant (diesel locomotive TGM - 4A)). Tests have shown the effectiveness and efficiency of this system. The use of abrasive magnetic powder increases the adhesion coefficient by 20 ... 30%. However, this system was not put into operation.

2.4. Partial braking of the driving wheels

In practice, light braking of locomotive wheels has become widespread [13, 14]. The tires are cleaned before long lifts. Partial braking can be used to reduce excess torque from a skidding wheelset. But this cleaning method cannot be considered effective, since additional wear of the brake pads and wheelset tires occurs. It should be noted that anti-slip braking can only be effective if locomotives have highly sensitive slip detection systems.

2.5. Mechanical cleaning of rails

Mechanical removal of dirt from the rolling surface is carried out using scrapers, brushes, elastic rollers located in front of a moving locomotive [15]. This method does not allow one to properly clean the contacting surfaces, the range of application is limited (no more than 15-20 km / h), low reliability in relation to the safety of train traffic.

The examples of cleaning rails using a strong jet of dried air [15], water or steam [16] are known. However, this method can only be used during summer. It is possible to clean the railheads [16] with a flame, it was supposed to heat up the rails in the most difficult sections, where there is a limitation of the weight norm of trains by adhesion. But this method is not profitable economically, as it requires huge energy costs.

2.6. Chemical cleaning of rails

Chemical cleaning of rails is based on the interaction of lubricant films, fatty acids with some chemical solvents. Various methods of chemical cleaning of rails were tested on a number of railways [17]. Experiments in which rails were treated with weak solutions of ether-caprylic acid showed that the adhesion coefficient increased to 30%. The rail surfaces were treated with an aqueous solution of polyamide resin and an alkali solution [18]. Effective chemical treatment of rails requires considerable time; it must be repeated after the passage of several trains. This limits the practical use of this method.

2.7. Electric spark cleaning of rails

The experiments related to electric spark treatment of rolling surfaces with high frequency current [19, 20] are known. With this method, a high-temperature effect on the treated surfaces occurs, while a high degree of cleaning is achieved. But there is a process of seizure of surfaces with pulling out of metal. The wear of the contacting bodies increases, which is explained by a change in the structure of the surface layers themselves, which is expressed in the formation of a thin layer of martensite with inclusions of pearlite and austenite. Since this method of cleaning is effective only on dry and clean rails, it did not go beyond the scope of experiments and did not find application in operation.

2.8. Plasma cleaning of rails

Much work has been carried out to increase the adhesion of the wheels to the rails by destroying the surface of the pollution films using plasma [21, 22]. The efficiency of using a plasma-cooled argon-hydrogen torch corresponded to an increase in the adhesion coefficient by about 0.1-0.2. However, this method, due to its complexity, has not found application on locomotives.

2.9. Cleaning with high energy sources

In [23], it is supposed to influence the substances of surface contaminants (especially oily or ice-covered rails) with an energy flow of microwave electromagnetic oscillations of the millimeter range (MW). In addition, it is proposed to use microwave heating with other cleaning methods (hot air, ultraviolet or infrared radiation). The positive qualities of this method include smooth regulation of heat flow, high speed, preservation of the material structure of the wheel and rails. The disadvantage is the need to create radiation only in the nearest radiation field.

2.10. Electric current in the contact zone

Currently, work aimed at studying the effect of an electric current passing through the wheel-rail contact on the adhesion coefficient is underway.

The article [24] presents the results of testing electric locomotives VL22M and VL23, which showed that when the motors were operated with a current of 250A, the adhesion coefficient was 0.25, and with current of 500A, it was 0.49.

2.11. Magnetic flux in the contact zone

The introduction of a magnetic field into the contact zone [25, 26] made it possible to increase the adhesion coefficient by 22%. However, the physical nature of the magnetic field effect currently has no general explanation. This is how the magnetoplastic effect is known [26]. The essence of this effect is to weaken the interaction of dislocations with obstacles. This effect is manifested in the change: the speed of the macroscopic flow in the surface layer; creep; yield point; internal friction. The study of the magnetic field influence on the tribological characteristics of the friction pairs of wheels with rails has not been studied properly and requires further research.

2.12. Friction surface modifiers

In a number of countries (USA, Canada, France), the friction modifiers Centrac VHPF, HPF, LCE are used to improve the parameters of frictional contact [27–29]. The new generation of friction modifiers is usually used in the form of a rod and is applied to the wheel using lubricators installed on locomotives in liquid form or using hand-held applicators. Modifiers like HPF improve friction conditions. However, each modifier cannot meet a wide range of environmental and track management requirements. When modifiers with a friction coefficient greater than 0.2 are used, problems associated with negative friction, wheel seizure, and dynamic braking arise [28, 29].

3. Conclusion

Many issues regarding the increase in adhesion are still unresolved. This is due to the fact that when wheels friction is on rails, mechanical, electrical, vibration, thermal, chemical, and magnetic processes occur simultaneously. A multivariate analysis of all these phenomena is practically impossible and, therefore, it is advisable to be limited to a small set of factors that could characterize the tribological system under study.

References

[1] Kragelsky I V 1977 *Fundamentals of calculations for friction and wear* (Moscow: Mashinostroyeniye) p 526

[2] Lucas F M da Silva and Öchsner A 2018 *Handbook of Adhesion Technology* (Luxembourg: Springer International Publishing AG) p 1805

[3] Goryacheva I G 1998 *Contact Mechanics in Tribology* (Luxembourg: Springer Science+Business Media Dordrecht) p 346

[4] Popov V L 2010 *Contact Mechanics and Friction. Physical Principles and Applications* (Luxembourg: Springer Heidelberg Dordrecht) p 362

[5] Rabinowicz E 2013 *Friction and Wear of Materials,* 2nd ed (Hoboken: John Wiley & Sons, Inc) p 336

[6] Shalygin M G 2019 Surface Parameters with Different Treatment Technologies and Their Effect on Wear *J. Frict. Wear* **40** 213–217

[7] Suslov A G and Shalygin M G 2019 A wear model for subrough surface contacts of martensitic steels *Proc. of the Institution of Mechanical Engineers, Part J: J. of Engineering Tribology* **223** 10

[8] Shalygin M G and Kuznetsov S V 2019 Technology to Increase Hydrocarbon Wear Resistance of Friction Units of Hydrogen Aggregates *MATEC Web Conf.* **297** 09001

[9] Xu K, Xu G and Zheng C 2015 Novel Determination of Wheel-Rail Adhesion Stability for Electric Locomotives *Int. J. of Precision Engineering and Manufacturing* **16(4)** 653-660

[10] Buckley-Johnstone L E 2017 *Wheel/Rail Contact Tribology: Characterising Low Adhesion Mechanisms and Friction Management Products,* PhD thesis (Sheffield: University of Sheffield) p 185

[11] Krymskii M D 1995 Magnetic properties of a powder magnetic-abrasive tool *Powder Metallurgy and Metal Ceramics* **33** 33–36

[12] Kumari C and Chak S K 2018 A review on magnetically assisted abrasive finishing and their critical process parameters *Manufacturing Rev.* **5**

[13] Savitski D, Schleinin D, Ivanov V and Augsburg K 2018 Robust Continuous Wheel Slip Control with Reference Adaptation: Application to Brake System with Decoupled Architecture *2018 IEEE Transactions on Industrial Informatics* 99

[14] Lewis S R, Riley S, Fletcher D I and Lewis R 2017 Optimisation of a Railway Sanding System: Adhesion Tests *Int. J. of Railway Technology* **6(2)** 87-98

[15] Neil R F, Arun M and Sudheer A P 2018 Design, modelling and fabrication of railway track cleaning bot *Procedia Computer Science* **133** 526-536

[16] James J, Wilson J, Jetto J and Thomas A 2016 Intelligent track cleaning robot *IEEE Int. Conf. on Mechatronics and Automation* 332-337

[17] Andrews H I 1955 The adhesion of electric locomotives *J. of the Institution of Electrical Engineers* **1** 785–803

[18] Kosikov S I 1967 *Frictional properties of railway rails* (Moscow: Nauka) p112

[19] Fichaux H and Moore D F 1968 Improvement in rail-wheel adhesion by spark-discharge *Wear* **11** 51-67

[20] Stavitskii B I 2010 Glimpses of the History of Electrospark Machining of Materials *Surface engineering and applied electrochemistry* **46** 178–191

[21] Balwanz W W 1979 Plasma Cleaning of Surfaces, in: *Surface Contamination* (Boston: Springer) pp 255-269

[22] Swan J, Candy M, Radoiu M and Richardson G 2019 Microwave generated plasma railway track treatment *AMPERE* 144-151

[23] Dincer I, Hogerwaard J and Zamfirescu C 2016 *Clean Rail Transportation Options* (Boston: Springer International Publishing) p 223

[24] Rudyakov V Z 1962 Increasing the carrying capacity of electric locomotives *Railway transport* **2** 28-31

[25] Antipin D, Bondarenko D and Vorobiev V 2020 Possibilities of Simulating Magnetic Fields in a Wheel-Rail System *2020 Int. Russian Automation Conf., RusAutoCon* 20007691
[26] Antipin D Y, Korchagin V O and Maslov M A 2020 Determination of Magnetization Efficiency of Wheel-Rail Contact Zone *Lecture Notes in Mechanical Engineering* 669–675
[27] Foo C T, Omar B and Jalil A S 2018 A review on recent wheel/rail interface friction management *J. of Phys.: Conf. Ser.* **1049** 012009
[28] Waara P 2006 *Lubricants Influence on Wear in Sharp Rail Curves,* PhD thesis (Luleå: Luleå University of Technology) p 36
[29] Mashkovich O I 1998 Optimization of the interaction of wheel and rail due to friction *Rail abroad. Sar. IV.* **5** 4-8

Laboratory stand for studying the interaction of rubbing bodies in a magnetic field

D Ya Antipin, V I Vorobiev, A I Gorelenkov and D A Bondarenko

Bryansk State Technical University, 7, Boulevard 50 let Oktyabrya, Bryansk, 241035, Russia

E-mail: unit@tu-bryansk.ru

Abstract. The paper presents the basic principles for the implementation of experimental studies of the tribological characteristics of friction pairs, taking into account the effect of a magnetic field on them. It has been established that modern experimental studies are carried out on laboratory units using powerful tools to analyze the data obtained, which makes it possible to study the structures of systems by changing the properties of materials under the influence of external factors and wear processes. Taking into account the influence and further research in tribo-conjugated metal bodies (for example, a wheel-rail) of such factors as magnetostriction, external magnetic field (constant, alternating, pulsed), current (direct, alternating, pulsed) passing through the contact, acting both separately, and in various combinations, the design of an experimental laboratory installation was proposed and implemented. The installation allows testing with two types of friction: sliding and rolling with sliding. To measure the friction force (friction coefficient), a 6-wire bridge circuit is used, which is described in the article. The paper shows a diagram of a noise-immune strain-gauge circuit of the installation, which allows a significant level of interference during research. Due to the fact that the friction coefficient with a change in temperature can vary over a wide range, the proposed installation implements a method for automatic temperature compensation in the contact zone of tribological samples. The use of thermocouples is proposed as temperature sensors. The signals from the thermocouples of the installation are transmitted to a measuring device consisting of an analog-to-digital converter and a personal computer. The proposed laboratory setup allows for a wide range of tribological testing studies and monitoring of processes in the friction zone exposed to the magnetic field.

1. Introduction

Experimental studies of the processes occurring in the contact of a wheel with a rail are carried out on roller rigs, in real operating conditions and in laboratory facilities. When using laboratory research, it is easier to follow the principle formulated by V.D. Kuznetsov [1]: "The task of scientific research is to investigate the phenomenon not in all its complexity, but in a simplified form, when of all the factors influencing this phenomenon, only one changes, and the rest remain unchanged."

In [2] it is indicated that about 50% of the tests described in the international journal "Wear", carried out over 13 years, were carried out on laboratory facilities and only 17% - on real machine parts. This confirms the importance of laboratory research, especially since at present powerful tools for analyzing friction surfaces have been developed, which makes it possible to study the structures of systems by changing the properties of materials under the influence of external factors and wear processes. In general, the tribosystem model is shown in Fig. 1.

The allocation of a system allows, firstly, to find its place in the system hierarchy as a subsystem and, secondly, to reduce the number of elements and, accordingly, factors that affect the tribosystem behavior, considering those factors that have a minor effect on the functioning and interconnections of the system's elements.

Back in 1845, Faraday studied the influence of a magnetic field on a number of substances, and after many experiments he wrote: "Now we can assume that all substances are subject to the action of magnetic forces in the same way as the action of gravity, electrical forces and cohesion forces. However, all bodies do not exhibit such magnetism, which is characteristic of iron" [3]. In 1926 E. Herbert described the effect of changing the properties of ferromagnetic materials and their subcultures when exposed to an external magnetic field [4]. The physical nature of the magnetic field effect in the contact area of metal bodies currently has no general explanation. Thus, a number of authors consider the effect of a magnetic field on the martensitic transformation in steel from the point of view of thermodynamic calculation [5].

The authors of [6, 7] explain the increase in the service life of the cutting tool as a result of magnetostrictional and magnetodispersive hardening of steel.

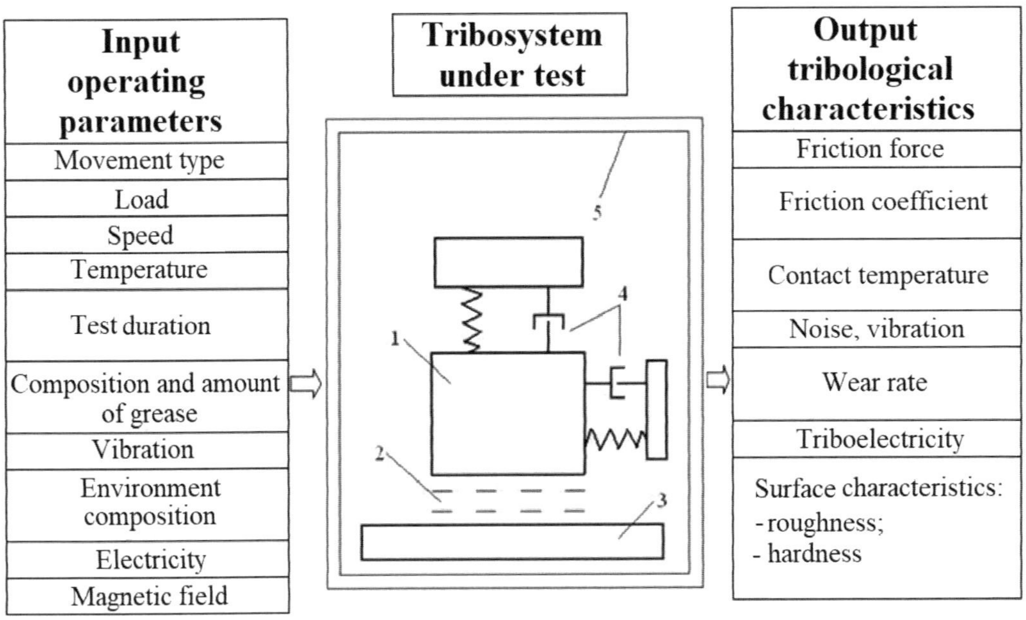

Figure 1. Tribosystem model: 1,3 - friction pair elements, 2 - oxide films, lubricant (third body), 4 - elastic-dissipative bonds, 5 - medium

Some hypotheses are based on the presence of a magnetic field in the friction zone. The authors of [6] assume that electromagnetic induction processes can cause changes in the properties of a magnetized instrument. In [8], it is assumed that the increase in the durability of magnetized parts is due to the interaction of excited hydrogen molecules appearing in the contact zone of rubbing surfaces with a magnetic field, which allows for the rapid removal of molecules and atomic hydrogen from the contact zone [9, 10].

In [9, 10], it is indicated that, when exposed to a magnetic field, the location of point defects changes, leading to the appearance of dipoles. The interaction of a magnetic field with dipoles leads to the decomposition of defect complexes.

In [11, 12], it was suggested that during friction of magnetized parts under conditions of boundary lubrication, a decrease in the wear rate may be associated with an improvement in the properties of lubricants under the influence of a magnetic field.

Japanese researchers Hayashi S., Takahashi S., Yamamoto M. [13] noted that an alternating magnetic field facilitates plastic deformation and associated this effect with the characteristics of ferromagnetic materials.

The article [14] considered the effect of an alternating magnetic field on the friction and wear of the XC48 ferromagnetic steel. The magnetic field was created by an alternating current with a frequency of 50 Hz. The experiments were carried out for sliding steel-on-steel contact. When the magnetic field strength changed from 0 to 8 kA / m, the friction coefficient increased from 0.16 to 0.23. In addition, it is noted that there is a decrease in wear, an increase in microhardness, and an increase in tribooxidation of friction surfaces under the action of an external magnetic field.

The introduction of a magnetic field into the contact zone, according to the authors of [15], made it possible to slightly (up to 22%) increase the adhesion coefficient of the wheel to the rail. The authors associate this increase with the increase in the pressure of the wheel on the rail, as well as with physical phenomena occurring on the friction surface. At the same time, it is not specified exactly what phenomena occur in the contact zone. The authors believe that the cost of copper necessary to create a magnetic field will not pay off such a slight increase in the adhesion coefficient.

Nevertheless, the study of the magnetic field influence on the friction surface is of interest in the theoretical aspect and in the field of application to engineering problems. Magnetoplastic effect [16], the essence of which is to weaken the interaction of dislocations with obstacles is known. This effect is manifested in the change: the speed of the macroscopic flow in the surface layer; creep; yield point; internal friction. However, to date, there is no comprehensive explanation of this effect.

In [17] the results of a study to determine the tribological characteristics of a wheel-rail friction pair when exposed to an external constant magnetic field are presented. It is noted that the coefficient of friction increases with increasing magnetic field strength. There is some improvement in the contact surface, in particular a decrease in roughness. It is established that the effect of an electromagnetic field, depending on the level of tension and the state of the intermediate medium, can be both the strengthening of the frictional bonds and their softening.

The work [18, 19] analyzes the points of view of a number of authors on the influence of various factors on the process of adhesion of wheels to rails. The considered factors were divided according to their importance into three groups: main, significant, insignificant. It is indicated that the performed ranking of factors is rather arbitrary. One of the reasons is the presence of incompletely studied factors that do not attract the attention of researchers at the moment due to the difficulties of their theoretical and experimental analysis. So among the insignificant factors is the electric current flowing through the contact. However, in [20], the results of tests of electric locomotives VL22M and VL23 are given, which show that when motors are operating with a current of 250A, the adhesion coefficient is 0.25, and with a current of 500A - 0.49. The influence of an external magnetic field on the processes occurring in the contact zone of metal bodies is not mentioned at all.

In [21, 22], it is noted that the capabilities of the magnetic-pulse processing technology consist in the fact that during the strengthening of parts, one can vary not only the field strength, but also the polarity, duration and amplitude of pulses, as well as the holding time after processing, which is emphasized in some cases after magnetic processing and holding, demagnetization is necessary, which, as a rule, does not reduce the effect of magnetic hardening.

2. Results

Investigation and evaluation of the phenomena occurring in tribo-conjugated metal bodies (for example, a wheel-rail) exposed to such factors as magnetostriction, external magnetic field (constant, alternating, pulsed), current (direct, alternating, pulsed) passing through the contact, acting separately , and in different combinations, demanded to develop an experimental installation, the general view of which is shown in Fig. 2.

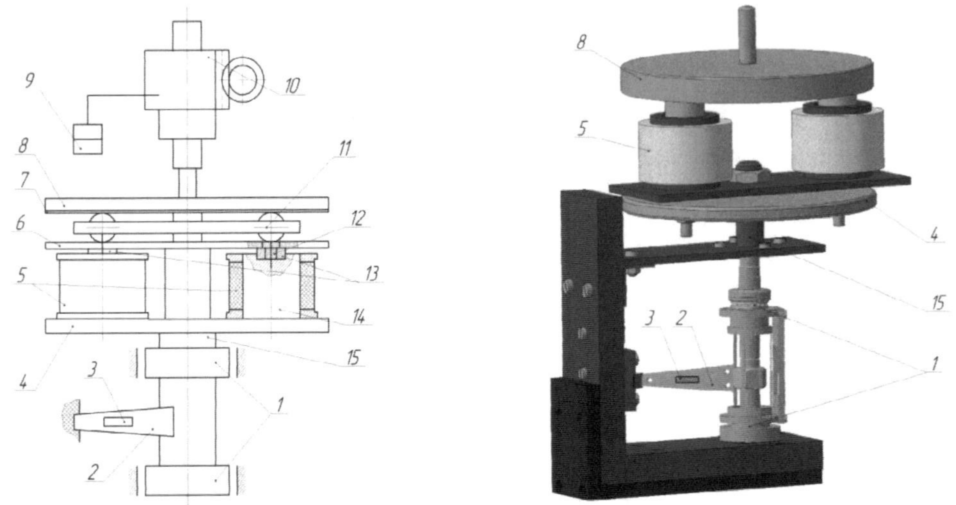

Figure 2. Installation for studying the tribological characteristics of a friction pair when exposed to a magnetic field: a) in the form of friction - rolling, b) in the form of friction - sliding

The installation is assembled on the basis of a drilling machine. On the shaft 15, containing two bearing assemblies 1, a support disk 4 is installed, on which the coils 5 are located. Coils 5 are located on the casings of 14 of samples 13. Above the cylindrical samples 13, a lower disk sample 6 is installed, on which an indenter 11 consisting of ball bearings, evenly spaced around the circumference of the separator is located. In the center of the separator there is a hole through which it is put on the shaft 15. An insulating sleeve is installed between the shaft 15 and the separator. A textolite mandrel with a holder 8 and a replaceable test disk 7 is inserted into the chuck of the drilling machine 10. The vertical load is set by the device 9. The machine's electric motor is powered by a static frequency converter, which allows smoothly changing the peripheral speed of the holder 8 over a wide range.

To measure the friction force (friction coefficient) a strain gauge 2 with a strain gauge 3 is attached to the shaft 15 through an insulating gasket. To determine the temperature in the contact zone of the rubbing surfaces in samples 13 and disc 6, thermocouples are installed. Samples are made of grade 60 tire steel.

The installation allows testing with two types of friction: sliding and rolling with sliding. In the first case, a disk sample 6 and an indenter 11 are removed from the shaft 15 (Fig. 2 (b)). In the second case, instead of the clips 14 with samples 13, holders are installed. Between the upper ends of holders and the lower end of the disc 6, a slight air gap is created. Electromagnetic coils 5 are attached to the holders.

When studying the effect of magnetostrictive processes on the characteristics of friction pairs, samples 13 were preliminarily magnetized in the magnetic field of a solenoid. The magnetic field strength during pole magnetization was determined by the expression [23]:

$$H = 78\frac{1.256nI}{\sqrt{l^2 + D^2}}, \tag{1}$$

where I – amperage, A; n – number of solenoid turns, pcs; l – solenoid length, cm; D – average diameter of the solenoid turns, cm.

The magnetization of samples 13 made of tire steel occurred in the range of saturation magnetic induction B∈ [1.42 ... 1.50] at a strength H∈ [12 · 103… 15 · 103].

When investigating the effect of an external magnetic field on the tribological characteristics of a friction pair, electromagnetic coils 15 were used. A storage battery was used to create a constant magnetic field; for alternating field of different frequency - static converter; for pulsed magnetic field - a capacitor of 600-800 μF.

The transmission of magnetic lines of force through the contact zone in the rolling friction mode is carried out as follows: the magnetic field passes through the first coil holder 5, the air gap, the lower disc 6, the balls of the indenter 11, the upper disc 7, the balls of the indenter, the lower disc 6, the air gap, the second coil holder 5, support disk 4 (Fig. 2, a).

In the sliding friction mode, the magnetic lines of force pass through the first holder 14 with the sample 13, the upper disk 7, the second holder with the sample 13 and the supporting disk 4.

To measure the friction force (friction coefficient), a 6-wire bridge circuit is used, shown in Fig. 3.

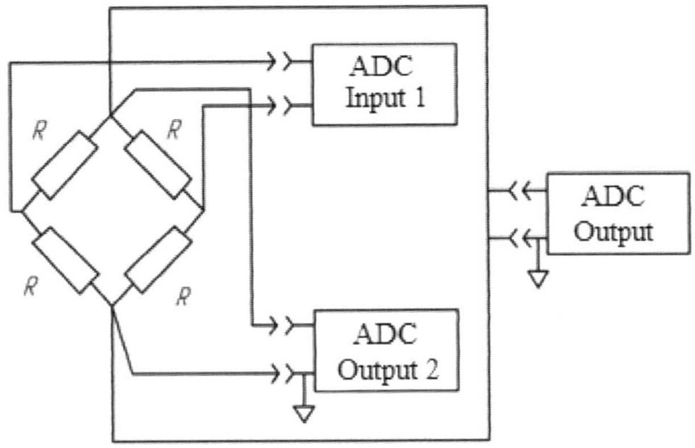

Figure 3. Connection diagram for strain gauges.

To implement this connection diagram, two input channels of an analog-to-digital converter (ADC) are required. The first channel in (Fig. 3) designated as "ADC Input 1" is used to measure the excitation voltage $U_{exc} = U_{gen} + U_{int}$, where U_{gen} – generator voltage, at DC or generator operating voltage at alternating current; U_{int} – voltage of the interferences induced by the electric motor or electromagnetic coils. It should be noted that U_{exc} should be measured directly on the connection diagram, which is especially important when using long wires, since in this case, when the measuring point is removed, the induced interference voltage drops. This fact should be taken into account when using connectors that allow U_{exc} measurement directly on the connector, in this case $U_{exc} = U_{gen}$. In this case, the interference immunity of the circuit will be minimal.

The second ADC channel in (Fig. 3) designated as "Input 2 ADC" is used to measure the total voltage U_{sum} directly on the measuring part of the bridge, while $U_{sum} = U_{gen} + U_{int} + U_{meas}$. After the data on both channels are taken, the difference between the voltage values U_{sum} and U_{exc} is calculated and the measurement signal $U_{meas} = U_{sum} - U_{exc}$ is extracted.

An important role in reducing the influence of interference in the measuring circuit is played by the measuring circuit in Fig. 4. To create a antiinterference strain gauge circuit, it is necessary to conditionally divide the sensor leads into the right - "P" (+) and left - "L" (-) (Fig. 4, a), and then, when they are connected to the bridge circuit (Fig. 4, b), it is necessary to connect the left lead of one sensor to the left lead of the other, and the right lead to the right. Thus, a contour is formed in which the EMF induced in different parts of the contour will have opposite directions and the same values, and, accordingly, exclude each other. The signals from the strain gauges are fed directly to the LA-2 USB data acquisition device connected to a computer, manufactured by "Rudnev-Shilyaev" CJSC.

Figure 4. Anti-interference strain gauge circuit diagram: a - connection of two sensors with signs of induced interferences; b - connection of anti-interference sensors

It is known that the coefficient of friction with a change in temperature can drop, increase, and even pass through one or two minima. Temperature measurement at the points of contact of the samples is carried out using thermocouples located in the immediate area of the contacting bodies. In this case, it is important to keep the temperature of the exposed ends constant, as changing it leads to errors. Of all the known methods (baths with melting ice, thermostats, etc.). the most convenient method is to automatically compensate for temperature changes in the free ends. Fig. 5 shows a diagram illustrating a method for automatically compensating the temperature of the exposed ends.

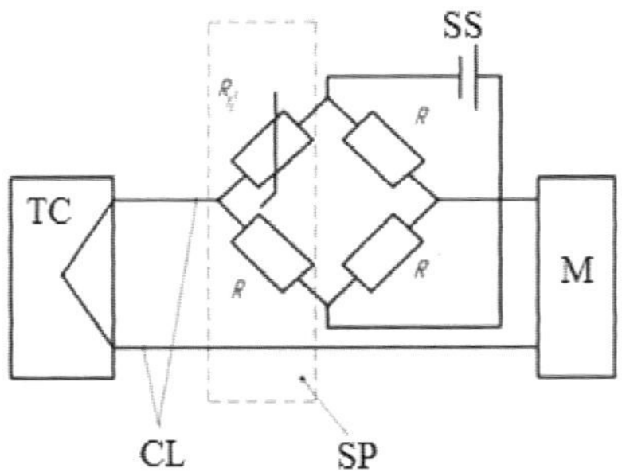

Figure 5. Thermocouple temperature compensation circuit

Bridge R is connected in series with the thermocouple TC, in one of the arms of which there is a thermistor $R_{t_0}^0$. The bridge output voltage is proportional to the temperature of the exposed ends of the thermocouple, since $R_{t_0}^0$ is located directly at the point where the exposed ends are soldered. This voltage is subtracted from the EMF of the thermocouple, resulting in automatic temperature compensation of the exposed ends.

The exposed ends of the thermocouples were soldered on special pads (SP). Compensating wires of the gearbox are used as supply wires. The signals from the thermocouples go to the meter M, which consists of an ADC and a computer. The circuit is powered from a stabilized source SS.

3. Conclusion

The above analysis of publications shows the contradictions in the explanation of the principles of magnetostriction, external magnetic field on the tribological characteristics of friction pairs, the absence of a unified theory explaining the essence of the phenomena occurring in the friction zone when exposed to a magnetic field. Thus, it is necessary and urgent to develop a laboratory testing facility that allows a whole range of tests and monitoring of processes in the friction zone exposed to the magnetic field.

References

[1] Kuznetsov V D 1947 *Solid State Physics* (Tomsk: Krasnoe Znamya) p 39
[2] Menezes P, Ingole S P, Nosonovsky M, Kailas S V and Lovell M R 2013 *Tribology for Scientists and Engineers* (New York: Springer-Verlag) p 948
[3] Laughlin D and Hono K 2014 *Physical Metallurgy* 5th Edition (Amsterdam: Elsevier) p 2960
[4] Hankins G A 1929 A Synopsis of the Present State of Knowledge of the Hardness and Abrasion Testing of Metals with Special Reference to the Work Done during the Period 1917–27, and a Bibliography *Proceedings of the Institution of Mechanical Engineers* **116** 317–387
[5] Zolotarevskii I V, Loskutov S V, Man'ko V K and Seidametov S V 2009 Effect of a magnetic field on the reverse and forward martensitic transformations in the Fe-33% Ni alloy *The Physics of Metals and Metallography* **108** 139–146
[6] Postnikov C H, Goldina A F and Tarakanov V N 1974 Some physical aspects of magnetic tool processing. In the book. Questions of electrophysics of friction and cutting *Proceedings of the Gorky Polytechnic Institute* **30** 27-35
[7] Postnikov S N 1975 *Electrical phenomena during friction and cutting* (Gorky: Vyatka book publishing house) p 280
[8] Garkunov D N, Suranov G I and Koptyaeva G B 1982 On the nature of increasing wear resistance of parts and tools by magnetic processing *Friction and wear* **3** 327–330
[9] Gorbunov N, Kovtanets M, Prosvirova O and Garkushin E 2012 Adhesion control in the system of "wheel-rail" *Transport problems* **7** 15–24
[10] Wolfer W G 2012 Fundamental Properties of Defects in Metals *Comprehensive Nuclear Materials* **1** 1–45
[11] Dehghani A, Khalaj S, Alireza A, Tehrani F and Mohammadi A 2017 Effects of magnetic assistance on improving tool wear resistance and cutting mechanisms during steel turning *Wear* **384–385** 1–7
[12] Ma L,Wang X, Liang Z, Liu Y and Zhang D 2019 Machining mechanism in pulsed magnetic treatment for micro-tools *The Int. J. of Advanced Manufacturing Technology* **101** 2391–2406
[13] Hayashi S S, Hayashi S and Takahashi M 1971 *J. Phys. Soc. Japan* **30** 381
[14] Chin K J 2001 Tribological behavior and surface analysis of magnetized slicing contact XC48 steel/ XC48steel *Wear* **250** 470-476
[15] Antipin D Y, Vorobiev V I and Maslov M A 2019 Improving the construction of magnetic clutch amplifiers of locomotive wheels with rails *Lecture Notes in Mechanical Engineering* **0 (9783319956299)** 2331-2338
[16] Morgunov R B 2004 Spin micromechanics in the physics of plasticity *UFN* **174** 131–151
[17] Antipin D Y, Korchagin V O and Maslov M A 2020 Determination of Magnetization Efficiency of Wheel-Rail Contact Zone *Lecture Notes in Mechanical Engineering* 669-675
[18] Luzhnov Yu M 2003 *Coupling of wheels with rails (nature and patterns)* (Moscow: Intekst) p 144
[19] Suslov A G and Shalygin M G 2016 The interrelation of the surface subroughness of martensitic steels with their granularity the quality of mechanical processing *AIP Conf. Proc.* **1785** 040085
[20] Rudyakov V Z 1962 Provisions for increasing the weight standards *Railway transport* **2** 28-31
[21] Malygin B V 1989 Magnetic hardening of tools and machine parts *Mechanical Engineering* p 112
[22] Shalygin M G 2016 Correlation between Nano-Roughness with Grains of Austenitic Steel and Machining Quality *J. Fundam. Appl. Sci.* **8(3S)** 2835-2841
[23] Moore P O 2012 *Nondestructive Testing Handbook* 3rd Edition **10** p 604

IOP Conf. Series: Materials Science and Engineering 1111 (2021) 012012 doi:10.1088/1757-899X/1111/1/012012

Study of the flammability of vegetable oils in the combustion chamber of diesel engines

V A Markov[1], S N Devyanin[2], V G Kamaltdinov[3] and B Sa[1]

[1] Power Engineering Faculty, Bauman Moscow State Technical University, 5, 2-ya Baumanskaya Str., Moscow, 105005, Russia
[2] Tractors and Cars Department, Russian State Agrarian University - Moscow Timiryazev Agricultural Academy, Moscow, 127550, Russia
[3] Motor Transport Faculty, South Ural State University, 76, Lenina Ave., Chelyabinsk, 454080, Russia

Email: saboven@bmstu.ru

Abstract. Promising biofuels for diesel engines are vegetable oils and fuels derived from them. The advantages of using biofuels in internal combustion engines were considered. The problems of using vegetable oils in diesel engines have been analyzed. It is noted that the main issues of using vegetable oils are their high viscosity and their worst flammability in the combustion chamber of diesel engines. The flammability indicators of vehicle fuels in the combustion chamber of diesel engines were considered. The most important among them is the cetane number of the fuel. The factors influencing the cetane number of vegetable oils have been determined. The dependences of the cetane number of vegetable oils on the flash point in a closed-cup, the content of oleic acid, and the content of saturated fatty acids have been obtained and analyzed. These dependences have been compared with a multi-factorial dependence of the cetane number on the basic triglyceride contents.

1. Introduction

One of the urgent problems in modern engine manufacturing is the need to find alternatives to petroleum diesel fuel (DF) and replace it with vehicle fuels produced from renewable raw materials. Recently, more and more fuels derived from vegetable oils (VOs) have been used as fuels for diesel engines [1, 2]. In different parts of the world, such as Europe, the United States, the countries of Central and Latin America, and Southeast Asia, these fuels are considered to be real alternatives to petroleum fuels. Moreover, these biofuels are used in various industries - road transport, agriculture, railway transport, and energy [3, 4]. This is due to the simplicity and environmental friendliness of obtaining VOs, the potential to improve emission performances of diesel engines fueled with these fuels, the relatively low cost of VOs, and their acceptable flammability in the combustion chamber of diesel engines.

Although there have been some problems in the operation of diesel engines using VOs and their derivatives (high viscosity, low cetane number, high freezing point, and high coking property), studies on the performance of diesel engines using these biofuels and their mixtures with other fuels, in particular, with petroleum DF, have continued [5-13]. Biofuels are expected to account for 4-6% of total vehicle fuel consumption by 2030 [3].

There are two possible ways to use VOs and their derivatives as vehicle fuels - centralized and decentralized production of biofuels [3]. The centralized production of vehicle fuels from VOs is to process VOs into esters (methyl, ethyl, butyl), which are then used as diesel engine fuels. The decentralized production is to use "pure" VOs or their mixtures with petroleum DF as vehicle fuel.

In the analysis of the possibilities of using various VOs to produce biofuels, rapeseed oil (RO) is usually considered [1-3]. At the same time, other kinds of VO are grown in Russia. The structure of VO production in Russia is as follows: sunflower oil (SuO) accounts for 86.84%, soybean oil (SO) accounts for 7.96%, RO accounts for 4.84%, mustard oil (MuO) accounts for 0.11%, and other oils, such as corn oil (CO), linseed oil (LO), account for 0.25% [2].

One of the problems arising in adapting diesel engines to run on VOs and their derivatives is the difference in physicochemical properties between VOs and petroleum DF. Some physicochemical properties of the most significant VOs for Russia are given in Table 1. It should be noted the high viscosity of VOs and their worst flammability in the combustion chamber (CC) of diesel engines. Therefore, if the cetane number (CN) of petroleum DF is 45 or more according to GOST 305-2013 (Diesel fuel. Technical condition), the CN of VOs usually ranges from 33 to 36 (see Table 1) [3]. This leads to a longer ignition delay, a higher rate of pressure rise in the CC, and more nitrogen oxide emissions from the exhaust gas.

Table 1. Physicochemical properties of petroleum DF and VOs

Physicochemical properties	DF	RO	SuO	SO
Density at 20 °C, kg/m^3	839	914	920	916
Kinematic viscosity, mm^2/s at:				
20 °C	3.8	75.0	65.2	68.7
40 °C	3.3	34.6	30.7	32.0
100 °C	2.5	8.1	7.4	7.7
Cetane number	45	36	34	33
Autoignition temperature, °C	280	318	320	321
Amount of air required to burn 1 kg of fuel, kg	14.5	11.9	12.4	12.4
Lower calorific value, H_U, MJ/kg	42.5	37.3	37.1	37.0
Weight content, %				
C	87.0	78.0	77.6	77.5
H	13.0	10.0	11.5	11.5
O	0	12.0	10.9	11.0

In order to assess the fuel flammability in the CC of a diesel engine and the dynamics of the subsequent combustion process, various indicators are used: the cetane number (CN), the ignition delay period τ_d, the maximum combustion pressure p_z, the maximum rate of pressure rise $(dp/d\varphi)_{max}$ and the average rate of pressure rise $(dp/d\varphi)_{ave}$ during combustion (φ - crank angle), the ratio of pressure rise during combustion $\lambda=p_z/p_c$ (p_c - pressure at the end of the compression stroke), etc. [1, 3]. The most important of these indicators is the CN of fuels. The higher the CN, the shorter the ignition delay period and the more smoothly the fuel-air mixture burns.

CN is determined as the volume fraction of cetane (n-hexadecane $C_{16}H_{34}$, the CN of which is taken as 100) mixed with α-methylnaphthalene (its CN is 0). By mixing these two components in different proportions, it is necessary to obtain the mixture with the same ignition delay period as the test fuel under the same conditions. The percentage of cetane in the mixture corresponds to the CN of the test fuel. This procedure for determining the CN of vehicle fuel was adopted in the first version of ASTM D613 issued in 1941. New versions of ASTM D613 (since 1962) do not use α-methylnaphthalene, but 2,2,4,4,6,8,8-heptamethylnonane that is sometimes called HMN or isocetane with a CN of 15. The mixture of n-hexadecane and α-methylnaphthalene is still used in GOST 3122-67 (Diesel fuels. Method for determining the cetane number). There were several reasons for excluding α-methylnaphthalene from ASTM D613. First, it readily forms peroxides that change the CN of mixtures based on it. In addition, α-methylnaphthalene has an unpleasant smell, possesses carcinogenic properties, and is difficult to obtain in a pure form.

The objectives of the work are to study the flammability of different VOs in the combustion chamber of diesel engines, to determine the factors affecting the CN of VO, and to determine the most significant factors. The following eleven VOs were investigated: rapeseed oil (RO), sunflower oil (SuO), soybean oil (SO), safflower oil No. 1 (SaO1), safflower oil No. 2 with high oleic oil (SaO2), corn oil (CO), sesame oil (SeO), peanut oil (PO), linseed oil (LO), cotton oil (CoO), and crambe oil (CrO).

2. Results

VOs are 95% to 97% composed of triglycerides (esters of glycerol and fatty acids), as well as monoglycerides and diglycerides. Acyglycerides, in the same way, contain molecules of different fatty (carboxylic) acids that are associated with the molecule of glycerol $C_3H_5(OH)_3$ [3, 14]. The fatty acid composition of VO includes both unsaturated fatty acids (linoleic, oleic, linolenic acids) and saturated acids (palmitic, stearic, arachidic, myristic acids). Table 2 shows the fatty acid composition of VO studied in [15, 16]. The physicochemical properties of these VOs are given in Table 3.

Table 2. Fatty acid composition of VOs

Fatty acid, composition formula, conditional composition formula	Oils										
	RO (No.1)	SuO (No.2)	SO (No.3)	SaO1 (No.4)	SaO2 (No.5)	CO (No.6)	SeO (No.7)	PO (No.8)	LO (No.9)	CoO (No.10)	CrO (No.11)
Saturated fatty acids											
Myristic, $C_{14}H_{28}O_2$, C 14:0	0	0	0	0	0.34	0	0	0	0	0	0
Palmitic, $C_{16}H_{32}O_2$, C 16:0	3.49	6.08	11.75	8.60	5.46	11.67	13.10	11.38	4.92	28.33	2.07
Stearic, $C_{18}H_{36}O_2$, C 18:0	0.85	3.26	3.15	1.93	1.75	1.85	3.92	2.39	2.41	0.89	0.70
Arachidic, $C_{20}H_{40}O_2$, C 20:0	0	0	0	0	0.23	0.24	0	1.32	0	0	2.09
Behenic, $C_{22}H_{44}O_2$, C 22:0	0	0	0	0	0	0	0	2.52	0	0	0.80
Lignoceric, $C_{24}H_{48}O_2$, C 24:0	0	0	0	0	0	0	0	1.23	0	0	1.12
In total	4.34	9.34	14.90	10.53	7.78	13.76	17.02	18.84	7.33	29.22	6.78
Unsaturated fatty acids											
Oleic, $C_{18}H_{34}O_2$, C 18:1	64.40	16.93	23.26	11.58	79.36	25.16	52.84	48.28	19.70	13.27	18.86
Linoleic, $C_{18}H_{32}O_2$, C 18:2	22.30	73.73	55.53	77.89	12.86	60.60	30.14	31.95	18.03	57.51	9.00
Linolenic, $C_{18}H_{30}O_2$, C 18:3	8.23	0	6.31	0	0	0.48	0	0.93	54.94	0	6.85
Erucic, $C_{22}H_{42}O_2$, C 22:1	0.73	0	0	0	0	0	0	0	0	0	58.51
In total	95.66	90.66	85.10	89.47	92.22	86.24	82.98	81.16	92.67	70.78	93.22

Table 3. Physicochemical properties of VOs

Properties	Vegetable oils										
	RO (No.1)	SuO (No.2)	SO (No.3)	SaO1 (No.4)	SaO2 (No.5)	CO (No.6)	SeO (No.7)	PO (No.8)	LO (No.9)	CoO (No.10)	CrO (No.11)
Density at 38 °C, kg/m^3	911.5	916.1	913.8	914.4	902.1	909.5	913.3	902.6	923.6	914.8	904.4
Viscosity at 38 °C, mm^2/s	37.0	33.9	32.6	31.3	41.2	34.9	35.5	39.6	27.2	33.5	53.6
Cetane number	37.6	37.1	37.9	41.3	49.1	37.6	40.2	41.8	34.6	41.8	44.6
Autoignition temperature, °C	246	274	254	260	293	277	260	271	241	234	274

The preliminary analysis of the data in Tables 2 and 3 showed that the CN (according to ASTM D613) of the studied VOs depends on their fatty acid composition, primarily on the content of saturated fatty acids and unsaturated oleic acid in VOs. Another factor affecting the flammability of VOs in the conditions of the CC of diesel engines, characterized by CN, is their flash point determined in accordance with ASTM D93. The methodology of this standard involves the determination of the flash point of petroleum products in a closed-cup in a temperature range from 40 to 360 °C by using a manual or automatic Pensky-Martens instrument. It is also possible to determine the flash point of biodiesel by this method, possessing fire safety with handling fuels.

One of the factors affecting the CN of VOs is the content of saturated fatty acids in them. This is due to the fact that saturated fatty acids have a relatively low autoignition temperature. The two most important saturated fatty acids in VOs are palmitic and stearic acids (see Table 2). Their autoignition temperatures determined by GOST 12.1.044-89 (Fire and explosion safety of substances and materials) are 210 and 320 °C, respectively. However, the autoignition temperatures of linolenic and erucic fatty acids (unsaturated fatty acids) are 470 and 454 °C, respectively.

In order to estimate the relationship between the CN and the content of saturated fatty acids in VOs (C_{SFA}), the data in Tables 2 and 3 were used. Figure 1 shows the CN and the content of saturated fatty acids for nine VOs in Tables 2 and 3 (nine experimental points). However, the data for two VOs with abnormal content of unsaturated fatty acids were not used - high oleic SaO2 (No. 5) contains 79.36% of oleic acid, and high erucic CrO (No. 11) contains 58.51% of erucic acid (the content of this fatty acid for other VOs in Tables 2 and 3 is close to zero). With the use of the data in Figure 1 and the least square method, an approximation of the dependence CN $= f(C_{SFA})$ is carried out. The expression describes the fitted line plotted in Figure 1:

$$ECN = 23.55 \times C_{SFA} + 35.6, \quad (1)$$

where ECN is the predicted cetane number.

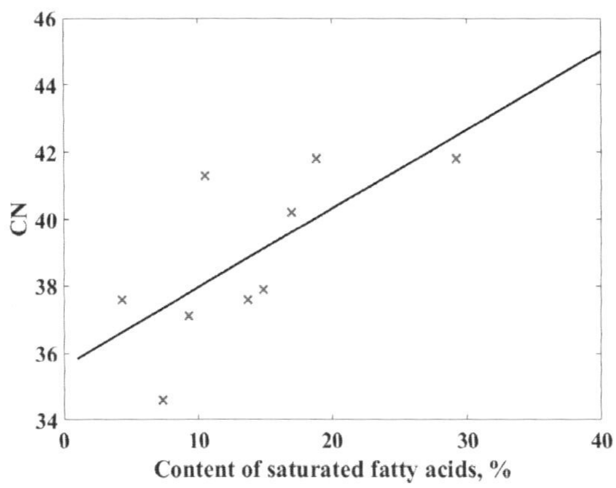

Figure 1. The relationship between CN and C_{SFA} (the points – experimental data in Tables 2 and 3; the straight line – predicted CN).

The correlation coefficient between the CN of the studied VOs and C_{SFA} was equal to R = 0.69 (the coefficient of determination - $R^2 = 0.48$), which indicates the presence of a correlation between CN and C_{SFA}.

Another factor affecting the CN of VOs is the content of oleic acid in them (C_{OLA}). Among the main unsaturated fatty acids composing vegetable oils, oleic acid has the lowest autoignition temperature. The autoignition temperature of oleic acid determined according to GOST 12.1.044-89 (Fire and explosion safety of substances and materials) is 271 °C. And the autoignition temperature of linolenic and erucic acids are 470 and 454 °C, respectively. The data for eight VO (No. 2-9 in Tables 2 and 3) were used to evaluate the correlation between the CN of VOs and C_{OLA} shown in Figure 2. At the same time, the data for three VOs with abnormal content of saturated fatty acids were not used – RO (No. 1) with a minimum content of saturated fatty acids (4.34%) among the studied VOs, CoO (No. 10) with a maximum content of saturated fatty acids (29.22%), and CrO (No. 11) with an abnormal content of erucic acid (58.51%).

With the use of the data in Figure 2 and the least square method, an approximation of the dependence CN $= f(C_{OLA})$ is carried out. A linear relationship has been obtained as:

$$ECN = 15.34 \times C_{OLA} + 34.64. \quad (2)$$

The correlation analysis of CN= f (C_{OLA}) confirmed the presence of a fairly close correlation relationship between these parameters - the correlation coefficient between CN of the studied VOs and C_{OLA} was R = 0.81 (coefficient of determination - $R^2 = 0.66$).

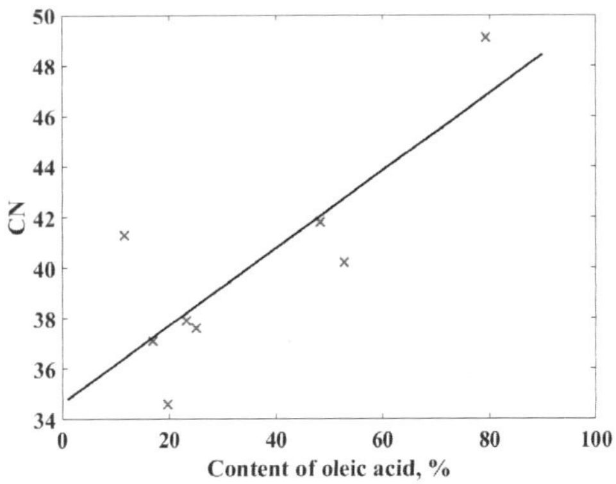

Figure 2. The relationship between CN and C_{OLA} (the points – experimental data in Tables 2 and 3; the straight line – predicted CN).

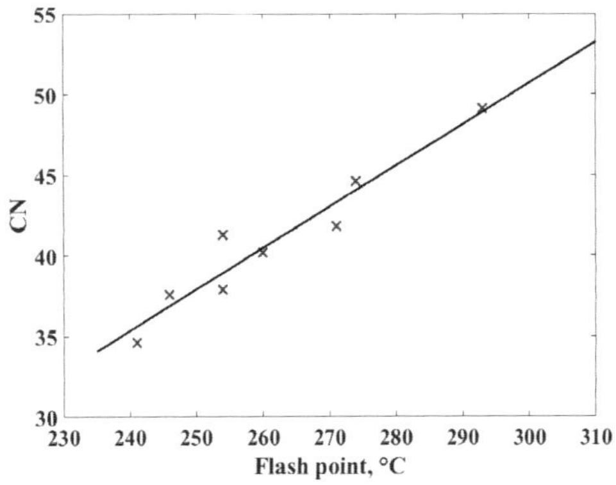

Figure 3. The relationship between CN and T_{FP} (the points – experimental data in Tables 2 and 3; the straight line – predicted CN).

As a third parameter associated with the CN of the studied VOs, their flash point (T_{FP}) in a closed-cup determined in accordance with ASTM D93 was taken into account. In order to analyze the relationship between the CN of VOs and their T_{FP}, eight points corresponding to VOs No. 1, 3-5, 7-9, and 11 in Table 2 and 3 were plotted in Figure 3. At the same time, data for VOs, which have an abnormal content of non-saturated fatty acids, were not used - SuO (No. 2) with a linoleic acid content of 73.73%, CO (No. 6) with a linoleic acid content of 60.60%, and CoO (No. 10) with a minimum content of unsaturated fatty acids equal to 70.78%.

Based on the data in Figure 3 and the least square method, an approximation of CN = f (T_{FP}) by a straight line was carried out. An analytical expression for CN = f (T_{FP}) has been obtained and has the following form:

$$ECN = 0.2552 \times T_{FP} - 25.88. \quad (3)$$

By conducting a correlation analysis of CN = f (T_{FP}), the correlation coefficient of this dependence was obtained and equal to R = 0.96 (the coefficient of determination - R^2 = 0.92). The high value of the correlation coefficient indicates a very close correlation relationship between the CN of VOs and their flash point.

3. Discussion

A large number of works have been devoted to the problem of using various VOs as additives to petroleum DF - there are more than 300 publications from 2015 to the present in the Scopus database [2]. The possibilities of using petroleum DF blended with various VOs - RO, SuO, SO, palm oil, Jatropha oil, Castor oil, Karanja oil as diesel engine fuels were reviewed in [12]. The performance of a DI-diesel engine fueled with petroleum DF, RO, and RO methyl ester was investigated in [17]. The possibility of operating a diesel engine on petroleum DF blended with biofuels at a concentration from 0 to 100% has been analyzed. The possibility of using LO as an additive to petroleum DF was investigated in [18]. The content of this additive was 30 and 50%. These publications indicate the urgency of studying the possibilities of using VOs as diesel engine fuels.

A number of publications [18, 19, 20] are devoted to analytical and experimental studies of the flammability of biofuels in the CC of diesel engines, including studies of the CN of biofuels, such as VOs.

The relationship between CN and the ignition delay τ_d in a constant volume combustion chamber was investigated for reference fuels (mixtures of hexadecane and 2,2,4,4,6,8,8-heptamethylnonane at different concentrations of these components) in [21]. The obtained relationship between CN and τ_d as determined in this chamber can de be expressed as:

$$ECN = 38.07\left(\tau_d - 0.35\right)^{-0.6484}. \tag{4}$$

And then, Freedman et al. [21] measured the ignition delays of different triglycerides in this chamber and calculated their CN by formula (4), as shown in Table 4.

Table 4. Estimated cetane number of triglycerides in [21], derived cetane number of triglycerides, and constant *D*

Conditional formula of triglycerides	Ignition delay, ms	Estimated cetane number	Derived cetane number
TG 16:0	1.96	89	88.89
TG 18:0	1.98	85	85.02
TG 18:1	2.45	45	44.47
TG 18:2	2.94	32	34.22
TG 18:3	3.85	23	24.47
TG 21:1	-	-	48.47
Non-basic triglycerides TG 14:0, TG 20:0, TG 22:0, TG 24:0	-	-	84.03
constant *D*			-43.41

In order to predict CN of VOs based on the CN of triglycerides, a multi-factorial dependence of the CN of VOs on the content of basic triglycerides in them has been proposed. This relationship can be expressed by taking into account the contribution of the basic triglycerides:

$$ECN = \sum_i DCN_i \times C_i + DCN_0 \times C_0 + D \times C_{USA} \times C_{SA}, \tag{5}$$

where C_i is the concentration of the *i*-th basic triglyceride; DCN_i is the derived CN of the *i*-th basic triglyceride; C_0 is the concentration of the group of non-basic triglycerides (i.e., triglycerides with fatty acids C 14:0, C 20:0, C 22:0, and C24:0); DCN_0 is the derived CN of the group of non-basic triglycerides; C_{USA} is the concentration of triglycerides with unsaturated fatty acids; C_{SA} is the concentration of triglycerides with saturated fatty acids; *D* is a constant. With the use of the least square method, the values of *DCN* in this formula were calibrated for VOs No. 2-4, No. 7, No. 9, No. 10 (in Tables 2 and 3), containing only fatty acids C 16:0, C18:0, C 18:1, C 18:2, and C18:3. After that, the obtained *DCN* for the basic triglycerides (except for TG 21:1) and the data on the fatty acid composition for all eleven VOs (Tables 2 and 3) were used to determine *DCN* of TG 21: 1, DCN_0 of the group of non-basic triglycerides and the constant *D*. The obtained values of these parameters are shown in Table 4.

The comparison of the CN of VOs calculated by the formula (5) with the experimental data in Tables 2 and 3 is shown in Figure 4.

Table 5 shows the regression parameters of the fitted dependences predicting CN in the study. The fitted dependence of the CN of VOs on their flash point in a closed-cup has the highest degree of fitting (prediction accuracy) - CN = *f* (T_{FP}) with a determination coefficient R^2 = 0.92. The dependence of the

CN on the content of oleic acid in VO has a relatively high prediction accuracy - CN $= f(C_{OA})$ with a determination coefficient $R^2 = 0.66$. The dependence of the CN on the content of saturated fatty acids in VO has the prediction accuracy - CN $= f(C_{SFA})$ with a determination coefficient $R^2 = 0.48$. The multi-factorial dependence of CN with the content of triglycerides in biofuel has a determination coefficient $R^2 = 0.53$. However, the multi-factorial dependence is more complicated to use than the other dependences predicting the CN of VOs.

Table 5. Regression parameters of the dependences predicting CN

Regression parameters	Value			
	CN $= f(C_{SFA})$	CN $= f(C_{OA})$	CN $= f(T_{FP})$	CN $= f(C_i, C_0, C_{USA}, C_{SA})$
Determination coefficient R^2	0.48	0.66	0.92	0.53
Maximum absolute deviation	3.22	4.89	2.36	5.7
Maximum relative deviation $\tilde{\Delta}_{max}$, %	7.88	11.84	5.71	11.61
Average absolute deviation	1.46	1.79	0.95	1.99

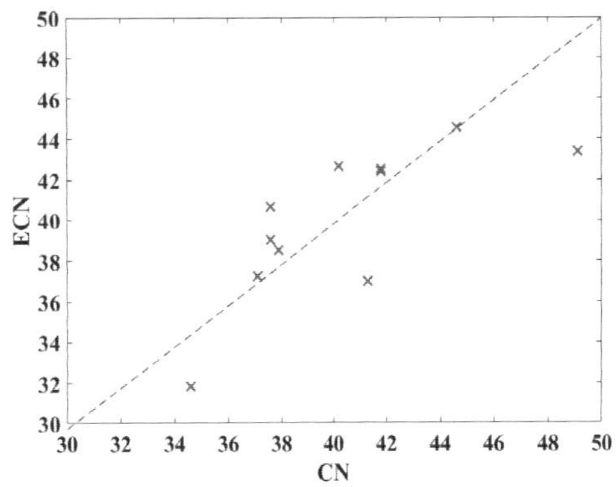

Figure 4. The comparison of the CN calculated by the formula (5) (ordinate axis) with the experimental data in Tables 2 and 3 (abscissa axis).

4. Conclusion

One of the problems of using VOs as diesel engine fuels is their lower flammability compared to petroleum DF. The flammability indicators of vehicle fuels in the CC of diesel engines were considered. The most important among them is the CN. The dependences of the CN of VOs on their flash point in a closed-cup, the content of oleic acid, and saturated fatty acids in oils have been analyzed. Corresponding fitted expressions for these dependences have been obtained. Compared with the multi-factorial dependence of CN on the triglyceride contents, single-parameter dependences have the highest accuracy and convenience to use. The fitted dependence of CN on flash point has the highest prediction accuracy with a determination coefficient of 0.92. The dependence of the CN on the content of saturated fatty acids has the lowest prediction accuracy with a determination coefficient of 0.48.

References

[1] L'otko V, Lukanin V N and Khachiyan A S 2000 *The use of alternative fuels in internal combustion engines* (Moscow: MADI)
[2] Markov V A, Kamaltdinov V G, Zykov S A and Sa B 2020 Optimization of the Composition of Blended Biodiesel Fuels with Additives of Vegetable Oils *Int. J. Energy a Clean Environ.* **20** 303–19
[3] Markov V A, Devyanin S N, Semenov V G, Bagrov V V and Zykov S A 2019 *Motor fuels produced from vegetable oils* ed V A Markov (Riga: Lambert Academic Publishing)
[4] Kamenev V F, Shatrov M G, Terenchenko A S and Karpukhin K E 2014 *Thermal engines of electrical and heat supply installations using biofuels* (Moscow: MADI)
[5] Vallejo Maldonado P R 2008 *Energy-saving technologies and alternative energy* (Moscow:

RUDN)
[6] Vasiliev I P 2009 *The influence of vegetable fuels on the environmental and economic performance of a diesel engine* (Lugansk: East Ukrainian National University named after V. Dahl)
[7] Parsadanov I V 2003 *Improving the quality and competitiveness of diesel engines based on an integrated fuel and environmental criterion* (Kharkov: KhPI Publishing House)
[8] Knothe G, Krahl J and Van Gerpen J 2015 *The biodiesel handbook* (Urbana: AOCS Press)
[9] Myo T 2008 *The Effect of Fatty Acid Composition on the Combustion Characteristics of Biodiesel* (Japan: Kagoshima University)
[10] Spessert B M, Arendt I and Schleicher A 2004 Influence of RME and Vegetable Oils on Exhaust Gas and Noise Emissions of Small Industrial Diesel Engines *SAE Tech. Pap.* 2004-32–0070
[11] İlkiliç C and Yücesu H S 2005 Investigation of the effect of sunflower oil methyl esther on the performance of a diesel engine *Energy Sources* **27** 1225–34
[12] Mat S C, Idroas M Y, Hamid M F and Zainal Z A 2018 Performance and emissions of straight vegetable oils and its blends as a fuel in diesel engine: A review *Renew. Sustain. Energy Rev.* **82** 808–23
[13] Hazar H and Sevinc H 2019 Investigation of the effects of pre-heated linseed oil on performance and exhaust emission at a coated diesel engine *Renew. Energy* **130** 961–7
[14] Alimentarius K 2007 *Fats, oils and derivatives* (Moscow: West World)
[15] Goering C E and Daugherty M J 1982 Energy accounting for eleven vegetable oil fuels *Trans. ASAE* **25** 1209–15
[16] Goering C E, Schwab A W, Daugherty M J, Pryde E H and Heakin A J 1982 Fuel properties of eleven vegetable oils *Trans. ASAE* **25** 1472–7
[17] Raman L A, Deepanraj B, Rajakumar S and Sivasubramanian V 2019 Experimental investigation on performance, combustion and emission analysis of a direct injection diesel engine fuelled with rapeseed oil biodiesel *Fuel* **246** 69–74
[18] Velmurugan K and Gowthamn S 2012 Effect of cetane improver additives on emissions *Int. J. Mod. Eng. Res.* **2** 3372–5
[19] Canoira L, Alcántara R, Torcal S, Tsiouvaras N, Lois E and Korres D M 2007 Nitration of biodiesel of waste oil: Nitrated biodiesel as a cetane number enhancer *Fuel* **86** 965–71
[20] Tat M E 2011 Cetane number effect on the energetic and exergetic efficiency of a diesel engine fuelled with biodiesel *Fuel Process. Technol.* **92** 1311–21
[21] Freedman B, Bagby M O, Callahan T J and Ryan III T W 1990 Cetane numbers of fatty esters, fatty alcohols and triglycerides determined in a constant volume combustion bomb *SAE Trans.* 153–61

The deterministic component of the traffic flow intensity

M G Boyarshinov and A S Vavilin

Perm National Research Polytechnic University, 29, Komsomlsky ave., Perm, 614990, Russia

E-mail: atm@pstu.ru

Abstract. The information obtained from enforcement complexes of traffic rules allows determining in real time the traffic flow intensity over long periods of time. The analysis of the obtained dependencies leads to the conclusion that there is a periodic deterministic component in the characteristic of the traffic flow intensity. The using of the smoothing procedure allows one to define the deterministic and stochastic components, which are qualitatively different on workdays and weekends. Statistical indicators of stochastic distributions of traffic flow intensities and random components isolated from them are determined. The estimates of the correspondence of the obtained curves to the normal law of probability distribution were carried out. It is shown that the deterministic components of the traffic flows intensity both on working days and on weekends are similar in shape and location of maximum and minimum values. It seems appropriate to use deterministic components for predicting traffic flows, controlling the operation of traffic lights, monitoring equipment operation, as well as in the reconstruction, design and construction of roads and road facilities.

1. Introduction

Due to the implementation of video recording equipment on the road network, it is now possible to take into account and analyze on the real-time basis significant volumes of information on the evolution of traffic flow characteristics for different time intervals on individual road sections, intersections, to form correspondence matrices, etc. To analyze traffic flows, the intensity indicator [1, 2] is usually used, which shows the number of cars passing through a conditional section of a road section per unit of time. In the present study the time dependence of the traffic flow intensity on one of the busy roads in the city of Perm is analyzed. The selected section of the two-lane road is equipped with the software and technical measuring complex (STMC) "Azimut 2" [3] that allows one to measure the speed of vehicles in the control point, average speed of the vehicles, drive-through time of the section, etc. Figure 1 shows the time dependences of the daily and weekly traffic flow intensity on the selected road, which were obtained as a result of processing the incoming data.

2. Materials and methods

Figure 1, a shows that the daily traffic flow rate is a random process. Meanwhile, Figure 1, b displays that the curves of the traffic flow intensity, which were obtained for several days in a row, are similar. Traffic intensity dependencies for working days (Monday-Friday) have two peak values (approximately from 7:00 to 9:00 and 16:30 to 18:30) and one peak value on weekends (Saturday-Sunday, approximately from 13:00 to 18:00).

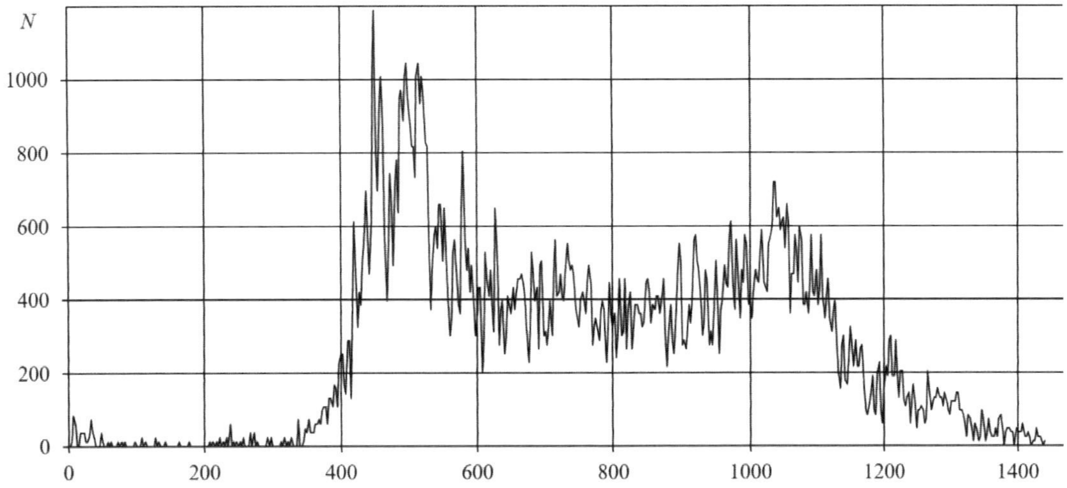

a

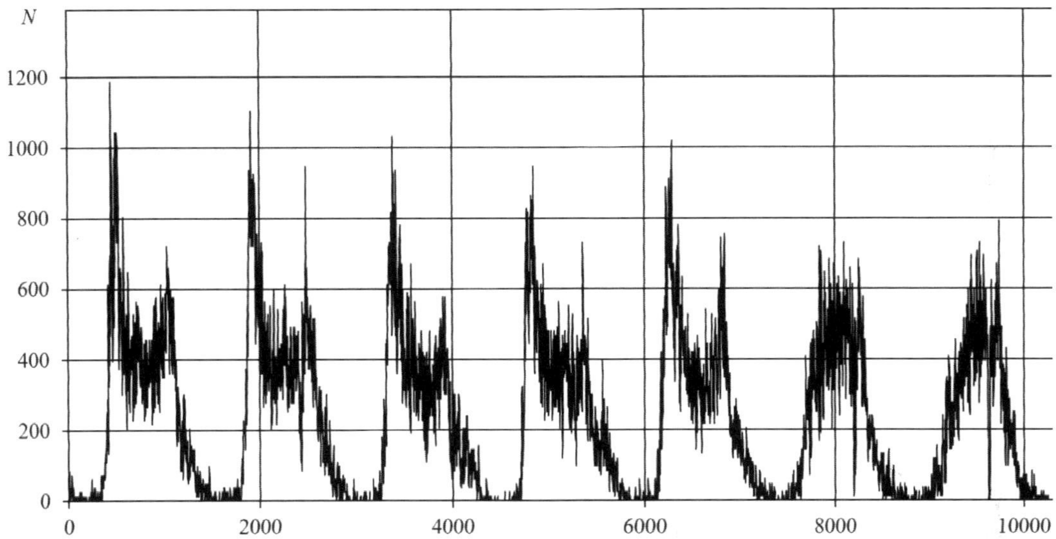

b

Figure 1. The traffic flow intensity N (vehicles/h) dependencies on time t (min) daily (Monday, a) and weekly (from Monday to Sunday, b)

Table 1. Statistical indicators of the traffic flow intensity (Monday)

Indicator	Value for N	Value for N_S
Average value, vehicles per hour	274.64	2.11
Median, vehicles per hour	264.00	–4.32
Standard deviation, vehicles per hour	242.71	113.98
Variation index, %	88.37	53.98
Asymmetry	0.84	1.07
Kurtosis	0.34	6.99
Pearson's statistics	226.9	27464.5
Kolmogorov's statistics	0.7987	1.1606

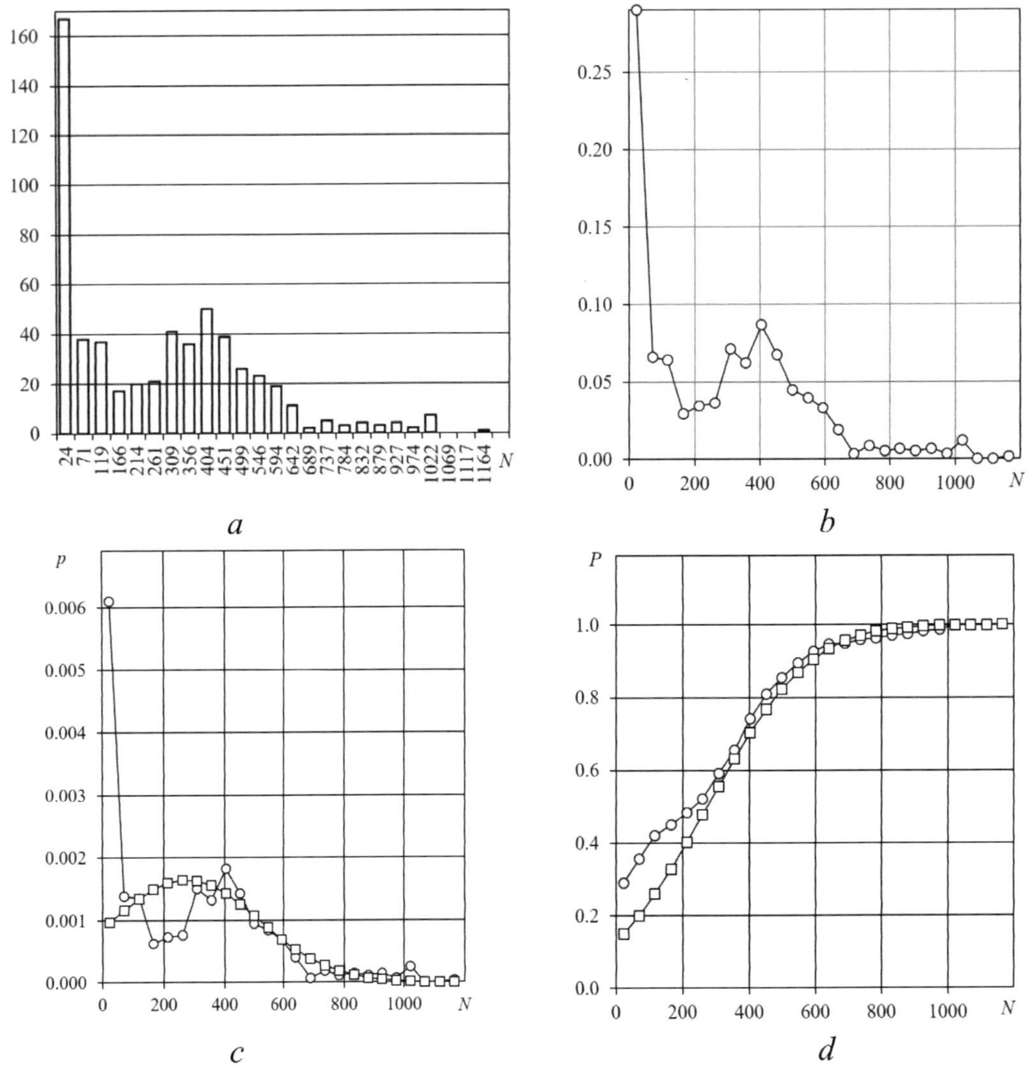

Figure 2. Statistical indicators of the traffic flow intensity N on a working day (Monday): a - histogram, b - frequency, c - experimental (–o–) and theoretical (Gaussian, –□–) curves of probability distribution density p, d - experimental (–o–) and theoretical (–□–) curves of probability distribution P

Figure 2 shows a histogram, frequencies, experimental and theoretical (normal law of distribution is considered) curves of the probability distribution P and its density p. Experimental and theoretical probability distribution functions and its density differ significantly.

For a quantitative assessment of the admissibility of the hypothesis about the normal law of distribution of the traffic flow intensity on the road under consideration, Kolmogorov's statistics $D_M = \underset{k=1,M}{\underline{max}}|F_M(N_k) - F(N_k)|$ and Pearson's criteria $\chi^2 = \sum_{k=1}^{m} (\nu_k - Mp_k)^2/Mp_k$ were used [4, 5]. Here $F_M()$, $F()$ are empirical and hypothetical distribution functions respectively; N_k are the values of the elements of the time series (Figure 1) and P_k – probability of being these values in the corresponding intervals of the groupings determined from the hypothetical density distribution; ν_k are the empirical frequencies of the sample points falling into the corresponding grouping intervals, m is the number of intervals, $\sum_{k=1}^{m} \nu_k = M$.

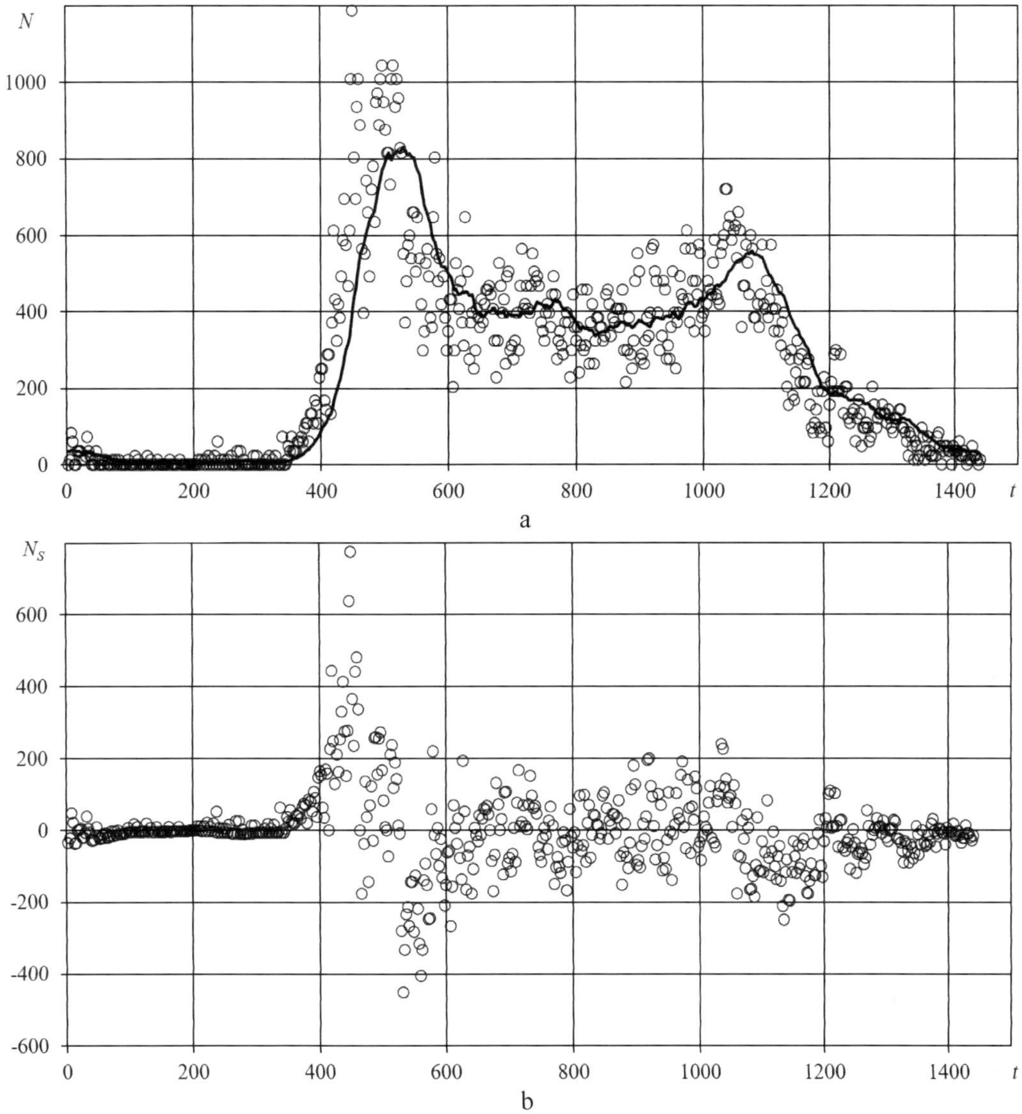

Figure 3. Time dependence of the traffic flow intensity N (a, markers «o», working day, Monday, vehicles per hour), the smoothed (deterministic) component of the intensity (a, solid curve) and the deviation (NS) of the intensity from the smoothed curve (b, markers)

Pearson's statistics for the considered time series gives a value of 226.9, while the significance level of 0.01 corresponds to $\chi^2_{0.01} = 43.0$; therefore, the hypothesis of the normal distribution of the traffic flow intensity should be rejected. Kolmogorov's statistics for the same time series gives a value of 0.7987, while the significance level of 0.01 corresponds to the critical value $t_{0,01}$ = 0.317, so the hypothesis of a normal distribution of traffic intensity also should be rejected.

3. Discussion

The assumption is made that at each time point the traffic flow intensity N can be represented as the sum $N = N_D + N_S$, where N_D is the deterministic (smoothed) component of the traffic flow intensity, N_S is the deviation of the traffic flow intensity from the smoothed curve (stochastic component).

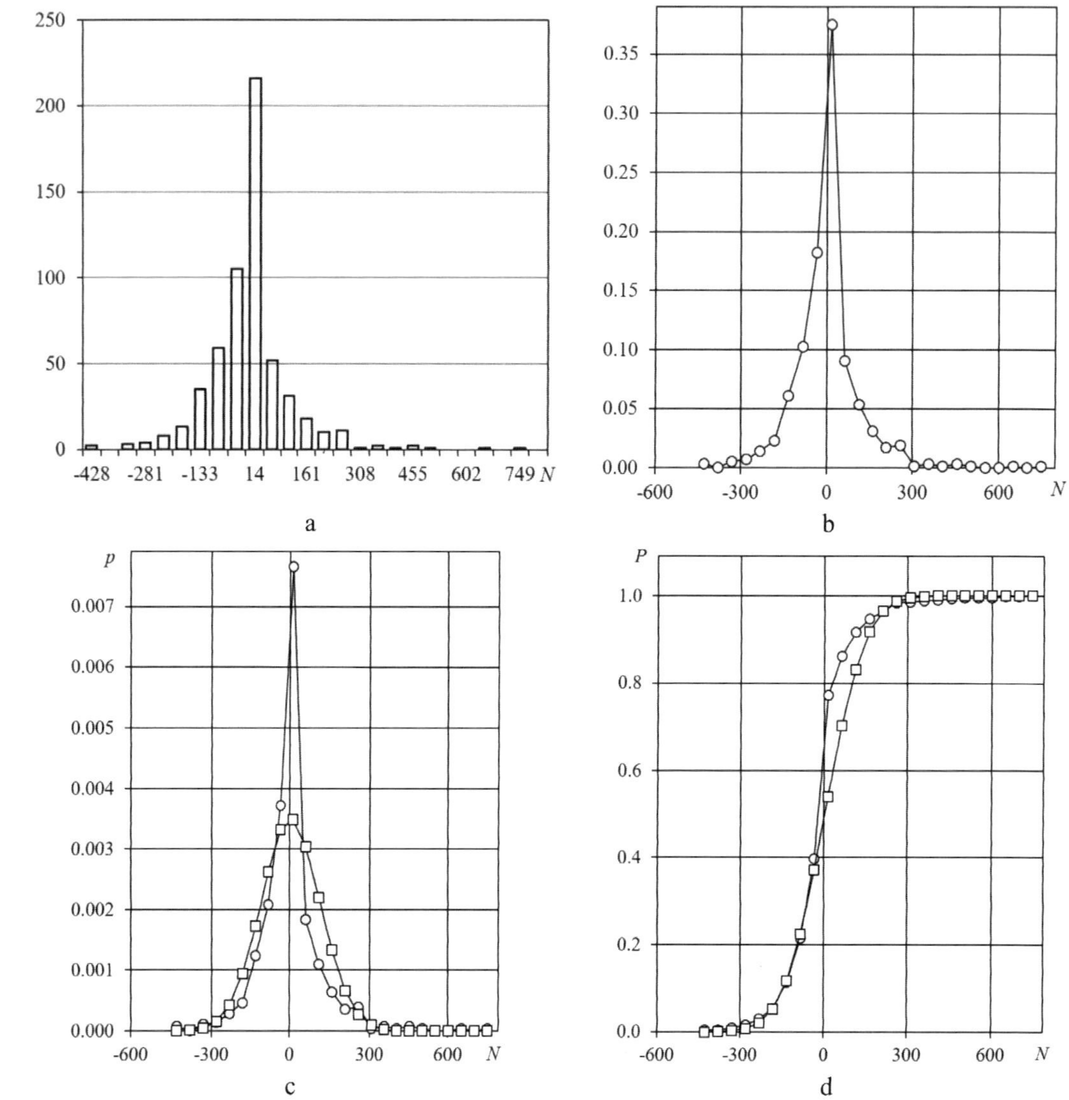

Figure 4. Statistical indicators of the deviation N_S of intensity from the smoothed curve on a working day (Monday): a – histogram, b – frequencies, c – experimental (–o–) and theoretical (–□–) curves of probability distribution density p, d – experimental (–o–) and theoretical (–□–) curves of probability distribution P

Table 2. Traffic flow intensity statistical indicators (Sunday)

Indicator	Value for N	Value for N_S
Average value, vehicles/h	203.72	0.73
Median, vehicles/h	168.72	–3.36
Standard deviation, vehicles/h	197.49	89.26
Variation index, %	96.94	121.51
Asymmetry	0.67	–0.74
Kurtosis	–0.83	3.93
Pearson's statistics	205.8	5651.5
Kolmogorov's statistics	0.9945	0.8480

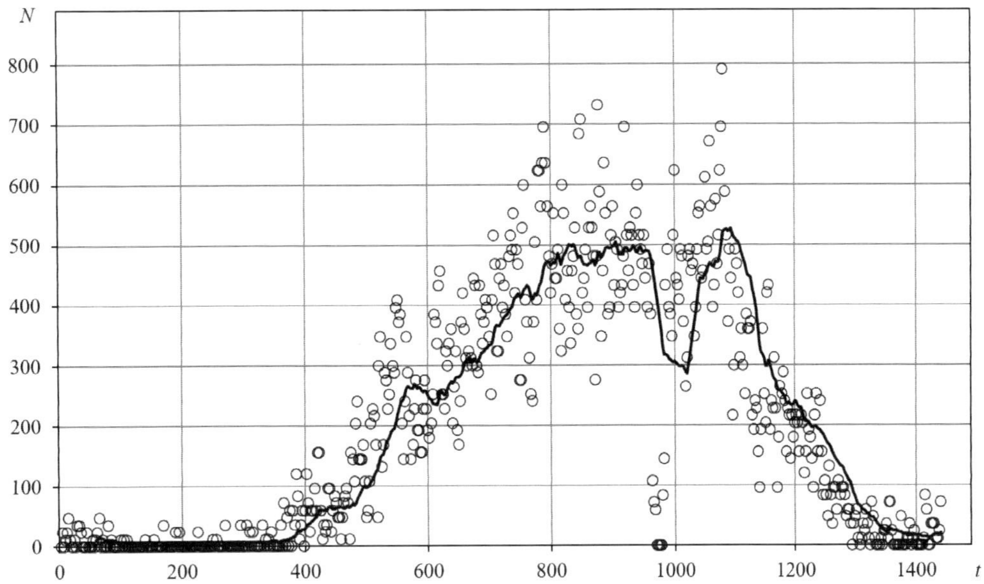

a

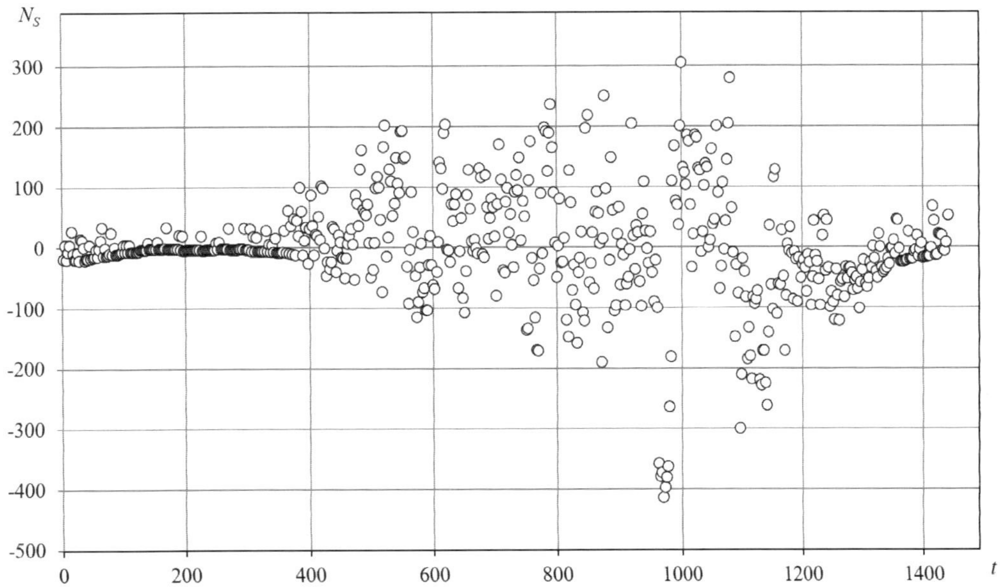

b

Figure 5. Time t dependence of the traffic flow intensity N (a, markers «o», weekend, Sunday, vehicles per hour), the smoothed (deterministic) component of the intensity (a, solid curve) and the deviation N_S of the intensity from the smoothed curve (b, markers)

To construct the deterministic component of the traffic flow intensity, the smoothing procedure by the moving average method is used [6]. The smoothed traffic intensity curve is shown in Figure 3, a.

The time dependence of the deviation N_S of the traffic flow intensity N on the smoothed curve N_D is shown in Figure 3, b, its statistical indicators are given in Table 1. Figure 4 shows a histogram, frequencies, experimental and theoretical curves of the distribution P of the probability and its density p for the deviation of N_S. Pearson's statistics for the time series N_S gives a value of 27464.5 with the same value $\chi^2_{0.01} = 43.0$; Kolmogorov's statistics for the same time series is 1.1606 with a critical value $t_{0.01} = 0.317$.

Both criteria require rejecting the hypothesis of the normal distribution of the deviation N_S of the traffic flow intensity from the smoothed curve. Further, the intensity of the traffic flow during the last day of observation is considered (Sunday, Figure 5, *a*, highlighted with markers). Figure 5, *b* shows the time dependence of the deviation N_S of the Sunday traffic flow intensity from the smoothed curve N_D. Statistical indicators of the distribution of the intensity N and deviation N_S of this traffic flow are given in Table 2.

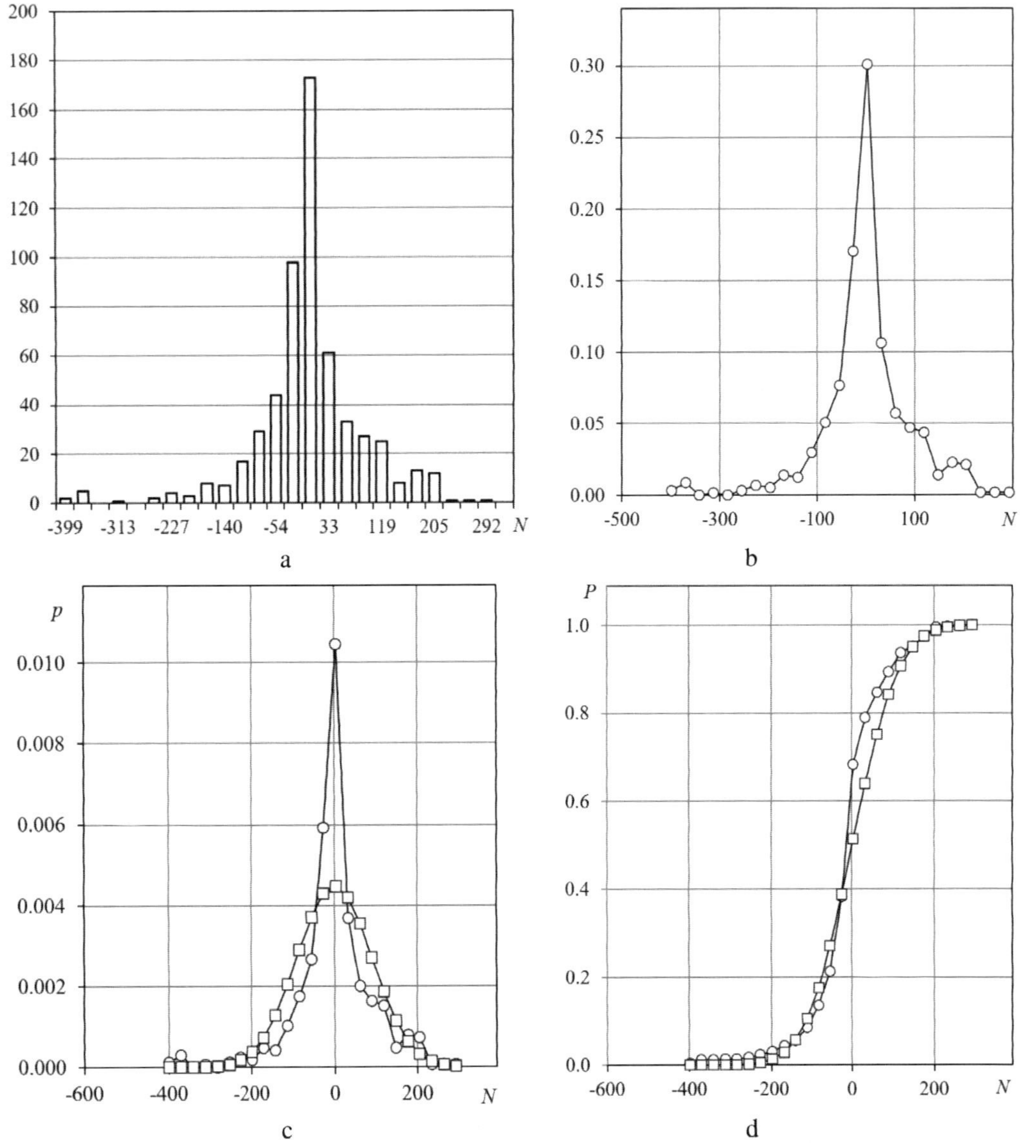

Figure 6. Statistical indicators of the deviation N_S of intensity from the smoothed curve on a weekend (Sunday): a - histogram, b - frequency, c - experimental (–o–) and theoretical (–□–) curves of the density p of the probability distribution, d – experimental (–o–) and theoretical (–□–) curves of the probability distribution P

Figure 6 shows a histogram, frequencies, experimental and theoretical curves of the probability distribution P and its density p for the deviation N_S of the Sunday traffic flow intensity. The experimental and theoretical curves of the probability distribution function P and its density p, as in the previous case, differ significantly.

Pearson's statistics for the considered time series gives a value of 5651.5, Kolmogorov's statistics for the same time series gives a value of 0.8480, meaning the hypothesis of a normal distribution of traffic intensity should be rejected in this case as well.

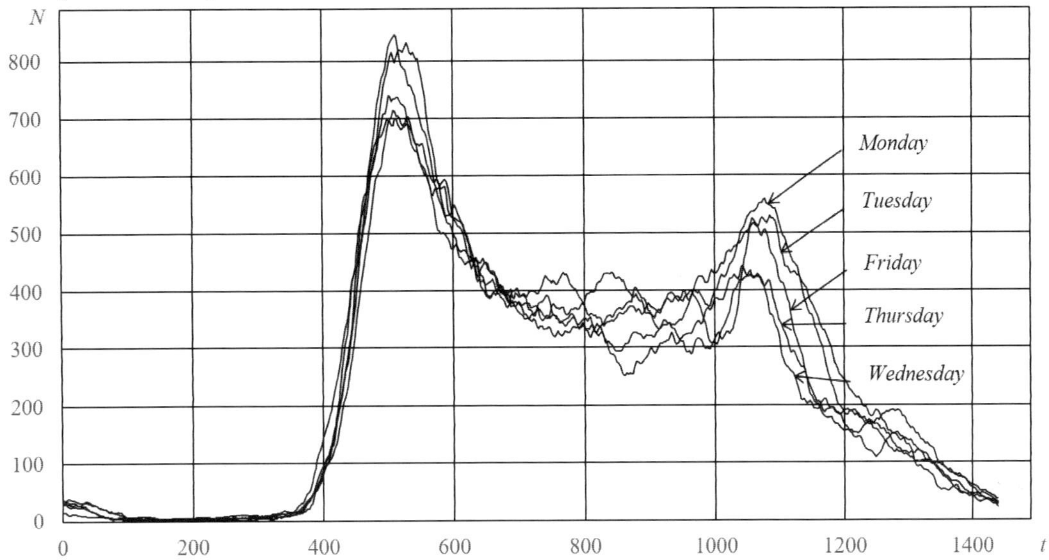

a

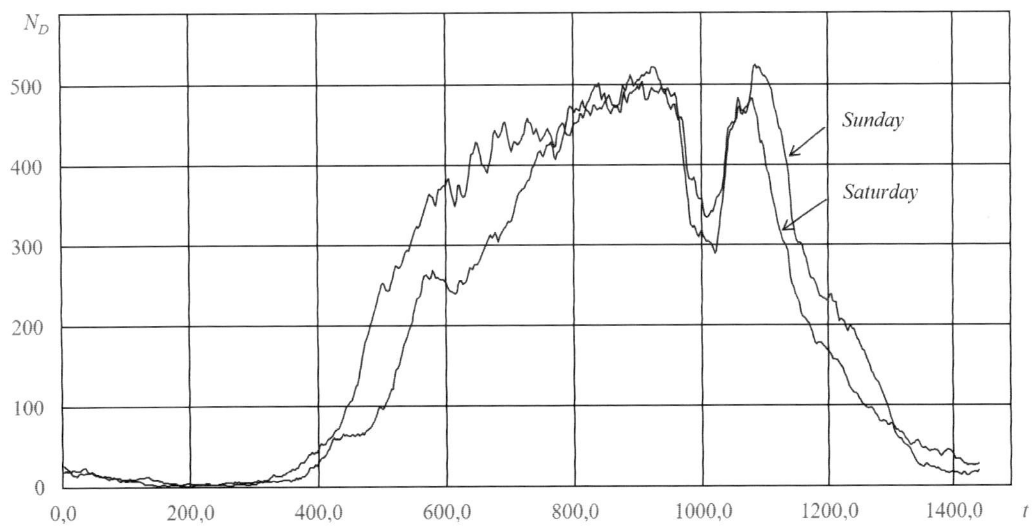

b

Figure 7. Deterministic components N_D of the traffic flows intensity N on working days (a) and weekends (b)

Figure 7 shows the deterministic components of the intensity of traffic flows on weekdays and weekends which were constructed using the algorithm described above. The results presented in this Figure show that the deterministic components N_D both on workdays (Figure 7, *a*) and on weekends

(Figure 7, *b*) are similar and can be used to analyze the traffic flow intensity of road transport in urban areas road networks.

4. Conclusion

The data obtained by the "Azimut 2" complex made it possible to determine the time dependences of the traffic flow intensity continuously throughout the week. An approach is proposed to define deterministic and stochastic components (Figures 3 and 5) of the traffic flow intensity, which are qualitatively different on workdays and weekends.

Statistical indicators of the distributions of the intensities and the random components isolated from them were determined. It is shown that on workdays and weekends, the deterministic components of the traffic flow intensity are similar in shape and location of maximum and minimum values. In all studied cases the distributions of random components cannot be recognized as complying to the normal law of distribution.

It is advisable to use the deterministic component to predict the intensity of traffic flows, control the algorithms for the operation of traffic lights, identify anomalies in the operation of equipment, use as input data for the reconstruction, design and construction of roads and road facilities.

References

[1] Donald R D 1972 *Traffic flow theory and control* (New York: McGraw-Hill Book Company)
[2] Silyanov V V 1977 *Traffic flow theory in road design and traffic management* (Moscow: Transport)
[3] STMC "Azimut 2" *Road safety technology*. Retrieved from: https://tbdd.ru/node/162
[4] Smirnov N V and Dunin-Barkovskiy I V 1969 *A course in the theory of probability and mathematical statistics for technical applications* (Moscow: Nauka)
[5] Ivchenko G I and Medvedev Yu I 1984 *Mathematical statistics* (Moscow: Vischaya Shkola)
[6] Greshilov A A, Stakun V A and Stakun A A 1997 *Mathematical methods for forecasts making* (Moscow: Radio i Svyaz)

IOP Conf. Series: Materials Science and Engineering 1111 (2021) 012014 doi:10.1088/1757-899X/1111/1/012014

Minimizing geometric tests in CAAP-systems

A N Bozhko

CAD Department, Bauman Moscow State University, 5/1, 2-ya Baumanskaya str., Moscow, 105005, Russia

E-mail: bozhkoan@bmstu.ru

Abstract. The article is devoted to modeling geometric obstacles in the assembly of complex products. The problem of minimizing the number of geometric tests is considered. Two assertions about geometric inheritance in the assembly of technical systems are introduced. Based on these assertions, a game-theoretic model has been developed. The model made it possible to minimize the number of geometric tests. It is a game of decision-maker and nature by coloring the vertices of an ordered set in two colors. A classification of ordered sets is proposed. The computational experiment on coloring of ordered sets of each class using different criteria is carried out. The experiment made it possible to determine the optimal criteria for the coloring ordered sets of each class. The use of these criteria allows minimizing the labor intensity of geometric analysis in computer-aided assembly planning.

1. Introduction

When designing the assembly process of a complex technical system, it is necessary to take into account the geometric obstacles. In publications on computer-aided assembly planning (CAAP), this essential condition for the existence of an assembly operation and an assembly process is called geometric solvability [1]. To solve this problem various approaches based on collision detection methods are proposed [2]–[5]. In this paradigm, mobile and stationary parts of the scene are set. Their shape is approximated using simple geometric shapes: spheres, parallelepipeds, convex hulls, polytopes, and so on. A trajectory of the mobile object is specified. In the process of movement, the intersection of the mobile and stationary shells is checked [2].

This approach is widely used in computer games, CAD-systems and geometric modeling systems. Experience has shown that it is ineffective for the analysis of obstacles in the assembly of complex products. Parts of machines and mechanical devices can have a rather complex shape. The approximation does not allow obtaining the high accuracy often required during assembly. The methods of collision detection can be used as auxiliary means. They are often used to model assembly processes in virtual reality systems [6]–[8].

The application of motion planning methods in CAAP is discussed in the papers [9]–[13]. In these articles, synthesis of a possible assembly plan is set as the problem of holonomic motion planning. The mounting part is represented by the point in the configuration space of assembled product. Search for a possible path of the point in multidimensional space is carried out, mainly, by means of probabilistic methods (Probabilistic roadmap method, PRM and Rapidly-exploring random trees, RRT). The methods of this group do not take into account the structural constraints (connections and mating) of assembly operations. All project alternatives are considered acceptable. In general, the number of geometric tests is $(n-1)!$, where n is the number of parts.

Methods for analyzing geometric obstacles based on projections of parts are described in [14] and [15]. If the projections do not intersect, the parts do not create a mutual obstacle in the direction that is perpendicular to the selected projection plane. Otherwise, you must select another plane and repeat this geometric test. This approach takes into account only the straight-line movements of the assembly elements and the position of the projection planes must be selected by the expert. The most severe restrictions on the movement of a part are imposed on the parts that are in close proximity to it. Various methods of modeling local geometric obstacles are offered in articles [16]–[19]. In this group it is necessary to note the method of *db*-graphs. It was offered in [18]. This method is used in several CAAP-systems, for example STAAT. The methods of local obstacle analysis give only necessary conditions of geometric solvability. They make it possible to identify obviously irrelevant design alternatives.

In several CAAP methods, the function of the geometrical solver is performed by an expert [21]–[24]. In the classic article [21] a method based on ordering the mechanical connections of a product is proposed. This information is provided by an expert, who should take into account the geometric constraints on the movement of parts. Various variants of rational organization of the expertise are described in papers [22], [23]. In article [24] the expert should choose a direction of disassembly and determine the geometric obstacles in the given direction of disassembly. Any procedures of geometrical solvability based on expertise are characterized by high labor intensity and low reliability.

The considered methods of geometric obstacles analysis have two main disadvantages:

1. The mechanical structure of a product (connections and matings) imposes strict constraints on the possible assembly sequences. In the models considered above, these limitations are not taken into account.
2. The problem of minimization of geometric tests is not considered.

Let us consider a model free of the specified lacks.

2. Game-theoretic model of geometric solvability. Problem statement

We introduce the necessary definitions and notations. We denote by $X = \{x_i\}_{i=1}^{n}$ the set of parts of a product *X*.

Definition 1. A connected subset of parts $Y \subseteq X$ is called an *s*-set if *Y* can be assembled independently.

The author's works [25], [26] propose a constructive way of generating *s*-sets. *S*-sets serve as a mathematical description of the connected and geometrically defined (coordinated) sets of parts, which can be assembled independently.

Definition 2. A geometric situation (*G*-situation) is a pair (*Y*, *x*) such that *Y* and $Y \cup \{x\}$ are *s*-sets.

The *g*-situation is a mathematical description of the product fragment for which a geometric test gives interpreted results.

Definition 3. A *g*-situation (*Y*, *x*) is called resolved if there is a movement of the part *x* realizing *s*-set $Y \cup \{x\}$. The situation is called unresolved, otherwise.

Figure 1 (a) shows a simple product and its two *g*-situations: (b) – resolved and (c) – unresolved.

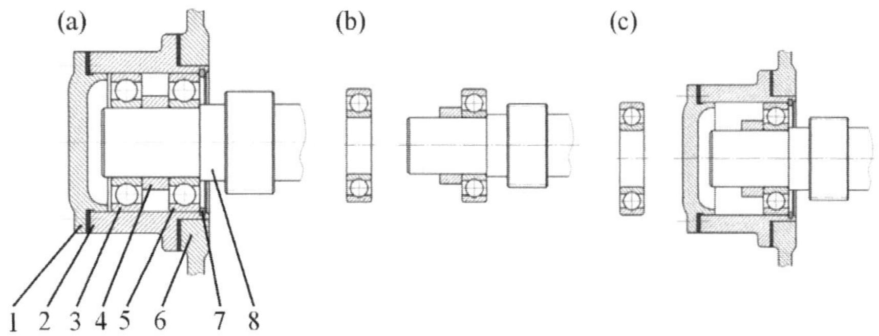

Figure 1. A simple product (a), a resolved *g*-situation (b), an unresolved *g*-situation (c).

Let (A, x) and (B, x) be two *g*-situations associated with the mounting of part x.

Assertion 1. If the situation (A, x) is resolved, then any situation (B, x), $B \subseteq A$, is also resolved.

Indeed, if *s*-set A does not contain geometric obstacles for the mounting of part x, then they cannot be in the smaller *s*-set B.

Assertion 2. If the situation (A, x) is unresolved, then any situation (B, x), $A \subseteq B$ is also unresolved.

In fact, if A has geometric obstacles that prohibit movement of the part x, then *s*-set B contains these obstacles. It means that there is no movement of part x which implements *s*-set $B \cup \{x\}$.

Denote $GS(x)$ – the set of all *g*-situations in a product X related to the mounting of a part x. We introduce a non-strict order on $GS(x)$ by the following relation $(A,x) \leq (B,x) \Leftrightarrow A \subseteq B$. This turns GS into an ordered set $(GS(x),\leq)$. Figure 2 (a) shows the gearbox, and figure 2 (b) shows the set of all *g*-situations related to the mounting of bearing 7.

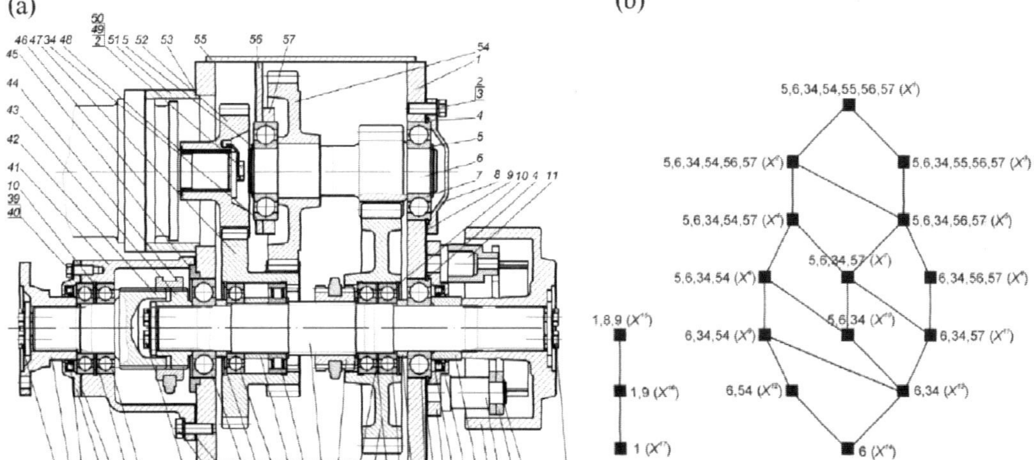

Figure 2. The gearbox (a), the set $(GS(7),\leq)$ (b).

We represent the set $(GS(x),\leq)$ using the Hasse diagram. The vertices of the diagram representing the unresolved *g*-situations will be shown in black. The vertices corresponding to the resolved *g*-situations will be shown in white.

Analysis of the geometric obstacles during the assembly can be set as a non-antagonistic game of decision-maker with nature. This game is played according to the following rules. An ordered set is given, presented in the form of Hasse diagram. Initially, all the vertices of this set have no colors. At each step of the game decision-maker chooses unpainted vertex A, nature selects a color of this vertex. If the color of vertex A is white, then all the vertices of the ordinal ideal $I(A)=\{B\in GS(x)\mid B\leq A\}$ get white (assertion 1). If the color of vertex A is black, then all the vertices of the ordinal filter $F(A)=\{B\in GS(x)\mid B\geq A\}$ are colored black (assertion 2). It is required to color all the vertices of the ordered set $(GS(x),\leq)$ in the minimum number of moves.

We denote this game by $\Gamma(GS)$. In meaningful terms, the vertex is a test configuration; the color of the vertex is the result of testing this configuration for geometric solvability. Obviously, a rational strategy in the game $\Gamma(GS)$ will minimize the number of necessary geometric tests.

3. The game $\Gamma(GS)$ as a decision-making problem

The optimal game solution for arbitrary ordered sets seems to be a very difficult, hardly solvable, problem. We will consider the game $\Gamma(GS)$ as a decision-making problem under conditions of uncertainty. Let the decision-maker have no a priori information about the possible colors of the vertices of an ordered set GS. Therefore, at every step of the game, we consider the answers of nature to be equally probable. We represent the game in the form of a payment matrix (table 1).

Table 1. The payment matrix of game $\Gamma(GS)$.

	X^1	X^2	...	X^n
Y_1	x_{11}	x_{12}	...	x_{1n}
Y_2	x_{21}	x_{21}	...	x_{2n}

In this table X stands for the decision-maker (the first player), Y represents nature. X^j, $j=\overline{1,n}$, – the first player moves; Y_1, Y_2 – the answers of nature (the colors of vertices); x_{ij} – the winning of the first player with his own move X^j and the answer of nature Y_i. The winning of the first player is equal $|F(A)|$ (cardinality of ordinal filter), if the vertex A is black. The winning is equal $|I(A)|$ (cardinality of ordinal ideal) if the vertex A is white.

4. Classification of ordered sets

To find a rational strategy in game $\Gamma(GS)$ we divide the class of ordered sets GS into 13 subclasses, whose representatives are fundamentally different in their structural characteristics. Figure 3 shows the classification of ordered sets (OS).

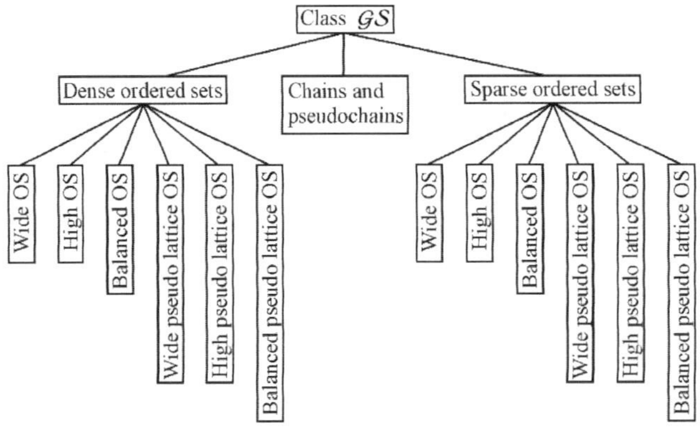

Figure 3. The classification of the ordered sets.

Definition 4. The height $h(P)$ of an ordered set P is the length of the maximum chain in P [27].

Definition 5. The width $w(P)$ of an ordered set P is the maximum cardinality of an antichain in P [27].

Definition 6. An ordered set is said to be dense if in the Hasse diagram of this set, the number of edges r is much greater than the number of vertices n, that is, $r >> n$. If $r << n$, then the ordered set is sparse.

Definition 7. An ordered set P is called high if $h(P) >> w(P)$. The ordered set P is called wide if $h(P) << w(P)$. The ordered set P is called balanced if $h(P) \approx w(P)$.

Definition 8. An ordered set is called pseudo-lattice if it has the greatest and least elements.

Definition 9. A connected ordered set P is called a pseudo-chain if most of the vertices of this set belong to the maximal chain, that is $h(P) \approx |P|$.

Examples of pseudo-chains are shown in figure 4 (d).

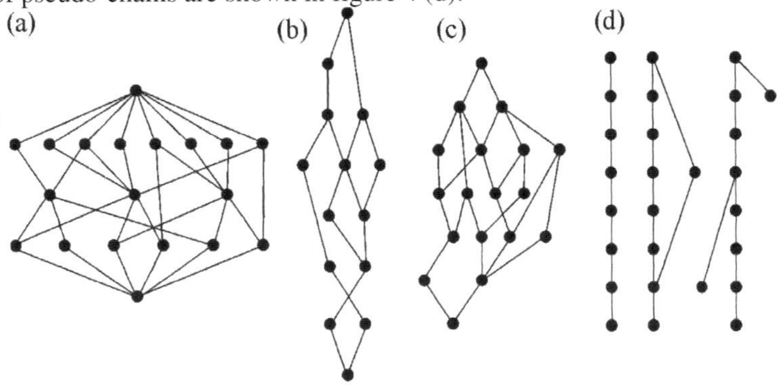

Figure 4. Pseudo-lattice ordered sets: wide (a), high (b), balanced (c); pseudochains (d).

5. Decision-making criteria

In each of the subclasses of GS the main criteria for decision-making under uncertainty were tested [28].

The Wald criterion (WLD). For the payment matrix of a general form it is given by the expression $\max_j \min_i x_{ij}$. For game $\Gamma(GS)$ it takes the form $\max_j \min\left(x_{1j}, x_{2j}\right)$.

The Bayes-Laplace criterion (BL). For the payment matrix of a general form, it is given by the expression $\max_j z_j$, $z_j = \sum_{i=1}^{n} x_{ij} p_{ij}$, where p_{ij} is the probability of the response of nature Y_i to the move of decision-maker X_j. For game $\Gamma(GS)$ it takes the form $\max_j z_j$, $z_j = \dfrac{x_{1j} + x_{2j}}{2}$.

The Savage criterion (SVG). For the payment matrix of a general form, it is given by $\min_j \max_i r_{ij}$, $r_{ij} = \max_p x_{ip} - x_{ij}$, where r_{ij} is an element of the loss matrix, which is equal to the difference between the maximum element in the i-th row and the element x_{ij}. For game $\Gamma(GS)$ the criterion takes the form $\min_j \max\left(r_{1j}, r_{2j}\right)$.

The multiplication criterion (MPL). In general, it is given by the expression $\max_j \prod_i x_{ij}$, for $\Gamma(GS)$ – $\max_j \left(x_{1j} \times x_{2j}\right)$.

IOP Conf. Series: Materials Science and Engineering 1111 (2021) 012014 doi:10.1088/1757-899X/1111/1/012014

The Hodge-Lehman criterion (HL) is a synthetic criterion that is built on the basis of the WLD and BL criteria. For game $\Gamma(GS)$ it has the form $\max_j\left(\beta\times\frac{x_{1j}+x_{2j}}{2}+(1-\beta)\times\min\left(x_{1j},x_{2j}\right)\right)$, where $\beta, 0\le\beta\le 1$, is a priori reliability coefficient.

The Hurwitz criterion (HW) is a synthetic criterion expressing a compromise between the points of view of extreme optimism and pessimism. For game $\Gamma(GS)$, it takes the form $\max_j\left(\alpha\times\max\left(x_{1j},x_{2j}\right)+(1-\alpha)\times\min\left(x_{1j},x_{2j}\right)\right)$, where $\alpha, 0\le\alpha\le 1$, is a priori confidence coefficient.

6. Computational experiment

We have developed special software to implement the computational experiment. The program creates an ordered set of random structure in a given subclass (figure 4) and implements a game session on the coloring of the created ordered set in two colors using the specified criterion. For this purpose, the best unpainted vertex is selected. If there are several vertices with the same criterion value, any of them is selected with equal probability. With the probability $\frac{1}{2}$ this vertex is assigned a color: white or black. After the choice of a color the program paints unpainted vertices of the ordered set by the rules of game $\Gamma(GS)$. The given sequence of actions repeats until all vertices of the ordered set are painted. One coloring session is performed with the help of one criterion. The program remembers the number of selected vertices.

The computational experiment was performed on a computer system with the following characteristics: Windows 7, CPU Core i5 with a clock speed of 2700 MHz, RAM 16GB and hard disk 1TB.

Tests were performed for the following criteria: WLD, BL, SVG, MP, HL and HW. Three values of the reliability coefficient were selected for HL $\beta=\frac{1}{3},\frac{1}{2},\frac{2}{3}$. Criterion HW was tested for two values of the confidence coefficient $\alpha=\frac{1}{3},\frac{2}{3}$ since HW = BL for $\alpha=\frac{1}{2}$.

The computational experiment was performed for all criteria in each of the subclasses of GS. For each subclass in the category of sparse ordered sets is made 100,000 sessions of game $\Gamma(GS)$. In the category of dense ordered sets – 60,000. This is due to the fact that the coloring dense ordered sets required large computational resources. 200,000 tests were performed for pseudo chains.

The program remembers the number of vertices (N_i) that the criterion selected to color the entire ordered set in each test. After the tests were completed, these numbers were averaged $Navr=\sum_{i=1}^{n}N_i$. The results for the subclass of dense balanced ordered sets are shown in table 2 and figure 5. H_t is the height of the ordered set. The best results correspond to lower *Navr* values.

Table 2. The results of the computational experiment in the subclass of dense balanced ordered sets.

H_t	WLD	BL	SVG	HW1/3	HW2/3	HL1/3	HL1/2	HL2/3	MPL
100	1093.6	620.2	1097.5	640.7	632.6	756.9	655.6	636.9	747.6
200	3835.2	2110.8	3954.3	2200.5	2105.6	2400.2	2402.6	2165.5	2566.4
300	7898.9	4408.8	7900.4	4600.6	4457.6	4744.3	4707.6	4558.3	5329.9
400	12993.5	7081	13022.3	7140.2	6897.3	7809.8	7488.3	7313.7	8537.5
500	19400.8	10493.7	19676.1	13042.3	10236.1	11906.2	12390.1	10906.3	12664.7

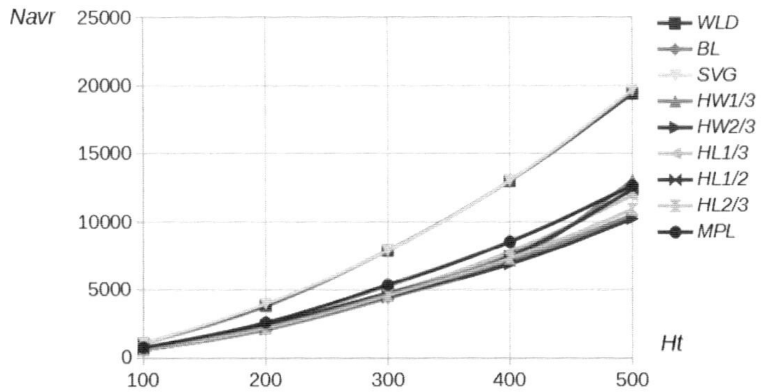

Figure 5. The results of the computational experiment in the subclass of dense balanced ordered sets.

The computational experiment has shown that in all subclasses of ordered sets, except for pseudo chains, the best results are given by the criteria of Hurwitz $\left(\alpha = \frac{2}{3}\right)$ and Bayes-Laplace. The Wald and Savage criteria were the most ineffective. The difference between the best and worst results grows as a low-degree polynom depending on the height of the ordered set. For pseudo chains, the results of all the criteria were approximately equal.

7. Conclusion

The article deals with the modeling of geometric obstacles in the assembly of complex products. The analysis of modern publications on CAAP is performed. It is shown that the problem of minimizing the number of geometric tests is not discussed in these papers. A game model of geometric solvability is proposed. It is a non-antagonistic game with nature to color the vertices of an ordered set. A rational strategy in this game makes it possible to minimize the number of necessary geometric tests in the computer aided design of assembly processes. The game is set as a problem of decision-making in conditions of uncertainty. The computational experiment was performed to determine the best coloring criterion for ordered sets. The special program has been developed that creates a random ordered set and colors it using a given criterion. The experiment showed that the best results are given by Hurwitz and Bayes-Laplace criteria. Using these criteria can significantly reduce the complexity of geometric modeling in CAAP systems.

References

[1] Delchambre A 2012 *Computer-aided Assembly Planning* (New York: Springer Science) DOI: 10.1007/978-94-011-2322-8
[2] Ericson Ch 2005 *Real-Time Collision Detection* (Elsevier) DOI: 10.1016/b978-1-55860-732-3.50007-3
[3] Eng T-H, Ling Z K, Olson W and Mclean Ch 1999 Feature-based assembly modeling and sequence generation *Computers & Industrial Engineering* **36** 17–33. DOI: 10.1016/s0360-8352(98)00106-5
[4] Miller J and Hofman R 1989 Automatic assembly planning with fasteners *Robotics and Automation Proc. IEEE Int. Conf.* **1** 69–74. DOI: 10.1109/ROBOT.1989.99969
[5] Hoffman R A 1991 Common sense approach to assembly sequence planning *Computer-Aided Mechanical Assembly Planning* **148** 289–313. DOI: 10.1007/978-1-4615-4038-0_12
[6] Chen T 2010 Virtual assembly of mechanical components and collision detection *Int. Conf. on Electronics and Information Engineering* **1.** DOI: 10.1109/ICEIE.2010.5559671

[7] Kim H S et al 1995 A collision detection method for real time assembly simulation *Proc. IEEE Int. Symp. on Assembly and Task Planning.* DOI: 10.1109/ISATP.1995.518799
[8] Jayaram S et al 1999 VADE: a virtual assembly design environment *IEEE computer graphics and applications* **19(6)** 44–50. DOI: 10.1109/38.799739
[9] Weiwei W, Kensuke H and Kazuyuki N 2018 Assembly sequence planning for motion planning *Assembly Automation* **38(2)** 195–206. DOI: 10.1108/AA-01-2017-009
[10] Halperin D, Latombe J C and Wilson R 2000 General Framework for Assembly Planning: The Motion Space Approach *Algorithmica* **26(3–4)** 577–601. DOI:10.1007/s004539910025
[11] Chang Hsuan and Tsai-Yen Li 1995 Assembly maintainability study with motion planning *Proc. of 1995 IEEE Int. Conf. on Robotics and Automation* **1.** DOI: 10.1109/ ROBOT.1995.525415
[12] Sundaram S, Remmler I and Amato N 2001 Disassembly sequencing using a motion planning approach *Proc. 2001 ICRA. IEEE Int. Conf. on Robotics and Automation* **2.** DOI: 10.1109/ROBOT.2001.932818
[13] Bonert M, Shu L H and Benhabib B 2000 Motion planning for multi-robot assembly systems *Int. J. of Computer Integrated Manufacturing* **13(4)** 301–310. DOI: 10.1080/095119200407660
[14] Romney B, Godard C, Goldwasser M and Ramkumar G An efficient system for geometric assembly sequence generation and evaluation. Retrieved from: http://robotics.stanford.edu/~latombe/cs326/2003/class15/romney.pdf
[15] Pan Ch, Smith Sh and Smith G 2005 Determining interference between parts in CAD STEP files for automatic assembly planning *J. of Computing and Information Science in Engineering* **5(1)** 56–62. DOI: 10.1115/1.1861473
[16] Krishnan S and Sanderson A 1991 Reasoning about geometric constraints for assembly sequence planning *Robotics and Automation* **1** 776–782. DOI: 10.1109/ROBOT.1991.131680
[17] Su Q A 2009 Hierarchical approach on assembly sequence planning and sequences analysing *Robotics and Computer-Integrated Manufacturing* **25(1)** 224–234. DOI: 10.1016/j.rcim.2007.11.006
[18] Wilson R and Latombe J C 1994 Geometric Reasoning about Mechanical Assembly *Artificial Intelligence* **71(2)** 371–396. DOI: 10.1016/0004-3702(94)90048-5
[19] Srinivasan H and Gadh R 2002 A non-interfering selective disassembly sequence for components with geometric constraints *IIE Transactions* **34(4)** 349–361. DOI: 10.1080/07408170208928875
[20] Ghandi S and Masehian E l 2015 Review and taxonomies of assembly and disassembly path planning: problems and approaches *Computer-Aided Design* **67–68** 58–86. DOI: 10.1016/j.cad.2015.05.001
[21] Bourjault A 1988 Methodology of Assembly Automation: A New Approach *Robotics and Factories of the Future '87* 37–45. DOI: 10.1007/978-3-642-73890-6_6
[22] De Fazio T and Whitney D 1987 Simplified generation of all mechanical assembly sequences *Robotics and Automation, IEEE J.* **l(3)** 640–658. DOI: 10.1109/JRA.1987.1087132
[23] Wilson R 1995 Minimizing user queries in interactive assembly planning *IEEE Transactions on Robotics* **11(2)** 308–312. DOI: 10.1109/70.370514
[24] Zha X, Lim S and Fok S 1998 Integrated Knowledge-Based Assembly Sequence Planning *Int. J. of Advanced Manufacturing Technology* **14(1)** 50–64. DOI: 10.1007/BF01179417
[25] Bozhko A 2018 Math Modeling of Sequential Coherent and Linear Assembly Plans in CAD Systems *Global Smart Industry Conf. (GloSIC)* DOI: 10.1109/GloSIC.2018.8570090
[26] Bozhko A N 2019 Hypergraph model for assembly sequence problem *IOP Conf. Ser.: Mater. Sci. Eng.* **560(1)**. DOI: 10.1088/1757-899X/560/1/012010
[27] Roman S 2008 *Lattices and Ordered Sets* (Springer Science) DOI: 10.1007/978-0-387-78901-9
[28] Yoe Ch 2019 *Principles of Risk Analysis. Decision Making Under Uncertainty* (Taylor & Francis Group) DOI: 10.1201/9780429021121

Factors affecting the tribotechnical properties of sintered materials

O T Cherney, E G Skachkova, A A Permovskiy, Zh V Smirnova and L I Kutepova

Minin Nizhny Novgorod State Pedagogical University (Minin University), Nizhny Novgorod, Russian

E-mail: fiolet1975@mail.ru

Abstract. The article considers the issues of production of powder antifriction materials based on iron by methods of pressing and rolling powder mixtures. The analysis of influence of composition of materials, technological conditions of their production and other factors on the structure, mechanical and tribotechnical properties of powder sliding bearings was carried out. During the operation of the machine, its elements, subjected to various influences, change in state, size and properties. The effectiveness of developments in the field of tribotechnical materials is determined by the properties of these materials, as well as the speed and breadth of their industrial use. The effect of porosity of samples on their tribotechnical properties and hardness after sintering and additional heat treatment is studied. The results of tribotechnical tests of iron-copper-graphite sintered sulfided compositions, not alloyed and doped with phosphorus, as well as materials including additional lubricant components: graphite, molybdenum disulfide or calcium fluoride, are presented. On the basis of tests, it was found that in conditions of liquid friction, alloying with phosphorus has a decisive effect on increasing wear resistance, and without supplying lubricant to sintered samples with impregnated oil, wear is more than an order of magnitude higher compared to the conditions of liquid friction.

1. Introduction

To obtain plain bearings, porous powder antifriction materials are widely used, capable of operating in a wide range of loads and rotational speeds both under conditions of forced lubrication and, in some cases, without it. Powder materials have a number of advantages over traditional cast materials. The advantages of porous bearings such as pores in the powder material contribute to the formation of a stable oil film on the friction surface.

The lubricant accumulated in the pores after the bearing impregnation operation allows ensuring its operation in conditions of lack or in some cases forced lubrication. Powder bearings have relatively low cost and lack of deficiency of initial components. Powder bearings have high operational reliability and simple manufacturing technology.

Many scientists and developers took part in the research of this direction. The merit of the revival of powder metallurgy and the transformation into a special technological method of pressing and sintering platinum powder belongs to Russian scientists P.G. Sobolevsky and V.V. Lyubarsky who worked in 1827. The pressing and sintering of platinum powder takes place by dissolving in tsarist vodka, depositing ammonium chloroplatinate, calcining to obtain platinum powders, their subsequent pressing, sintering or hot pressing [1].

2. Materials and methods

We conducted a study on a friction machine MI-1M. Seven sintered compositions and cast bronze material were compared. Various enterprises and workshops of the automotive industry took part in the manufacture of samples.

The following materials were used to make parts. Iron powder PZHV 3-4-160-460-34 is marked as "Zh". Graphite powder GC -2 or GC -3 is marked as "Gr". Copper powder of the grade PM -1 is marked as "D". Sulfur powder is marked as "K". Molybdenum disulfide powder is labeled "Ms". The milled ferrophosphore domain is marked as "P". Ground melting spar with calcium fluoride content of not less than 97% in the final product is marked as "Fk". Tin powder is marked as brands PO-1, PO-E, PO-2. Numbers after letters indicate the mass fraction of elements in percent. Additional treatment after sintering is paroxidation (PO).

3. Study of the tribotechnical characteristics of materials

The performance of antifriction sintered materials is positively affected by such technological operations as sulfidation and oxidation, which are widely used in workshops and in powder metallurgy. Roller samples were tested according to the established mode when friction in oil. The test results are shown in Table 1.

Table 1. Properties of typical material grades depending on the version of their manufacturing technology

Material Tag	Properties		Additional Processing Operation	Friction in oil p = 25 kgf/cm^2, 0.8 m/s		
	Hardness (HV)	Porosity P, %		Wear and tear μm\hour	Relative wear	friction coefficient
ZhGr0.3D3K0.3	95.16	16.0	-	1.380	2.59	0.056
ZhGr0.3D3K0.3P0.2	64.8	19.7	-	0.532	1.00	0.080
ZhGr0.5D3	125.5	15.9	-	Samples did not pass the test (teases)		
ZhGr0.5D3	131.0	16.7	Sulfided impregnation with sulfur, roasting T = 600 ^{0}C	0.612	1.16	0.050
ZhGr0.5D3	82.0	15.9	Sulfided sulfur impregnation, roasting T = 550 ^{0}C	0.800	1.29	0.050
ZhGr0.3	123.0	17.0	Steam oxidation T = 550 ^{0}C	4.20	7.89	0.090
Cast bronze Br0F1	66.0	-	-	3.2	6.01	0.020

In the accepted mode of friction tests, the least wear belongs to sintered material of the ZhGr0.3D3K0.3P0.2 grade, obtained using the technology of one of the automobile plants; its relative wear is taken as a unit.

Compared to a material that does not contain phosphorus, their resistance is more than double. Some authors also point to an increase in wear resistance of iron-based sintered materials containing phosphorus.

Sintered materials that do not contain antitussive components (sulfur, phosphorus) show unsatisfactory tribotechnical properties.

Oxidation (for example, superheated steam treatment) markedly restores the antifriction properties of the sintered materials, but not as significantly as sulfidation carried out by impregnation with sulfur followed by annealing or by addition of sulfur to the original powder mixture.

Cast bronze OF-1 has the lowest coefficient of friction, and in terms of wear resistance, it even exceeds, for example, a sintered material such as oxidized ZhGr0.3D3-PO [2].

Serviceability at the given load of the bearing is strongly dependent on external friction conditions in the sense of supply of lubricant to the friction zone into the friction, liquid, boundary and dry. The degree of this constraint is different for each material tag. Table 2 shows examples from the results of the tribotechnical tests of iron-copper-graphite sintered sulfided compositions not alloyed and doped with phosphorus. There are materials including additional lubricant components: graphite, molybdenum disulfide or calcium fluoride.

From the table it follows that:

- in liquid friction conditions, phosphorus doping has a decisive effect on increasing wear resistance. The presence of solid lubricants in the composition of the material in this friction mode has no significant effect on improving operability;
- under conditions of boundary friction, i.e. without supply of lubricant to sintered samples, they were only impregnated with oil. Wear compared to the conditions of liquid friction is more than an order of magnitude, but samples containing solid calcium fluoride in their composition have less wear. It is sufficient to have only 1% to surpass an alternative material containing 1.5% graphite and 4% molybdenum disulfide in the charge;

Table 2. Tribotechnical characteristics of sintered samples depending on lubrication conditions

Tag designation	Wear, W μm\hour	Friction coefficient
	In an oil bath	
ZhGr0.2D3K0.3	1.4	0.08
ZhGr0.2D3K0.3R0.2	0.46	0.08
ZhGr0.2D3K0.3R0.2Ms3	0.61	0.07
ZhGr0.2D3K0.3R0.2Fk4	0.44	0.07
	Self-lubricating mode of oil impregnated sample (boundary friction)	
ZhGr1.5D3K0.3Ms4	10.31-13.75	0.10
ZhGr0.2D3K0.3R0.2Fk1	8.7-9.1	0.05-0.06
	Dry friction of samples (without oil impregnation)	
ZhGr0.2D3K0.3R0.2Fk1	51	0.33
ZhGr0.2D3K0.3R0.2	76	0.25
ZhGr0.3DK0.3R0.2Ms3	74	0.30
ZhGr0.3DK0.3R0.2Fk4	41	0.24
After nitrocementation and hardening wear, W = 34 μm\hour		

- in the absolutely dry friction mode, wear increases even more (by one or even two orders of magnitude) compared to boundary friction. Materials that do not contain dry lubricant have the greatest wear. The best values for materials that contained calcium fluoride in an amount of 1-4%. Molybdenum disulfide is less effective [3].

Thus, calcium fluoride in dry friction conditions provides high performance of bearings.

Noting the role of calcium fluoride as an effective lubricant, certain limitations on its content by other evaluation factors should be taken into account. Although increasing its content to 4% still allows slightly increasing the wear resistance of the bearing under dry friction conditions, the mechanical strength of the sintered composition is noticeably reduced due to weakening of the metal frame. Table 3 shows this [4].

Thus, it can be seen from Table 3 that the optimum content of calcium fluoride in the material should not exceed 1%.

Based on the studies conducted, the sintered material of the brand ZhGr0.3D3K0.3R0.2Fk1 is widely used in the manufacture of sintered parts operating under dry friction. One of these parts was the packing ring of the rear axle of the truck [5].

Among the factors, affecting the performance of the sintered bearing article, there is the nature of the microstructure.

Table 3. Effect of calcium fluoride additives on the mechanical characteristics of the sintered material ZhGr0.3D3K0.3R0.2

Type of tests	Units of measure	Content of calcium fluoride in % (wt.)					
		0	1	2	3	4	5
Tensile strength	kgf/mm^2	25.9	25.3	23.2	22.7	20.9	18.6
Specific elongation	%	2.93	4.1	3.10	3.10	2.70	2.60
Hardness (HV)	kgf/mm^2	90.4	94.4	89.4	87.6	89.9	92.4

It is well known that the medium of microstructures of iron-based materials not subjected to thermal hardening is preferred in terms of wear resistance to pearlite microstructure. However, there are certain technological difficulties in machining materials with a lamellar pearlite or pearlite structure with excessive cementite.

In order to improve workability, annealing on the microstructure of the granular pearlite is usually used, which allows a sharp improvement in cutting conditions.

The thermal annealing of sintered articles has become traditional. But as studies have shown, annealing on the microstructure of granular perlite reduces mechanical properties. At the same time, the doping of low carbon compositions with phosphorus provides a wear resistance superior to that of materials with a granular perlite structure [1, 6]

This is evidenced by the results (Table 4) of the comparison of materials of the grades ZHGr1.7DZK0, 3P0.2 (annealing at 710-740 °C within 6 hours) and ZHGr0.3DK0.3P0.2 (not subjected to annealing). It is interesting to note that both brands contain 0.2% phosphorus. The effectiveness of phosphorus as an alloying element is found only in low-carbon material due to the poor solubility of phosphorus at a high carbon content in the iron.

Table 4. Comparative values of low-carbon and high-carbon (annealed) antifriction materials

Grade, type of microstructure, type of heat treatment	Carbon, C %	Physical and mechanical parameters				Tribotechnical properties	
		Porosity P, %	(HV), hardness	Ultimate strength kgf/mm^2	Specific elongation, %	Wear W, µm\hour	Friction coefficient
ZhGr1.7D3K0.3R0.2 annealing on granular pearlite structure within 6 hours	0.9	22.8	65.9	19.0	3.2	0.494	0.05
ZhGr1.3D3K0.3R0.2	0.10	21.6	63.6	22.5	3.5	0.410	0.08

Thus, the use of phosphorus-doped sintered material instead of the use of pearlite annealing material requiring additional annealing after sintering, from the point of view of production technology, is economically advantageous [8]. In addition, the wear resistance of the phosphorous-doped ferrite sintered material is noticeably higher not only according to the results of tests on a standard friction machine, but also according to the mileage tests of real products - thrust washers of car pivots, as shown in Table 5.

The operational wear resistance of the second low-carbon material is 1.3-1.4 times is higher than that of the first granular pearlite annealed material of the microstructure, which is used for sintered automotive parts.

The use of the low-carbon material with phosphorus for parts such as guide bushings of engine valves is most promising [9].

In fact, the maximum operating mode on the surface of the valve sleeves is full of the exhaust, the temperature reaches 500-550°C. The ordinary motor lubrication loses its necessary properties. In addition to sulphides, calcium fluoride is introduced into the material of the sintered bushings as an effective solid lubricant [10]. Standard bushings with a microstructure of granular pearlite did not contain such lubricant. The results of tests of sleeves of two materials ZhGr1.5D3K0.3 (standard with annealing) and ZhGr0.3D3K0.3R0.2Fk1 (1% with fluoric calcium) are given in table 6.

Table 5. Test results of "washer" with different microstructures

Sintered material grade	Description of microstructure	Porosity of part, %	Car mileage in thousands of km	
			Mean values	Confidence limit of 90%
ZhGr1.7D3K0.3R0.2	After additional annealing: grainy pearlite sulphide and phosphide eutectic grains along grain boundaries	15	186	105
ZhGr1.3D3K0.3R0.2	Solid solution of copper and phosphorus in iron sulfides and rare grains of cementite	15	239	145

Table 6. Comparative wear of guide bushings of the motor engine exhaust valve (bench tests)

Materials	Test mode	Description of microstructure	Average maximum wear of bushings for 100 hours.
Standard for Part ZhGr1.5D3K0.3	60 hours run-in 100 hours run-in - part manufacturing quality test	Granular pearlite, sulfides	19.85
Advanced for Part ZhGr0.3D3K0.3R0.2Fk1	40 hours of wear test (total 200 hours with a maximum power of 2800 rpm)	Solid solution of copper and phosphorus in iron, individual grains of sulfides, cementite and calcium fluoride	16.95

4. Conclusion

The parts that were tested on two engines of medium-duty trucks were explored.

According to the average data, the wear of standard exhaust valve bushings is 17% more than that with a ferrite microstructure.

It has been confirmed that the prospect of using the material ZhGr0.3D3K0.3R0.2 with a calcium fluoride additive is more economical and it is wear resistant.

References

[1] Arzamasov B N and Cherepakhin A A 2013 *Materials science: textbook for universities* (Moscow: Academy)
[2] Cherney O T and Morozov D L 2019 Powder antifriction materials in the automotive industry *Integration of information technologies into the system of professional and additional education. Collection of articles on the materials of the VI Regional Sci. and Pract. Conf.* pp 36-39
[3] Sorokin G M 2000 *Tribology of steels and alloys: textbook for universities* (Moscow: Nedra)
[4] Smirnova Z V, Vaganova O I, Cherney O T, Mukhina M V and Kutepova L I 2020 The process of protecting metals from corrosion *Int. J. of Emerging Trends in Engineering Res.* **8(5)** 1728-1731
[5] Smirnova Z V and Cherney O T 2020 Laser alloying technology when changing the processing speed *Int. J. of Emerging Trends in Engineering Res.* **8(4)** 1196-1198

[6] Barkhatov N A, Revunov S E, Smirnova Z V, Cherney O T and Mukhina M V 2020 Recovery of gaps in records of geomagnetic observatories by neural network *Int. J. of Emerging Trends in Engineering Res.* **8(4)** 1243-1246
[7] Bataev A A and Bataev V A 2002 *Composite materials: textbook for universities* (Novosibirsk: NSTU)
[8] Bobovich B B 2009 *Non-metallic structural materials: textbook* (Moscow: MGIU)
[9] Filippov A A, Pachurin G V, Goncharova D A, Mukhina M V, Katkova O V, Chaikina Z V, Smirnova J V and Tsapina T N 2019 Resource efficient preparation of high quality rolled stock for motor vehicles fastening *IOP Conf. Ser.: Mater. Sci. Eng.* 012078
[10] Pachurin G V, Filippov A A, Goncharova D A, Nuzhdina T V, Matveyev Yu I, Mukhina M V and Kutepova L I 2019 Effect produced by corrosive environment on fatigue resistance of automotive structural steel *IOP Conf. Ser.: Mater. Sci. Eng.* 012037

Analysis of narrowband data transfer technologies on the Internet of Things (IoT)

Kh A Daudov[1], A M Patrusova[2] and V V Nadrshin[3]

[1] Chechen State University, 17a, Dudaev Boulevard, Groznyy, 364015, Russia
[2] Bratsk State University, 40, Makarenko str., Bratsk, 665709, Russia
[3] Irkutsk National Research Technical University, 83, Lermontova str., Irkutsk, 664074, Russia

E-mail: ibr024@mail.ru

Abstract. The article provides an overview and comparative analysis of narrowband data transfer technologies on the Internet of Things. For analysis, a method of comparing the parameters of each of the reviewed technologies based on data from domestic and foreign literature, scientific articles and publications was chosen. This comparative analysis helps to identify a more efficient and preferable technology.

1. Introduction

The concept of the Internet of Things appeared in 1999 as the prospect of connecting a huge number of physical objects (sensors) through radio frequency identification for interaction between themselves and with the external environment, that is, this is a new stage in the development of the Internet, which connects the objects around us into a computer network. They exchange information among themselves and work without human intervention and in real time. In fact, this is the Internet, exciting the real world.

Filling the concept with a variety of technological content and introducing practical solutions for its implementation since the 2010s are considered as a stable trend in information technologies. The development of the Industrial Internet of Things (IIoT) should be discussed separately [13]. The range of possible data transfer technologies covers all possible means of wireless and wired networks.

For wireless data transmission, such qualities as efficiency in low-speed conditions, fault tolerance, adaptability, and self-organization play a particularly important role in building the Internet of Things.

In terms of transmission range, Internet of Things communication technologies can be divided into short-distance technologies and long-range technologies (WAN, wide area network technologies). The first are represented by protocols such as Zigbee, Wi-Fi, Bluetooth, Zwave, etc.

Their typical use is "smart homes", "smart manufacturing" or stand-alone industrial scanning systems for difficult or unsafe work environments [14]. The latter are used in low-speed data services and are usually defined as Low-power Wide-area Network (LPWAN) technologies. At the same time, the development of LPWAN technologies is especially promising.

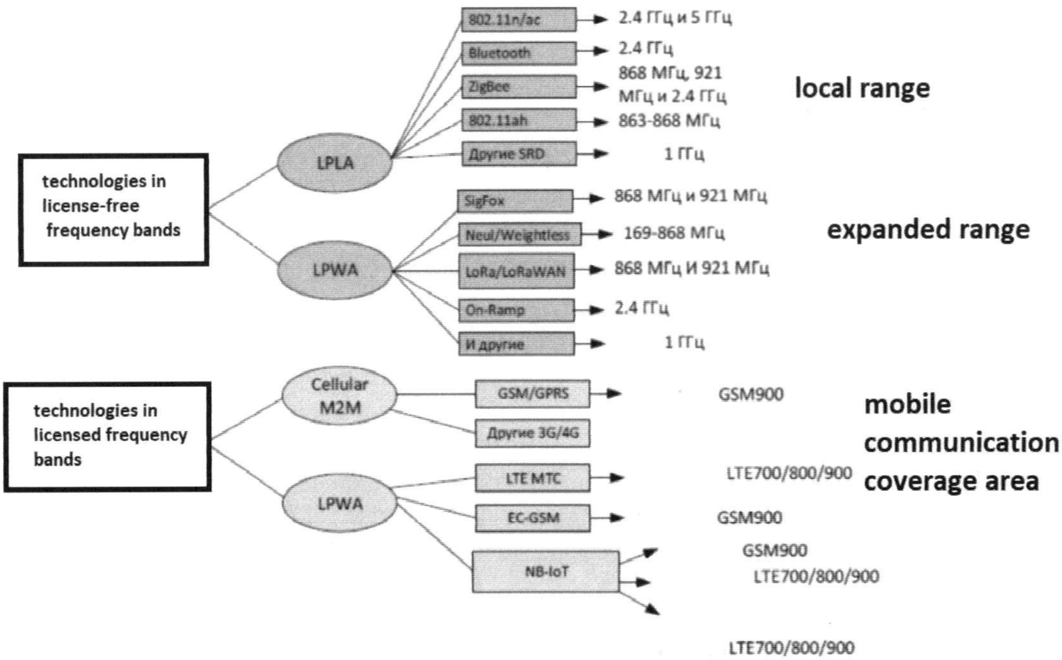

Figure 1. IoT technologies and their use in the radio frequency spectrum

In terms of frequency spectrum licensing, LPWAN technologies can be divided into two categories: technologies operating in unlicensed and licensed frequency bands (Figure 1).

The first category is represented by Lora, Sigfox and other technologies, most of which are not licensed.

The second category is represented by some relatively old 2G/ 3G cellular technologies (such as GSM, CDMA, WCDMA, etc.), LTE and LTE-A technology, which support various terminal categories. Standards for these permitted spectrum communication technologies are mainly developed by international standardization organizations such as 3GPP and 3GPP2.

Due to the rapid growth of low data rate of the Internet of Things services, LPWA technology is becoming increasingly popular in industry, and its market share is gradually increasing. According to a report by Hequan Wu at the 2016 China Internet of Things Conference, intelligent IoT applications can be divided into three categories depending on the requirements for data transfer speed in 2020 [1].

Table 1. Distribution of indicators by smart IoT connection technology in 2020

Distribution by 2020	categories	how to connect to the network
10%	high data rate (more than 10 Mbit/s)	3G:HSPA/EVDO/TDS 4G:LTE/LTE-A WiFi 802.11
30%	average data rate (less than 1 Mbit/s)	2G:GPRS/CDMA2K1X MTC/eMTC
60%	low data rate (more than 100 Kbit/s)	NB-IoT SigFox LoRa Zigbee

1. High data rate. The data rate is higher than 10 Mbps. Available access technologies: 3G, 4G and Wi-Fi. Mainly used in live television, electronic healthcare, automotive navigation and entertainment system, etc. The expected market share for this type of application is IoT 10%.

2. Average data rate. The data rate is below 1 Mbps. Available access technologies: 2G and MTC/eMTC. Such applications are, for example, a POS machine, a smart home and 8 M2M feedback. The expected market share for such applications is IoT 30%. However, in the future 2G M2M will be gradually replaced by MTC/eMTC technology.

3. Low data rate. The data rate is below 100 kbps. Available access technologies are NB-IoT, SigFox, LoRa, and short-range wireless technologies such as ZigBee. Applications are mainly used in LPWA technologies, including sensors, intelligent metering, product tracking, logistics, parking, and intelligent agriculture. The expected market share for such applications is IoT 60%. [1]

Since this article analyses narrowband data transfer technologies in IoT, let us look at LORA, SigFox, and NB-IoT technologies in more detail.

2. Technology analysis

LORA Technology. The developers of LoRa Alliance believe that LoRa technology has significant advantages over WiFi and cellular networks, due to the ability to deploy machine-to-machine (M2M) connections up to 20 km at a speed of up to 50 kbps, and also has a minimum power consumption that provides several years of battery life on a single battery. The scale of applications of this technology is large: from home automation and the Internet of things to industry and smart cities [2, 7, 11].

Figure 2 describes the architecture of LoRaWAN networks 1 [6, 8].

The classic LoRaWAN network consists of the following elements: end nodes, gateways, network server and application server.

The End-Node is designed to perform control, monitoring, and measurement functions. It contains a set of necessary sensors and control elements. They usually have battery power. Nodes include data transmission for only a certain period of time (usually 1-5 seconds), at the end of which two time windows are opened to receive data. The rest of the time, the end-node transceiver is either inactive or in the receive state, depending on the device class (A, B, or C).

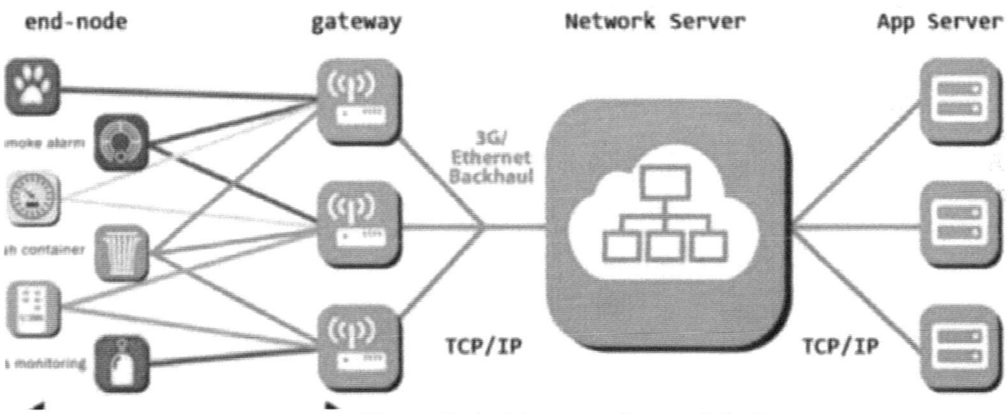

Figure 2. Architecture of network LoRa

The device accepting data from final devices by means of a radio channel and transferring them to transit network - the LoRa (Gateway/Concentrator) Gateway. Transit networks can be Ethernet, WiFi, cellular networks and any other telecommunication channels. The gateway and endpoints form a star network topology. Often, this device contains multi-channel transceivers for processing signals in several channels simultaneously or even, several signals in one channel. Accordingly, several such devices provide network coverage and transparent bidirectional data transmission between the end nodes and the server. Network Server is designed for network management: scheduling, rate adaptation, storage and processing of received data. The Application Server can remotely monitor the operation of the target nodes and collect the necessary data from them.

3. Technology SigFox

SIGFOX is a private company that aims to create a worldwide network specially designed for devices (Internet of Things, IoT). The technology allows data transmission over long distances at low transmitter power and low battery capacity.

The network is great for simple and stand-alone devices that send a small amount of data to this network. So the SIGFOX network is similar to the cellular infrastructure (GSM and GPRS-3G-4G), but is more energy efficient and at the same time less expensive. SIGFOX uses Ultra Narrow Band (UNB) based on radio technology to connect devices to the wide area network. The use of UNB is a key factor in ensuring a very low transmitter power level that will be used during the state of maintaining a reliable data connection. The network operates in existing unlicensed bands (The industrial, scientific and medical, ISM) on a global scale and coexists in these frequencies with other radio technologies without the problem of network overlap or bandwidth problem. In Europe, the 868.8 MHz band is widely used (as defined in ETSI and CEPT), and in the USA 915 MHz (as defined by the FCC). The coverage area of SIGFOX is about 30-50 km in rural and urban areas. Usually, if there is more difficulty and noise, the range decreases from 3 to 10 km. Figure 4 shows the topology of the SIGFOX network [9, 10].

The overall network topology was designed to provide a scalable, high-performance network with very low power consumption, while maintaining a simple and easy star topology based on mesh infrastructure.

Figure 3. Network topology SIGFOX

In SIGFOX, nodes can be used in two configurations:

• P2P mode: direct communication between nodes (LAN interface);

• Hybrid mode: SIGFOX/ P2P (P2P GW in SIGFOX network).

SIGFOX technology is aimed at low cost devices where a wide coverage area is required. There are a number of applications that need this wireless technology. Areas in which SIGFOX networks can be used include:

• homes and consumer goods;
• energy communications;
• transport - including technical management;
• remote monitoring and control;
• retail point of sale, storage of updates, etc.;
• security.

The standard has a number of advantages over other basic LPWAN network technologies:

• large coverage area;

• high penetrability;
• up to 20 years of sensor operation from 2 AA batteries, i.e. ultra-low power consumption;
• low cost.

Like all technologies of the modern world, the energy-efficient SIGFOX network, unfortunately, also has negative characteristics:

• low data rate;
• dependence on cellular infrastructure;
• limited noise immunity;
• no SIGFOX networks in Russia [3].

4. Technology NB-IoT

The narrow-band Internet of things (Narrow-Band Internet of Things, NB-IoT or LTE Cat-NB1) is the large-scale technology of energy efficient network of distant range (LPWA) offered 3GPP for collecting and data processing and intended for intellectual applications with a low speed of data transmission. Typical applications are smart instrument readings and intelligent environmental monitoring.

NB-IoT provides massive connections, ultra-low power consumption, wide coverage area and bidirectional connection between control plane, and signaling and data plane.

To create networks for the Internet of Things in Russia according to the standard NB-IoT according to the decision of the State Commission on Radio Frequencies (GKRCH) of December 28 2017; SCRF No. 17-44-06, radio frequency bands allowed 453-457,4 MHz and 463-467,4 MHz, 791-820 MHz, 832-862 MHz, 880-890 MHz, 890- 915 MHz, 925-935 MHz, 935-960 MHz, 1710-1785 MHz 1805-1880 MHz, 1920-1980 MHz, 2110-2170 MHz, 2500-2570 MHz and 26200-2690 MHz [4, 12]

In order to operate the communication network according to the NB-IoT technology, a frequency band of 180kHz must be allocated in the radio frequency resource, which corresponds to one LTE radio network resource unit (EUTRAN). 3GPP defines the three modes of operation shown in Figure 4 [3]:

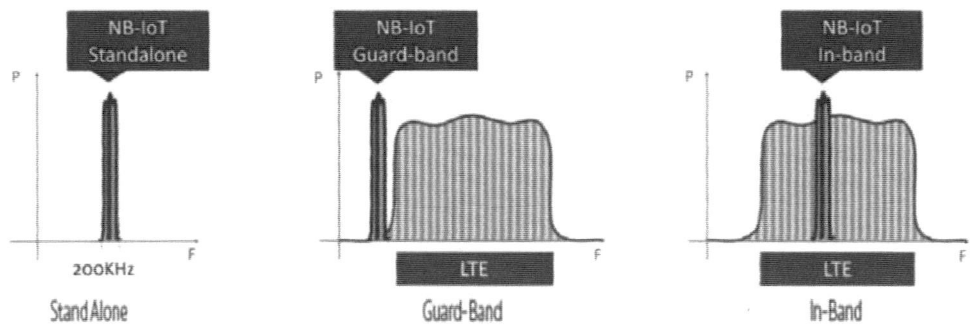

Figure 4. Three operating modes supported by NB-IoT

NB-IoT - stand-alone mode. Autonomous mode of operation, which uses an independent frequency band that does not overlap with the LTE frequency band: one frequency interval of 200 kHz is allocated in the GSM band, which allows you to organize a NB-IoT radio channel with two protective intervals of 10 kHz.

NB-IoT - guard-band mode. The mode of operation in the protective band of the LTE network, in which one resource unit is allocated in the protective interval outside the operating band used to provide LTE services; in particular, the protective band for the LTE range with a width of 20MHz is 1MHz on each side, for the LTE range with a width of 10MHz - 500 kHz.

NB-IoT - in-band mode. The LTE band mode, which uses the LTE band for deployment and requires 1 Physical Resource Block (PRB) of the LTE band for deployment.

The most common is in-band operation. It is widely used in Europe, where networks with NB-IoT in-band are deployed by telecommunications companies Vodafone, Deutsche Telekom, Telecom Italia Mobile and others.

Despite the prevalence of the in-band mode of NB-IoT use, it has disadvantages. Thus, the narrowband in-band signal NB-IoT occupies a frequency of 180 kHz and in LTE networks can interfere with physical resource units (PRBs). In this mode, the NB-IoT takes away the functional resources of the cellular networks.

The NB-IoT network is shown in Figure 6, which shows that it consists of 5 parts:

1. Terminal NB-IoT. Devices IoT in all industries have access to the NB-IoT network if the appropriate SIM card is installed;

2. Base station NB-IoT. This generally applies to base stations that have already been deployed by carriers and support all three types of deployment modes mentioned previously;

3. The core of the NB-IoT network. Through the core of the network, the base station NB-IoT can connect to the cloud platform NB-IoT;

4. Cloud Platform NB-IoT. The cloud platform NB-IoT processes various services and directs results to a vertical technical center or terminals NB-IoT;

5. Vertical technical center. Receives NB-IoT service data and stores it in its own center, and also takes control of NB-IoT terminals.

Coverage Levels In most cases, NB-IoT is used to transmit mini-packets, and it is NB-IoT difficult to provide a long-term and continuous indication of channel quality change, so NB-IoT uses different coverage levels instead of a dynamic line adaptation scheme.

There are three types of coverage levels, including normal coverage, reliable coverage and extreme coverage, which correspond to minimum communication losses of 144, 154 and 164 dB, respectively [5].

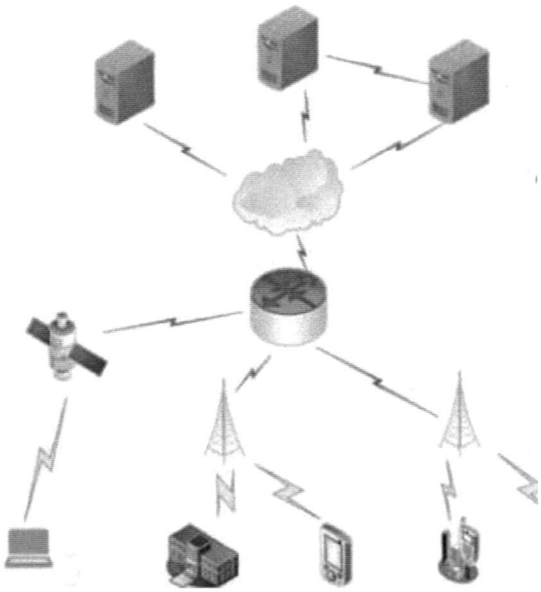

Figure 5. Network NB-IoT

The original goal of the NB-IoT was 50.000 connections per cell.

Based on preliminary calculations and estimates, the current version of the NB-IoT mainly meets the requirements. However, the ability to achieve the goal in practice depends on the service model of the NB-IoT terminal in the cell, so further tests and assessments are needed.

5. Conclusion

This paper provides a comparative analysis of narrowband technologies in IoT, such as NB-IoT, LoRa, and SigFox. Comparison results are shown in **Table 2**.

Table 2. Comparison between NB-IoT, LoRa and SigFox

point	NB-IoT	LoRa	SigFox
power supply level	Low (10 years of battery life)	Low (10 years of battery life)	very much Low (20 years of battery life)
communications budget	164 Db	157 Db	149 Db
bandwidth	180 kHz (licensed)	125 kHz, ISM	100 Hz, ISM
Frequency	700/800/900/1800 MHZ end any LTE frequency range	858,8 MHZ (Europe), 915 MHZ (USA) 433 MHZ (Asia)	868,8 MHZ (Europa), 915 MHZ (USA)
Dplnost	up to 11 kilometers in the city	up to 2.5 kilometers in the city	up to 1 kilometer in the city
Number of connections	50000 on the sector	40000 on the sector	50000 on the sector
Speed	up to 250 kbps	0,3-50 kbps	100-600 bit/s
Scalability	high	average	low
Coating (vs. GSM)	more by 20 db	more by 20 db	more by 11 db
Expansion	supports reconfiguration of existing FDD or GSM LTE networks	inconveniently	inconveniently

Based on comparative analysis, it can be concluded that the standard NB-IoT a more promising technology, since it has a number of advantages:

• high density of connections (50 thousand devices IoT per sector of base station);

• provision of maximum communication range in urban conditions (up to 11 km) due to communication budget up to 164 dB;

• optimal speed (up to 250 kbps) for trouble-free operation of devices that transmit a small amount of information of the same type with a certain frequency;

• cheap use and maintenance of NB-IoT when deployed at existing frequencies.

References

[1] Nolan K E, Guibene W and Kelly M Y 2016 An Evaluation of Low Power Wide Area Network Technologies for the Internet of Things *Int. Wireless Communications and Mobile Computing Conference (IWCMC)*

[2] Kirichek R and Kulik V 2016 Long-Range Data Transmission on Flying Ubiquitous Sensor Networks (FUSN) by Using LPWAN Protocols *Communications in Computer and Information Science* **678** 442–453. DOI: 10.1007/978-3-319-51917-3_39.

[3] Retrieved from: https://www.3gpp.org/release-13

[4] 2018 Decision of the State Commission on Radio Frequencies at the Ministry of Communications of Russia dated December 28, 2017 "On the use of radio frequency bands by electronic means of the LTE standard and its subsequent modifications in the NB-IoT mode (RCRCH decision No. 17-44-06)," January 24, 2018

[5] Pilar A-M, Ameigeiras P, Prados J, Navarro Ortiz J and Lopez-Soler M J 2017 NarrowBand IoT Data Transmission Procedures for Massive Machine Type Communications *IEEE Network*

[6] Gimranov R R, Kirichek R V and Shpakov M N 2015 Technology of machine-to-machine interaction LoRa *Information technologies and telecommunications* **2(10)** 62-73. Retrieved from: http://www.sut.ru/doci/nauka/review/2-15.pdf

[7] Klyuev R V, Bosikov I I, Tavasiev V Kh and Gobeev M A 2020 Improving the energy efficiency of transport equipment in ore mining *IOP Conf. Ser.: Mater. Sci. Eng.* **918** 012123. DOI: 10.1088/1757-899X/918/1/012124

[8] Bosikov I I, Klyuev R V, Tavasiev V Kh and Gobeev M A 2020 Influence of transport and road complex on the natural-technical system *IOP Conf. Ser.: Mater. Sci. Eng.* **918** 012124. DOI: 10.1088/1757-899X/918/1/012223

[9] Nolan K E, Guibene W and Kelly M Y 2016 An Evaluation of Low Power Wide Area Network Technologies for the Internet of Things *Int. Wireless Communications and Mobile Computing Conference (IWCMC)* pp 439–444. DOI: 10.1109/IWCMC.2016.7577098.

[10] Retrieved from: https://www.saures.ru/

[11] Sidorenko A A, Dmitriyev P N and Ivanov V V 2020 A study of gas drainage methods efficiency in Kotinskaya mine in Russia *ARPN J. of Engineering and Applied Sciences* **15(4)** 530-535

[12] Klyuev R, Bosikov I, Gavrina O, Madaeva M and Sokolov A 2021 Improving the energy efficiency of technological equipment at mining enterprises *Advances in Intelligent Systems and Computing* **1258** 262-271. DOI: 10.1007/978-3-030-57450-5_24

[13] Makhovikov A B, Katuntsov E V, Kosarev O V and Tsvetkov P S 2019 *Digital transformation in oil and gas extraction. Innovation-Based Development of the Mineral Resources Sector: Challenges and Prospects: 11th conf. of the Russian-German Raw Materials* pp 531-538

[14] Kosarev O V, Tcvetkov P S, Makhovikov A B, Vodkaylo E G, Zulin V A and Bykasov D A 2019 Modeling of industrial IOT complex for underground space scanning on the base of Arduino platform *Topical Issues of Rational Use of Natural Resources – Proc. of the Int. Forum-Contest of Young Researchers* pp 407-412

Calculation of shock absorber parameters of torsional vibrations of screw pumps' rod string

A Yu Davydov, V A Petrov and L M Zaripova

Ufa State Petroleum Technological University, Branch of the University in the City of Oktyabrsky, 54a, Devonskaya St., Oktyabrsky, Republic of Bashkortostan, 452607, Russian Federation

E-mail: alex-dy@yandex.ru, info@of.ugntu.ru

Abstract. Torsional vibrations of the rod string occur during the operation of surface-driven screw pumping units, which reduces the reliability of the units. It is proposed to include a torsional vibration damper in the rod string to maintain stability. The calculation of the parameters of the shock absorber is presented.

1. Introduction

The use of screw pump installations with a surface drive and having a low rotor speed is a promising method of operating wells with low flow rates and high-viscosity oil. During operation of screw pump installations with a surface drive, oscillatory processes of downhole equipment occur, which reduce the durability of the installation elements, increase the frequency of rod string breakage and wear of the friction surfaces of the screw and the rubber stator lining. Oscillatory processes during operation of screw pump installations with surface generally drive negatively affect the operating parameters of screw pump installations. Elimination of vibration processes in the rod string layout suggests including a torsional vibration damper [1].

Determination of the optimal parameters of the torsional vibration damper is associated with the formulation of the problem of the torsional vibrations of the downhole equipment of the screw pump installations with a surface drive, arising from the falling characteristics of the friction coefficient of the pump screw on the stator rubber cage. The equations of the viscoelastic model describe the model of the operation of a torsional vibration damper, structurally made with an elastic element made of rubber. Figure 1 shows the design layout [2].

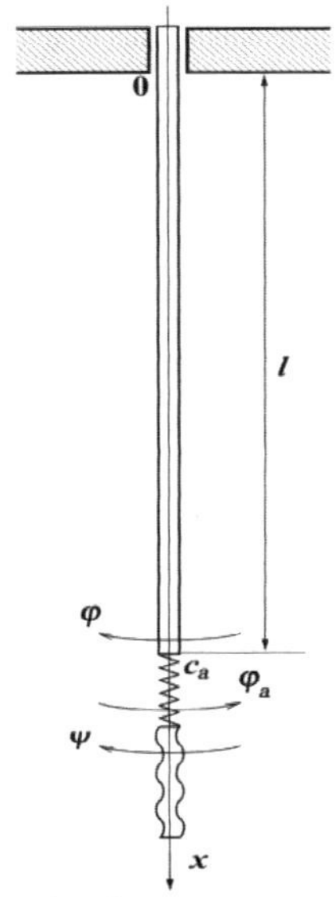

Figure 1. Scheme for calculating the optimal parameters of the torsional vibration damper when installed above the pump screw

2. Materials and methods

The following system of equations describes the above problem:

For a rod string:

$$\frac{\partial^2 \varphi}{\partial t^2} = a^2 \frac{\partial^2 \varphi}{\partial x^2} - 2\nu \frac{\partial \varphi}{\partial t}, \tag{1}$$

For the shock absorber:

$$I_a \frac{d^2 \varphi_a}{dt^2} + C_a\left(\varphi_a - \varphi|_{x=l}\right) + h_a\left(\frac{d\varphi_a}{dt} - \frac{\partial \varphi}{\partial t}|_{x=l}\right) = -M_{kp} \tag{2}$$

Initial conditions:

$$\psi(t)|_{t=0} = \varphi_a(t)|_{t=0} = \varphi(t,x)|_{t=0} = 0, \tag{3}$$

$$\frac{d\varphi_a}{dt}|_{t=0} = \frac{\partial \varphi}{\partial t}|_{t=0} = \frac{d\psi}{dt}|_{t=0} = 0. \tag{4}$$

Border conditions:

$$\frac{\partial \varphi}{\partial t}|_{t=0} = \omega_0 = const \tag{5}$$

$$GI_p \frac{\partial \varphi}{\partial x}|_{x=l} = C_a\left(\varphi_a - \varphi|_{x=l}\right), \tag{6}$$

$$M_{kp} = 4r_v^2 l_v P \frac{0,07}{\frac{\partial \psi}{\partial t} r_v + 0,28} \arccos\left(\frac{r_v - \delta}{r_v}\right) + \\ + \frac{32\rho_{zh} geDT}{\pi^3 \eta_g \eta_o}\left[250\pi^2 - 7\left(\frac{\partial \psi}{\partial t}\right)^2 (eDT)^2\right]. \tag{7}$$

where φ_a is the shock absorber twist angle, I_a is the moment of inertia of the shock absorber, C_a is the stiffness of the elastic element of the shock absorber, h_a is the viscous friction coefficient of the shock absorber, l is the length of the rod column, ψ is the screw twisting angle.

Equation (6) is a necessary condition for matching equations (1) and (2).

The equations obtained above can be solved by the grid method, i.e. by replacing differential equations with the corresponding difference equations [3].

The region of integration (0…t; 0…l) can be replaced by a set of equally spaced nodes (t_i; x_j)

$$t_i = i\Delta t (i = 0,1,...m); \quad (8)$$

$$x_j = j\Delta x (j = 0,1,...n), \quad (9)$$

where Δt is the partitioning step in time, Δx is the partitioning step along the column length.

The grid function u^i_j can replace the function $\varphi(t, x)$:

$$u^i_j = \varphi(t_i, x_j) = \varphi(i\Delta t; j\Delta x). \quad (10)$$

The function $\varphi_a(t)$ can be replaced by the function v^i:

$$v^i = \varphi_a(t_i) = \varphi_a(i\Delta t). \quad (11)$$

The function $\psi(t)$ can be replaced by the function w^i:

$$w^i = \psi(t_i) = \psi(i\Delta t). \quad (12)$$

Difference analogues replace the equations:

$$\frac{u^{i+1}_j - 2u^i_j + u^{i-1}_j}{\Delta t^2} = a^2 \frac{u^i_{j+1} - 2u^i_j + u^i_{j-1}}{\Delta x^2} - 2v \frac{u^{i+1}_j - u^i_j}{\Delta t}; \quad (13)$$

$$I_a \frac{v^{i+1} - 2v^i + v^{i-1}}{\Delta t^2} + C_a\left(v^i - u^i_n\right) + h_a\left(\frac{v^i - v^{i-1}}{\Delta t} - \frac{u^i_n - u^{i-1}_n}{\Delta t}\right) = -M_{kp} \quad (14)$$

The initial conditions will take the form:

$$u^0_j = v^0 = w^0 = 0, \quad (15)$$

$$\frac{u^1_j - u^0_j}{\Delta t} = \frac{v^1 - v^0}{\Delta t} = \frac{w^1 - w^0}{\Delta t} = 0; \quad (16)$$

Border conditions:

$$\frac{u^i_0 - u^{i-1}_0}{\Delta t} = \omega_0, \quad (17)$$

$$GI_p \frac{u^i_n - u^i_{n-1}}{\Delta x} = C_a\left(v^i - u^i_n\right), \quad (18)$$

$$M_{kp} = -4r_v^2 l_v P \frac{0,07}{\frac{w^i - w^{i-1}}{\Delta t} r_v + 0,28} \arccos\left(\frac{r_v - \delta}{r_v}\right) - \frac{32\rho_{zh} geDT}{\pi^3 \eta_g \eta_o}\left[250\pi^2 - 7\left(\frac{w^i - w^{i-1}}{\Delta t}\right)^2 (eDT)^2\right] \quad (19)$$

An algorithm was developed for solving these equations, and a computer program was written in the Delphi programming language, which allows obtaining the results of calculations in graphical form and the form of a table [4, 5].

3. Results

Below are the results of solving the equations with the following parameters: shear modulus of the rod material G=8.06·10^8 N/m^2; material density of rods ρ_{sh}=7815 kg/m^3; screw tension δ = 0.11 mm; diameter of rods d_{sh} = 0.022 m; screw diameter d_v = 0.06 m; the length of the rod string L_{sh} = 1000 m; screw length L_v = 2 m; resistance coefficient ν = 2.1; operating pressure of the pump P_n = 6·10^6 Pa; drive rotation speed n = 100 rpm; stiffness of the elastic element of the shock absorber C_a = 2800 N·m / rad; the coefficient of viscous friction of the shock absorber h_a = 0.88 N·m·s / rad.

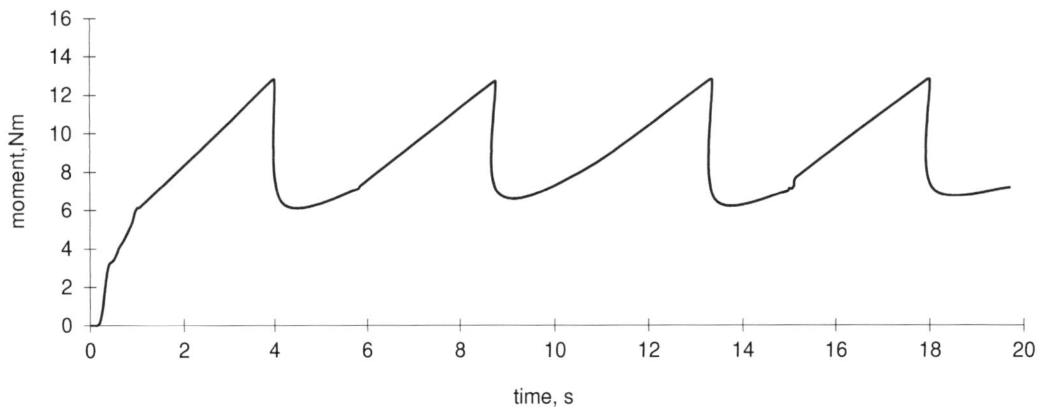

Figure 2. Dependence of torque on time when a torsional vibration damper is included in the rod string assembly when installed above the pump screw

The magnitude of the non-uniformity of the rod string torque when the torsional vibration damper above the pump screw was installed in the rod string assembly was m = 0.71. This indicator is 2.5 times lower than the unevenness of the torque without a shock absorber [6, 7].

Figure 3 shows the design layout with the shock absorber installed directly under the polished stem.

The problem is solved using a system of equations:

For the rod string:

$$\frac{\partial^2 \varphi}{\partial t^2} = a^2 \frac{\partial^2 \varphi}{\partial x^2} - 2\nu \frac{\partial \varphi}{\partial t}, \tag{20}$$

For the shock absorber:

$$I_a \frac{d^2 \varphi_a}{dt^2} + C_a(\varphi_a - \psi) + h_a\left(\frac{d\varphi_a}{dt} - \frac{d\psi}{dt}\right) = M_a \tag{21}$$

Initial conditions:

$$\psi(t)\big|_{t=0} = \varphi_a(t)\big|_{t=0} = \varphi(t,x)\big|_{t=0} = 0, \tag{22}$$

$$\frac{d\varphi_a}{dt}\Big|_{t=0} = \frac{\partial \varphi}{\partial t}\Big|_{t=0} = \frac{d\psi}{dt}\Big|_{t=0} = 0. \tag{23}$$

Border conditions:

$$GI_p \frac{\partial \varphi}{\partial x}\Big|_{x=0} = C_a(\varphi_a - \psi). \tag{24}$$

$$GI_p \frac{\partial \phi}{\partial x}\Big|_{x=l} = -4r_v^2 l_v P \frac{0,07}{\frac{\partial \phi}{\partial t}\Big|_{x=l} r_v + 0,28} \arccos\left(\frac{r_v - \delta}{r_v}\right) - \\ -\frac{32\rho_{zh} geDT}{\pi^3 \eta_g \eta_o}\left[250\pi^2 - 7\left(\frac{\partial \phi}{\partial t}\Big|_{x=l}\right)^2 (eDT)^2\right]. \tag{25}$$

where ψ is the angle of rotation of the polished rod, $d\psi / dt = \omega_0 = \text{const}$, the remaining designations are the same as in the previous problem.

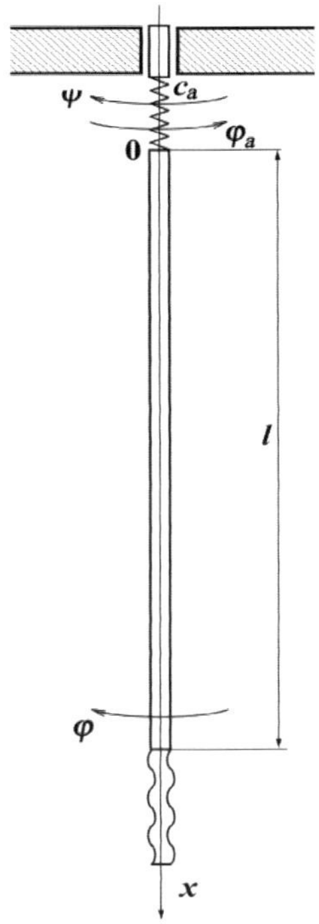

Figure 3. Scheme for determining the optimal parameters of the shock absorber of torsional vibrations of a rod string when installed under a polished rod

Equation (24) is an indispensable condition for matching equations (20) and (21).

The equations obtained in this way can also be solved by the grid method, i.e., by replacing the differential equations with the corresponding difference equations [8, 9].

The system of equations is solved using an algorithm similar to the algorithm of the previous problem.

Solution results for the same values: stiffness of the elastic element of the shock absorber C_a = 2800 N·m/rad; the coefficient of viscous friction of the shock absorber h_a = 0.88 N·m·s/rad.

The magnitude of the torque irregularity (Fig. 4) when installing a torsional vibration damper in the rod string arrangement directly above the pump screw was m = 0.73, which is 2.4 times lower than the torque irregularity without a shock absorber [10, 11].

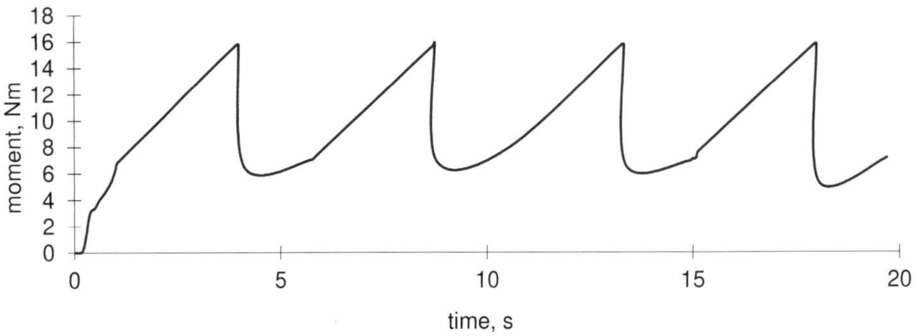

Figure 4. Dependence of torque on time when a torsional vibration damper is included in the arrangement when installed under a polished rod

4. Conclusion

The presented study allows concluding that the inclusion of a screw pump with a surface drive of a torsional vibration damper in the rod string arrangement leads to a decrease in the torque unevenness by 2.4-2.5 times.

References

[1] Davydov A Yu 2018 Utilization of expert systems for screw pump sets with surface drive management *Advances in Engineering Research (AER)* **157** 129-133

[2] Davydov A Yu 2019 Monitoring of the operation of surface-driven screw pumping units using expert control systems *J. of Phys.: Conf. Ser.* **1333(4)** 042005. DOI: 10.1088/1742-6596/1333/4/042005.

[3] Mukhametshin V V 2018 Bottomhole formation zone treatment process modelling with the use geological and geophysical information *IOP Conf. Ser.: Earth Env.* **194(2)** 022024. DOI: 10.1088/1755-1315/194/2/022024

[4] Khabibullin M Ya and Suleimanov R I 2019 Choice of bottom-hole filter construction for viscous fluid filtration *Advances in Engineering Research (AER)* **157** 272-275

[5] Kozikhin R A, Daminov A M, Fattakhov I G, Kuleshova L S and Gabbasov A Kh 2018 Identifying the efficiency factors on the basis of evaluation of acidizing of carbonate reservoirs *IOP Conf. Ser.: Earth Env.* **194(6)** 062013. DOI: 10.1088/1755-1315/194/6/062013

[6] Suleimanov R I, Gabdrakhimov M S, Khabibullin M Y, Zaripova L M and Vasilyeva E R 2018 The study of hydraulic hammer device in drilling tool assembly in hydraulical rotary drilling *Int. J. of Engineering and Technology* 7(2) 28-30

[7] Bailey M P and Ondrey G 2015 Twin- and triple-screw pumps for a variety of fluids *Chemical Engineering* **122(10)** 40

[8] Brandt J U, Lewerenz J, Müller-Link D and Reichwage M 2015 Development of a multiphase down-hole twin-screw pump and its use in multiphase-operated transport networks *Oil Gas European Magazine* **41(3)** 153-156

[9] Muhammed A and Childs D W 2016 Vibration modeling and experimental results of two-phase twin-screw pump *J. of Engineering for Gas Turbines and Power* **38(9)** 092601

[10] Bi H, Wu M and Zhang X 2016 Design of parameters optimization system for screw pump well *Int. J. of Simulation: Systems, Science and Technology* **17(25)** 6

[11] Duque J LR, Nakashima C Y and De Oliveira Junior S 2016 Modeling and simulation of a twin screw multiphase pump, considering gradual last chamber opening and liquid recirculation between suction and discharge *Proc. of the Int. Offshore and Polar Engineering Conf.* pp 279-290

Computer simulation of heat transfer processes in aircraft landing gear braking mechanisms

E A Deren, A E Volkhonsky and D V Konoval

Moscow Aviation Institute (National Research University), 4, Volokolamskoe highway, Moscow, 125993, Russia

E-mail: eaderen@mai.education

Abstract. Landing gear brakes – multi-disc brakes, as the responsible mechanism of the aircraft, must have the lowest possible weight and dimensions, high performance, increased service life and reduced life cycle cost. Solving these problems will reduce the design and development cycle of multi-disc chassis brakes. To perform the calculations, a computer mathematical model based on the use of the finite element method is proposed, replacing the classical version of the calculation based on currently used mathematical models, taking into account the heating of the aircraft landing gear braking mechanism, which will increase the speed, accuracy of calculations and reduce their labor intensity. We have developed a computer mathematical model of heat transfer processes occurring in the friction elements of multi-disc landing gear braking systems at the second stage of landing using the COMSOL Multiphysics program. Based on the calculations performed using this model, it is possible to take into account the influence of various system parameters such as the number of friction disks, their diameter and thickness, the medium in the interstitial space, the size of the solver grid on the maximum heating temperature of the brake mechanism, the cooling rate, the calculation time and the calculation error. On the basis of the obtained experimental data, we can conclude about the preferable construction of the brake mechanism. For example, it is possible to produce friction plates, steel with application to the surface of the composite friction material, allowing it to maintain the necessary friction and thus reduce the dimensions and maximum temperature, wet-disc brakes. The results of this work can be applied not only in the aviation industry, but can also be used to calculate the mechanisms of multi-disc oil-loaded brakes, which are used in agricultural, forestry, construction, quarry and other special equipment. In future work, the developed model will allow carrying out calculations taking into account the changes of material properties depending on temperature, refining the geometry of the brake mechanism, taking into account periods of reverse thrust engines and aerodynamic brakes.

1. Introduction

The design of the chassis of the modern mainline aircraft is subject to the mandatory requirements imposed by the aviation regulations AR25 [1] (paragraph 25.735. "Brakes and braking systems"), which are fundamental when creating it and are the main source data for its mathematical modeling.

Disc brakes of the aircraft landing gear are the most important systems that ensure the safe completion of the flight and are a complex set of mechanical devices connected by the necessary connections [2-7]. This complex should allow you to optimize the braking process in conditions of changing a large number of different parameters. Consideration of the aircraft braking process during the post-landing run is fundamental in the design, development, and improvement of brakes and braking systems. Based on the analysis of physical processes occurring at the time of wheel braking, it is possible to determine the main design parameters of the friction discs of brakes, increase their energy consumption, which will reduce the braking distance, braking time, etc.

In the design of the landing gear wheels of the modern aircraft, disc brakes are used to reduce the length of the run when landing, which consume up to 95% of the kinetic energy [2]. Rotating and friction discs of such brakes must be made of materials with increased friction properties to increase energy consumption. The temperature on the surface of the friction elements during heavy braking can exceed 1000°C, which imposes restrictions on the operating modes of the aircraft. Therefore, the design and selection of disk materials is a responsible task, because they must provide heat removal from the friction surfaces, which would not lead to overheating of the entire structure during prolonged braking.

Mathematical modeling aims to provide opportunities to study the influence of various parameters of the chassis design on the braking process, the choice of its rational design. And it minimizes the braking distance on the runway when designing and certifying the parts for the compliance with AP25 standards during normal operation and design failures.

A model was developed, considering the processes of heat transfer in the friction elements of disc brakes at the second stage of landing.

The work was performed in the universal numerical simulation environment COMSOL Multiphysics, designed for modeling and calculating scientific and engineering problems based on partial differential equations (PDE) by the finite element method (FEM). PDE coefficients are set as physical properties and conditions, and their conversion to coefficients of mathematical equations occurs automatically. This program also supports loading the already created construction geometry from all common programs, such as SIEMENS NX, CATIA, SolidWorks, and AutoCAD.

The presented mathematical model allows studying the braking process when various parameters of its elements change within a wide range, as well as the parameters of the aircraft itself, the runway condition, etc.

2. Analysis of current research and publications

To describe the processes of heat transfer at the interface of solids, we can derive an equation from the localized heat balance equation written in the form for a spatial frame:

$$\rho\frac{\partial E}{\partial T}+\rho\boldsymbol{u}\nabla E+\nabla(q+q_r)=\sigma{:}D+Q. \tag{1}$$

It shows that changes in internal energy over time are balanced by internal energy convection, thermal conductivity, radiation, mechanical stress dissipation, and additional bulk heat sources. In the entry for a solid, the right-hand side is thermoelastic damping and accounts for thermoelastic effects in solids [8].

By means of transformations, we can derive the equation we are looking for that describes the processes of heat transfer at the interface of solids:

$$\rho C_p\left(\frac{\partial T}{\partial t}+\boldsymbol{u}_{trans}\nabla T\right)+\nabla(q+q_r)=-\alpha T{:}\frac{\partial S}{\partial t}+Q \tag{2}$$

or

$$\rho C_p\frac{\partial T}{\partial t}+\rho C_p\boldsymbol{u}_{trans}\nabla T+\nabla q=Q_{ted}+Q \tag{3}$$

where ρ – density [kg/m^3]; C_p – specific heat at constant pressure [j/(kg·K)]; T – absolute temperature [K]; $\boldsymbol{u}_{trans}$ – translational velocity vector [m/s]; $q=-k\nabla T$ – heat flow due to the conduction effects [W/m^2]; k – thermal conductivity [W/(m·K)]; q_r – heat flow due to radiation effects [W/m^2]; α – coefficient of thermal expansion [1/K]; S – second Piol–Kirchhoff stress tensor [Pa]; Q – constant that characterizes additional heat sources [W/m^3].

Thermoelastic damping that takes into account thermoelastic effects in solids:

$$-\alpha T{:}\frac{\partial S}{\partial t}=Q_{ted}. \tag{4}$$

By default, the boundary condition for all heat transfer interfaces is written as:

$$-nq=0. \tag{5}$$

This boundary condition means that there is no heat flow across the boundary, i.e. the temperature gradient across the boundary is zero. For this to be true, the temperature on one side of the border must be equal to the temperature on the other side. Since there is no temperature difference at the border, heat cannot be transferred through it. As a default boundary condition, it can only be applied to external borders; but when added manually, it can also be applied to internal borders.

In our model, we do not consider the landing gear braking mechanism as a thermally insulated system, but due to the presence of air blowing the brake, with a speed in the first approximation equal to the speed of the aircraft, heat exchange with the environment occurs due to external forced convection. To do this, we redefined the boundary condition for all heat transfer interfaces and wrote it as:

$$-nq = h(T_{ext} - T) \tag{6}$$

$$h = \begin{cases} 2\dfrac{k0.3387\sqrt[3]{Pr}\sqrt{Re_L}}{L\left[1+\left(\dfrac{0.0468}{Pr}\right)^{\frac{2}{3}}\right]^{0.25}} \; if \; Re_L \leq 5\cdot 10^5 \\ 2\dfrac{k}{L}\sqrt[3]{Pr}[(0{,}037Re_L)^{0.8} - 871] \; otherwise \end{cases} \tag{7}$$

where T_{ext}– external temperature [K]; $Pr = \frac{\mu C_p}{k}$ – Prandtl number; $Re_L = \rho\frac{\vartheta L}{\mu}$ – Reynolds number; k – air thermal conductivity [W/(m·K)]; L – characteristic size, in our case – the radius of the brake disc [m]; μ – dynamic viscosity [Pa·s]; ϑ – speed of the plane [m/s]; C_p – specific heat at constant pressure [J/(kg·K)]; ρ – air density [kg/m^3]; h – average heat transfer coefficient for the plate under external forced convection.

In addition to forced convective heat exchange, it is necessary to take into account that bodies whose temperature exceeds the environment have heat exchange by radiation. To account for this in our mathematical model, we redefined the boundary conditions for all external surfaces and wrote it as:

$$-nq = \varepsilon\sigma\left(T_{amb}^4 - T^4\right) \tag{8}$$

where T_{amb} – ambient temperature; $\sigma = 5{,}670367(13)\cdot 10^{-18}$ – Stefan–Boltzmann constant [W/(m^{2}K^4)]; ε – degree of blackness of the body.

It is worth noting that redefining the boundary conditions for the external boundaries of the body under study in order to account for heat exchange caused by the effects of forced convection and radiation do not contradict each other and the presence of one condition does not cancel the other. That is, the presence of convective heat exchange does not exclude the presence of heat exchange by radiation.

When the disks of the aircraft landing gear brake system are brought into contact, due to their design, i.e. the alternation of fixed friction disks that play the role of brake pads with rotating disks that are analogs of the rotor, the kinetic energy of the aircraft movement is converted into thermal energy of friction. To take into account the presence of thermal contact, i.e. heat exchange between disks, in our mathematical model, we need to assign boundary conditions for the contact surfaces in the form:

$$\begin{cases} -n_d(-k_d\nabla T_d) = -h(T_u - T_d) + rQ_b \\ -n_u(-k_u\nabla T_u) = -h(T_u - T_d) + (1-r)Q_u \end{cases} \tag{9}$$

where u – index denoting the upper side of the thermal contact surface; d – index denoting the lower side of the thermal contact surface.

These equations characterize the dependence of heat fluxes at the upper and lower boundaries on the temperature difference, and also have a correlation estimate of the joint conductivity on two contacting surfaces:

$$h = h_c + h_g + h_r\,. \tag{10}$$

The joint conductivity h consists of three parts: the conductivity of the narrowing h_c from the contact spots, the conductivity of the gap h_g due to the presence of an external environment (air) in the interstitial space, and the radiation conductivity h_r, since at the microscopic level, the contact is made with a finite number of points.

Microscopic surface asperites (irregularities) are characterized by their average height $\sigma_{asp,u}$ and $\sigma_{asp,d}$ and the average slope $m_{asp,u}$ and $m_{asp,d}$ for the upper and lower surfaces, respectively. To make further calculations, it is necessary to introduce the concepts of root-mean-square values σ_{asp} and m_{asp}, which can be calculated using the following formulas, respectively:

$$\sigma_{asp} = \sqrt{\sigma_{asp,u}^2 + \sigma_{asp,d}^2} \tag{11}$$

$$m_{asp} = \sqrt{m_{asp,u}^2 + m_{asp,d}^2}. \tag{12}$$

The constriction conductivity was calculated using the Cooper–Mikich–Jovanovich correlation, which is valid for isotropic rough surfaces and was formulated using a model that assumes plastic deformation of surface roughnesses [9]. However, this model does not calculate or store plastic deformations of asperites. This means that, despite the fact that plastic deformation of asperites (irregularities) is assumed, this contact model has no memory. For example, if the load is applied twice, the thermal contact is identical in both cases. The Cooper–Miki–Jovanovich correlation says that the constriction conductivity depends on the root-mean-square values of asperites and the pressure load at the contact point:

$$h_c = 1.25k_{contact}\frac{m_{asp}}{\sigma_{asp}}\left(\frac{p}{H_c}\right)^{0.95} \tag{13}$$

where H_c – microhardness of the softer material; p – contact pressure; $k_{contact} = \frac{2k_u k_d}{k_u + k_d}$ – average harmonic value of contact conductivities.

If k_u or k_d (contact conductivities of the upper and lower surfaces, respectively) is not isotropic, they are replaced with their normal conductivities.

The relative pressure p/H_c can be estimated either by specifying the H_c value directly or by using the following relations:

$$\frac{p}{H_c} = \left(\frac{p}{c_1\left(1.62\frac{\sigma_{asp}}{\sigma_0}m_{asp}\right)^{c_2}}\right)^{(1+0.071c_2)^{-1}}. \tag{14}$$

The coefficients c_1 and c_2 are the Vickers correlation coefficient and the size index, respectively, and σ_0 is 1 μm. For materials with Brinell hardness between 1.3 and 7.6 GPa the Vickers correlation coefficients can be calculated from the following relations:

$$\frac{c_1}{H_0} = 4 - 5.77\frac{H_B}{H_0} + 4\left(\frac{H_B}{H_0}\right)^2 - 0.61\left(\frac{H_B}{H_0}\right)^3 \tag{15}$$

$$c_2 = -0.37 + 0.422\frac{H_B}{c_1} \tag{16}$$

where H_B – Brinell hardness; $H_0 = 3.178$ GPa.

It is worth noting that if we were to calculate the rough surface and assume the presence of elastic deformations of surface roughnesses, then it would be necessary to use the Mikich correlation to determine the constriction conductivity:

$$h_c = 1.54k_{contact}\frac{m_{asp}}{\sigma_{asp}}\left(\frac{\sqrt{2}p}{mE_{contact}}\right)^{0.94}. \tag{17}$$

Here $E_{contact}$ is an effective young's modulus for a contact satisfying the relation:

$$\frac{1}{E_{contact}} = \frac{1-\nu_u^2}{E_u} + \frac{1-\nu_d^2}{E_d} \tag{18}$$

where E_u and E_d – Young's modules of two contacting surfaces; ν_u and ν_d – Poisson coefficients.

The gap conductivity must be taken into account if the medium (liquid) in the interstitial space has a high temperature conductivity or there is a large contact pressure. The correlation of the gas gap of parallel plates implies that the liquid in the interstitial space is a gas (air) and defines h_g as:

$$h_g = \frac{k_g}{Y + M_g} \tag{19}$$

where k_g – gas conductivity; Y – average thickness of the separation; M_g – gas parameter, equal to:

$$M_g = \alpha\beta\Lambda, \qquad \Lambda = \frac{k_B T_g}{\sqrt{2}\pi D^2 p_g} \tag{20,21}$$

where α – local thermal contact parameter; β – gas parameter (1.7 for air); Λ – average path length of a gas molecule; k_B – Boltzmann's constant; D – average diameter of the gas particle; p_g – gas pressure; T_g – temperature gap is equal to:

$$T_g = \frac{T_u + T_d}{2}. \tag{22}$$

The average separation thickness Y is a function of the contact pressure p. For small values of p (about 0 Pa) Y goes to infinity, because no contact occurs. For large values of p is greater than $H_0/2$ in the Cooper–Mikich–Jovanovich model and greater than $H_0/4$ in the elastic Mikich model – Y decreases to zero, this means that the contact is considered perfect.

At high temperatures above 600 degrees Celsius, radiation conductivity must be taken into account. The gray-diffusion parallel plate model defines h_c as:

$$h_c = \frac{\varepsilon_u \varepsilon_d}{\varepsilon_u + \varepsilon_d - \varepsilon_u \varepsilon_d}\sigma\left(T_u^3 + T_u^2 T_d + T_u T_d^2 + T_d^3\right) \tag{23}$$

or

$$h_r(T_u - T_d) = \frac{\varepsilon_u \varepsilon_d}{\varepsilon_u + \varepsilon_d - \varepsilon_u \varepsilon_d}\sigma\left(T_u^4 - T_d^4\right) \tag{24}$$

$$h_r(T_d - T_u) = \frac{\varepsilon_u \varepsilon_d}{\varepsilon_u + \varepsilon_d - \varepsilon_u \varepsilon_d}\sigma\left(T_d^4 - T_u^4\right). \tag{25}$$

To describe thermal friction in our mathematical model, we divided the heat of friction Q_b into $rQ_b$and $(1 - r)Q_b$ in the heat contact zone. If both bodies are identical, then r and $(1 - r)$ are 0.5, so that the heat of friction is evenly distributed between the two surfaces. In general, when both bodies are made of different materials, the partition coefficient may not be equal to 0.5. the value of r and $(1 - r)$ is determined from the Sharron ratio:

$$r = \frac{1}{1 + \xi_d}, \qquad \xi_d = \left(\frac{\rho_u C_{p,u} k_u}{\rho_d C_{p,d} k_d}\right)^{0.5} \tag{26,27}$$

$$(1 - r) = \frac{1}{1 + \xi_u}, \qquad \xi_d = \left(\frac{\rho_d C_{p,d} k_d}{\rho_u C_{p,u} k_u}\right)^{0.5}. \tag{28,29}$$

If k_u or k_d (contact conductivities of the upper and lower surfaces, respectively) are anisotropic, they are replaced with their normal conductivity values.

3. Mathematical modelling

To test the performance of the mathematical model, we created a simplified geometry of the brake discs of the main landing gear struts in the SIEMENS NX 10 program. After that, we loaded it into the COMSOL Multiphysic program, pre-configuring the 3D space dimension, physics-thermal conductivity in solids, and the time-dependent solver. Next, we set the following properties of materials for rotating and friction disks: density, thermal conductivity, heat capacity at constant pressure, and the degree of blackness of the body. To describe the model, it is necessary to specify the following system parameters: the speed of the aircraft to braking, acceleration, deceleration, radius of wheel landing gear, the weight of the aircraft on landing, the ambient temperature, time of braking and the friction disks. The next step is to set the speed functions of the aircraft, considering the deceleration to be equidistant, and the speed

IOP Conf. Series: Materials Science and Engineering 1111 (2021) 012018 doi:10.1088/1757-899X/1111/1/012018

of the aircraft before and after deceleration to be constant. At the same time, we set the existence of the second derivative and the size of the smoothing zone to 0.2. This is necessary for the existence of the acceleration function, which we constructed as a derivative of the velocity function.

After that, we set the motion function of the rotating brake discs of the landing gear, redefined the boundary conditions by setting the microhardness, pressure in the contact zone, and heat release intensity, taking into account that not all the kinetic energy of the aircraft's movement is absorbed by the landing gear braking system. The last step in setting up a mathematical model is to set a grid, given that the smaller the grid elements, the more accurate the calculations are, but the more time they require. For example, all other things being equal, the solution time for the coarser grid sizes was 3 minutes, and for the extremely fine sizes it was 4 hours, 17 minutes, and 5 seconds.

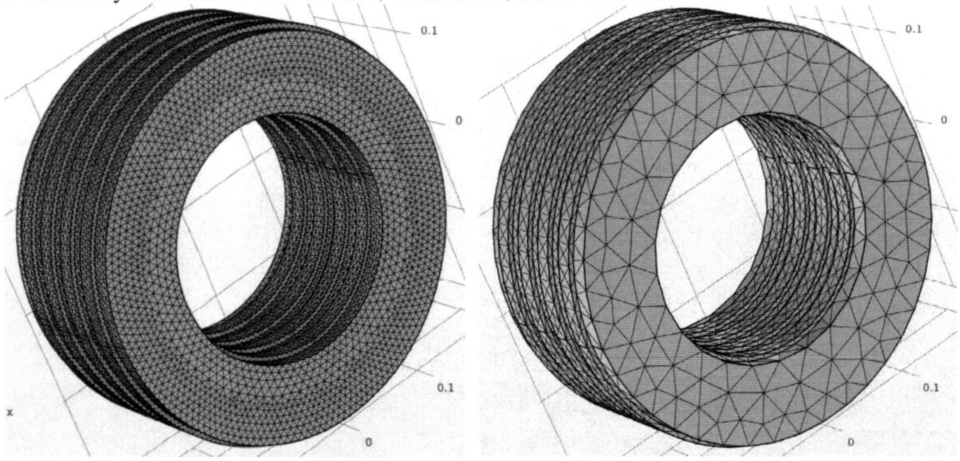

a b

Figure 1. Automatically generated grids with settings: a – extremaly fine; b – coarser.

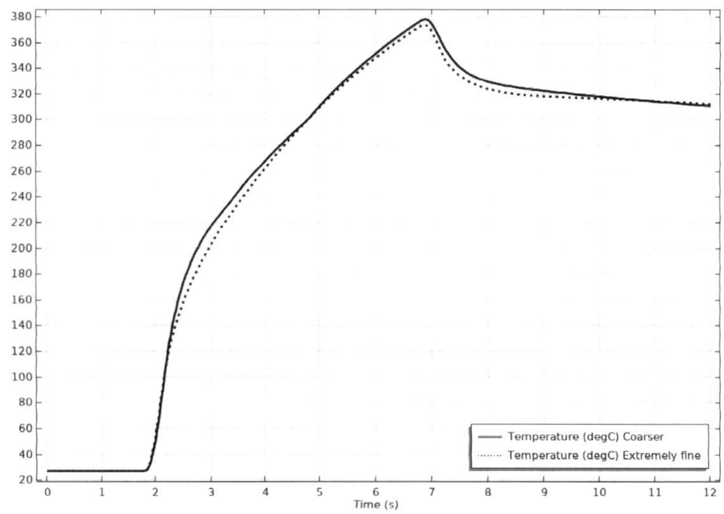

Figure 2. Graphs of changes in the maximum heating temperature over time depending on the solver grid settings

4. Results and Discussion

The proposed model considers the processes of heat transfer in the friction brakes at the second stage of aircraft landing. Also, by changing various parameters, we qualitatively tracked their impact on the system [10, 11]. Analyzing the obtained graphs of changes in the maximum temperature from time to time of the disks of the braking system of the aircraft landing gear, we can draw the following conclusions.

When choosing a grid solver, we need to find a compromise between the desired accuracy of calculations and the available time that we can or want to spend on calculations.

It is quite clear that as the number of friction disks increases, the maximum heating temperature decreases, but with each new disk added, the effect that it has an the overall decrease in the heating temperature becomes smaller. Therefore, when designing, it is necessary to find a balance between the allowed maximum heating temperature and the dimensions/weight of the aircraft chassis.

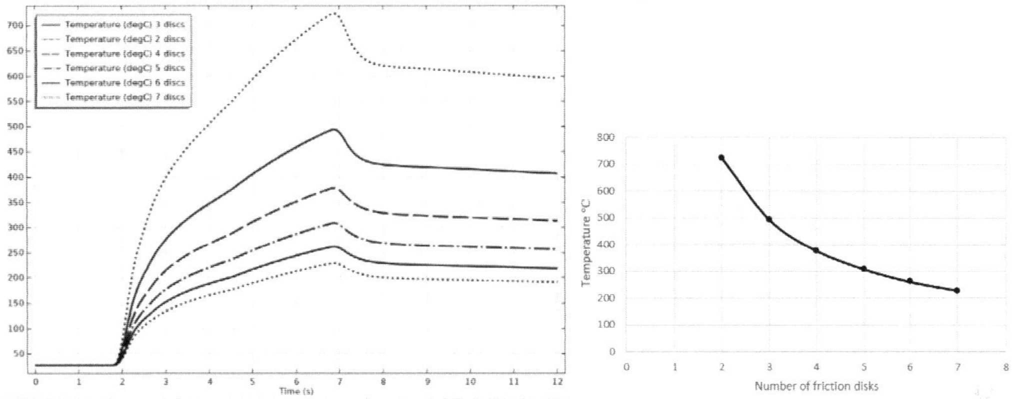

a b

Figure 3. Graphs: a – changes in the maximum heating temperature over time depending on the number of friction discs; b – dependences of the maximum heating temperature on the number of friction discs

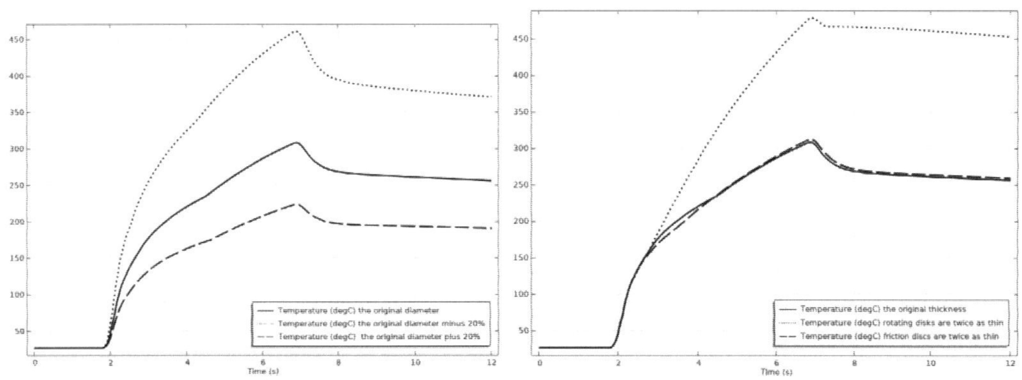

a b

Figure 4. Graphs of changes in the maximum heating temperature of the chassis brake system disks depending on: a – diameter of the disks; b – thickness of the movable and friction disks

Reducing the diameter of the brake discs increased the maximum temperature by 49.7%, while increasing the diameter by 20% reduced the maximum temperature by only 27.4%. Therefore, it is not always effective to go by increasing the diameter of the brake discs to reduce the temperature – the more we increase the diameter, the more the dimensions/weights of the chassis increase and the less the maximum temperature decreases. You need to find a balance between these characteristics.

When the thickness of the friction disks is reduced 2 times, the maximum temperature value increases slightly, but when the thickness of the movable disks is reduced by the same number of times, the maximum temperature increases significantly (Fig. 2). Due to the fact that mobile disks are made of steel heat capacity and thermal conductivity at constant pressure which is much higher than that of cast iron, from which the friction discs are made. If you use composite materials, replacing cast-iron discs with steel ones that will be coated with friction composite material, we can reduce the overall weight of the structure and reduce the maximum heating temperature of the brake discs.

The environment in which the brake mechanism is located has a significant impact on the maximum heating temperature and cooling rate. Provided that all other parameters are constant, and only the environmental parameters change, the maximum heating temperature of the brake mechanism in water will be lower, and cooling will go faster. When braking in transformer oil, the maximum temperature is also lower than that in air, and cooling is faster, but still not comparable to the indicators in water. This has found application, for example, in motor graders. In their balancing cart, multi-disc oil-loaded brakes are installed on the shaft connecting the driven gear of the on-board gearbox to the wheel hub.

5. Conclusion

Our mathematical model allows us to optimize the dimensions and weight of the braking mechanism, select materials so that when braking, the maximum heating temperature is in the range of acceptable values, which prevents rapid wear of the structure and increases the life cycle of the aircraft.

Further refinement of our model will be due to the fact that part of the kinetic energy at the landing stage of the aircraft is absorbed, for example, by the engine thrust reversal or brake pads, and, as a result, part of the kinetic energy absorbed by the brake is not constant and changes during braking. When taking into account the period of operation of the engine thrust reverse and brake flaps, it is necessary to specify the functions of acceleration of braking and speed of the aircraft. The geometry of the braking mechanism will be improved, taking into account the use of composite materials in the design, the fuel consumption during the braking process. This leads to changes in the mass of the aircraft, as well as the properties of materials such as thermal conductivity, heat capacity at constant pressure, microhardness, coefficient of friction, density, and the degree of blackness of the body, which change depending on temperature.

References

[1] 1994 *Aviation regulations. Part 25. Airworthiness standards for transport category aircraft* (Moscow: M Gromov LII) 237 p
[2] Kokonin S S, Kramarenko E I and Matveenko A M 2007 *Fundamentals of designing aircraft wheels and brake systems* (Moscow: Moscow Aviation Institute) p 264
[3] Antonyuk V E, Goman A M, Zabolotskii M M and Rudyi V V 2009 *J. of Friction and Wear* **30(5)** 329–336
[4] Polyakov P A, Tagiev R S, Fedotov E S, Polyakova E A and Moskalenko M B 2020 *Bull. of Irkutsk State Technical University* **24(1)** 64-76
[5] Tarasik V P 2015 *Bull. of the Belarusian-Russian University* **3(48)** 68-78
[6] Pogosyan M A, Liseitsev N K, Strelet D Y, Kiselev V A, Tarasov A Z, Dolgov O S, Dudar E N, Timoshenko V P, Eremin N A, Penkov E A et al 2018 *Aircraft design*, ed Pogosyan M A (Moscow: Moscow Aviation Institute) p 864
[7] Anfinogenov S B and Balakin V A 2004 *Bull. of the Gomel State Technical University named after P. O. Sukhoi* **3(16)** 9–16
[8] Mauthorgin A 1999 *Thermomechanics of Nonlinear Irreversible Behaviors: An Introduction* p 391
[9] Cooper M, Mikic B and Yovanovich M 1967 *Int. J. of Heat and Mass Transfer* **12(3)** 279–300
[10] Deren E A and Tikhonova A Y 2020 *Gagarin readings – 2020: Coll. of abstracts* pp 115-117
[11] Deren E A and Konoval D V 2020 *Youth. Technic. Cosmos: Book of abstracts of the twelfth all-Russian youth STC* **1(4)** 91-100

Using the wave algorithm for optimal route construction

K Zhigalov[1], E Klochkova[2], T Ya Druzhinina[3], M Yu Vakhrusheva[4] and D K-S Bataev[5]

[1] V.A. Trapeznikov Institute of Control Sciences of Russian Academy of Sciences, Moscow, Russia
[2] Moscow University of the Ministry of Internal Affairs of the Russian Federation n.a. V. Ya. Kikotya, Moscow, Russia
[3] Irkutsk National Research Technical University, 83, Lermontova str., Irkutsk, 664074, Russia
[4] Bratsk State University, 40, Makarenko str., Bratsk, 665709, Russia
[5] Integrated Research Institute named after Kh.I. Ibragimov, Russian Academy of Sciences of the Chechen Republic, 21A, Staropromyslovsky ave., Grozny, 364051, Russia

E-mail: kshakalov@mail.ru

Abstract. Integration of modern routing algorithms in logistics is aimed at minimizing costs. The formation of the most effective solutions must take into account the complexity and the variety of types of tracks. It is advisable to optimize the route only after localizing the square of movement, analyzing the terrain, developing the infrastructure of the sector, as well as taking into account the technical characteristics of the declared vehicle and the construction algorithm. The study of the principles of the hierarchical approach in the formation of the route justifies the use of the wave algorithm in the formation of the optimal route on a digital map that meets multifactor criteria: determining the direction of movement of an object by sector, calculating the resources spent on overcoming a specific area of terrain and displaying the results of movement-minimizes costs when building routes.

1. Introduction

The task of finding optimal routes depending on the level of transport costs is becoming more urgent every year. New algorithms for solving this problem are constantly being proposed and developed [1]. The tasks of transport logistics are extremely diverse. Examples of such tasks can be planning long-distance cargo transportation with regard to ensuring its quality and safety, planning the movement of vehicles in the city depending on the load of routes, managing the movement of specialized transport, when the optimal criterion is to minimize the time of arrival at the destination, etc [2-4].

Despite numerous studies in this area, the task of tracing (modeling) routes continues to be relevant and one of the most time-consuming. This is primarily due to the fact that there is a wide variety of implementations of different routes, for each of which it is necessary to apply specific optimization criteria and restrictions [5, 6]. From a mathematical point of view, route modeling is the most difficult task of choosing the optimal solution from a huge number of options.

This article discusses the problem of planning an optimal route of the vehicle in the general case. On the map of the area, you need to determine the route from the start point to the destination point. In this case, the path is laid taking into account the technical characteristics of the vehicle and the possibility of moving on roads with different characteristics.

Mathematically, a graph model can be used to describe the existing road network. The graph structure may change as a result of various factors. The vertices are a set of intersection points of highways (intersections, interchanges, etc.), and arcs are a set of roads. Each of the edges will have its own weight, which depends directly on the road itself.

As a rule, a graph describing the structure of a road network has a significant number of vertices and edges, some of which do not meet the requirements of resource consumption. Thus, it is possible to reduce the calculations performed by first excluding them from consideration.

2. Terrain map elements

When talking about building a route, you can't help but focus on the maps you use. To implement the algorithm, we will use digital maps, which are a coded set of spatial objects – material or abstract objects of the real or virtual world, and at the same time its digital representation (model), containing its location and a set of properties (characteristics, attributes). To organize the display of spatial information, a layered principle is used, the essence of which is to form thematic layers that include all the variety of information about the displayed territory. Layer-by-layer data organization assumes that layers in space do not have breaks, and there is some information everywhere, at least "there is no data about the presence or absence of objects". In this case, the layers do not necessarily have to correspond exactly to the thematic division.

Data on electronic vector maps has the following structure:

- passport data about the map sheet (scale, projection, coordinate system, rectangular and geodetic coordinates of the corners of the sheet, and so on);
- metric data of map objects (coordinates of objects in the area);
- semantic data of map objects (various properties of objects);
- information about conditional signs (rules for displaying map objects).

Such a variety of types of localization of map objects is necessary for a more visual description of terrain objects. Electronic map objects can be combined into vector map layers based on some characteristics. A description of the types of vector map objects, semantic characteristics (properties, attributes) of objects, layers that combine objects, and symbols used in forming an electronic map on graphic devices is stored in the digital classifier (resource file) of the electronic map. Description of object types and semantic characteristics contains information about the encoding system (classification) of objects, characteristics, and their values.

Using digital maps allows you to get various information about the earth's surface-terrain, type of soil, etc., necessary for assessing the patency of certain sections of the route.

3. The description of the wave algorithm

To solve the problem of finding the optimal route, we plan to use a wave algorithm, the initial data for which is a digital map of the area, presented in the form of a raster model of spatial data – matrix $Z=z(x, y)$.

Each cell corresponds to a square plot on the terrain, taking into account the scale (the coordinates of the plot are determined by its central point and are equal to (x, y)). To describe the earth's surface, a generalized cross-country indicator of this section is used. It depends on the parameters of a specific area of terrain (altitude, waterlogging, the presence of natural and artificial barriers, etc.) and is stored for each element of the matrix.

The set of center points of cells in a grid is a grid that supports accurate calculations of the cross-country capability of a given terrain. We will search for the shortest path from the destination point to the final one using a wave algorithm. From the starting point, a so-called wave is formed, which can propagate in all directions. Each element (cell) that received the primary wave generates a secondary wave, and these elements later become the source for the third wave, etc.

To account for the movement of the wave, each cell visited by the wave is marked as passed. The wave propagation is possible only for cells that have not yet been passed and are "passable".

The movement of the wave continues until the destination is reached, after which the route is laid (according to the coordinates of the passed cells on the plane). If the wave has passed all the available cells, and the final destination is not reached, then we can say that it is impossible to lay the desired route.

The wave can move in one of eight directions. At the same time, it is necessary to take into account that moving along the diagonal of the site is more expensive and it is necessary to introduce corrections for such movement. We will assume that the coefficient that takes into account the ratio of costs for a direct intersection of the cell to the diagonal is 2/3 (assuming that the area that we use for routing is divided into square elements). An example of constructing a wave propagation matrix is shown in Figure 1.

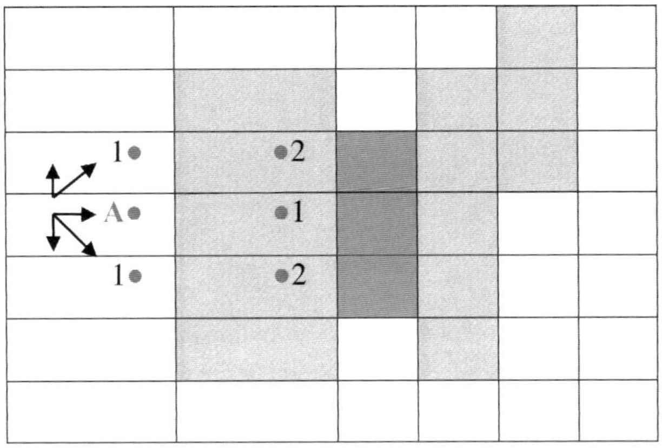

Figure 1. Example of building a wave propagation matrix (Propagation of the first wave front)

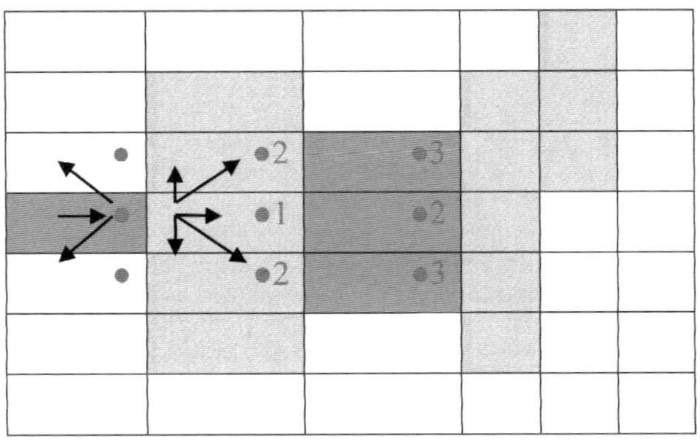

Figure 2. Example of building a wave propagation matrix. (Wave propagation matrix)

The second stage of the algorithm is the direct construction of a possible route for the vehicle, which must take into account the following conditions:

- the route is calculated for user-defined vehicle specifications;
- the route is built taking into account the requirements for road surface and road quality (only paved roads, the possibility of using unpaved roads, etc.);
- traffic can only be carried out on permitted areas of the terrain.

When building a route, we assume that the transition between adjacent cells is the transition between the central points of these sections. We will determine the cost of overcoming the site:

zs=g(S)+h(S), where g(S) is the actual cost of overcoming a certain section S by a given technical means, and h(S) is a function that allows you to estimate the cost of the best sequence of actions from this route point to the destination point.

As an example, we suggest using the following estimate, which depends on the geometric distance from the starting point to the destination.

$$h(S) = R(x\ ,y\ ,x_{\kappa}, y_{\kappa}) \cdot \langle g(S) \rangle = \left(\sqrt{(x_{\kappa} - x)^2 + (y_{\kappa} - y)^2} \right) \cdot \langle g(S) \rangle$$

The total cost of traversing the route is determined by the sum of the costs for all the areas involved: $Q = \sum_{i=1}^{N-1} z_i$, where N is the number of route points. When building the optimal route, you must choose the direction of further movement at each step. The result of the selection at each step is denoted by SC, i.e. the control of all movements can be described as a matrix of step controls C: C=(c1, c2, c3, ….st-1). Any constructed route can be described using the step control matrix C and the cost matrix for overcoming it Q.

4. Dijkstra's algorithm in C++ and python

For C++ we must enter the following data types:

- array of visited vertices (visited);
- path matrices (w);
- array of minimal paths (D);
- maximum size of the " int " type (INT_MAX);
- and the source number (st).

The source code has the following view:

```
void Dijkstra(int n, int st)
{
  vector<vector<int >> w;
  w.resize(n);
  for (int i=0;i<n;i++)
    w[i].resize(n);
  bool visited[n];
  int D[n];
  for(int i=0;i<n;i++)
  {
    D[i]=w[st][i];
    visited[i]=false;
  }
  D[st]=0;
  int index=0,u=0;
  for (int i=0;i<n;i++)
  {
    int min=INT_MAX;
    for (int j=0;j<n;j++)
    {
      if (!visited[j] && D[j]<min)
      {
        min=D[j];
        index=j;
      }
    }
    u=index;
    visited[u]=true;
    for(int j=0;j<n;j++)
    {
```

```
            if (!visited[j] && w[u][j]!=INT_MAX && D[u]!=INT_MAX && (D[u]+w[u][j]<D[j]))
            {
              D[j]=D[u]+w[u][j];
            }
          }
        }
        cout<<" The cost of the path from the initial vertex to the rest (Dijkstra's Algorithm):\t\n";
        for (int i=0; i<n; i++)
        {
          if (D[i]!=INT_MAX)
            cout<<st<<" -> "<<i<<" = "<<D[i]<<endl;
          else
            cout<<st<<" -> "<<i<<" = "<<" the route is not available "<<endl;
        }
      }
```

For Python we must enter the following data types:
- the number of vertices (N);
- the number of the starting vertex (S);
- adjacency matrix of the original graph (matrix).
The source code has the view:

```
def Dijkstra(N, S, matrix):
  valid = [True]*N
  weight = [1000000]*N
  weight[S] = 0
  for i in range(N):
        min_weight = 1000001
        ID_min_weight = -1
        for i in range(N):
              if valid[i] and weight[i] < min_weight:
                    min_weight = weight[i]
                    ID_min_weight = i
        for i in range(N):
              if weight[ID_min_weight] + matrix[ID_min_weight][i] < weight[i]:
                    weight[i] = weight[ID_min_weight] + matrix[ID_min_weight][i]
        valid[ID_min_weight] = False
  return weight
```

NavMesh technology, which implements navigation mechanisms for AI agents, cannot be ignored. The source code that defines the navigation logic is presented as:

```
using UnityEngine;
using UnityEngine.AI;
public class Target : MonoBehaviour
{
  private NavMeshAgent[] navAgents;
  public Transform targetMarker;
  private void Start ()
  {
   navAgents = FindObjectsOfType(typeof(NavMeshAgent)) as NavMeshAgent[];
  }
  private void UpdateTargets ( Vector3 targetPosition )
  {
   foreach(NavMeshAgent agent in navAgents)
```

```
    {
      agent.destination = targetPosition;
    }
  }
  private void Update ()
  {
    if(GetInput())
    {
      Ray ray = Camera.main.ScreenPointToRay(Input.mousePosition);
      RaycastHit hitInfo;

      if (Physics.Raycast(ray.origin, ray.direction, out hitInfo))
      {
        Vector3 targetPosition = hitInfo.point;

        UpdateTargets(targetPosition);
        targetMarker.position = targetPosition;
      }
    }
  }
  private bool GetInput()
  {
    if (Input.GetMouseButtonDown(0))
    {
      return true;
    }
    return false;
  }
  private void OnDrawGizmos()
  {
    Debug.DrawLine(targetMarker.position, targetMarker.position + Vector3.up * 5, Color.red);
  }
}
```

5. Using the wave algorithm for constructing the optimal route

The work of the algorithm for finding the optimal path can be presented in the following form. We start the movement from the starting point A. Since the algorithm is just beginning its work, it is necessary to immediately add the starting point to the so-called open list of cells that will be processed in the future. Initially, there is only one item in this list, but in the future it will be updated. This list is intended for storing a list of cells located along a possible path and through which the route will be laid in the future. This is a list of cells that will be subjected to additional verification.

When you select the direction of travel from the starting cell, the elements located next to it are selected. In this case, if they are passable or conditionally passable, they are added to the open list, otherwise the algorithm ignores them. For the selected elements, we make a note that point A is their "parent cell". This information is necessary in order to be able to build an optimal route in the future.

After that, the starting point is moved from the open list to the "closed list". The "closed list" will contain cells that are no longer subject to verification. The next step is to select one of the neighboring cells that were previously placed in the open list. The criterion for selecting an element is the minimum total cost of overcoming this area. After that the whole process is repeated again and in general can be described as follows:

a) select the item with the lowest cost value (z_{ij}) and make it the current one;

b) move the selected item from the open list to the closed list;

c) check each of the eight neighboring elements using the following procedure.

Check the elements remained in the closed list, if such elements are found, then ignore them. Next, check the elements for possibility; if we find impassable elements, then we ignore them, and for being in the open list, if we find elements that are not included in the open list, then they are entered in it. For these elements, we calculate the costs of $g(S)$, $h(S), z_S$.

We check the items for being in the closed list, if such items are found, we ignore them. Next, we check the elements for possibility, if we find impassable elements, then we ignore them, and for being in the open list, if we find elements that are not included in the open list, then they are entered in it. For these elements, we calculate the costs of $g(S)$, $h(S), z_S$.

If the elements are included in the open list, then we check whether the path through this cell will not be cheaper, for comparison, we use the real cost of overcoming the section- $g(S)$. A lower value of $g(S)$ means that the route will be cheaper. We select this element to include it in the calculated route, and then recalculate the costs for this cell $g(S)$, z_S. We remember the choice made in the matrix of step controls C.

d) the algorithm ends its work if an element where the destination point is located, the so-called target cell, is added to the open list, in which case we can say that the optimal route is built. Or the open list is empty, and the target cell is not reached. In this case, we say that it is impossible to build a route and there is no path.

e) the route found in accordance with this algorithm is saved using the matrix of step controls. Moving from the target point in the opposite direction, passing from each point to its parent, we reach the starting point. This will be the desired path.

The presented algorithm for finding the optimal route requires significant computational costs when implementing it, which makes it much more difficult to use it in real time. To improve the algorithm, you can use hierarchical planning, when the search for the optimal path begins with a rough approximation, and as the solution is implemented, the result is refined.

The hierarchical approach is based on using a hierarchically ordered set of maps. For each card, the maximum and minimum values of the cost function are stored. The set starts with a rough representation of the surface, the maps in the hierarchy are a more detailed (detailed) version of the previous one, and so on. Similarly, there is a complete list of all maps, in which the last one is detailed enough to solve the task.

Then the solution of the problem of finding the shortest path from A to B using a hierarchical approach can be formulated as follows.

Any C cell that is a higher-level map spans multiple cells at a lower level of detail. For each cell C, the highest and lowest value of the price function of any lower-level map cell is stored.

1. The shortest path from A to B, calculated based on the worst (highest) costs of each part of the map. This value is the upper bound of the cost of the optimal path from A to B.

2. For each cell C, the shortest path from A to C and the shortest path from C to B are calculated, using the best (lower) costs. The sum of these two values is the lower bound of the shortest path from A to B.

If the shortest path passing through C is longer than the shortest path that is guaranteed to exist, then it is obvious that the optimal path does not pass through C. Therefore, in a further search, the cell C and all its subcells can be excluded from further consideration. Then all cells that were not excluded are replaced with a more detailed version and the whole process is repeated.

It is interesting to note that this method usually excludes all nodes that are not included in some kind of ellipse formed by cell boundaries around A and B. When A and B are close, almost the entire map is excluded from further consideration at the early stages of consideration, since the ellipse between these points is extremely small.

6. Conclusion

There are different approaches to building routes, depending on the specific tasks and technical characteristics of the data center. In general, the described approach (using the wave algorithm) allows reducing the load not only on the operational recording device (RAM), but also the final calculation time (route construction).

References

[1] Mazakov E B, Matrokhina K V and Trofimets V Y 2021 *Traffic management at the enterprises of the mineral industry, in: Advances in raw material industries for sustainable development goals* (London: CRC Press) p 528
[2] Klyuev R, Madaeva M and Umarova M 2020 Mathematical modeling of specific power consumption of electrolyzers *Int. Ural Conf. on Electrical Power Engineering (UralCon)* pp 356-361. DOI: 10.1109/UralCon49858.2020.9216247
[3] Klyuev R, Turluev R and Khadzhiev A 2020 Analysis of power quality and pumping units operation's optimization *Int. Ural Conf. on Electrical Power Engineering (UralCon)* pp 324-329. DOI: 10.1109/UralCon49858.2020.9216254
[4] Klyuev R, Bosikov I, Gavrina O, Madaeva M and Sokolov A 2021 Improving the energy efficiency of technological equipment at mining enterprises *Advances in Intelligent Systems and Computing* **1258** 262-271. DOI: 10.1007/978-3-030-57450-5_24
[5] Afanas'ev A S, Egoshin A M and Alekseev S V 2018 Justification of logistical approach application in road safety management *IOP Conf. Ser.: Earth Env. Sci.* **194** 072001
[6] Afanasyev A S, Egoshin A M and Alekseev S V 2019 The choice substantiation method of road construction equipment for mining enterprises *J. of Phys.: Conf. Ser.* **1399** 33021

Experimental study for choosing an active substance in a microcracks detection system in the turbine blade

M S Grinkrug[1,2], M Kara Balli[1], J I Tkacheva[1] and N A Novgorodov[1]

[1] General Physics Department, Komsomolsk-on-Amur State University, 27 Lenin Prospect, Komsomolsk-on-Amur, Khabarovsk Territory, 681013, Russian Federation

E-mail: grin@knastu.ru

Abstract. In this work, an experiment was carried out that allows choosing an active substance in the system for early detection of microcracks in the blades of gas turbine engines. The aim of the work is to compare several active substances by measuring the signal generated by each of them at the same temperatures and concentrations of the active substance. A current signal is obtained by applying a voltage between two electrodes placed in a stream of hot gas. The strength of the electric current arising between the electrodes in the hot gas flow depends on the ionization of the gas caused by the active substance. The current signal is the main criterion for choosing an active substance. A study was also carried out on the effect of gas temperature on the strength of the signal caused by the active substance. This experiment is one of the main stages in the creation of a system for detecting microcracks in gas turbine blades.

1. Introduction

In gas turbine engines, the most loaded elements are the rotor blades of the first turbine stages. They operate at high temperatures and stresses from centrifugal forces and bending moments. Also, variable aerodynamic forces act on it. These factors lead to the risk of crack formation in them. Developing cracks are a hazard when the engine is running and can lead to an emergency situation. Timely crack detection is an important task in aircraft engine diagnostics. Currently, various methods are used for this, described in [1–6]. However, all these methods are applied in stationary conditions on the ground during engine inspections.

In works [7–9], a method for detecting microcracks in the blades of operating turbomachines is proposed and described. The proposed technology works as follows. Several thin-walled capsules with an active substance are located in the blade. At the origin of the crack, when it reaches the capsule with the active substance, the thin-walled shell of the capsule breaks and the active substance enters the flow path. After that the active substance is detected by the appearance of an electric current between the electrodes. The increase in the current is recorded with an ammeter.

2. Materials and methods

In the system for early detection of microcracks, the active substance is considered one of the most important factors affecting the success of this system and its effectiveness, since the magnitude of the recorded signal depends on it. Therefore, it is necessary to test several materials and select the most suitable one, which gives a clear and accurate indication for the formation of microcracks in the body of the turbine blade. One of the main conditions that were taken into account is the possibility of ionization of these materials at high temperatures. And when the active substance passes between the electrodes, an electric current appears in the signal detection circuit, because of the attraction of various ions to the electrodes due to the high electric field, created by a large voltage difference. An experiment was carried out to select a suitable active substance. The scheme diagram for the experiment is shown in Figure 1.

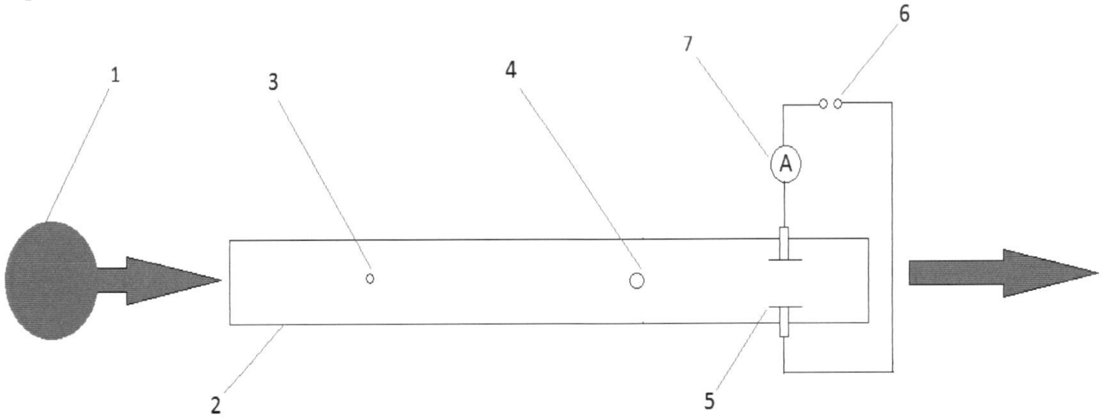

Designations: 1 - gasoline burner; 2 - pipe; 3 - place of active substance input; 4 - thermocouple; 5 - electrodes; 6 - voltage source; 7 – micrometer

Figure 1. Schematics of the experiment

The experimental stand consists of a stainless steel pipe 2 with a diameter of d = 8 cm with a thick wall h = 1 mm. A gasoline burner 1 is installed at one end of the pipe to give a hot gas flow. It can give a gas flow with a temperature of up to 800^0C. A thermocouple 4 is installed in the pipe to measure the temperature of the gas flow and electrodes 5. The pipe has a hole 3 for introducing an active substance into the gas flow. A constant voltage is supplied to the electrodes from the source 6, which can be regulated. A micrometer 7 is connected to the electrodes to register the output currents. The active substance is added to the flow part through hole 3 at the initial section of the pipe to ensure its uniform mixing with the hot gas flow. The gas flow temperature is measured by a pre-calibrated thermocouple 4 connected to a voltmeter. The value of the gas temperature was determined according to the equation $T = 11.48 \cdot V - 14$, obtained during calibration, where V is the voltage on the voltmeter.

Electrodes 5 were connected to a voltage source 6 (2000 V). When the gas flow with the introduced substance passes between two electrodes, different ions move between the two electrodes, which causes an electric current, which is measured by the micrometer 7. The image of the experimental stand is shown in Figure 2.

Figure 2. Experimental stand

We carried out six different active substances containing alkali and alkaline earth metals: potassium chloride (KCl), potassium monophosphate (KH_2PO_4), calcium nitrate ($Ca(NO_3)_2$), magnesium sulfate ($MgSO_4$), potassium nitrate (KNO_3) and sodium bicarbonate (NaHCO3). The experiments were preliminarily carried out at a temperature of 600 °C, at the same concentration of all substances, which was 15 grams of active substance per 100 grams of solvent (water), injected into the flow path. This concentration was chosen because it ensures complete dissolution of all these substances.

The volume of the injected substance was 2 cm^3 per second. The increase in the current ΔI between the electrodes was observed for 2-3 seconds. After that, the current dropped to its original value.

Electric current ΔI was measured each time when a new active substance was added. The experiment was repeated for each active substance at least seven times to ensure the results, and each time, the tube and electrodes were cleaned of any remaining active material that might have deposited on the walls of the tube and electrode.

3. Results and discussion

Table 1 and Figure 3 show the obtained results.

Table 1. Experimental results

Active substance	KCl	KH_2PO_4	$Ca(NO_3)_2$	$MgSO_4$	KNO_3	$NaHCO_3$
ΔI, μA	4	0.5	3	0.8	16	3.5

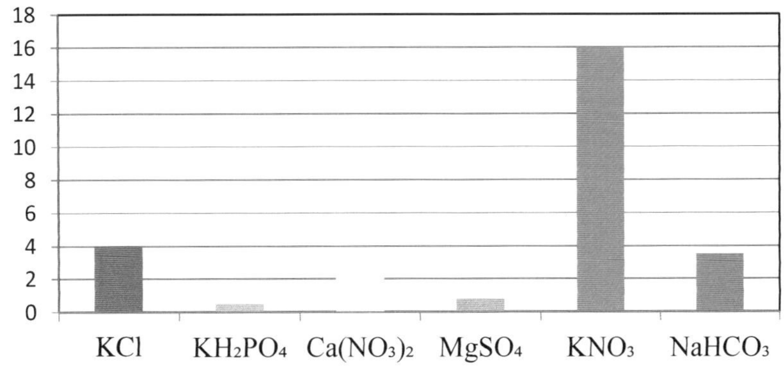

Figure 3. The magnitude of the recorded current for various active substances

Note that (KNO_3) turned out to be the most effective than other materials, since the current was at least four times higher than the other substances. This allows us to select (KNO_3) as an active substance in the early detection of microcracks in a turbine blade.

It was shown in [10] that an increase in temperature affects the durability of gas turbine blades. It leads to an acceleration of crack growth and a decrease in the strength properties of blade materials.

Therefore, the influence of the gas flow temperature on the current obtained during signal registration was investigated. In this case, a concentration of 36 grams (KNO_3) in 100 grams of water was chosen, since it provides the highest degree of solubility of this active substance. The experiment was carried out at four different values of temperature and voltage between the electrodes of 2000 V. The obtained results are presented in Table 2 and in Figure 4.

Table 2. Experimental results

Temperature (^{0}C)	500	600	700	800
ΔI, μA	2,5	21	500	2300

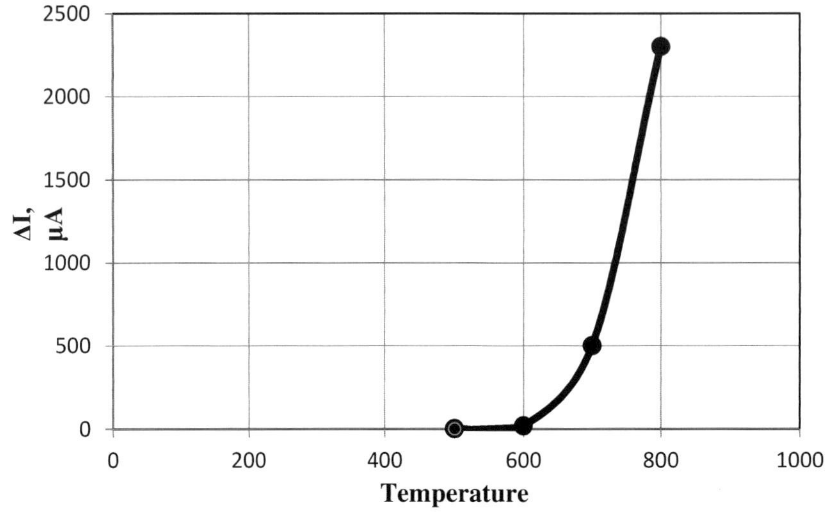

Figure 4 - Dependence of the current between the electrodes on the temperature of the gas flow

We can see from the results obtained that the temperature has a great influence on the strength of the resulting current.

4. Conclusion

Based on the research results, it can be concluded that:

1) it is advisable to use potassium nitrate as an active substance, since it gives the maximum value of the received signal;

2) the gas flow temperature is a significant factor influencing the signal value. With an increase in temperature from 600 to 800 ^{0}C, the signal value increases 110 times.

References

[1] Rani S, Agrawal A K and Rastogi V 2019 Vibration analysis for detecting failure mode and crack location in first stage gas turbine blade *J. of Mechanical Science and Technology* **33(1)** 1–10. DOI 10.1007/s12206-018-1201-x

[2] Yong Jia, Rui-min Zhang, Wei Zhang, Guo-feng Jin, Ji-fang Liu, Gan Tian and Zheng-wei Yang 2018 Rapid Detection of Cracks in Turbine Blades Using Ultrasonic Infrared Thermography *IOP Conf. Ser.: Mater. Sci. Eng.* **389** 012023. DOI: 10.1088/1757 899X/389/1/012023

[3] Rongqiao Wang, Kanghe Jiang, Fulei Jing and Dianyin Hu 2016 Thermomechanical fatigue failure investigation on a single crystal nickel superalloy turbine blade, in: *Engineering Failure Analysis* (Elsevier) DOI: 10.1016/j.engfailanal.2016.04.016. Hal-01324007.

[4] Li Tian, Yan Hai, Zeng Qingyue and Yin Qin 2019 Non-destructive testing Techniques based on Failure Analysis of Steam Turbine Blade *IOP Conf. Ser.: Mater. Sci. Eng.* **576** 012038. DOI: 10.1088/1757-899X/576/1/012038.

[5] Mevissen F and Meo M 2019 *A Review of NDT/Structural Health Monitoring Techniques for Hot Gas Components in Gas Turbines* (Materials Research Centre, Department of Mechanical Engineering, University of Bath, Bath BA2 7AY, UK)

[6] Rau E, Bamberg J, Burchards J, Berwig P and Voelkl W2016 High-resolution crack detection on turbine blade roots by the use of eddy current and ultrasonic Rayleigh waves *19th World Conf. on Non-Destructive Testing*

[7] Grinkrug M S and Popovskiy A V, RF Pat. No. 2168724 C1. *Method for the diagnosis of incipient defects* (6 October 2001)
[8] Grinkrug M S, Mohamad K B and Tkacheva Yu I 2019 Method of detecting microcracks in the blades of operating gas turbine engines *Youth and science: topical problems of fundamental and applied research: mater. of the II All-Russian nat. sci. conf. of students, post graduate students and young scientists (Komsomolsk-on-Amur, April 08-12, 2019*) (Komsomolsk-on-Amur state university) part 3, pp 226-229
[9] Grinkrug M S, Balli M K, Tkacheva J I and Novgorodov N A 2020 An experimental bench for testing the cracks detecting technology in the blades of working aircraft engines *IOP Conf. Ser.: Mater. Sci. Eng.* **734** 012022. DOI: 10.1088 / 1757-899X / 734/1/012022
[10] Błachnio J, Bogdan M and Zasada D 2017 Increased temperature impact on durability of gas turbine blades *Maintenance and Reliability* **19(1)** 48–53. DOI: http://dx.doi.org/10.17531/ein.2017.1.7.

Change of exploitation parameters of a vessel after the switch to the usage of LNG as the main fuel

L V Ivanov, A Y Baranov, V A Tsvetkov, A V Vasilenok and A M Andreev

ITMO University, 9, Lomonosova st., Saint Petersburg, 191002, Russia

E-mail: lvivanov@itmo.ru

Abstract. The study presents the analysis of the changes in the fuel autonomy and the effective range performed by the vessels which underwent modernization in order to use LNG as the main fuel. In this article various variants of modernization were considered. It is suggested that the project core geometry of the self-propelled tankers and tugboats should not be altered. The change in the parameters of the fuel autonomy after placing the cryogenic fuel tanks of different constructions was estimated. The various operating modes of cargo tanks of different constructions were observed. Besides that, there are two possible operating modes, particularly the drainage mode and the drain-free mode. It is proved that the tug boats' fuel autonomy is reduced 2 times. The usage of the boil-off-gas in the self-propelled LNG carrier's engine compensates for the decrease in the fuel autonomy. Thus, it depends on the distance between the final consumer of LNG and the loading satellite terminal of the cryogenic product whether to opt for using self-propelled or tug and barge transport.

1. Introduction

Liquefied natural gas (LNG) is an environmentally friendly, energy efficient and fairly cheap type of fuel. In the regions with developing production of LNG, liquefied gas may become a viable alternative to the traditional fuels such as diesel fuel (DF) and coal. As a general rule, the water transport is used to deliver LNG in industrial amounts. It is reasonable to modernize vessels that deliver LNG so that they can consume LNG as the main type of fuel. It may reduce the shipping cost by means of the reduction of the shipping leg.

Construction of new gas carrying vessels requires big sums of money. The construction project of a gas carrier vessel includes engineering of the vessel hull. The usage of the existing engineering projects adapted to the navigation conditions of a particular water basin can significantly reduce the cost of the development of the gas carrier project. The modernization of existing LNG carriers makes it possible to build the LNG carriers' fleet in the shortest possible time.

Traditionally, in the arrangement of the delivery of oil products a large volume of shipping is performed by means of tug and barge transportation [1], therefore, they are considered in the article on the equal basis with self-propelled vessels.

The analysis is carried out basing on the example of the Russian Federation where the manufacturing facilities are located in the northern parts of Siberia. The specific location of the LNG sources of terminals Sabetta and Utrenniy in the gulf of Ob requires the usage of river-sea vessels for shipping.

2. Technical solutions for modernization

The most widespread types of vessels were selected to undergo the modernization. Those ships will be employed in LNG shipping. They are tugboats with the power of 600-900 ehp and the self-propelled low-tonnage oil tankers with the total displacement tonnage of 3,680 and 3,216 tons. While modernizing the vessels, it is proposed to modernize the propulsion system of the vessel to use LNG as fuel. It is suggested that the propulsion unit should be changed in order to function as the gas-diesel in the dual-fuel mode. It is also advised that the cryogenic fuel tank should be installed as the replacement for the traditional marine diesel oil (MDO) or Heavy Fuel Oil (HFO) tanks. The modernization is expected to be carried out without changing the geometry of the hull. This kind of modernization is considered in SWOT analysis in the Yangtze shipping case study [2]. The authors characterize the increase in the volume of a fuel tank as a weakness. But since the Chinese LNG bunkering infrastructure is better developed than the infrastructure of Siberian Russian rivers, the fuel autonomy loss is not big of a problem. During the exploitation of the dual-fuel power plant, it is necessary to install the tank for the pilot diesel fuel. The placement of the diesel tanks is supposed to be on the upper deck.

When the self-propelled tankers undergo modernization, the cargo tank is required to be changed in order to be able to store LNG. The basic product oil tanks can be replaced with the independent LNG tanks. Various LNG tank systems are known. They differ in volume utilization, weight, type of thermal insulation and its thickness. Moreover, LNG tanks can be exploited in various modes. The boil-off gas is removed from the tanker in the drainage mode. In the drain-free mode LNG vapour is stored inside the tank. That causes the increase in the internal pressure which is also known as the pressure build-up. Membrane tanks which are used for sea shipping are not suitable for this modernization project, since this kind of cargo containment systems is the integral and inseparable part of the vessel's hull.

The most important criterion for choosing the cargo containment system for the modernization is the compactness. The highest parameters of useful utilization of space are characteristic of prismatic tanks type A (LNT A-BOX) and type B (IHI SPB) [3-4]. Those tanks are not designed to store LNG at high internal pressure. Cylindrical type C tanks can be used at high pressure, but they poorly integrate into the existing geometry and, therefore, they have the lowest level of compactness of placing in the vessel's hull.

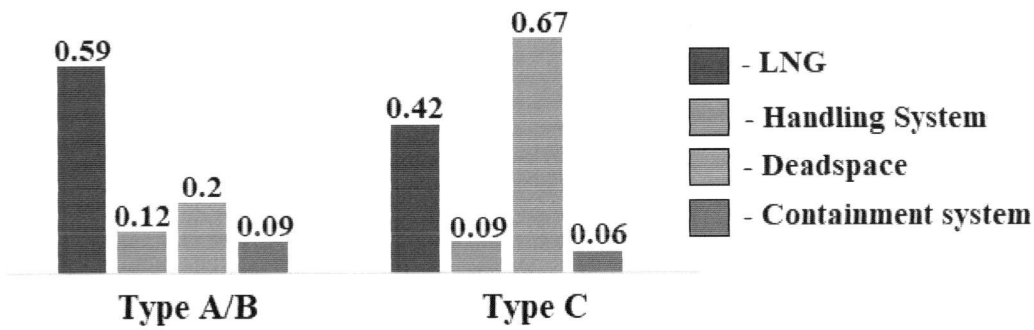

Figure 1. Volume utilization coefficient for fuel tank types A, B and C.

Type B tanks have lower specific weight than that of type A tanks, since there is no secondary barrier in the construction of type B tanks that allows the latter to be fitted into the original deadweight of the oil tanker. For tug and barge transportation it is rational to use type C tanks which are installed on barge-platforms [1,5]. By varying the initial filling level of the tank such tanks can be used in drain-free mode without exceeding the maximum allowable internal pressure and loss of LNG [6].

It is possible to calculate the volume of cryogenic LNG fuel tanks after modernization.

Table 1. The volume of the fuel tanks after the modernization

Vessel type	Fuel tanks volume before the modernization, m^3	Tanks volume with type C fuel tanks installed, m^3	Tanks volume with type A/B fuel tanks installed, m^3
Tanker P-77	99	41.6	58.7
Tanker 621	103.6	43.5	61.5
Tugboat 1741A	65.3	38.8	27.4
Tugboat 758B	80.8	48	33.9
Tugboat 07521	158	93.7	66.3

In the case of modernizing the self-propelled tankers, the volume of the transported LNG after the modernization can be determined by the relation of the volume of displacement and the total bulk volume of the tanks. For each type of cargo containment systems, it is necessary to take into consideration the parameters of the vessel on which the tanks are installed. For A type tanks this relation is:

$$a = \frac{L \cdot B \cdot D}{W} = \frac{195.3 \cdot 30 \cdot 9}{45000} = 1.17 \quad (1)$$

where L - length overall, m; B - beam, m; D is draught, m; W - total volume of cargo containment system, m^3.

For B type tanks the relation can be calculated in the similar way. In this calculation the overall dimensions of the tanker "Polar Spirit" were used:

$$a = \frac{238.94 \cdot 40 \cdot 8.8}{88100} = 0.95.$$

The total volume of cryogenic cargo tanks after the modernization of Lenaneft P-77 is going to be:

for type A:

$$W_A = \frac{108{,}6 \cdot 14{,}8 \cdot 2{,}5}{1{,}17} \approx 3400 \text{ m}^3; \quad (2)$$

for type B:

$$W_B = \frac{108.6 \cdot 14.8 \cdot 2.5}{0.95} \approx 4200 \text{ m}^3 .$$

The quantity of appeared boil-off gas depends on the boil-off rate. BOR shows the amount of evaporating LNG per day. For tanks IHI SPB and LNT A-BOX the BOR index is 0.15 %/day. Therefore, the volume of BOG is equal to:

for type A:

$$BOG_A = W_A \cdot BOR_A = 3400 \cdot 0.0015 = 5.14 \text{ m}^3/\text{day}; \quad (3)$$

for type B:

$$BOG_B = W_A \cdot BOR_A = 4200 \cdot 0.0015 = 6.31 \text{ m}^3/\text{day}.$$

Type C tanks installed on barges can be efficiently used in the drain-free mode.

3. The analysis of changes in the degree of fuel autonomy and effective range of LNG shipping

Ship fuel autonomy according to capacity of cryogenic fuel tank after the modernization can be calculated by the formula:

$$A = \frac{V}{Q - BOG} \quad (4)$$

where Q is consumption of fuel LNG, m^3/day.

The consumption of fuel LNG for Lenaneft P-77 will be:

$$Q = \frac{M \cdot n \cdot q \cdot 24}{10^3 \cdot \rho_{LNG}} = \frac{660 \cdot 2 \cdot 156 \cdot 24}{10^3 \cdot 430} = 11.49 \text{ m}^3/\text{day}, \quad (5)$$

where M - main engine power, kW; n - quantity of main engines; q - fuel consumption, g/kWt*h; ρ_{LNG} - density of LNG, kg/m^3.

Evaporated boil-off gas is reasonable to supply to the vessel's engine. However, this mode of exploitation can be applied only to self-propelled vessels, since there is no technical possibility to transport boil-off gas from barge to tugboat [7]. In this case of LNG shipping, the endurance of tugboats according to the fuel capacity is:

Table 2. Fuel autonomy of tugboats after the modernization

The project of the vessel	Original autonomy, day	Autonomy after the modernization, day	
		Tank types A/B	Tank type C
Tugboat 1741A	9	6	4.2
Tugboat 758B	15	6.9	4.9
Tugboat 07521	20	10.4	7.4

After the transition to the gas engine fuel, the level of fuel autonomy of tugboats is significantly decreased. This is caused by the lower specific density of LNG. Besides that, part of the usable space of the hull can be taken by the thermal insulation of the fuel tanks.

Table 3. Fuel autonomy of self-propelled vessels after the modernization

The project of the vessel	Original autonomy, day	Autonomy after the modernization, day	
		Tank type A	Tank type B
Tanker P-77	15	9.2	11.3
Tanker 621	15	11.9	16.3

In the design of cryogenic tank it is possible to vary BOR level. This leads to the change in the amounts of boil-off gas generation. Volume of BOG supplied to the main dual-fuel engine as fuel helps to make an increase in the fuel autonomy of the gas carrying vessel.

Effective range of LNG shipping is estimated by means of considering conditions under which the full fuel autonomy potential is used. The total shipping comprises the condition of the distance that the loaded vessel has covered in order to reach the consumers. The distance is equal to to the distance covered by the loaded vessel to the initial point. This distance is the effective delivery range:

$$R_L = R_{Ul} = R \,, \tag{6}$$

where R_L - distance covered by a loaded vessel, km; R_{Ul} - distance covered by a vessel without load, km; R - effective delivery range.

The effective delivery range can be found according to the following equation:

$$R = \mathrm{A} \cdot x \cdot S_L \cdot 24 = A \cdot (1 - x) \cdot 24 \cdot S_{Ul} \tag{7}$$

where x - percentage of swimming in the loaded state; S_L - speed of a loaded vessel, km/h; S_{Ul} - speed of an unloaded vessel, km/h.

In this case the effective delivery range can be calculated for the modernization variants under study.

The speed of a tug is considered in the fully loaded condition with a barge's total capacity of 3000-5000 t. The unloaded vessel's speed is decreased, since the empty barges should be towed to the point of departure. The effective delivery range of self-propelled tankers is significantly bigger than that of the tugboats. All that allows the self-proplled tankers to deliver LNG to the most remote consumers.

Table 4. The effective delivery ranges

The project of the vessel	The speed of a loaded vessel, km/h	The speed of a vessel without load, km/h	The effective delivery range, km
Tanker P-77	19	35	3350
Tanker 621	18.5	35	4740
Tugboat 1741A	12	16.5	1002
Tugboat 758B	10	16.5	1028
Tugboat 07521	11	16.5	1655

4. Conclusion

After the modernization that allows vessels to use LNG as the main fuel and does not require changes in the vessel's geometry, the fuel autonomy has noticeably decreased. This decrease in the autonomy can be compensated by the usage of boil-off gas from the cargo containment system in the main dual-fuel engine. The system of this kind can be used only in self-propelled LNG carriers. By changing the parameters of the thermal insulation, the fuel autonomy can be varied. The effective LNG delivery range performed by self-propelled LNG carriers is 2.5-3 times wider than the one performed by tug and barge LNG transportation. Thus, it is efficient to deliver LNG to the remote consumers by means of self-propelled vessels. The parameters of thermal insulation should be chosen according to the distance to the most remote consumers.

References

[1] Seitz M 2015 *Masterplan for LNG on Rhine-Main-Danube Axis* (LNG Masterplan) Retrieved from: https://www.prodanube.eu/images/2015-06-10_LNG_Masterplan_Status-June_2015.pdf
[2] Chengpeng Wan, Yan X P, Di Zhang and Shanshan Fu 2015 Emerging LNG-fueled ships in the Chinese shipping industry: a hybrid analysis on its prospects *WMU J. of Maritime Affairs*. DOI: 10.1007/s13437-015-0080-6
[3] David Wu D 2019 LNT A-Box for Small & Mid scaled LNG carriers. Retrieved from: https://www.norwep.com/content/download/40633/297704/version/1/file/04+David+Wu+-+LNT+A-Box+in+Vietnam.pdf
[4] Nagata Y, Tanoue A, Kida T and Kawai T 2015 IHI SPB Tank for LNG fueled ships *IHI Engineering Review* **2(47)**
[5] Wursig G-M 2011 LNG fuel tank – benefits and challenges. Retrieved from: https://docplayer.net/23911750-Lng-fuel-tank-benefits-and-challenges-dr-gerd-michael-wursig-business-director-lng-fuelled-ships.html
[6] Theokokatos G, Dimetrellou S, Strntzali E and Pagonis D N 2015 *Desing of LNG storage and feeding system for an open type ferry*
[7] *IGC Code Int. Code for Construction and Equipment of Ships Carrying Liquefied Gases in Bulk* (MSC. 177(79))

Principles of protection, storage and movement of documents during electronic document flow within and outside the organization

Kh A Daudov[1], **A A Zinoviev** [2] **and Zh L Gavrilova**[3]

[1] Chechen State University, 17a, Dudaev Boulevard, Groznyy, 364015, Russia
[2] Bratsk State University, 40, Makarenko str., Bratsk, 665709, Russia
[3] Irkutsk National Research Technical University, 83, Lermontova str., Irkutsk, 664074, Russia

E-mail: ibr024@mail.ru

Abstract. Today, electronic document management refers to a documentation management system in which the entire array of created, transmitted and stored documents is supported using technologies on computers combined into a network structure, which provides the ability to form and maintain a distributed database.

1. Introduction
The number of documents depends on the legislative requirements for information created or edited in a given area and the functional purpose of this area.

It is advisable to distinguish two main document routes:

• movement of documents within the organization;
• movement of documents in communication with the outside world.

The internal movement of documents is carried out in various ways, one of the main ones is the transfer of data over local networks, which are currently distinguished by high data rates (the most common format is 100 Mbps).

Optimizing the movement of documents within an organization is closely related to the term Intranet. In the simplest sense, Intranet is the use of Internet technology in corporate systems, but in reality this concept is much more serious - Intranet carries with it a new philosophy of information management within the organization.

2. Results
The movement of documents when the organization communicates with the outside world includes a variety of connections with counterparties of the enterprise and state executive bodies, as well as with elements of the enterprise outside its limits (employees on a business trip). A characteristic feature of this part of the document flow should be its strict regulation:

• only specially authorized documents can be outbound;
• privacy regimes are established for outgoing documents;
• incoming documents are checked for false and destructive information.

The organization's communications with the outside world are carried out directly through an information intermediary - (Intranet, Internet, Extranet). Communication through an information broker is generally financially more profitable, but is often accompanied by significant time losses and requires increased information security requirements. Graphically, this interaction is represented in

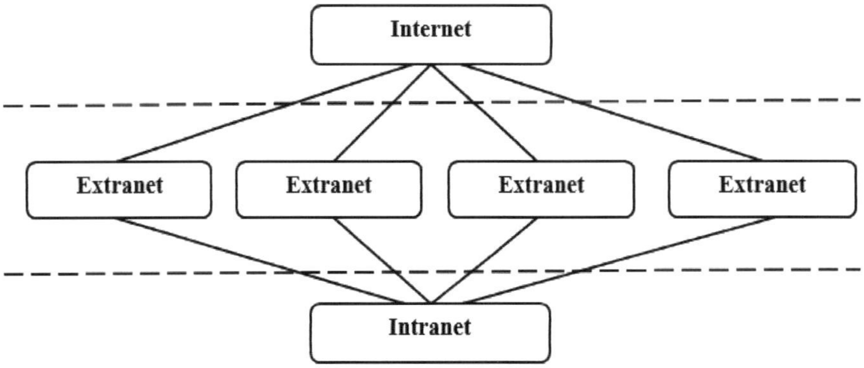

Figure 1. Relations of the organization with the outside world via Internet, Intranet, Extranet

Communication type selection is determined by the following basic parameters:

• territorial distance of communication facilities;

• intensity of document flow with the given object;

• the ability of this type of communication to ensure the timely passage of documents.

Depending on the distance and amount of documents sent, the following communication types are selected:

• by courier;

• normal telephone lines;

• leased lines;

• fiber optic connections;

• satellite channels, etc.

In organizing information repositories, the organization uses three main principles - centralization, decentralization and duplication.

In an isolated application of the centralization principle, it is intended to create a minimum number of information stores with a large amount of heterogeneous information stored in them and the ability to access this information of many users.

The principle of decentralized storage of information involves the creation of many information repositories, the number of which is oriented to the number of documents received and created in the organization [1, 2].

The principle of duplication involves the storage in each repository of information of all documents necessary for the implementation of a process.

Thus, it is not possible to state clearly the advantages or disadvantages of applying each principle in the design of a particular repository of information. - and considering the fact that the choice of an information distribution organization has a significant impact on the overall efficiency of the automated electronic document management system in the organization and, accordingly, on the performance of the organization itself - This choice should be based on mathematical methods that make it possible to find the most effective option for distributing information to warehouses.

When organizing the storage of documents in electronic form, some features should be provided.

• electronic storage of documents increases the risk of unauthorized access to them, as well as reproduction and copying of information stored in documents, so the requirements for ensuring security and tougher access to such documents should be increased;

• technical characteristics of the material on which the information is stored undergo some changes over time. It is known, for example, that the lifetime of magnetic carriers is lower than that of paper. The preservation of diskettes and other magnetic carriers is greatly influenced by the temperature and humidity regime, as well as the presence of electromagnetic radiation in storage places. Consequently, in addition to strengthening these factors, it is necessary to provide for regular checks on the safety of information, and in cases where there are doubts about its further safety, to copy information to other similar media.

It is necessary to take into account the costs of ensuring information protection when designing an automated electronic document management system in an organization, when assessing the cost of ensuring information protection, it is necessary to pay attention to the following indicators:

- number of documents circulating in the automated electronic document management system and stored in the archive;
- number of automated workplaces in the organization.

You should pay special attention to information protection technology: most of the information should remain protected from accidental or malicious access [3, 4]. Using modern cryptographic and other methods of information protection, it is most advisable to provide the following four levels of data protection:

- authentication;
- access control;
- digital signatures.

Authentication refers to reliable identification of the user, identification should be bidirectional - the servers on which data warehouses are to authenticate users (automated workstations), and users, in turn, should authenticate servers. Authentication must be applied every time the workstation connects to the data warehouse.

In the access control lists, it is necessary to determine which workstation has access to each of the resources, and what right to use them (drawing up documents, editing, deleting, etc.). The following levels of information access are most useful:

- custodian - has the right only to write information to data warehouses. This level of access is useful for data warehouses into which documents fall, for example, by e-mail;
- reader - this level of access allows you to only read information from the data warehouse;
- author - can create new documents and read already created, but cannot edit or modify documents in data warehouses;
- editor - has the right to read, write and edit documents in the data warehouse;
- manager - this type of access allows you to add or remove other users from the access control list.

Encryption involves encrypting or encoding information so that if it is accessed by an unauthorized person, it will not be possible for him to understand the content of the protected field. The four most common encryption levels are:

- message level - i.e. individual messages can be encrypted to read only one or more selected recipients;
- network layer. Encryption will not allow anyone to randomly "listen" or wedge into local network traffic, since its content will not make any sense to the intruder;
- field level. Databases can be configured so that certain document fields are encrypted (for example, for financial documents, the "amount" field), and only specified users could read them;
- database level. Encryption protects all documents stored in a particular repository of information from access by someone, with the exception of certain users who have the ID and password used in encryption [5-7].

The use of digital signatures is simply necessary in cases where the user needs to make sure that the information he has received has actually been sent by a certain sender. A digital signature provides such a guarantee and is an option for mutual authentication. In addition, this technology allows you to provide the user with a guarantee that the document has not been corrected during the transfer process.

3. Conclusion

The implementation of the electronic document management system requires solving the following key problems:

1) It is necessary to transform existing documents from a paper form to an electronic one. In an organization with a paper archive of documents, it is almost impossible to find electronic copies of them on the network to create an electronic version of the archive. Moreover, electronic copies could be modified by anyone, and it is almost impossible to verify this. Therefore, in this case, you need to create an electronic archive of existing documents from paper by scanning. If there is no archive, it is necessary to carry out a sufficiently long preparatory work. First, you need to achieve a single procedure for storing documents in catalogs. You should prohibit (including, if possible, by technical means) the storage of documents on local disks of computers. It is necessary that all users of the organization have on the network their directories on the network in which they store documents. This will greatly facilitate further migration of documents to the document management system. It is convenient when the workflow system allows you to maintain the versivity of the entire birth history of the document, the ability to view it and a short description of how one version differs from another or which version was given to the counterparty for approval.

2) When implementing the electronic document management system, there is a need to immediately analyze the possibility and need for its integration with the already existing automated management system. This is due to the fact that employees of a number of specialized organizations combine work with "ordinary" documents (for example, work order, timesheet, etc.) and engineering documents (drawings, etc.), and all this - when performing operational management functions. It is highly desirable that electronic document management systems contain special means for such integration.

3) The business is quite demanding in relation to the concept of responsibility both from the point of view of technology (medicine, the production of baby food, etc.) and the optimal execution of business processes and the preservation of trade secrets, which creates the need to document the implementation of each stage of the process. In such cases, the electronic document management system should allow a step-by-step analysis of each activity, even after several years; give a "cut" of a certain point in time.

4) Holding structures are characterized by a large territorial disunity (regional structure), a complex management hierarchy with many stages, sometimes a weak standardization of existing information technologies or a "zoo" systems. However, there are high requirements for data security and restricted access. It is necessary to draft a single functioning system.

References

[1] Vatolina M V 2011 *Document Management. Higher Education Series* (Rostov-on-Don: Phoenix) pp 76-116

[2] Efremov A 2008 *Information life cycle. Electronic archives of unstructured information. Electronic office. Technologies and procedures* pp 20-21

[3] Klyuev R V, Bosikov I I, Revazov V Ch, Krysanov K S and Dzeranov B V 2020 A comprehensive analysis of the properties of reservoir rocks to determine the productive oil horizons *IOP Conf. Ser.: Mater. Sci. Eng.* **952** 012017. DOI: 10.1088/1757-899X/952/1/012017

[4] Kortiev L I, Klyuev R. V, Totchiev F G, Vaneev S D and Babunska N I 2020 Temporary and operational transport roads in construction of transfer linear structures and their scientific and technical support *IOP Conf. Ser.: Mater. Sci. Eng.* **952** 012023. DOI: 10.1088/1757-899X/952/1/012023

[5] Kudryaev V A 2001 *Organization of work with documents* (Moscow: INFRA-M) 592 p

[6] Klyuev R V, Bosikov I I, Egorova E V and Gavrina O A 2020 Assessment of mining-geological and mining technical conditions of the Severny pit with the use of mathematical models *Sustainable Development of Mountain Territories* **3** 418-427. DOI: 10.21177/1998-4502-2020-12-3-418-427

[7] Petelin V G 2009 *Office work at the enterprise: Short course* (Moscow: Axis-89) 112 p

Assessment of the effectiveness of immune neural network

K Zhigalov[1,2], E O Pokhomchikova[3] and M V Sygotina[4]

[1] V.A. Trapeznikov Institute of Control Sciences of Russian Academy of Sciences, Moscow, Russia
[2] Kh. Ibragimov Complex Institute of the Russian Academy of Sciences (CI RAS), Russia
[3] Irkutsk National Research Technical University, 83, Lermontov str., Irkutsk, 664074, Russia
[4] Bratsk State University, 40, Makarenko str., Bratsk, 665709, Russia

E-mail: kshakalov@mail.ru

Abstract. Nowadays the information technology industry is facing a boom in the research of neural networks and related technologies. Networks are used almost everywhere. However, the question of their effectiveness is omitted. Although there is a growing need to move from quantity to quality. The paper considers a variant of the assessment of the effectiveness of the immune neural network with an example.

1. Introduction

During the development of information security system based on human immune system (working on the model of it), we faced the need to conduct research on the effectiveness of its work. Since this problem can be posed by other researchers, we will try to describe it as much as possible in this paper.

2. Mathematical base of assessment

For the experiment, we took the immune network we were developing, which should respond to all events that are not present in the information system. Decision-making on the presence of an "alien" information process $P^{al}(t)$ in the computer system of the workplace can be made upon the fact that the frequency f exceeds a certain threshold value f_t (4):

$$P(t) = \begin{cases} P^c(t), \text{if } f \le f_t, \\ P^{al}(t), \text{if } f > f_t. \end{cases} \quad (1)$$

Moreover, in accordance with the system identification method, any process P(t), which differs from all processes $P_n(t)$, $n=\overline{1, M}$, occurring at a given time, will be considered "alien".

The statistical probability Ver of the presence of an "alien" information process in a computer can be reflected by a value inverse to the number of coincidences of new information processes with the hosts of the artificial immune network (2).

$$Ver\left(P^{al}(t)\right) = 1 - Ver\big(P(t)\big), where$$

$$Ver(P(t)) = \frac{n^+}{n}, \tag{2}$$

where n^+ - the number of positive comparisons of information environment processes with hosts of the immune network, and n - the total number of events of the comparison of information environment processes with network hosts.

The number of hosts in the system corresponds to the number of processes that take place in the system. Generalizing the processes performed in accordance with the affinity criterion *q*, we reduce the number of hosts to a level sufficient to identify the process. For the convenience of processing, the hosts have attribute fields. The total number of hosts *Y* can be expressed through the formula (3):

$$Y = \sum_{i=0}^{n} F_i^{uniq}(q), \tag{3}$$

where $F_i^{uniq}(q)$ –the number of hosts in a given artificial immune network, selected in accordance with the affinity criterion *q*. The number of network hosts is the greater, the higher the *q* index and it depends on the number of operations performed at the workplace at a given time.

The monitoring system must function in real time. Therefore the speed of the comparison of the processes of the information environment with the hosts of the artificial immune network is a key factor in the response. The reduction of *q* parameter of the immune network affinity will reduce the number of hosts involved in the comparison, but will reduce the quality of the system's response to malicious processes. The increase in the parameter *q* will lead to the increase in the hosts of the system and therefore it will increase the time spent on processing incoming events.

To speed up the procedure of event processing, the comparison has exponential nature, from the most general feature to the most unique, which is described above when compiling a process card. For the analysis of the most general features, *q* parameter is not used, and therefore it does not affect the number of created hosts. Therefore, the time spent on the comparison of general features will grow linearly depending on the number of these features (4):

$$T = \sum_{i=0}^{n} F_i^{gen} * k_i, \tag{4}$$

where T - comparison operation for general features of information processes, F_i^{gen} – general features, k_i – the number of general features of a given type, i=$\overline{0,n}$ – feature reference.

In general, the parameter *T* can be represented as a constant. Then the assessment of the number of comparisons for parameters unique to information processes can be represented as (5):

$$O = T + k * \sum_{i=0}^{n} F_i^{uniq}, \tag{5}$$

where O – comparison operation in general, F_i^{uniq} – features unique for each host, k – number of unique features of a host, i=$\overline{0,n}$ – feature reference.

Unique features will not arise without general ones, since the comparison begins with them. In addition, each general feature depends on the previous one. On the other hand, when the degree of affinity is zero, the number of features for comparison will be equal to the number of general features. Thus, the formula for the calculation of the number of general features can be modified as follows (6):

$$O = \begin{cases} \sum_{i=1}^{n} F_i^{gen}(F_{i-1}^{gen}) * k_i + \dot{k} * \sum_{l=0}^{m} F_l^{uniq}(F_n^{gen}), l = \overline{0,m}, i = \overline{1,n} \\ F_i^{gen}(F_{i-1}^{gen}) = 0, if\ F_{i-1}^{gen} = 0; \\ F_i^{gen}(F_{i-1}^{gen}) = 1, if\ \exists\ F_i^{gen}\ \text{и}\ F_{i-1}^{gen} \neq 0 \\ F_l^{uniq}(F_n^{gen}) = 0, if\ F_n^{gen} = 0 \\ F_l^{uniq}(F_n^{gen}) = 1, if\ F_n^{gen} = 1 \end{cases}, \tag{6}$$

where $\dot{k}$ – the number of unique features.

Consequently, the number of testing will grow depending on the growth of indicators n and m, i.e. on the number of unique hosts in the system. For m=0, i.e. the absence of unique features of hosts, the number of checks will be limited to $n * \sum k$. In the case when m $\neq$ 0, the number of checks is limited to $n * \sum k + m$.

The graph of the function of the maximum number of testing will grow linearly depending on the growth in the number of general and unique features. According to the table for the generation of information about the process, the reference to general features i will change from 0 (if the name of a user who launched the process was not known earlier) to 3 ($i=\overline{0,3}$). The reference to unique features l will always be equal to 3, since if the process exists and found among general features, it will always have activity described by unique features.

The minimum number of testing will correspond to the number of general features of the first characteristic. The least commonly used general feature of a corporate computer network of user's workstation is the name of a user who started the process. In the pre-collected data, the number of users is 4. Accordingly, the minimum number of testing will be compared with 4 hosts containing different usernames. If the username does not exist, the system marks this process as alien and does not perform verification with further host fields. It was mentioned above that it was the field with the name of the user who started the process that the author chose as the first sign of the general characteristics of running processes.

3. Realization of experiment

In order t conduct the experiment, the data collected for the primary analysis were loaded into the prototype of the program for the formation of an artificial immune network. After processing data consisting of 4447 lines of the csv-file collected by the Process Monitor program, each of which described a process passing with the user's workstation, 634 hosts were formed in the system with an affinity coefficient of q = 0.8. Changing the affinity coefficient, we can observe the change in the number of hosts processing the data belonging to the system:

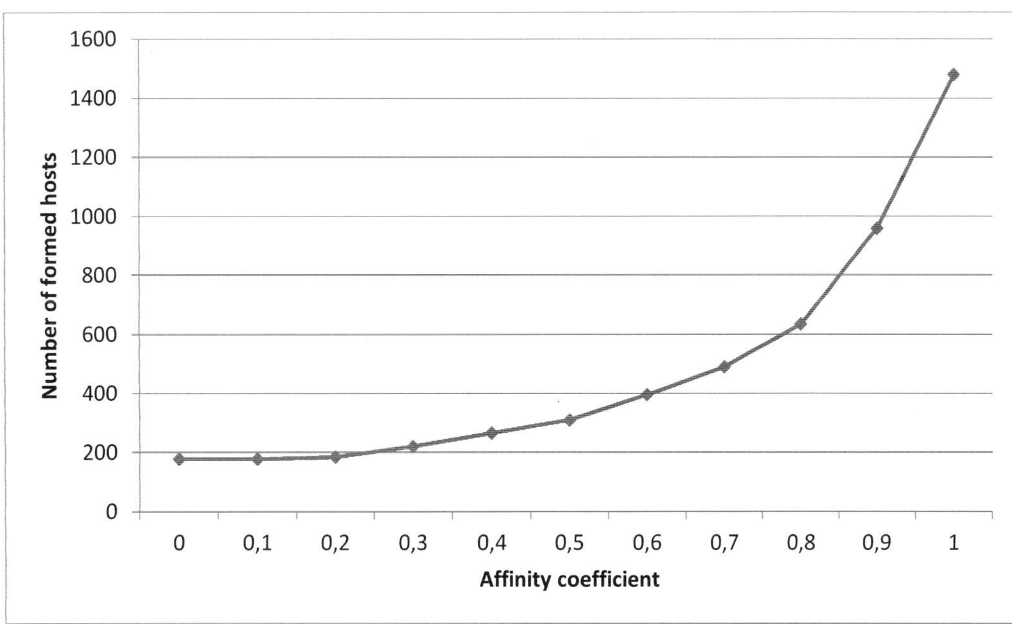

Figure 1. Graph of dependence of system hosts on the affinity coefficient

For the assessment of the effectiveness of the algorithm to detect an artificial immune network, the formed network was provided with data collected on the same workstation at a slightly different time. The collected data consisted of 1238 lines, each of which characterized the process taking place in the system. Since the data for detection was collected at a different time of a user's work, the control of the time when the process started was cancelled in the detection algorithm. However, it was planned to use this parameter during the current functioning of the network of hosts of the artificial immune network, since it allowed tracing the patterns of user activity depending on time.

The presented graph (Figure 1) shows that with the increase in the affinity coefficient, the number of formed hosts increases. This graph also indicates that processes continue to occur on user's workstation, corresponding to the processes reflected in the data collected in previous periods and used to build the network, which is logical. Similar experiments have been carried out in other industries [1-3].

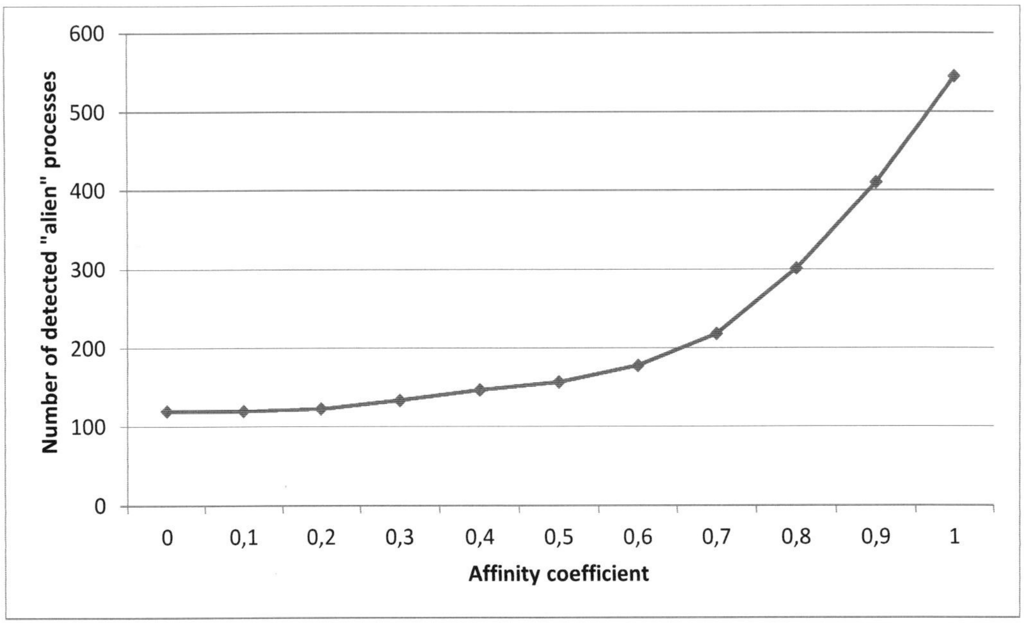

Figure 2. Graph of dependence of detection of anomalies of user processes depending on the affinity coefficient

This graph (Figure 2) shows that with the increase in the affinity coefficient, the number of processes detected as “alien” ones increases. This graph also indicates that processes continue to occur on user's workstation, corresponding to the processes reflected in the data collected in previous periods and used to build the network, which is logical.

In order to analyze the work of the immune network with abnormal data, the authors changed the data used for the initial training of the system. A total of 21 lines were changed to describe the file. The authors of the research did not deliberately change parameters such as the name of the process or the name of a user who started the process, but only touched upon those features that were unique for each process. The graph (Figure 3) of the dependence of the changes on the coefficient of affinity is presented below.

Figure 3. Graph of minor changes in incoming data

Due to the significantly larger number of processes occurring at the workstation of the employees of an organization than those presented for training the system, it is necessary to maintain a balance between the required accuracy of the correspondence of the processes performed at the workstation and the size of the detecting network. In addition, it is necessary to take into account that a user of a workstation can periodically perform actions that he has not performed before, which do not pose any threat to information security. To eliminate the redundancy of the system's response to “alien” information processes, as described earlier, the process weight coefficient was introduced which solves two problems at once.

First, the processes that are too rarely performed by a user are removed from the set of hosts of the immune network, which means that the triggering of these detectors will be quite rare and can be neglected. Secondly, the process weight coefficient is assigned to an “alien” process, and the system responds to it only when the weight gains a certain value. In tests that measure the response of the system to alien processes, the weight of the process is not taken into account, due to the very modest set of input data.

This system, already at the prototype stage, shows its ability to react to user actions, revealing the data that is different from the original. Having created an artificial immune network, trained by data collected over a sufficient period of time to observe a user of a workplace, a snapshot of the system will be obtained, reflecting the activities of an employee at his workplace.

4. Conclusion

The analysis of these experiments will reveal the effectiveness of the immune network or any other neural network, and hence the reasonability of further refinement of the system interfaces for its further industrial use. This will allow developers to more efficiently choose certain algorithms for the operation of neural networks for specific tasks.

References

[1] Klyuev R V, Bosikov I I, Revazov V Ch, Krysanov K S and Dzeranov B V 2020 A comprehensive analysis of the properties of reservoir rocks to determine the productive oil horizons *IOP Conf. Ser.: Mater. Sci. Eng.* **952** 012017. DOI: 10.1088/1757-899X/952/1/012017

[2] Kortiev L I, Klyuev R V, Totchiev F G, Vaneev S D and Babunska N I 2020 Temporary and operational transport roads in construction of transfer linear structures and their scientific and technical support *IOP Conf. Ser.: Mater. Sci. Eng.* **952** 012023. DOI: 10.1088/1757-899X/952/1/012023

[3] Klyuev R V, Bosikov I I, Tavasiev V Kh and Gobeev M A 2020 Improving the energy efficiency of transport equipment in ore mining *IOP Conf. Ser.: Mater. Sci. Eng.* **918** 012123. DOI: 10.1088/1757-899X/918/1/012124

Radio-frequencies technologies for local communication in cyber-physical environment

I M Daudov[1], Zh L Gavrilova[2] and T N Iakovkina[3]

[1]Chechen State University, 17a, Dudaev Boulevard, Groznyy, 364015, Russia
[2]Irkutsk National Research Technical University, 83, Lermontov str., Irkutsk, 664074, Russia
[3] Bratsk State University, 40, Makarenko str., Bratsk, 665709, Russia

E-mail: ibr024@mail.ru

Abstract. The paper is devoted to the Internet of Things which is becoming more popular and, in particular, to wireless connections and technologies. The paper considers two technologies that are currently most widely used on the Internet of Things: Bluetooth and Wi-Fi. The milestones of their development are considered, as well as advantages and features of each technology. Various devices that employ Bluetooth and Wi-Fi technologies and their interaction are described. The conclusion about the perspectives of development of Wi-Fi technology is made.

1. Introduction

The Internet of Things, which not only allows objects to interact with each other, but also provides interaction between objects and people, is becoming a reality and completely changing our world in many ways. Wireless connection of everyday items can automate data exchange and increase efficiency, which has a positive impact on the lives of people and organizations [1, 2]. The introduction of the Internet of Things (IoT) involves the use of wireless networks, which are present around us more and more every day.

Wireless networks mean networks in which two or more terminals can communicate without a wired connection. Wireless networks are based on radio wave connections instead of conventional cables. Each network has its own radio access technologies, for example, Bluetooth, Zigbee, Wi-Fi, WiMAX, 5G, etc. The wireless network categories are shown in Figure 1 [3].

There are two main types of radio access technologies - in licensed and unlicensed frequency spectra. The development of the Internet of Things has aroused interest in radio access technologies in an unlicensed frequency spectrum. Next, the focus will be on two technologies operating in an unlicensed spectrum of frequencies that are currently most widely used on the Internet of Things - Bluetooth and Wi-Fi. These technologies are also widely used in the Industrial Internet of Things, for example, in open-pit and underground mining operations, for the communication of autonomous drones for various purposes with the base station and among themselves [4].

Bluetooth technology is well suited for exchanging information over short distances between a large number of mobile devices (for example, to connect a smartphone to a car, headset or computer). Zigbee technology is also used at short distances.

IOP Conf. Series: Materials Science and Engineering 1111 (2021) 012024 doi:10.1088/1757-899X/1111/1/012024

Wi-Fi technology is better suited for transmitting large amounts of information over long distances. The following will focus on the latest developments in Bluetooth and Wi-Fi technologies. The development of Bluetooth technology has more than 20 years. The main steps in improving this technology are summarized in Table 1 [5-7].

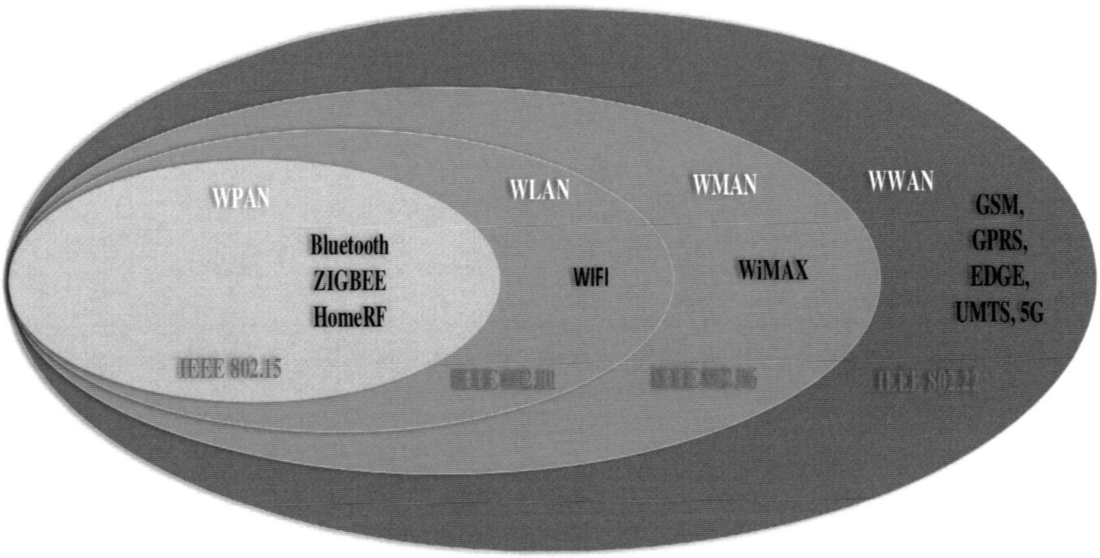

Figure 1. Wireless network categories and technologies

Table 1. Bluetooth Technology Milestones

Version	Date	Properties
1.0	1998	1 Mbps speed.
1.1	2002	Standardized IEEE 802.15.1.
1.2	2003	Frequency hopping.
2.0	2004	Optionally, increase the data rate to 3 Mbps.
2.1	2007	Encryption, authentication methods, power reduction.
3.0	2009	Improved power management.
4.0	2010	Low power mode, encryption.
4.1	2013	Coexistence with mobile wireless communication.
4.2	2014	Connection level privacy, IPv6 support profile.
5.0	2016	1, 2, or 3 Mbps, low power consumption, optional 802.11 feature for speeds up to 24 Mbps, mesh network.
5.1	2019	Determine the exact location of the device by analyzing the direction to connected devices. One of the devices must have an array of several antennas.
5.2	2020	Dynamic optimization of transmission power, timing of transmission channels.

2. Results

There are two types of Bluetooth devices: one is called Bluetooth Classic (it is used in wireless speakers, car infotainment systems and headsets), and the other is called Bluetooth Low Energy (BLE). These two types of Bluetooth devices are incompatible with each other. This means that the Bluetooth Classic device cannot directly communicate with the Bluetooth Low Energy device.

Therefore, some devices, such as smartphones, prefer to implement both types (sometimes it is called a Bluetooth device in double mode), so they can interact with both types of devices.

Bluetooth Low Energy (BLE), or Bluetooth Smart, is a low-power version of the well-known Bluetooth wireless technology. Low power consumption is achieved by changing the data protocol to create low duty cycle transmissions.

Bluetooth Low Energy is designed to reduce the power consumption of Bluetooth devices. BLE was introduced in 2010 with Bluetooth 4.0, and not with traditional Bluetooth radio, which consumes more energy.

With the introduction of Bluetooth 5.0 in 2016, all audio devices communicate through Bluetooth Low Energy, which means less power consumption and a longer battery life.

Since many IoT systems include small devices and sensors, BLE has become a more common protocol in IoT in terms of energy efficiency, which is very useful on the Internet of Things.

In 2017, the Bluetooth SIG Consortium (Special Interest Group) published the following three specifications of the Bluetooth mesh network:

- Meshy 1.0 profile - requirements of integrated mesh-network solution;
- Mesh 1.0 model specification with node functionality;
- Mesh 1.0 device properties.

A mesh network is a distributed, peer-to-peer, self-organizing network with a mesh topology. In a mesh network, all nodes are peer, each node is both a provider, router, and switch. According to the Mesh 1.0 specification, the maximum number of mesh network nodes can be 32,767, the maximum lifetime (TTL) is 127. If the message is received by the node and then retransmitted, the TTL is reduced by one. This prevents infinite cycles.

Features of mesh networks:

- The "intelligence" of the network means that when connected, each node automatically receives information about all other access nodes in the network and "finds out" its role;
- self-healing and self-adaptation. When a network is turned on, each device automatically determines the state of its neighbors and its role in the common topology. Therefore, when one of the nodes fails, the network is able to redirect data - redefine routes automatically;
- quick deployment. Deploying a mesh network does not require expensive infrastructure. Due to self-healing and self-adaptation abilities, this network is economical to operate;
- concept of network filling. Each incoming packet is sent through each outgoing communication, except for a reference to the parent of the message. The node automatically finds the shortest route.

Another wireless personal area network (WPAN) standard is IEEE 802.15.4. The standard defines only the physical and channel layers. The model was ratified in 2005. The goal of 802.15.4 and the protocols that are on it is an inexpensive WPAN with low power consumption. One of the latter is the IEEE 802.15 .4e specification ratified in 2012. The protocol operates in an unlicensed spectrum in three different radio frequency bands: 868.3 MHz, 915 (902-928) MHz, 2,405-2480 MHz. The 2.4 GHz band is the most commonly used. The radius of the 802.15.4 protocol with direct visibility is about 200 meters, in the room - about 30 m. The data transfer rate reaches 250 kbit/s.

The Zigbee protocol is a WPAN protocol based on IEEE 802.15.4 for low-power networks. It is Bluetooth-like, but designed for the Internet of Things. It consumes little and is designed to send small amounts of data (for example, turn on the light in the bedroom).

The Zigbee Alliance approved specification 1.0 in 2004. The protocol is a private and closed standard. The Zigbee network has three main components:

- a controller for forming and starting network functions. Once the network is formed, the controller can act like a router;
- the router is involved in routing multi-hop messages, assigning logical network addresses, and allowing nodes to join or leave the mesh network;

- the final device, for example, a light switch or a thermostat. It has sufficient functionality to communicate with the controller. Zigbee endpoints do not participate in routing.

The protocol operates in the 2.4 GHz band, but also at 868 MHz in Europe and 915 MHz in the United States.

Wi-Fi wireless LAN technology is developed on the basis of IEEE 802.11 protocols and provides high data transfer speed in a reliable and secure way. However, it consumes a lot of energy, and its range is relatively limited (35 m), more through obstacles, such as thick walls. To solve the problem in terms of high power consumption taking into account IoT applications, Wi-Fi 6, also called Wi-Fi ax, was released in 2019.

Wi-Fi ax improves the connectivity of objects, and at a speed of 10 Gb/s it is also used to more efficiently manage more devices with less power consumption.

The protocol was first developed in 1991 for cash registers. In 1999, the Wi-Fi Alliance was created (from the phrase Wireless Fidelity - wireless accuracy). The Wi-Fi logo is a trademark of the Wi-Fi Alliance. Equipment that complies with the IEEE 802.11 standard can be tested in the Wi-Fi Alliance, obtain the appropriate certificate and the right to apply the Wi-Fi logo.

The improvement of the IEEE 802.11 standard goes in the direction of increasing the transmission rate using the unlicensed frequency spectrum of 2.4/5 GHz (Table 2).

Table 2. Options for various Wi-Fi technology options.

Protocol IEEE 802.11	Date	Frequency, GHz	Bandwidth, MHz	Transmission rate min-max, Mbps	Range indoor/outdoor, m
802.11	1997	2,4	22	1-2	20 / 20
n (Wi-Fi 4)	2009	2,4/5	20 40	7,2 - 72,2 15 - 150	70 / 250
ac (Wi-Fi 5)	2013	5	20 40 80 160	7,2 - 96,3 15 – 200 32,5 – 433,3 65 – 866,7	35 / 35
ax (Wi-Fi 6)	2019	2,4/5	20, 40, 80, 160	450 – 10 000	35 / 35
be (Wi-Fi 7)	2024	7	320	30 000	35 / 35

3. Conclusion

The promising development of Wi-Fi 7 involves support for real-time applications, coexistence with the technologies of cellular mobile networks 4G/5G. Consortium 3GPP continues development and standardization of a radio network of access for the fifth generation 5G NR (New Radio) where one of functionality, as well as in 4G of LTE, is put work in not licensed frequency ranges.

References

[1] Makhovikov A B, Katuntsov E V, Kosarev O V, Tsvetkov P S 2019 *Digital transformation in oil and gas extraction. Innovation-Based Development of the Mineral Resources Sector: Challenges and Prospects* (ed. Litvinenko) - 11th conference of the Russian-German Raw Materials (ISBN: 978-0-367-07726-6) pp. 531-538

[2] Makhovikov A and Kryltcov S 2016 Development of secure and reliable communications system for corporate sector over mobile telecom networks. *Asian Journal of Information Technology* **15(24)** 5120-5125

[3] Li P 2020 *Architecture of the Internet of Things* (Moscow: DMK Press)

[4] Kosarev O V, Tcvetkov P S, Makhovikov A B, Vodkaylo E G, Zulin V A, Bykasov D A 2019 *Modeling of industrial iot complex for underground space scanning on the base of arduino platform.* Topical Issues of Rational Use of Natural Resources - Proceedings Of The International Forum-Contest of Young Researchers, pp. 407-412

[5] Silicon Labs BG22 Secure Bluetooth 5.2 SoC Promises 10-Year Life batteries. Archived 20.01.2020. The date on which the 20.08.2020 was contacted.

[6] What are we waiting for in Wi-Fi 7, IEEE 802.11be? Archived May 10, 2020. Date appeals 20.08.20.

[7] Asma Amraoui and Badr Benmammar 2014 *Réseaux de radio cognitive: Allocation des ressources radio et accès dynamique au spectre*

Simulation in the study of operating experience and application of the latest technology for the repair of rod plunger pumps

M L Galimullin, M Ya Khabibullin and R I Suleimanov

Ufa State Petroleum Technological University, Branch of the University in the City of Oktyabrsky, 54a, Devonskaya St., Oktyabrsky, Republic of Bashkortostan, 452607, Russian Federation

E-mail: minivaris02@mail.ru, info@of.ugntu.ru

Abstract. Insufficient attention has been paid to the issues of repair and increase of the service life of plunger borehole pumps used in the oil production industry. The oil company Bashneft aims to increase the life and the IRP of borehole pumps. For this purpose, a lot of work has been done to analyze downhole equipment failures across the entire stock of oil wells equipped with sucker rod pumps. To repair the plunger pair of sucker rod pumps, all the documents (a technical task and technical conditions for repairs) were developed and the technological process was prepared. For the fields NGDU Ufaneft and NGDU Krasnokholmneft, the best results were achieved by pumps with a landing group (0.06–0.11), and for NGDU Ishimbayneft - pumps with a landing group (0.03–0.06). When repairing a pump, one can choose a landing group (plunger-cylinder), which will increase the service life of a sucker rod pump that has previously worked out its service life.

1. Introduction

In the oil industry, the main stock of wells has been operated by sucker rod pumping units. The main operating element is a deep rod plunger pump. The economic efficiency of oil producing enterprises depends on the reliability of this unit. Therefore, improving the performance of downhole plunger pumps is a key task to reduce the cost of oil production [1].

2. Results

The reliability of borehole plunger pumps has increased over the past 10 years. If the inter-repair period t(IRP) did not exceed 300 days, currently it has reached 400 days or more. This is due to the use of advanced technologies in the manufacture of pumps and improvement of technology and quality of service during the operation of pumping units. Until now, insufficient attention has been paid to the issues of restoration and increase of the service life of plunger borehole pumps used in the oil production industry. The oil company Bashneft aims to increase the life and the IRP of borehole pumps. For this purpose, a lot of work has been done to analyze downhole equipment failures across the entire stock of oil wells equipped with sucker rod pumps. Data on failure modes of downhole equipment of one of the NGDU ANK "Bashneft" are summarized in Table 1 [2].

Table 1. Types of downhole equipment failures at NGDU "Tuimazaneft"

Types of failures	Number	%
Broken rods	185	23.3
Lapel rods	259	31.3
Polished stem flap	84	10.1
Breakage of tubing	4	0.5
Leakage of tubing	38	4.6
Pump failures due to wear of valve pairs and wear of a plunger pair	261	30.4
Well stock	831	100

Table 1 shows that pump failures account for 30.4% of the total number of failures. When considering the raised pumps, it was found that 72% of cases are worn or waxed valves and seized plungers, 18% - worn plunger pairs. Therefore, it was decided to repair pump plunger pairs (plunger-cylinder) and valve pairs (seat-ball) [3]. Economic calculations confirmed the inexpediency of repairing the valve pair, instead of replacing it with a new one. But a prerequisite for this is an incoming control and a tightness test on a vacuum tester. To organize the repair of the plunger pair of sucker rod pumps, all the documents (a technical task, technical conditions for repairs) were developed. To optimize repair costs, taking into account the large volume of accumulated pumps, RT-61701high-performance machines for honing (mechanical processing of cylinders), a stand for straightening cylinders, a plunger with a force of Q = 60 kN, a washing unit, a shaft furnace SShTsN-10- 60/9 for heat treatment of the cylinder, a machine for thermal spraying of the plunger, a pump test bench were installed [4].

Considering that the most critical unit of the pump is a plunger pair (cylinder-plunger), Tables 2, 3 show the technological processes of repairing the cylinder and plunger of a NN-44 submersible pump.

The sucker rod pump cylinder is one of the main parts that is of paramount importance. Therefore, operations 005, 006, 007 are performed on a unique honing machine with a bed length of 12 m, grade RT 6171. The final honing operation is performed with a honing head filled with 125^{x}12^{x}6 mm diamond stones LS6 80/63 100-M-2. The size of the allowance to be removed during finishing is 0.03 mm and the resulting roughness of the cylinder working surface is not less than R 0.32, which is checked with a special device manufactured by the Mitugoito company and by comparison with roughness samples [5].

Taking into account the 005 technological process, the honing operation restores the correct geometry (cylindricality) of the cylinder. In addition, there is a removal from its working surface along the entire length of ulcerative corrosion damage. Next, the thermal nitriding process is performed in the SShTsN10-60 / 9 shaft furnace. At the same time, 36 cylinders are nitride, and for quality control, witnesses are placed in each batch (metal samples from the same material); they are attached to the batch in three places along the length of the cylinders [6].

Nitriding is a process of chemical-thermal treatment, causing saturation of the metal surface with atomic nitrogen, which is formed during the dissociation of ammonia according to the scheme:

$$2NH_3 = 2N+6H. \quad (1)$$

The gas nitriding process can be carried out for parts made of structural and alloyed steels such as steel 40X, 40XH, 15XHMA, 38X2MYUA, 25X2M16, etc.

Nitriding is the final process of the heat treatment cycle used to increase the wear resistance and endurance limit of a rod pump cylinder.

Table 2. Technological process of sucker rod pump cylinder repair

No of the operation	Name of the operation	Name of equipment	Number and short description of transitions
001	Detecting	Detection table	Reveal a cylinder defect in accordance with TU-366570-001356 45-2000; if there are cracks, washes, reject the cylinder on the working surface
002	Washing	Washing unit O.ADx013109.00.000	Carry out a wash of the outer and inner surfaces of the cylinder until the complete removal of dirt and corrosion
003	Adjusting	Stand for straightening the cylinder ShGN 0595-158.000	Adjust the cylinder until the curvature is no more than 0.125/1500 mm or 0 . 3 mm over the entire length of the cylinder
004	Lathing	Screw-Vintorez machine 1465-5	Calibrate the thread M56x1 . 5-6 q at length L = 42s-2 sides
005	Honing	Honing machine RT-61701	1 - honing a hole with a diameter of 44.45 without unclamping the stones at a length of L = 5530 mm 2 - measure the inner diameter of the cylinder from 2 sides 3 – hone a hole with a diameter of 44.49, ensure the cylindricity of the channel
006	Heating	Shaft furnace SShTsN-10-60 / 9	1– collect 30 cylinders in a cassette 2 – nitride the surface "B" to a depth of h = 0.2-0.3 mm, achieve a hardness of 800 - 950 HV
007	Honing	Honing machine RT 61701	1 – hone the cylinder to a depth of h = 0.03 mm along the length L = 5530 m 2 – hone a hole 44.52
008	Locksmith	Marking table	1 - mark the font size PO-5 GOST 2930-82 by striking 2 - cylinder serial number 3 - nominal cylinder diameter 4 - date of cylinder repair (month, year)
009	Controlling	Control bench	1 - control of dimensions according to drawings 2 - roughness control should be ≤ Ra 32 3 - control of the hardness of the nitrided layer (on the sample) 800 - 980 HV

Taking into account the experience of chemical-thermal treatment of tractor engine cylinder liners at NPO VNIITMASH, PO Chelyabinsk Tractor Plant, and the OZNPO ANK Bashneft plant, a cylinder working surface hardness of up to 1100 HV was achieved. These results were achieved by alternating saturation and diffusion holding cycles. Nitriding was carried out at a temperature of 510-530°C in a two-stage regime. The two-stage supply of ammonia made it possible to reduce the duration of the heat treatment process by 15% and the consumption of ammonia by 40%; the data are given in Table 3. Then the pump plunger is repaired. The technological process of plunger repair (Table 4) represents the classic sequence of operations, which are carried out with thermal spraying.

Table 3. Two-stage heat treatment mode

Nitriding mode		HV	Effective layer thickness h_e, mm	Total layer thickness h_n, mm
Saturation period π_n>1.5	Diffusion period π_n<0.3			
520 °C 24 hours	520°C – 6 hours	1103	0.25	0.4
	560°C – 6 hours	1022	0.35	0.4
	600°C – 6 hours	1066	0.45	0.5
520–560 °C 6–18 hours	–	1048	0.3	0.45
	600°C – 6 hours	1003	0.5	0.6

Table 4. Technological process of sucker rod pump plunger repair

Operation number	Operations	equipment identification	Number and short description of transitions
005	Blanking	Control and location stand	Perform troubleshooting of the plunger according TU366570-00135645-2000. Reject in case of cracks of any nature and location, washing.
010	Adjusting	Straightening stand Q = 6 t	Adjust the plunger until a curvature of no more than 0.05 / 1200 mm is obtained.
015	Lathing	Screw-cutting lathe CA562 C200	1.Calibrate the M38 ˣ 1 thread. 5-6H with a tap along the length L = 30 from 2 sides Dor = 37. 026 2. Grind the outer surface to Ø 42 along the length L = 30, clean 3. Grind the outer surface to Ø 42. 4 at L = 1140 mm, clean
020	Shot blasting	Shot blasting machine OB-85	Blast the outer surface of the plunger
025	Thermal spraying	Spraying unit TRG-3, rotator	Spray the outer surface with powder PRNKh16SRZ TU14-1-3785-90 up to Ø 45.2 at L = 1140 mm
030	Floating	Burner G-2-04 GOST1077-79	Melt the sprayed surface to obtain the phenomenon of "fogging"
035	Lathing	Screw-cutting universal lathe SA-5622S-200	Groove 4 grooves keeping dimensions R 1; Ø39. 5; 5 and sizes 120; 320; 320 ˣ 3 = 960 mm.
040	Grinding	Round machine grinding 3M175	Grind the outer surface to Ø 44.5 over L = 1140 mm.
045	Locksmith	Marking table	Mark with the font size PO-3 GOST 2930-82, serial number, nominal diameter
050	Controlling	Control bench	1. Control the dimensions according to the drawing. 2. Surface roughness. 3. Control the hardness.

Wear-resistant powder PNx16S4R brand is used as a spraying. The powder consists of nickel and contains 16% chromium, 4% silicon and boron. The spraying process is composed of two stages: coating spraying and fusing. As a result of spraying, a hard (900-1000 HV) wear-resistant coating is formed on the plunger surface, which must be machined to obtain the required dimensions and roughness [7]. The technological process of plunger repair is completed by grinding the working surface of the plunger on a CA562C200 screw-cutting lathe for mating the plunger with the cylinder.

3. Discussion

The repaired parts, the cylinder and the plunger are subjected to the assembly operation, which is performed on a special stand (Fig. 1).

The assembled pump is delivered and placed for testing on a special bench. It is fixed with crimping yews in three places along the length of the pump, the piping of test pump 5 and the piping of receiving tank 9 are connected. The test is composed of three stages. At the first stage, the plunger is in position I, at the second stage - in position II and at the third stage in position III [8–10]. Using test pump 5, the pressure rises to P_{init} = 18.0 MPa and is maintained at t = 65 s, while the pressure should decrease to P_{fin} = 9.0 MPa; according to the volume of oil leakage through the annular gap (plunger-cylinder) at an oil temperature t = 20°C, the conjugation group is determined (Table 5).

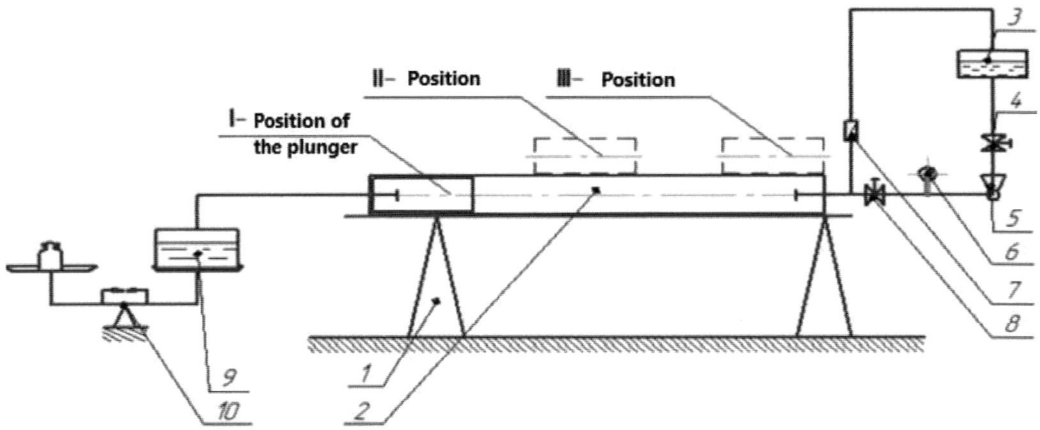

Figure 1. Test stand: 1 - test stand; 2 - sucker rod pump pump cylinder; 3 - container with oil M-10; 4, 8 - shut-off valves; 5 - LP pump for hydrotesting; 6 - manometer; 7 - safety valve; 9 - receiving container for oil; 10 - laboratory scales; I, II, III - the position of the plunger in the cylinder

Table 5. Parameters for determining the mating group of the sucker rod pump assembly during hydrotesting after overhaul

Group	1				2			
Conditional diameter of the pump, mm	P_{in}, MPa	P_{fin}. MPa	Test time sec	Leaking pump max – min, cm^3 / min at t=20°C	P_{in}, MPa	P_{fin}, MPa	Test time, sec	Leaking pump max – min, cm^3 / min at t=20°C
29	18.0	9.0	65	3.4–0.36	18.0	6.3	65	4.3–0.72
32	18.0	9.0	65	4.2–0.48	18.0	6.3	65	5.4–0.96
38	18.0	9.0	65	6.2–0.6	18.0	6.3	65	7.9–1.2
44	18.0	9.0	65	7.02–0.72	18.0	6.3	65	9.0–1.44
57	15.0	8.0	65	9.5–0.84	15.0	5.0	65	12.3–1.68
70	12.5	6.5	65	12.1–0.96	12.5	4.0	65	15.5–1.92
95	-	-	-	-	-	-	-	-

4. Conclusion

For the fields NGDU Ufaneft and NGDU Krasnokholmneft, the best results were achieved by pumps with a landing group (0.06–0.11), and for NGDU Ishimbayneft - pumps with a landing group (0.03–0.06). When repairing a pump, one can choose a landing group (plunger-cylinder), which will increase the service life of a sucker rod pump that has previously worked out its service life.

References

[1] Meirmanov A and Nekrasova I 2013 Mathematical models of a hydraulic shock *J. of mathematical analysis and applications* **408(1)** 76-90

[2] Haoran Zh, Yongtu L and Xingyuan Zh 2017 Sensitivity analysis and optimal operation control for large-scale waterflooding pipeline network of oilfield *J. of petroleum science and engineering* **154** 38-48

[3] Kuleshova L S, Kadyrov R R, Mukhametshin V V and Akhmetov R T 2019 Auxiliary equipment for downhole fittings of injection wells and water supply lines used to improve their performance in winter *IOP Conf. Ser.: Mater. Sci. Eng.* **560(1)** 012071. DOI: 10.1088/1757-899X/560/1/012071

[4] Korn G A and Korn T M 1984 *Mathematical Handbook for Scientists and Engineers: Definitions, Theorems, and Formulas for Reference and Review* (Moscow: Nauka)

[5] Heuzeroth F, Fritzsche J and Werzner E 2015 Viscous force - An important parameter for the modeling of deep bed filtration in liquid media *Powder technology* **283** 190-198

[6] Valeev A S, Kotenev Yu A, Mukhametshin V Sh and Sultanov Sh Kh 2019 Substantiation of the recovery of residual oil from low-productive and heterogeneous formations in Western Siberia by improving the waterflood system using gas and water-gas impacts *IOP Conf. Ser.: Mater. Sci. Eng.* **560(1)** 012204. DOI: 10.1088/1757-899X/560/1/012204

[7] Kuleshova L S, Kadyrov R R, Mukhametshin V V and Safiullina A R 2019 Design changes of injection and supply wellhead fittings operating in winter conditions *IOP Conf. Ser.: Mater. Sci. Eng.* **560(1)** 012072. DOI: 10.1088/1757-899X/560/1/012072

[8] Khabibullin M Ya 2019 Managing the reliability of the tubing string in impulse non-stationary flooding *J. of Phys.: Conf. Ser.* **1333** 052012. DOI: 10.1088/1742-6596/1333/5/052012

[9] Sun W and Mun-Hong H 2017 Forecasting and uncertainty quantification for naturally fractured reservoirs using a new data-space inversion procedure *European Assoc Geoscientists & Engineers Computational geosciences 15th Conf. on the Mathematics of Oil Recovery (ECMOR) (Amsterdam, Netherlands)* **21(5-6)** 1443-1458

[10] Khabibullin M Y 2019 Managing the processes accompanying fluid motion inside oil field converging-diverging pipes *J. of Phys.: Conf. Ser.* **1333** 042012. DOI: 10.1088/1742-6596/1333/4/042012

IOP Conf. Series: Materials Science and Engineering 1111 (2021) 012026 doi:10.1088/1757-899X/1111/1/012026

Optimization of traffic flows to increase the capacity of urban unsignalized junctions

P I Pospelov[1], S V Kostin[2], D S Martyakhin[1] and T K Komarova[1]

[1]MADI, 64, Leningradsky Ave., Moscow, 125319, Russia
[2]Committee on Architecture and Urban Planning of Moscow, Moscow, Russia

E-mail: tatianakkom@gmail.com

Abstract. Regular studies of traffic flows allow obtaining the actual parameters of the vehicle movement. Field observations using unmanned aerial vehicles help to optimize the research due to the wide capture field of the video camera, which allows comprehensively assessing traffic flows simultaneously utilizing all approaches to the intersection. When designing a road network, it is necessary to determine the capacity of all its elements. The characteristic elements are unsignalized intersections and junctions, the capacity of which largely depends on the intensity of the main and secondary directions, the behavior of drivers and the adoption by drivers of intervals between cars of the main traffic flow when entering the intersection.

1. Introduction

When designing the elements of the road network, it is necessary to ensure reliability and safety of traffic, as well as comfortable driving conditions for drivers characterized by the level of service.

A characteristic indicator of the intersection efficiency is its capacity during the design process.

Many factors affect the capacity values. Russian [2, 9, 10] and foreign researchers [4, 5, 6, 11, 12] note high influence of the traffic stream mix, traffic conditions, geometric parameters of intersections, as well as the difference in driver behavior on rural and urban roads.

The traffic stream mix and the dynamic characteristics of cars are fundamental factors that determine the capacity using the load equivalency factors [1, 3, 10]. The study of traffic flows on the example of Moscow space intervals [8] confirmed a decrease in the intervals between different types of cars in conditions of free movement compared to the study results of 1970s conducted for rural roads [10].

Foreign research [11, 12] also confirms that regular studies of traffic flows are necessary for a more accurate determination of road capability in modern traffic conditions. For example, the study of regulated intersections in Japan [12] is characterized by a systematic approach to the assessment of traffic flows by observing the same vehicles annually and by an integrated comparison after several years.

The issue of determining the boundary interval and the capacity of unsignalized intersections and junctions is relevant for a number of northern European countries [5], due to the fact that unsignalized intersections and junctions are one of the most common elements of the road network.

As a rule, the movement at unsignalized intersections and border intersections is determined by priority traffic signs for the adjacent direction. If there is no interference for drivers moving in the forward direction, except for cars moving in front of them, then the drivers of the secondary direction are forced to stop before intersection waiting for a sufficient interval in the main flow of vehicles to perform a safe left, right or forward maneuver.

The present study was conducted to determine the traffic stream mix in modern conditions and confirm changes in driver behavior using the example of boundary intervals.

2. Methods to determine the capacity of unsignalized intersections

The approach to determine the capacity of unsignalized intersections in Russia [2, 3] and abroad [1] is based on the probability that the drivers accept the secondary direction of intervals between cars of the main direction of movement (boundary intervals). Many authors [2, 4–6] note a high degree of influence of the drivers' behavior on the boundary interval.

There are two models of behavior of drivers depending on the traffic intensity at the intersection. The classic model of acceptance of the boundary interval by drivers in the conditions of a continuous queue of secondary direction is the basis for the method to determine the capacity in Russia, which is used in the current regulatory documents [2, 3] and in the guide for determining the traffic flow capacity [1].

The second model of behavior of drivers while turning is caused by a high tension in a situation of high congestion in the intersection. At the same time, the drivers of the secondary direction take smaller boundary intervals, and the drivers of the main direction are forced to reduce speed and give way to drivers of the secondary direction. A number of European studies [4–6] focus on the analytical comparison of the capacity values of unsignalized intersections depending on the adopted model of behavior of drivers by comparing experimental data and computer simulations.

The approach to determining the boundary interval accepted by the drivers of the secondary direction is based on the comparison of the probabilities of acceptance and deviation by the drivers of the intervals available for integration into the main stream and safe maneuver.

Due to the fact that it is impossible to constantly observe the continuous queue of cars at unsignalized intersections and junctions, the Raff method is adopted as a method for determining the boundary interval in this work [7].

3. Results

A number of unsignalized intersections in Moscow were chosen as the subject of the study. The observations were made under conditions close to favorable. Time periods of measurements are picked with good illumination, no precipitation and dry pavement. Traffic at all facilities is organized by priority signs for drivers.

Field observations were made using a quadcopter. Video shooting was performed near the intersection, while the drone hovered at the height from 50 to 100 m, which made it possible to record the simultaneous movement of all traffic flows when approaching the intersection [8]. The shooting lasted 15 minutes.

The data were collected during laboratory processing of video material, which made it possible to obtain the following database: the traffic stream mix, the intensity of traffic in the main and secondary directions of movement, the intervals between cars of the main and secondary directions of movement.

The study of the traffic stream mix showed a predominantly high proportion of motor vehicles – 80-90%. The results are shown in Figure 1.

Subsequent statistical processing of the database was carried out in order to determine the values and the probability of intervals between cars of the main direction of movement. The nature of the secondary direction is determined by the average distance between cars. The boundary intervals are also defined.

Figure 2 shows an example of comparison of distribution functions of cumulative frequency of received and rejected intervals according to Raff method [7].

A comparative analysis of the boundary intervals obtained by field observations and theoretical values currently used in Russia is given in Figure 3.

IOP Conf. Series: Materials Science and Engineering 1111 (2021) 012026 doi:10.1088/1757-899X/1111/1/012026

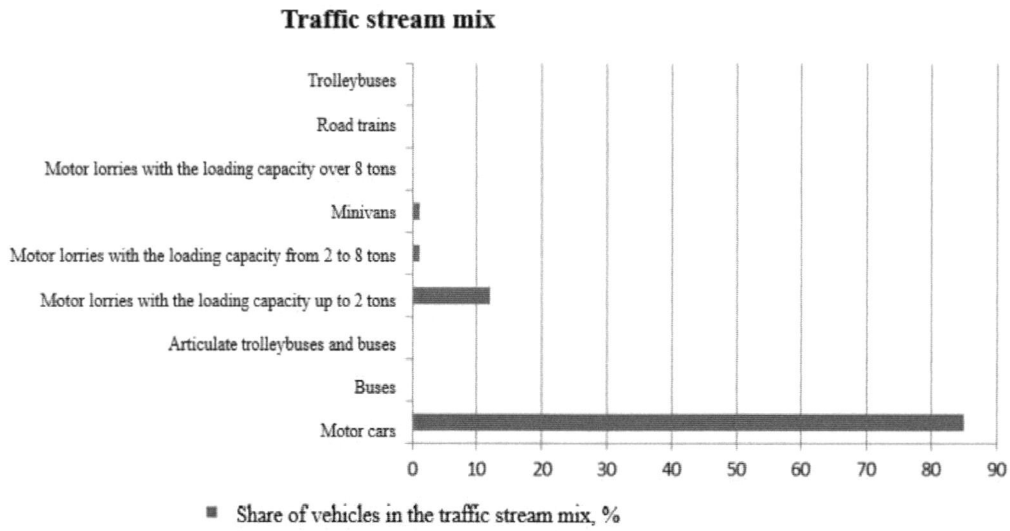

Figure 1. Traffic flow in Moscow

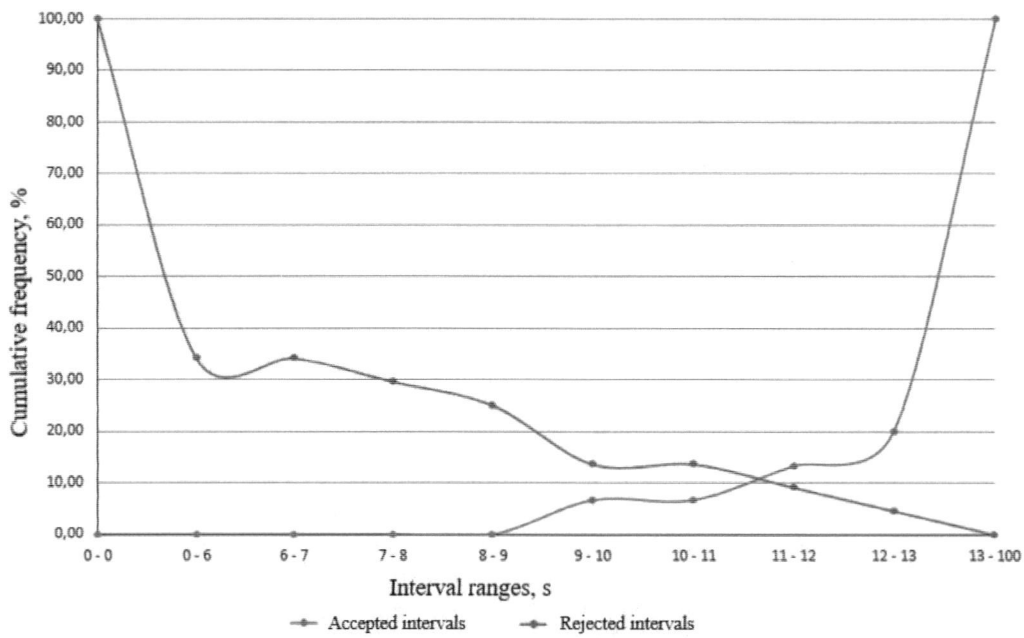

Figure 2. Definition of the boundary interval of a pair of directions for unsignalized left-turn junction

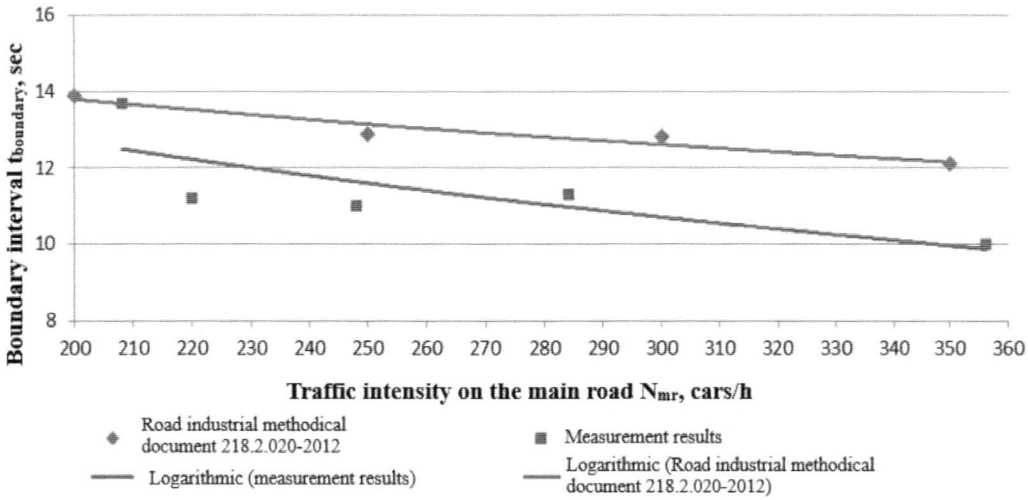

Figure 3. Dependence of the boundary interval on the intensity of the main direction when turning left

4. Conclusion

The capacity of unsignalized intersections depends on the composition of the flow, the intensity of the traffic stream mix and the probability of the drivers accepting boundary intervals of the secondary direction of movement.

During the study, the traffic stream mix, traffic intensity and the intervals between the cars of the main and secondary directions were determined. The values of the boundary intervals accepted by the drivers of the secondary direction at the entrance to the intersection are determined.

The values of boundary intervals for various planning solutions of unsignalized intersections will make it possible to clarify the capacity of unsignalized intersections and optimize the parameters of the road network.

The value of the boundary interval is influenced by a significant number of factors: geometric parameters of the intersection, number of lanes, intensity of the main and secondary directions, behavior of drivers at the entrance to the intersection, weather conditions, time of day, state of the road surface, equipment of intersections and organization of traffic, movement of pedestrians and cyclists.

References

[1] 2016 *Highway Capacity Manual* (Transportation Research Board, Washington, D.C.)
[2] Kostsov A V, Bakhirev I A, Borovik E N and Martyakhin D S 2017 *Transport layout of cities: Textbook* (Moscow: A-project)
[3] ODM 218.2.020-2012. *Methodological recommendations for assessing road capacity*
[4] Ruškić N and Mirović V 2021 Estimation of left-turn capacity at the unsignalized intersection *Simulation Modelling Practice and Theory* **106** 102170. DOI: https://doi.org/10.1016/j.simpat.2020.102170.
[5] Aakre E and Aakre A 2017 Modeling cooperation in unsignalized intersections *Procedia Computer Science* **109** 875-880. DOI: 10.1016/j.procs.2017.05.409.
[6] Pitlova E and Kocianova A 2019 Determination of priority stream volumes for capacity calculation of minor traffic streams for intersections with bending right-of-way *Transportation Research Procedia* **40** 875-882. DOI: 10.1016/j.trpro.2019.07.123.
[7] Raff M S and Hart J W 1950 *A Volume Warrant for Urban Stop Signs* (Eno Foundation for Highway Traffic Control, Saugatuck, Connecticut)
[8] Nemchinov D M, Martyakhin D S, Komarova T K and Pospelov P I 2020 Coefficients of traffic flow of cars when calculating the capacity of city streets *Highways* **9** 144-149

[9] Mikhailov A Yu and Golovnykh I M 2004 *Modern trends in the design and reconstruction of street and road networks of cities* (Novosibirsk: Nauka)
[10] Silyanov V V 1977 *Theory of traffic flows in the design of roads and the organization of traffic* (Moscow: Transport)
[11] Mehar Arpan, Chandra Satish and Senathipathi Velmurugan 2014 Passenger Car Units at Different Levels of Service for Capacity Analysis of Multilane Interurban Highways in India *Journal of Transportation Engineering* **140** 81-88. DOI: 10.1061/(ASCE)TE.1943-5436.0000615.
[12] Eri Aoyama, Keisuke Yoshioka, Sumio Shimokawa and Hirohisa Morita 2020 Estimating saturation flow rates at signalized intersections in Japan *Asian Transport Studies* **6** 100015. DOI: 10.1016/j.eastsj.2020.100015.

Reading and processing of analog signals of the software complex "Automated control system for the technological process of the associated production of concentrates of nanostructures MD1 and MD2"

V V Kondrat'ev, A D Kolosov, M V Konstantinova and E A Guseva

Irkutsk National Research Technical University, 83 Lermontov Street, Irkutsk, 664074, Russia

E-mail: kvv@istu.edu

Abstract. In this paper, the main order to automate the control of the technological process of the associated production of concentrates of nanostructures MD1 and MD2 is presented. The algorithm has been developed for processing incoming raw integer data into a real format in the parameter range, which is implemented in software and hardware at the level of the analog input module. The configuration of the algorithm is performed by specifying the input data using the program. The developed algorithms make it possible to implement an effective automated control system for further implementation in production.

1. Introduction

The main technological processes of silicon production are constantly evolving [1-5]. However, in addition to improving the main processes, it is necessary to develop the processing of waste from this production [6] and its automation [7-9]. Industrial waste from silicon production is used in construction [10-12] and other industries [13-15]. One of the main criteria for the applicability of the technologies being developed is the payback and economic efficiency for the business. One of the main ways to reduce costs is scaling up production and automating production processes. The efficiency of the developed automation systems, their algorithms and the minimization of hardware without losing the quality of the system operation are some of the key tasks of any production automation. The development processes of fundamentally new technologies require economical approaches to the development of automation systems to increase the investment attractiveness of the technological process as a whole.

2. Purpose and characteristics

This algorithm is designed to process incoming raw integer data into real format in the parameter range. The whole algorithm is implemented in software and hardware at the level of the analog input module OWEN MV110-8AS. The configuration of the algorithm is performed by specifying the input data using the M110 Configurator software.

It is used to process signals in a unified format (4-20 mA, 0-10 V, etc.) and signals from resistance thermometers, ohmic resistance, thermocouples. Input data are represented in the WORD, REAL types; output data in type WORD and REAL.

Information used

Table 1. Input data.

No.	Input information array name	Description	Data type	Range of values	Notes
1	ComF	Input filter type (common to all channels of the module)	WORD	0 .. 4	0 - filter off. 1 - 50 Hz 1st ord. 2 - 50 Hz 2 ord. 3 - 50 Hz 4 ord. 4 - 200 Hz 1 ord.
2	In-t	Connected sensor type	WORD	0 .. 4	0 - channel off. 1 - 4..20 mA 2 - 0..20 mA 3 - 0..5 mA 4 - 0..10 V
3	Ain.H	Upper limit of the measured parameter	REAL	$-3.402823e^{38}$ .. $3.402823e^{38}$	Default 20000.0
4	Ain.L	The lower limit of the measured parameter	REAL	$-3.402823e^{38}$ .. $3.402823e^{38}$	Default 0.0
5	dP	Decimal point offset	WORD	0..4	Default 0
6	Peak	Peak filter	WORD	0..200	Default 200 (range / sec)
7	in.Fd	Constant time for exponential filter	WORD	10..10000	Default 10(мs)
8	Out.F	Output filter type	WORD	0..16	0 - filter off. 1 - exponential. 2..16 - the length of the moving average

Solution results

Table 2. Output data.

No.	Input information array name	Description	Data type	Range of values	Notes
1	SRD	Signal processing output flags	WORD	0 .. 61455	0 - measurement is successful 61440 - Value knowingly wrong 61446 - Data not ready 61447 - The sensor is disabled 61450 - Measured value too big 61450 - Measured value too little 61453 - Open sensor 61455- Incorrect calibration factor.
2	Read	Output processed parameter value	REAL	-3.402823e[38] .. 3.402823e[38]	

Mathematical description

The ADC converts an analog signal into a digital code. The conversion frequency for each measurement channel is 1600 Hz.

Low Pass Filter:

- cutoff frequency, 2 kHz;
- slope of the frequency response in the far zone - minus 40 dB per decade

Peak filter: limiting the rate of change of the signal provides effective filtering of impulse noise. The operation of the signal change rate limiter is carried out in accordance with the formula:

$$Y_i = \begin{cases} X_{i-1} + P & \text{if} \quad (X_i - Y_{i-1}) > P \\ X_{i-1} - P & \text{if} \quad (X_i - Y_{i-1}) > -P \\ X & \text{else} \end{cases} \tag{1}$$

where Xi - signal at the input of the limiter at the current time; Y_i - signal at the limiter output; X_{i-1} - signal at the limiter input in the previous cycle (5 ms) of measurement; Y_{i-1} - signal at the output of the limiter in the previous cycle (5 ms) of measurement; P is the speed limit parameter calculated by the formula:

$$P = \frac{Peak \cdot \text{measurement range}}{200} \tag{2}$$

Rate limiting is controlled by the Peak parameter. The value of the limiting parameter is set in fractions of the change in the input signal relative to the measured range per second.

The input filter provides suppression of interference with frequencies that are multiples of the frequency of an industrial network of 50 Hz, increases the resolution of the meter by averaging and decreases the sampling frequency of the measured signal from 1600 to 200 Hz.

The control of the input filter for all eight measuring channels simultaneously is determined by the ComF parameter.

Output filter: The sampling frequency of the input and output of the output filter is the same and equal to 200 Hz. The output filter control determines the OutF parameter:

0 - output filter is disabled;

1 - filter with exponential impulse response is on (analogue of RC filter).

The filter has an attenuation of minus 3 dB at a frequency of 1 / 2nT (T is the filter time constant in ms, set by the in.Fd. parameter).

The frequency response of the filter has a slope of minus 20 dB per decade.

2 ... 16 - a first-order moving average filter with a length of, respectively, from 2 to 16. When the filter length is a multiple of four, the filter provides suppression of industrial frequency 50 Hz noise.

Coercion to measuring range:

Read = (Ain.H - Ain.L) / 1023 * Val + Ain.L

For more details on the operation of the input and output digital filters of the device, see the Help "Functioning of the input and output digital filters of the MV110-AC device" posted on the company's website: www.owen.ru.

Test case requirements

Input data are selected from valid ranges of values.

Table 3. Input data for testing.

No.	Input information array name	Selection for testing
1	ComF	3
2	In-t	0 or 1, 2-4 not used
3	Ain.H	Upper range limit
4	Ain.L	Lower range limit, Ain.L < Ain.H
5	dP	0
6	Peak	4
7	in.Fd	10
8	Out.F	8

Table 4. Output data.

No.	Input information array name	Test results
1	SRD	Should take values depending on the value of the current at the module input, which is in the range 0..24 mA: 0 - measurement is successful 61440 - Value knowingly wrong 61446 - Data not ready 61447 - Sensor disabled 61450 - Measured value is too big 61451 - Measured value is too little 61453 - Open sensor 61455 - Incorrect calibration factor
2	Read	The numerical value of the output value in real format, reduced to the range Ain.L..Ain.H, with the value SRD = 0, in other cases the value NAN (English Not-a-Number, "not number") is displayed

3. Conclusion

In order to automate the control of the technological process of the associated production of concentrates of nanostructures MD1 and MD2, an algorithm has been developed for processing incoming raw integer data into a real format in the parameter range, which is implemented in software and hardware at the level of the analog input module. The configuration of the algorithm is performed by specifying the input data using the program. The developed algorithms make it possible to implement an effective automated control system for further implementation in production.

References

[1] Rozhikhina I D, Nokhrina O I, Yolkin K S and Golodova M A 2020 Ferroalloy production: State and development trends in the world and Russia *IOP Conf. Ser.: Mater. Sci. Eng.* **866** 012004
[2] Yolkin K S, Yolkin D K, Nemarov A A, Sysoev I A and Karlina A I 2018 Conduct of reduction smelting of metallic silicon: Theory and practice *IOP Conf. Ser.: Mater. Sci. Eng.* **411** 012029
[3] Sivtsov A V, Yolkin K S, Kashlev I M and Karlina A I 2020 Processes in the Charge and Hearth Zones of Furnace Working Spaces and Problems in Controlling the Batch Dosing Mode during the Smelting of Industrial Silicon and High-Silicon Ferroalloys *Metallurgist* **64(5-6)** 396-403
[4] Strakhov V M, Surovtseva I V, Elkin D K, Elkin K S and Cherevko A E 2012 Low-ash carbon reducing agents for electrothermal silicon production *Coke and Chemistry* **55(5)** 172-175
[5] Nekrasov I V, Sheshukov O Y, Metelkin A A, Sivtsov A V and Tsymbalist M M 2016 Slag conditions in electrosmelting: A review *Steel in Translation* **46(6)** 435-442
[6] Kondratiev V V, Govorkov A S, Kolosov A D, Gorovoy V O and Karlina A I 2017 The development of a test stand for developing technological operation "flotation and separation of MD2. The deposition of nanostructures MD1" produce nanostructures with desired properties *Int. J. of Appl. Engineering Research* **12(22)** 12373-12377
[7] Ershov V A, Gorovoy V O, Nebogin S A, Govorkov A S and Suhanov A S 2019 System of process control for associated concentrate nanostructures production *J. of Phys.: Conf. Ser.* **1384** 012010
[8] Nebogin S A, Ershov V A, Gorovoy V O, Govorkov A S and Suhanov A S 2019 The main types of jigging machines structures *J. of Phys.: Conf. Ser.* **1384** 012034
[9] Ershov V A, Kondratiev V V, Karlina A I, Kolosov A D and Sysoev I A 2018 Selection of control system parameters for production of nanostructures concentrates *J. of Phys.: Conf. Ser.* **1118(1)** 012014
[10] Svintsov A P, Galishnikova V V and Stashevskaya N A 2020 Dataset on the effect of nano-modified additives of concrete mixes technological properties for winter concreting *Data in Brief* **31** 105756
[11] Galishnikova V V, Abdo Sh and Fawzy A M 2020 Influence of silica fume on the pervious concrete with different levels of recycled aggregates *Magazine of Civil Engineering* **93** 71-82
[12] Ponomarev A, Steshenko D and Rassokhin A 2018 Development of technology for production of fire-resistant nanocomposite constructional rebar and structural elements based on it *MATEC Web of Conferences* **245** 04001
[13] Konyuhov V Yu, Konstantinova M V and Gladkih A M 2019 Determination of restored units spectrum of equipment and development of the assembly unit repair method at industrial enterprises *J. of Phys.: Conf. Ser.* **1353** 012047
[14] Konyuhov V Yu, Gladkih A M, Galyautdinov I I and Severina Y D 2019 Economic aspects of green technologies *IOP Conf. Ser.: Earth Env.* **350** 012036
[15] Ivanchik N N, Balanovsky A E, Shtayger M G, Sysoev I A and Karlina A I 2018 Capability enhancement of production of activating fluxes for arc welding using ultradispersed products of silicon waste processing *IOP Conf. Ser.: Mater. Sci. Eng.* **411** 012035

Control of the servo drive and the regulating body of the software

M V Konstantinova, E A Guseva, V V Kondrat'ev and A D Kolosov

Irkutsk National Research Technical University, 83 Lermontov Street, Irkutsk, 664074, Russia

E-mail: kvv@istu.edu

Abstract. This paper presents control of the servo drive and the regulating body of the software complex "Automated process control system for the associated production of concentrates of nanostructures MD1 and MD2". The developed algorithm is designed to control a two-position valve with an electric drive by "open / close" commands. It allows to carry out: control in manual and automatic mode, control of extreme positions and excess torque on the drive shaft; blocking of work in automatic mode with a command to close; determination of the priority of the close command; delay when reversing to stop the drive.

1. Introduction

At present, environmental issues [1, 2] and including of industrial waste as additives in various materials [3-7] are becoming increasingly important. One of these is micro silica. It is a waste, the volume of which is tens of thousands of tons per year in many countries of the world. Currently, various methods of enrichment and processing of industrial waste have been developed. In the course of developing a technology, it is also necessary to think about its automation, using advanced technologies in the field of mechanical engineering, cybernetics and robotics. One of such technologies and examples of automation are [8-12], which must be solved in conjunction with the development of silicon production processes [13-16]. After developing the technology as part of the development of the automated process control system, we developed the servo drive control and the software regulator.

2. Servo control

This algorithm is designed to control a motorized ON / OFF valve with open / close commands. The algorithm allows:

- control in manual and automatic mode;
- control of extreme positions and excess torque on the drive shaft;
- blocking of work in automatic mode with a command to close;
- priority of the close command;
- delay in reverse to stop the drive.

Input data are represented in the BOOL and TIME types; output in type BOOL (Tables 1-4). The solution algorithm is presented in Figure 1.

Information used

Table 1. Input data.

No	Input information array name	Description	Data type	Range of values
1	MAN	Manual control	BOOL	TRUE, FALSE
2	MAN_OPEN	"Open" command in manual mode	BOOL	TRUE, FALSE
3	MAN_CLOSE	"Close" command in manual mode	BOOL	TRUE, FALSE
4	EN	Work permit command in automatic	BOOL	TRUE, FALSE
5	V_OPEN	mode	BOOL	TRUE, FALSE
6	V_CLOSE	Open command in automatic mode	BOOL	TRUE, FALSE
7	FO	Close command in automatic mode	BOOL	TRUE, FALSE
8	FC	End position signal "open"	BOOL	TRUE, FALSE
9	FAULT	Closed end position signaling	BOOL	TRUE, FALSE
10	T_OPEN	Signaling of excess torque on the drive shaft	TIME	T#1ms .. T#9999s
11	T_CLOSE	Opening time	TIME	T#1ms .. T#9999s
12	T_STOP	Stop time of the electric drive	TIME	T#1ms .. T#9999s

Solution results

Table 2. Output data.

No	Input information array name	Description	Data type	Range of values
1	TGT_STATE	Steady state command	BOOL	TRUE, FALSE
2	WORKING	Work signal	BOOL	TRUE, FALSE
2	STOPING	Stop signal	BOOL	TRUE, FALSE
4	FB ERR	Feedback error	BOOL	TRUE, FALSE
5	OPEN	Opening signal	BOOL	TRUE, FALSE
6	CLOSE	Close signal	BOOL	TRUE, FALSE

Test case requirements

Input data are selected from valid ranges of values.

Table 3. Input data for testing.

No	Input information array name	Selection for testing
1	MAN	TRUE - manual mode, commands EN V_OPEN V CLOSE are ignored, FALSE - automatic
2	MAN OPEN	On the front command to open in manual mode
3	MAN CLOSE	On the edge of the command to close in manual mode
4	EN	TRUE - automatic control is allowed, FALSE - V_OPEN V_CLOSE

IOP Conf. Series: Materials Science and Engineering 1111 (2021) 012028 doi:10.1088/1757-899X/1111/1/012028

Table 3. Input data for testing.

No	Input information array name	Selection for testing
		commands are ignored and the close command is set at the block level
5	V OPEN	On the front command to open in automatic mode
6	V CLOSE	On the edge of the command to close in automatic mode
7	FO	Signal from actuator: TRUE open end position, FALSE otherwise
8	FC	Signal from actuator: TRUE closed end position, FALSE otherwise
9	FAULT	Signal from the drive: TRUE excess torque on the shaft of the electric drive, FALSE otherwise
10	T OPEN	In the range of values
11	T CLOSE	In the range of values
12	T STOP	In the range of values

Solution Algorithm

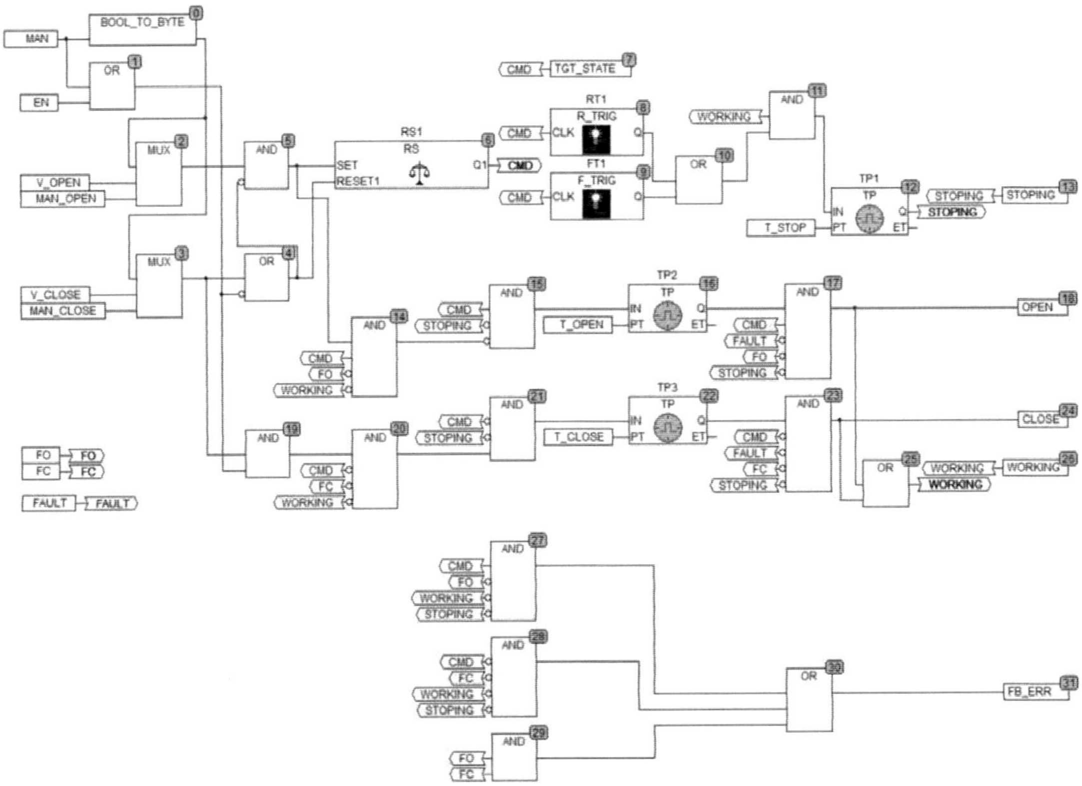

Figure 1. Solution algorithm.

Table 4. Output data.

No	Input information array name	Test results
1	TGT_STATE	Steady state of the command at the block level TRUE - opening, FALSE - closing
2	WORKING	When a control signal is issued: TRUE - open or close, FALSE - no control signal
3	STOPING	When the steady state of a command is changed at the block level while the drive is running: TRUE - after the change, the time T STOP has not passed, FALSE - has passed
4	FB_ERR	Feedback error: TRUE - during WORKING, the end position has not been reached during T_OPEN or T_CLOSE, FALSE - otherwise or after a new command has been given from the input
5	OPEN	Open signal: TRUE - when the steady state of the command at the block level changes to open for the time T_OPEN, FALSE - otherwise either FO or FAULT or STOPING
6	CLOSE	Closing signal: TRUE - when the steady state of the command at the block level changes to close for the time T_CLOSE, FALSE - otherwise, either FC or FAULT or STOPING

Here we see that the rational algorithm allows manual and automatic control, and mercury control is available when needed. The logical use is shown of control of extreme positions and excess torque on the drive shaft. An innovative blocking of automatic operation with a closing command is proposed. The priority of the close command is analyzed.

Reverse delay set to stop drive.

3. Results and discussion

This algorithm is designed to control a three-position regulator (PO) with open / close / stop commands with a built-in automatic stepping PI-controller.

It is used to control a three-position regulating body with the possibility of manual control and automatic regulation of various process parameters.

Input data are represented in the WORD, BYTE, REAL, INT types; the output is in type BYTE and REAL (Tables 5-8).

Information used

Table 5. Input data.

No	Input information array name	Description	Data type	Range of values
1	COM_RST	Resetting controller settings	Bool	
2	LMNS ON	Manual control	Bool	
3	LMNUP	"Open" command in manual mode	Bool	
4	LMNDN	"Close" command in manual mode	Bool	
5	LMNR_HS	End position signal "open"	Bool	
6	LMNR_LS	Closed end position signaling	Bool	
7	CYCLE	Block call time	Time	T#1ms .. T#9999s
8	SP_INT	Regulation task	Real	

Table 5. Input data.

No	Input information array name	Description	Data type	Range of values
9	PV_IN	Adjustable value (position PO)	Real	
10	GAIN	Gain	Real	0 .. 99999
11	TI	Integration time	Time	
12	T_STOP	Stop time of the electric drive	TIME	T#1ms .. T#9999s

Solution Algorithm

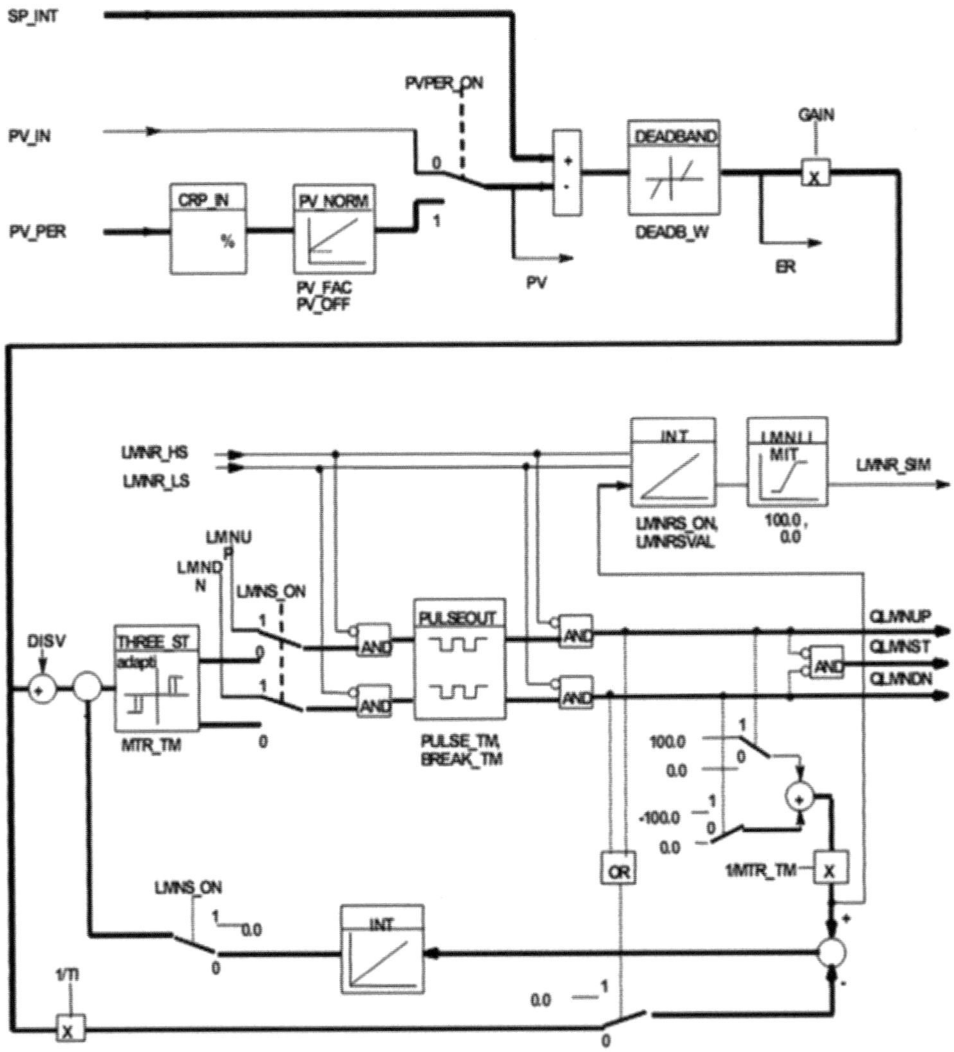

Figure 2. Solution Algorithm.

Solution results

Table 6. Output data.

No	Input information array name	Description	Data type	Range of values
1	QLMNUP	Opening signal	Bool	QLMNUP
2	QLMNDN	Close signal	Bool	QLMNDN
3	QLMNST	Stop signal	Bool	QLMNST
4	ER	The amount of mismatch (error)	Real	ER
5	PV	Adjustable value	Real	PV

Test case requirements
Input data are selected from valid ranges of values.

Table 7. Input data for testing.

No	Input information array name	Selection for testing
1	COM RST	False
2	LMNS_ON	Check operation with manual control - True, in automatic mode - False
3	LMNUP	When LMNS ON = True, True - command to open
4	LMNDN	With LMNS ON = True, True - close command
5	LMNR HS	True, False
6	LMNR LS	True, False
7	CYCLE	T#100ms
8	SP INT	In the range of values
9	PV IN	In the range of values
10	GAIN	In the range of values
11	TI	In the range of values
12	DEADB W	In the range of values

Table 8. Output data.

No	Input information array name	Test results
1	QLMNUP	With LMNS ON = True and LMNUP = True and LMNR HS = False: True – opening With LMNS_ON = False and LMNR_HS = False: True - opening according to the regulation law
2	QLMNDN	With LMNS ON = True and LMNDN = True and LMNR LS = False: True - close With LMNS_ON = False and LMNR_LS = False: True - closing according to the regulation law
3	QLMNST	With QLMNUP = False and QLMNDN = False: True - stop

Table 8. Output data.

No	Input information array name	Test results
4	ER	Numerical value - difference between PV IN - SP INT
5	PV	PV IN value

To check the adequacy of the developed algorithm, many examples and test problems were worked out, some of which are given above. After analyzing the test results, one can draw conclusions about the effectiveness of the developed control of the servo drive and the regulating body of the software.

4. Conclusion

In order to develop the including of industrial waste in subsequent production cycles and the production of additives on their basis, a system for their automation was created.

The developed algorithm is designed to control a two-position valve with an electric drive by "open / close" commands and allows to carry out: control in manual and automatic mode, control of extreme positions and excess torque on the drive shaft; blocking of work in automatic mode with a command to close; determination of the priority of the close command; delay when reversing to stop the drive.

References

[1] Konyuhov V Yu, Gladkih A M, Galyautdinov I I and Severina Y D 2019 Economic aspects of green technologies *IOP Conf. Ser.: Earth Env.* **350** 012036

[2] Konyuhov V Yu, Konstantinova M V and Gladkih A M 2019 Determination of restored units spectrum of equipment and development of the assembly unit repair method at industrial enterprises *J. of Phys.: Conf. Ser.* **1353** 012047

[3] Svintsov A P, Galishnikova V V and Stashevskaya N A 2020 Dataset on the effect of nano-modified additives of concrete mixes technological properties for winter concreting *Data in Brief* **31** 105756

[4] Galishnikova V V, Abdo Sh and Fawzy A M 2020 Influence of silica fume on the pervious concrete with different levels of recycled aggregates *Magazine of Civil Engineering* **93** 71-82

[5] Ponomarev A, Steshenko D, Rassokhin A 2018 Development of technology for production of fire-resistant nanocomposite constructional rebar and structural elements based on it *MATEC Web of Conferences* **245** 04001

[6] Gozbenko V E, Kargapoltsev S K and Karlina A I 2019 Synthesis and structure of sulfur-containing polymers based on polymer industrial waste applied for rail lubrication *IOP Conf. Ser.: Earth Env.* **229** 012021

[7] Gozbenko V E, Kargapoltsev S K and Karlina A I 2019 Environmental benefits of new industrial waste-based lubricant compositions *IOP Conf. Ser.: Earth Env.* **229** 012020

[8] Kondratiev V V, Nebogin S A, Sysoev I A, Gorovoy V O and Karlina A I 2017 Description of the test stand for developing of technological operation of nano-dispersed dust preliminary coagulation *Int. J. of Appl. Engineering Research* **12(22)** 12809-12813

[9] Ershov V A, Gorovoy V O, Nebogin S A, Govorkov A S and Suhanov A S 2019 System of process control for associated concentrate nanostructures production *J. of Phys.: Conf. Ser.* **1384** 012010

[10] Nebogin S A, Ershov V A, Gorovoy V O, Govorkov A S and Suhanov A S 2019 The main types of jigging machines structures *J. of Phys.: Conf. Ser.* **1384** 012034

[11] Ershov V A, Kondratiev V V, Karlina A I, Kolosov A D and Sysoev I A 2018 Selection of control system parameters for production of nanostructures concentrates *J. of Phys.: Conf. Ser.* **1118** 012014

[12] Nekrasov I V, Sheshukov O Y, Metelkin A A, Sivtsov A V and Tsymbalist M M 2016 Slag conditions in electrosmelting: A review *Steel in Translation* **46(6)** 435-442

[13] Sivtsov A V, Yolkin K S, Kashlev I M and Karlina A I 2020 Processes in the Charge and Hearth Zones of Furnace Working Spaces and Problems in Controlling the Batch Dosing Mode during the Smelting of Industrial Silicon and High-Silicon Ferroalloys *Metallurgist* **64(5-6)** 396-403
[14] Yolkin K S, Yolkin D K, Nemarov A A, Sysoev I A and Karlina A I 2018 Conduct of reduction smelting of metallic silicon: Theory and practice *IOP Conf. Ser.: Mater. Sci. Eng.* **411** 012029
[15] Strakhov V M, Surovtseva I V, Elkin D K, Elkin K S and Cherevko A E 2012 Low-ash carbon reducing agents for electrothermal silicon production *Coke and Chemistry* **55(5)** 172-175
[16] Rozhikhina I D, Nokhrina O I, Yolkin K S and Golodova M A 2020 Ferroalloy production: State and development trends in the world and Russia *IOP Conf. Ser.: Mater. Sci. Eng.* **866** 012004

Bluetooth based technology for industrial personnel local positioning

I M Daudov[1], M N Orobey[2] and I V Ignatev[3]

[1]Chechen State University, 17a, Dudaev Boulevard, Groznyy, 364015, Russia
[2]Irkutsk National Research Technical University, 83, Lermontov str., Irkutsk, 664074, Russia
[3] Bratsk State University, 40, Makarenko str., Bratsk, 665709, Russia

E-mail: ibr024@mail.ru

Abstract. The article presents the implementation of radio beacon functions in Bluetooth technology. The architecture of Bluetooth Low Energy technology is considered. The stages of development of Bluetooth technology from version 4.0 to version 5.2 are given. Advertising channels used by the beacon in the frequency range 2402-2480 MHz are shown. Functions and application of Bluetooth beacons are described.

1. Introduction

To ensure the safety of operations in dangerous industrial environment it is crucial to track the location of personal. There are number of various options to track person in the indoor-environment [1]. The most inexpensive and affordable tracking system is one that based on common devices as smartphones. Here the option with Bluetooth Low Energy beacons is discussed.

Bluetooth Low Energy Technology (BLE-Bluetooth Low Energy), sometimes called Bluetooth Smart, is a lightweight subset of classic Bluetooth that was introduced by the Bluetooth SIG consortium (Special Interest Group) as part of the basic Bluetooth 4.0 specification. This is Wireless Personal Area Network (WPAN). There are many wireless protocols available for engineers and product designers, but the main feature of BLE is the ability to communicate with very low power consumption. Currently, this is by far the easiest way to develop applications for interacting with any modern mobile platform (iOS, Android, Windows phones, etc.). The BLE architecture is shown in Fig. 1 [2, 3].

The BLE stack consists of two main architectural components: a host (master) and a controller. Each of them contains different stack layers.

Physical Layer (PHY). The transmitter uses GFSK (Gaussian frequency shift keying) modulation and operates in an unlicensed 2.4 GHz frequency band. Using this PHY layer, BLE provides a data rate of 1 Mbps to 3 Mbps (Bluetooth v5.0). It uses a frequency hopping transceiver. Two variants of the PHY level are indicated, namely uncoded and encoded data. The time division duplex (TDD) topology is used in both PHY modes.

The Link Layer is responsible for advertising, scanning, and creating/maintaining connections. The role of BLE devices is changed in peer-to-peer and broadcast modes. In general, Direct Test Mode is part of the Link Layer.

The Host Controller Interface (HCI) defines a series of commands that the host can use to communicate with the controller and events that the controller uses to communicate with the host. A significant part of BLE intelligence is implemented by the controller. This allows the master to stay asleep longer and wake up to the controller signal.

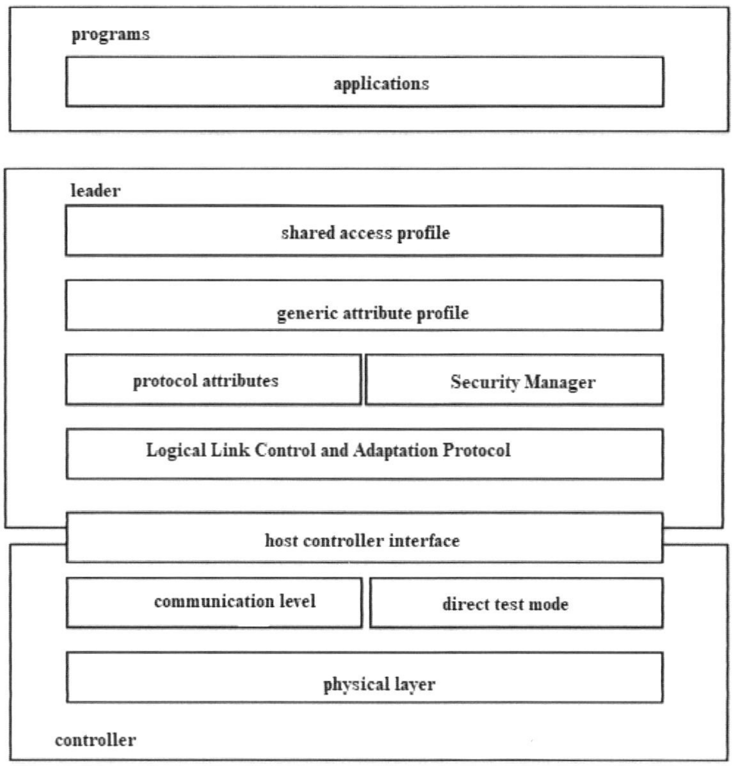

Figure 1. Architecture Bluetooth LE

The Logical Link Control & Adaptation Protocol (L2CAP) is responsible for the multiplexing protocol, flow control and segmentation, and the reassembly of data blocks.

Attribute Protocol (ATT) attributes are one of the main mechanisms by which applications of connected Bluetooth devices communicate with each other using PDUs defined by the protocol and procedures in higher-level specifications. This level allows the BLE device to provide specific data items or attributes.

Security Manager (SMP). This level of Security Manager provides methods for interfacing devices and distributing keys. It offers services for other protocol stack layers for secure connection and communication between BLE devices.

General Attribute Profile (GATT). This layer is the service structure that defines the routines to use the ATT. Data are exchanged between two BLE devices using these subroutines. Applications and/or profiles will use GATT directly.

Generic Access Profile (GAP). This layer directly interacts with the application layer and/or the profiles contained therein. It manages the device discovery and services associated with connecting BLE devices, and supports the launch of security features.

Applications. The layer controls the device discovery and services associated with connecting BLE devices. The levels of the BLE protocol stack communicate with applications and profiles as needed. Bluetooth applications interact with profiles. A profile defines vertical interactions between layers as well as peer-to-peer interactions between specific layers between devices. All profiles/applications run on top of the GAP/GATT layers of the BLE protocol stack.

BLE is a wireless personal network technology developed and supported by the Bluetooth Special Interest Group (SIG) consortium. To evaluate BLE beacons, it is important to understand the difference between classic Bluetooth and Bluetooth Low Energy. Classic Bluetooth consumes more power and transmits data over long distances, which is suitable for headsets and speakers. Bluetooth Low Energy transmits less data in a smaller frequency range, therefore consuming much less energy. BLE beacons transmit small amounts of data at regular intervals.

The development of Bluetooth technology has more than 20 years. There are different versions of Bluetooth. Version 4.0 and higher is called BLE. The last of the series is currently version 5.2. The main stages of improvement of this technology, starting with version 4.0, are summarized in Table 1 [4].

Table 1. Main stages of Bluetooth technology development

Version	Date	Properties
4.0	2010	Low power mode, encryption.
4.1	2013	Coexistence with mobile wireless communication.
4.2	2014	Connection level privacy, IPv6 support profile.
5.0	2016	1, 2, or 3 Mbps, low power consumption, optional 802.11 feature for speeds up to 24 Mbps, mesh network.
5.1	2019	Determine the exact location of the device by analyzing the direction to connected devices. One of the devices must have an array of several antennas.
5.2	2020	Dynamic optimization of transmission power, timing of transmission channels.

One of the most important aspects of Bluetooth Low Energy is advertising. BLE announcements are also important for beacons, which have become popular for determining the location of objects. By using BLE for advertising, one can reduce power consumption, accelerate connectivity, and increase reliability.

Advertising communication occurs when a BLE device sends packets to all surrounding devices. The receiving device may then act on this information or connect to obtain additional information. Basically, this is what beacons do: they simply transmit packets using advertising channels 37, 38 and 39, as shown in Figure 2 [5].

Channels 37, 38 and 39 are used only for sending advertising packets. The rest is used to exchange data when connecting (this part is not the subject of our topic). The 2.4 GHz spectrum for Bluetooth is from 2402 MHz to 2480 MHz. BLE uses 40 channels with a width of 1 MHz, numbered from 0 to 39.

Channels 37, 38 and 39 are distributed over a 2.4 GHz spectrum. If only one advertising channel is blocked, the other channels are likely to be free because they are separated by a bandwidth of several MHz. The long distance between advertising channels helps BLE better cope with interference from Wi-Fi, classic Bluetooth, microwave ovens and other devices to ensure the success of advertising.

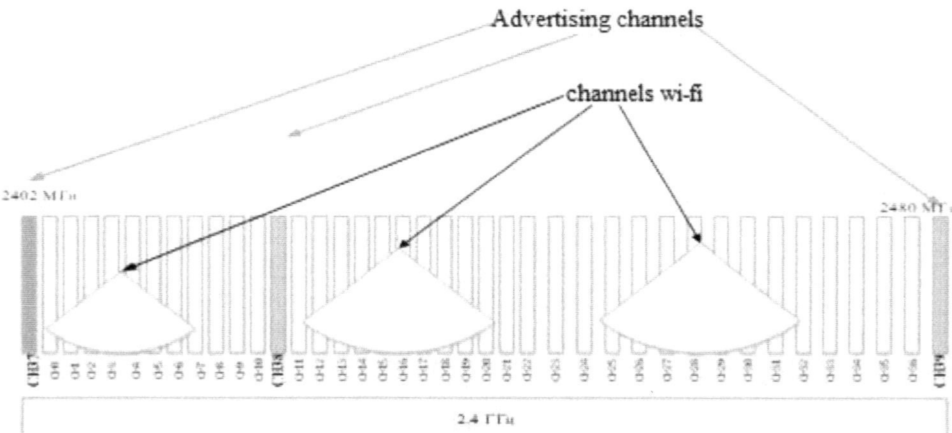

Figure 2. Advertising channels 37, 38 and 39

During advertising, the BLE device transmits the same packet through all 3 advertising channels, one after the other. A central device looking for devices or beacons will listen to these channels for advertising packages, which helps it detect nearby devices.

When the BLE device is in advertisement mode, advertisement packages are regularly sent to each advertisement channel. The time interval between a set of packets is fixed with a random delay. The interval is specified between a set of 3 packets. The larger the interval between advertising tags, the less accurate the location of the moving target will be.

2. Bluetooth Beacons Operating Principle

The transmitted data from the low-power Bluetooth device is formatted according to the Bluetooth core specification and consists of the components shown in Figure 3 [6].

The preamble (1 byte) is used to synchronize the transmitter and receiver. This is followed by the Access Address (4 bytes). On all advertising channels, it is the same. Next, a data packet (PDU) (2 to 39 bytes). In version 5.0, the length of the data packet is increased to 257 bytes. Three bytes of checksum (CRC) follow at the end of each advertisement packet.

Although Bluetooth 5.0 is known for its longer range and faster data transfer rates, the Bluetooth v5 promotional extensions extend the technology's capabilities. Instead of simply sending ad data through all 3 advertising channels (as shown in Figure 2), BLE allows you to combine ad packages together and use the remaining 37 channels that did not previously transmit ad data. This is used to obtain advertising data even in conditions of high interference.

Bluetooth 5 also allows the ad package to contain up to 257 bytes of data, which is much more than the 39 packets that were possible in Bluetooth 4.0, 4.1, and 4.2 [7].

In the Internet of Things (IoT - Internet of Things), one of the tasks is to estimate the cost of transmission in terms of power. Depending on transmitter power, Bluetooth devices are divided into three classes:

- Class 1 devices have a maximum output power of 100 mW (20 dBm) and provide communication range up to 100 meters;
- devices of class 2 have power up to 2.5 mW (4 dBm) and provide communication range up to 10 meters;
- devices of class 3 have power up to 1 mW (0 dBm) and communication range up to 1 meter.

The device IoT typically has a limited number of PDUs that it can send before the battery reaches a point where it cannot connect until electricity is restored or replenished. Suppose that the protocol iBeacon is declared in the interval of 600 ms, the packet length is 34 bytes, and the device (IoT) uses a battery CR2025 with a nominal capacity of 240 mA at 3 V. The beacon electronics consume 50 μA at 3 V.

We can now predict the lifetime of the beacon and transmission efficiency:

- power consumption of one beacon is 50 μA × 3B = 0.150 mW;
- the amount of information in bytes per second is 34 × (1 s/600 ms) × 3 channels = 170 bytes/s;
- the amount of information in bits per second is 170 bytes/s × 8 = 1360 bits/s;
- energy per bit is 0.150 mW/(1360 bps) = 0.110 μJ/bit;
- energy costs per advertisement are 0.110 μJ/bit × 34 bytes × 8 bits/byte = 29.92 μJ/advertising;
- energy stored in the battery: 240 mAh × 3 V × 3.6 s = 2592 J;
- battery life is (2592 J × (1,000,000 μJ/J))/((29.92 μJ/ad) ×(1 advertisement/0.6 s)) × 0.7 = 36385027 = 421 days = 1.15 years.

When an Android or iOS device is within reach of a Bluetooth beacon, it recognizes the signal sent by the beacon and can perform contextual tasks depending on what it receives. The beacon itself does nothing but transmit a signal. It does not receive data from the mobile device. Because it works in one direction thanks to the new Bluetooth 4.0 specification introduced in 2010 [8].

Bluetooth beacons are not really the Bluetooth SIG standard. Instead, they are what might be called "pseudo-standards" or formalized formats for radio-bus applications led by a major provider or group of companies. There are three main market standards for beacons: Apple's iBeacon, Google's Eddystone and Radius Networks' AltBeacon. All three pseudo-standards use the BLE broadcast methodology to place advertising packets on BLE channels 37, 38, and 39 to avoid traffic conflict with Wi-Fi in the unlicensed 2.4 GHz band for industrial, scientific, and medical (ISM) applications. Let us briefly consider the main differences between them [9].

Standard iBeacon. Apple was one of the first adherents of iBeacon technology in 2013. iBeacon defines a 30-byte packet to be broadcast at 100 ms intervals. A protocol developed by Apple that allows applications on smartphones to scan beacons in a certain range and display content when detected. Recommended when companies want to run Bluetooth marketing campaigns through their own apps. The Beaconstac SDK is an easy way to enable marketing and location analysis through a iBeacon-compatible BLE network.

The Eddystone standard was developed by Google Transnational Corporation in 2015. It can support four types of packages, Eddystone-URL, Eddystone-UID, Eddystone-EID, and Eddystone-TLM, until iBeacon supports these packages. The Eddystone-UID package is very similar to Apple's iBeacon, supports additional telemetry data with Eddystone-TLM.

Eddystone-URL is a single resource location. This frame allows the receiver to display web content based on the location of the beacon signal. To activate content, the app does not need to be installed. The content has a variable length and applies unique compression schemes to reduce the size of the URL to a limit of 17 bytes.

Eddystone-UID is a unique 16-byte beacon identifier with a 10-byte namespace and a six-byte instance. It uses the Google beacon registry to return attachments.

Eddystone-EID is a short-lived identifier for beacons requiring a higher level of security. There is no fixed namespace and identifier, identifiers are constantly rotating and require a permitted application for decoding. It uses the Google beacon registry to return attachments.

Eddystone-TLM broadcasts telemetry data about the beacon itself (battery level, time since inclusion, amount of advertising). It broadcasts along with URIs or URL packets.

Standard AltBeacon. Network Radius launched its beacon in 2014. It is an open source machine with a wide range of open devices with different types of beacon applications.

All three standards use the BLE broadcast mechanism to transmit advertising packets over the BLE channel. They use channels 37, 38, and 39 to avoid a Wi-Fi traffic conflict.

3. Scope of Bluetooth Beacons

Any Bluetooth mobile phone can be used to detect and view BLE devices. One of the most useful beacon services is that they can be used for advertising wherever there are people: in shopping centers, shops, parks and even at events. To use it, there must be a beacon-compatible device [5].

Bluetooth beacons will have a transformative effect on interaction with the physical world. They provide context awareness based on proximity, using the technology that most people in the world have - a smartphone with applications. The use of wireless technology to detect approximation is not news, but with the introduction of low-power Bluetooth features in 2010, beacons are now deployed on a wide scale. This is an important and significant technology for the IoT [10] and IIoT especially [11].

Bluetooth information can be environmental data (temperature, atmospheric pressure, humidity, etc.), micro-location data (asset tracking, retail, etc.) or orientation data (acceleration, rotation, etc.). These devices are used to transmit data using Bluetooth Low Energy signals. This technology can be used for interaction between autonomous drones, or to create a reference network for correcting the location of drones in unattended or abandoned mine workings [12].

Note that with Bluetooth capability, beacons can locate customers and send messages to their smartphones. Beacons are safe in the sense that they allow access only to applications and websites authorized by the user. This means that the tags only transmit information and provide assistance that has been authenticated by the user, providing user convenience [13-15].

4. Conclusion

The introduction of the Internet of Things contributes to the emergence of new technologies and services. One such technology is Low Power Bluetooth (BLE-Bluetooth Low Energy), which allows the use of BLE beacons using the latest advances in cellular mobile endpoints.

The article discusses the BLE architecture, the principles of BLE-beacons, provides an example of calculating the life of the beacon battery, and discusses the scope of BLE-beacons. The depth of consideration of BLE technology is designed for managers involved in advertising distribution.

References

[1] Bragin D, Kostyuchenko E, Faerman V, Kobzev A, Sidorov I 2021 Comparison of technologies of local patient positioning *International Journal of Advanced Research and Technology* **12(1)** 362-386.

[2] Townsend K, *Introduction to Bluetooth Low Energy.* Retrieved from: Https://learn.adafruit.com/introduction-to-bluetooth-low-energy

[3] *Woolley M, Bluetooth Core Specification Version 5.2.* Retrieved from: https://3pl46c46ctx02p7rzdsvsg21-wpengine.netdna-ssl.com/wp-content/uploads/2020/01/Bluetooth_5.2_Feature_Overview.pdf

[4] *Silicon Labs BG22 Secure Bluetooth 5.2 SoC promises a 10-year battery life.* Retrieved from: https://cnx-software.ru/2020/01/20/silicon-labs-bg22-secure-bluetooth-5-2-soc-obeshhaet-10-letnij-srok-sluzhby-batarei/

[5] *BLE ADVERTISING PRIMER.* Retrieved from: https://www.argenox.com/library/bluetooth-low-energy/ble-advertising-primer/

[6] *Why is the Bluetooth beacon so attractive to marketers??* Retrieved from: Https://www.mokosmart.com/ru/bluetooth-beacon/

[7] Lindh J, *Bluetooth low energy Beacons.* Retrieved from: Http://file.elecfans.com/web1/M00/00/05/pIYBAFnLi6KABs7RABvYrQep_v4502.pdf
[8] Li P 2020 *Architecture of the Internet of Things* Moscow: DMK Press
[9] *What are Bluetooth beacons? All you need to know is.* Retrieved from: https://ru.gadget-info.com/19060-what-are-bluetooth-beacons-everything-you-should-know
[10] Lance L, Bluetooth BLE Beacon Standards from iBeacon, Eddystone, and AltBeacon. Retrieved from: Https://www.silabs.com/community/blog.entry.html/2016/04/25/bluetooth_ble_beacon-IGMb
[11] Makhovikov A B, Katuntsov E V, Kosarev O V, Tsvetkov P S 2019 *Digital transformation in oil and gas extraction. Innovation-Based Development of the Mineral Resources Sector: Challenges and Prospects* (ed. Litvinenko) - 11th conference of the Russian-German Raw Materials (ISBN: 978-0-367-07726-6), pp. 531-538
[12] Kosarev O V, Tcvetkov P S, Makhovikov A B, Vodkaylo E G, Zulin V A, Bykasov D A 2019 *Modeling of industrial iot complex for underground space scanning on the base of arduino platform. Topical Issues of Rational Use of Natural Resources* - Proceedings Of The International Forum-Contest of Young Researchers, pp. 407-412
[13] *Developing Beacons with Bluetooth low energy (BLE) Technology*. Retrieved from: http://pages.silabs.com/rs/634-SLU-379/images/Whitepaper-Developing-Beacons-with-Bluetooth-Low-Energy-Technology.pdf
[14] *What is a Bluetooth beacon?* Retrieved from: Https://www.beaconstac.com/what-is-a-bluetooth-beacon
[15] *Beacon Technology - What is, How does it Work and it's Uses.* Retrieved from: Https://blog.sagipl.com/beacon-technology/

Automation of technological processes in mechanical engineering

E P Kozlova, S N Kuznetsova, E V Romanovskaya, N S Andryashina and E P Garina

Minin Nizhny Novgorod State Pedagogical University, 1, Ulyanova str., Nizhny Novgorod, 603002, Russia

E-mail: elka-a89@mail.ru

Abstract. This article discusses trends of production development in the engineering industry. The need to control automation processes at various stages of product manufacture has been identified. It was concluded that automation leads to increasing the efficiency of production, economic activities and quality of products, as well as making and introducing new products. On the basis of Russian and foreign experience the key stages of realization of automatic control system for technological processes were pointed out. Using the example of Business Units, the stages of implementing the latest automated equipment were considered. The project team conducted a survey of the business unit and determined that automation of the casting process in the foundry should be carried out. As a result, it was proposed to introduce an automatic cutting and grinding complex into the production process. These innovations made it possible to reduce the enterprise expenses by reducing the number of employees, as well as reduce the percentage of defects. Moreover, it became possible to solve a number of other significant problems: a decrease in the waste percentage, the need to meet the world requirements for the result of processing, elimination of the consequences of difficult working conditions (reduction in the number of accidents, noise levels and concomitant diseases). The results can be used not only for the considered objects of research, but also for organizations of other industries in order to increase the efficiency of activities through the introduction of automation of the control system of technological processes.

1. Introduction

In today's dynamic environment, industrial production has a significant impact on the economic environment. According to Rosstat, the current state of industrial development by August 2020 is at the stage of restoring industrial production, and this trend has generally remained. The index to the corresponding period of the previous year was 92,8% after 92,0% in July and 90,6% – on June. At the same time, compared to the previous month, production increased by 0,5%, while in July it amounted to 0,9% (with the elimination of the seasonal factor). According to the Center for macroeconomic analysis and short-term forecasting (CMASF), the index of the corresponding period of the previous year is set at 95,6% after 94,9% in July and 94,3% in June; with the elimination of seasonality, the growth index in August by July was + 1,4% (Figure 1).

Structurally, the tendency of the trend in August changed slightly. If in May-July the restoration of output in the affected industries was almost frontal, then in August in a noticeable part of them it stopped or even there was some backsliding. It was especially significant in two sectors, non-ferrous metallurgy, where export supplies sat, as well as in the production of cars, where manufacturers in July clearly overestimated the speed of recovery in consumer demand. New and significant positive developments in August are the beginning of the oil production restoration and the acceleration of the increase in gas production which began a month earlier.

It should be noted that the current situation is characterized by the continued excess of production of the last year relative to consumer goods of everyday demand. At the same time, the key point is the scale of external tourism that has significantly decreased, and therefore it is not worth counting on stabilizing this situation in the face of declining incomes.

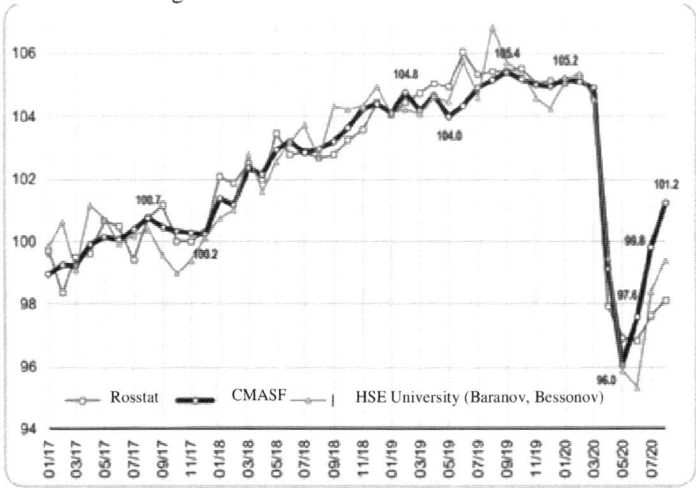

Figure 1. The volume of industrial production according to Rosstat, assessment of CMASF and National Research University «Higher School of Economics» (seasonality eliminated, 2017 = 100) [1]

On the basis of the proposed statistics, it could be concluded that there is a volatility of results in most areas of the industrial sector. To stabilize and dynamically develop the industrial complex as a whole, the sustainability of specific enterprises should be developed. One of the largest industries is mechanical engineering. It establishes a basic level of scientific and technological development. Due to this industry, other areas are being supplied. One of the characteristics of the machine-building complex is the profound specification of the production and significant scale.

Trends in the production development of engineering enterprises include a range of measures: providing production lines with the latest equipment, implementing control systems and automation of management. In the engineering industry, in the context of automation, it is necessary to pay attention to all kinds of factors. For example, from an engineering point of view, there is a difficulty in automating the entire production cycle. Most often this is not necessary, it could be sufficient to decide on the automation implementation at individual stages of the process.

In the foundry, for example, the charge loading process, temperature control and metal pouring into the molds are automated. Various sensors associated with the controller, feedback and remote control can minimize human involvement in this hazardous and harmful production [2].

In the machine-building complex, consisting of more than 7,5 thousand units of enterprises, one of the forms of blanks production is foundry. According to experts, at the moment in Russia there are about 1,1 thousand functioning foundry enterprises, which have significant volumes of castings and are also engaged in the production of equipment and materials for foundry production. Manufactured products (cast parts made of black alloys and non-ferrous ones) are most often used in the industry. Each industry sets its respective characteristic requirements for castings in terms of nomenclature, mechanical and operational properties, type of alloy, mass of castings, and, accordingly, type of technological processes and equipment.

2. Materials and methods

Automation of production processes in the engineering industry is a process of control, control and regulation of the production process, where part of the work is carried out not by man, but by automatic devices. One of the difficulties of mechanization and automation processes in the engineering industry is especially the use of continuous technology.

One of the basic factors of automation of technological processes and production in the engineering industry is the introduction of an automated process control complex, within which control takes place without direct human impact. This system includes various technological processes: continuous, individual and combined [3].

The implementation of the automated process control system (APCS) is carried out in stages and includes many stages. Based on the experience of domestic and foreign authors, the process of APCS implementing at the stage could be divided (Figure 2).

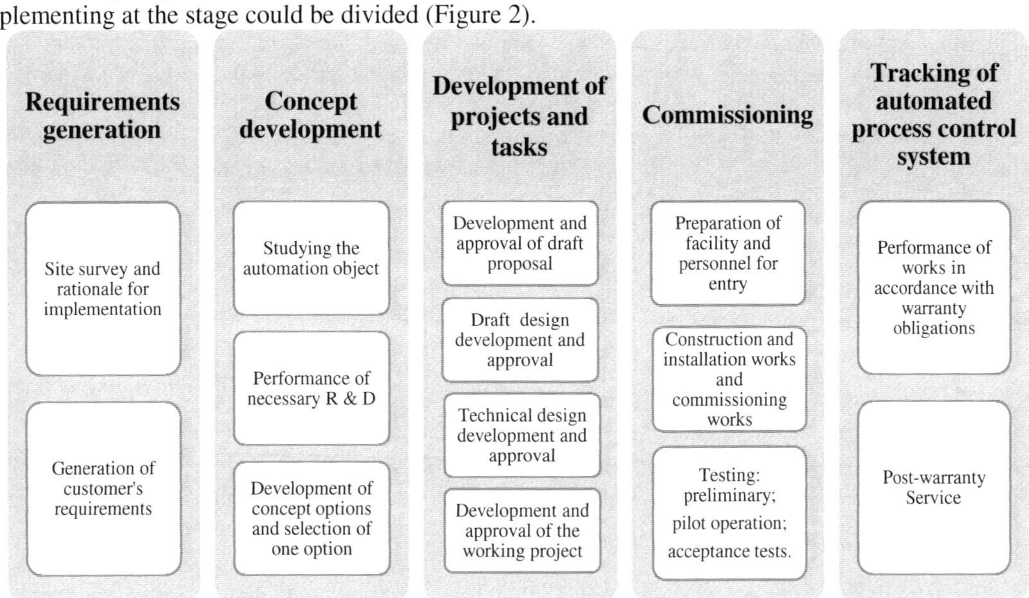

Figure 2. Key stages of implementation of automated process control system

The key task of automation is to improve the quality characteristics of the implemented process. For the engineering industry, the basic factors for increasing production efficiency are increased productivity, reduced working time, the greatest accuracy and stabilization of operations. It is possible to solve tasks due to automation of technological processes [4].

Modern enterprises use automation everywhere, and this is not a factor depending on the size of the company. In mechanical engineering, a significant number of various operations are used in the production and manufacture of hundreds of thousands of parts, such as casting, forging, welding; multidisciplinary machining; heat treatment, multilevel assembly operations, etc.

The modification and reconstruction of foundry enterprises also consists in the gradual introduction of new environmentally friendly processes and materials, modern equipment capable of carrying out the casting process at a high quality level, while meeting the standards existing in this industry [5]. Only a small part of modern Russian enterprises direct funds for the modernization of foundry, which contributes to improving the rate of quality of cast blanks and increasing labor productivity. Other companies are only on the way to building a flexible production process system; there are problems of such automation, ensuring the continuity of the equipment process chain. These steps allow you to carry out reassignment in the production of a wide range of castings.

3. Results

The practical use and efficiency of production automation are considered by the example of a machine-building enterprise; we designate it as a Business Unit. In this work, the implementation of systems and automation of management stages is presented in a separate fragment, since the central element of automation of production processes is the proposed automation project. In this article we focus on the first three stages of system and automation of management «development of projects and tasks».

As part of the automation project, it is necessary to take into account standard tasks to ensure production efficiency. To solve such problems, we will use the proposed stages of the system and automation of management [6].

At the first stage «Formation of requirements», the project team conducted a survey of the business unit and established the need to automate the process of processing the cast billet RF98MM-6303-CD «Crankshaft» in the foundry.

During the treatment of experimental batches, it was revealed that most of the defects after machining are at the top of the casting, so it was proposed to increase the allowance on all main and connecting rod necks from above by 1 mm, from 2 to 3 mm (Figure 3).

It becomes necessary to develop such a project and purchase an automated complex, which will be removed 1 mm on all main and connecting rod necks from above from the connector line (9 places, highlighted in red).

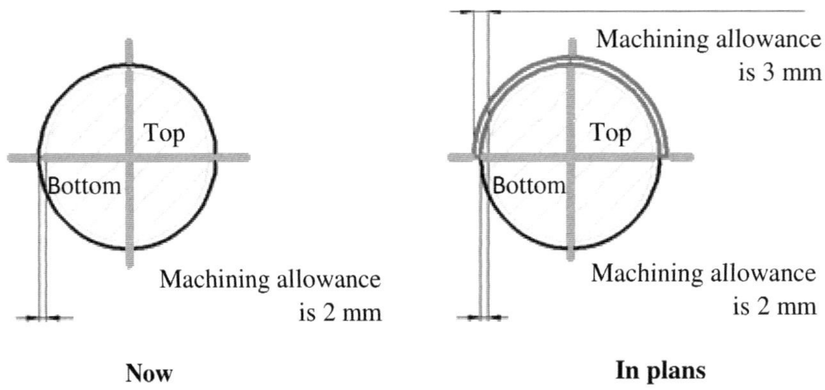

Figure 3. Requirements for the cast workpiece process

At the next stage, «Concept development», the automation object was studied.

The following means have been put forward to solve them:

- automation of the production process based on the introduction of new modern high-quality equipment: automatic refining complex;
- reduction of machine time due to application of implemented equipment;
- reduction of the number of employees per procurement processing activity.

At the third stage «Development of projects and tasks» it is necessary to develop a technical assignment and a preliminary design. At this stage, new progressive technological processes are being developed, the transition to which should solve the three-unit problem «quality-timing-costs». At this stage, the «sketch» expresses the best option, which allows eliminating the risks of making incorrect expensive decisions. It was decided to replace manual labor with automated equipment [7].

Manual execution of this operation entails the presence of a large number of employees, the percentage of waste is quite high, the possibility of scrap is very high (at the stage of grinding the seam), and it is necessary to meet the requirements for the treatment result that set the quality standards for the engineering industry. Another disadvantage of manual processing of a part is difficult working conditions (a large percentage of accidents, the level of noise isolation is quite low, concomitant diseases are possible). New processes require the selection of appropriate equipment.

The working group considered a proposal for the installation of an automatic cutting and grinding complex. P.S. Autogrinding Limited (PSAG), UK. This automatic cleaning complex is specially designed for foundry (Table 1).

Table 1. Equipment specification

Characteristic	Description
Number of axes	7
Operating range	700 D x 350 width
Nominal casting weight	50 kg
Total loading power (casting + gripping)	70 kg
Electrical switching requirements	17.3 KW
Pneumatic Requirements	3.5 l/cycle
Cutting disk spindle	
Power	4.5 KW drive
Speed	3500 turns/min
Typical disk sizes	355 D x 12 width
Grinding spindle	
Power	3.7 KW
Speed	15.000 turns/min
Typical Disk Sizes	50 D x 15 width
Air offtake	Not determined
Volume	60 m³/min
Nozzle size	D200
Dimensions	2000 width x 3425 length x 2305 height Weight 2450 kg

This automatic cut-off complex has the following characteristics:

- Compact design allows 1 operator to work on 2 or more machines.
- Diamond disc technology.
- Durable grinding disc - up to 6 months; more than 25 years of successful experience, more than 2000 machines put into operation.
- Speed of reconfiguration.
- Ease of operation and programming.
- Processing of grinding dust, no contamination.

A technical design was developed to determine the quantity of equipment and, after approval by all members of the project team, it was approved. As part of the task, technological processes were established, the loading of machine tools was determined. At the next stage of this stage, a working design was developed, which reflects the scheme of optimal arrangement of equipment, taking into account workshop traffic flows [8].

One of the important stages of this stage is the design and preparation of a feasibility study for the proposed complex implementation.

In accordance with the technical assignment, the wage fund for the cleaning site was 14,035,910 rubles. Wages in the industry are on average 23,000 rubles, after taxes are paid.

To calculate the number of people per shift: 14,035,910/369,600 = 38, that is, about 19 people are needed to execute work.

Taking into account this number of personnel and the volume of castings, it turns out one person must process 30 castings per minute (or spend about 40 seconds on one casting) [9].

Based on the experience of the business unit in the foundry, about 49 people per shift are needed to organize the grinding area for processing operations in accordance with the technical one. Labor costs for the year are on average 40,000,000 rubles.

With the complete automation of the cleaning site by Koyama complexes, the cost of acquiring the machine will be 167 850 000 rubles. The introduction of the new complex makes it possible to reduce the number of employees for the production of castings. The economic impact of the reduction in the number of workers is shown in table 2.

Table 2. Calculation of the rate of cost reduction for the Wage Fund

Name	Project implementation phase		Savings (+), increases (-), people	Average salary, rub.*	Savings (+), increase (-), rub per year
	Before implementation	After implementation			
Number of staff	100	14	86	22 000	37 877 496
Full investment repayment period					4.4 years

* after taxes are paid

During the implementation phase of the machine, the company understands how the expected results can be achieved. These stages of system and automation of management could be tested during the implementation of other projects at the enterprise [10].

The fifth stage is already the performance of works in accordance with warranty obligations and post-guarantee maintenance.

4. Conclusion

At the moment, only a small number of company executives are trying to maintain the usual and outdated production technologies. The head of this Business Unit believes that too conservative views unreasonably believe that automation is an economically unprofitable activity. A correct management decision in this area will be able to make significant changes to the usual situation of the enterprise. A competent leader understands that the modern situation in the industry dictates its own rules, forcing to conduct activities flexibly, not forgetting about the automation of modern production.

References

[1] Official website of the Center for macroeconomic analysis and short-term forecasting. Retrieved from: http://www.forecast.ru/default.aspx

[2] Filipov A I 2019 Automation of the control system in the foundry, in: *Problems and prospects for the development of Russia: a youth view in the future* pp 159-161

[3] Dulger A G and Pechenkina L S 2018 Automation of technical preparation of foundry on the example of Voronezh enterprises *Casting and metallurgy* 83-84

[4] Official site of the corporate disclosure center. Retrieved from: http://www.e-disclosure.ru/portal/files.aspx?id=859&type=3

[5] Kuznetsova S N, Kuznetsov V P, Kozlova E P, Potashnik Y S and Tsymbalov S D 2020 Transformational Period of Russian Development in the Digital Economy *Lecture Notes in Networks and Systems* **91** 663-669

[6] Kozlova E P, Kuznetsov V P, Garina E P, Romanovskaya E V and Andryashina N S 2020 Methodological Bases of the Assessment of Sustainable Development of Industrial Enterprises (Technological Approach), in: Popkova E and Sergi B (Eds) *The 21st Century from the Positions of Modern Science: Intellectual, Digital and Innovative Aspects. ISC 2019. Lecture Notes in Networks and Systems* **91** (Springer, Cham)

[7] Popkova E G and Sergi B S 2018 Will Industry 4.0 and Other Innovations Impact Russia's Development? In: Sergi B S (Ed) *Exploring the Future of Russia's Economy and Markets: Towards Sustainable Economic Development* (Bingley, UK: Emerald Publishing) pp 51-68

[8] Popkova E G, Popova E V and Sergi B S 2018 Clusters and Innovational Networks Toward Sustainable Growth, in: Sergi B S (Ed) *Exploring the Future of Russia's Economy and Markets: Towards Sustainable Economic Development* (Bingley, UK: Emerald Publishing) pp 107-124

[9] Smirnova Zh V, Vaganova O I, Chaykina Z V and Golubeva O V 2020 Calculation of metallurgical processes during welding *Int. J. of Emerging Trends in Engineering Research* **8(5)** 1529-1534

[10] Smirnova Z V, Vaganova O I, Cherney O T, Bystrova N V and Golubeva O V 2020 Kinematic study of the cam mechanism by the method of diagrams *Int. J. of Emerging Trends in Engineering Research* **8(7)** 3358-3361

Principles of organization of technological service systems to ensure their reliability

V N Kurochkin[1], N A Shcherbak[2], and S L Nikitchenko[3]

[1]South Federal University, Graduate School of Business, 105/42, Bolshaya Sadovaya st., Rostov-on-Don, Russia
[2] Moscow State University of Technology and Management named after K.G. Razumovsky (First Cossack University), 73, Zemlyanoy Val st., Moscow, Russia
[3]Azov-Black Sea Engineering Institute-Branch of the Don State Agrarian University, 21, Lenina st., Zernograd, Russia

E-mail: Valentin952@mail.ru

Abstract. The article analyzes technological systems, which are the production basis of organizations that provide services and produce goods, and service subsystems. In order to study the principles and methods of organizing services to ensure the reliability of the technological system, it is necessary to consider four types of technological systems in the agro-industrial complex: technological systems of plant growing, livestock and other agricultural and processing companies; technological systems of plant-growing processes, animal husbandry, processing of vegetable and animal raw materials; technological systems of production units; and technological systems of production units of enterprises. The systematic approach and the monographic method were used; a systematic analysis of the service system was carried out. It was established that the reliability of a technological system can be improved by organizing technical services and relying on the following principles: creation of a parallel system; organization of a reliability management system; the entry of reliability management into the quality management system, risk management and the general structure of enterprise management; analysis of the types and consequences of failures of the technological system; formation of redundancy; state reliability management; optimization of the structure of the subsystem of technical service by modeling their functions and developing tools for providing technical services.

1. Introduction

Technological systems are the production basis for organizations that provide services and produce goods. The concept "technological system" is defined as "a set of interconnected means of technological equipment, production items and performers in technological processes or operations under the regulated production conditions" [1]. The subject of the service industry is a service that is characterized by intangibility, since most services are actions and works that are inseparable from their sources; services are heterogeneous by quality, short-lived and cannot be registered as property objects.

A system is an emergent set of elements united by the system-forming factors, possessing general systemic properties of stability, reliability and efficiency. These properties allow the system to compete in fairly complex conditions of the changeable external environment, to maintain its structure and hierarchy throughout the entire life cycle.

The structure of a system is the mutual relationship of its subsystems and elements, determined by functional and parametric connections. Subsystems share some functional or structural characteristics of the technological system of a higher level. An element of the technological system is its part conventionally taken as indivisible at this stage of research. The system hierarchy includes four levels of technological systems: operations, processes, production divisions and organizations as a whole. In addition to a structure and a hierarchy, the system is characterized by organization and technology, composition of fixed assets, parameters and conditions.

An organization is an internal structure, methods and ways of achieving goals.

Technology is a sequence of operations performed during the operation of a system.

System parameters can be input and output, the latter characterize the efficiency, reliability, stability, quality of a system.

The reliability is "a property to keep values of all parameters characterizing the ability to perform required functions in the specified conditions." It is a complex indicator that, depending on the industry and the situation, may include reliability, durability, maintainability and preservation (the property of a system to continuously maintain an operable state until the end of its life cycle), adaptability to the restoration of an operable state and preservation after storage or forced long-term downtime [2]. Service systems allow you to maintain operability, reduce the failure rate, restore operability and save it during the periods of non-use, for example, during quarantine.

The efficiency depends on organization, methods and principles of ensuring the operable state of the serviced technological system.

The goal of the research is to study the principles and methods for ensuring the reliability of the technological system.

2. Materials and methods

The research methods are monographic, system, and functional types of analysis of systems. A number of research institutes, including the Institute of Mechanical Engineering of the Russian Academy of Sciences, and industry research institutes, were involved in improving the reliability of technological systems. An operating state of the system is "the state of the technological system, in which the value of at least one parameter of productivity of the technological system does not correspond to the requirements established in the normative-technical and (or) design and technological documents", when the inoperative state of the corresponding parameters do not meet the established requirements [1]. Failures of technological systems are divided into functional, with a complete loss of the ability to work, and parametric, in which the system continues to function, but the efficiency and quality are partially reduced. There are failures of service parameters (parametric), quality, productivity, and cost. In case of cost failures, the organization becomes uncompetitive, and cannot sell its products or services at a profit, its financial stability decreases to an unstable state. Failures can be caused both by internal and external factors.

3. Results and discussion

An analysis of the technological systems showed that its service should be aimed at increasing reliability as a complex property, characterized by the probability that there will be no failures of quality of manufactured products. The utilization rate of the technological system, the probability of performing work and the productivity retention factor will be within acceptable limits. The probability of producing the i-th product is determined by formula

$$P_{з_i}(t) = P\{V_i(t) \geq V_{0_i}\}, \quad (1)$$

where $V_i(t)$ и V_{0_i} is actual and set volumes of products *i* of required quality produced for period t [1].

The utilization factor characterizes the relative proportion of time during which the vehicle is in a working state, taking into account all types of downtime.

Therefore, the service organization should be aimed at reducing non-productive downtime caused by functional and parametric failures of the technological system.

For example, a grain processing enterprise is a technological system of the service industry serving the agro-industrial complex. In turn, it has its own service system for ensuring operational reliability - a reliability management system (Fig. 1).

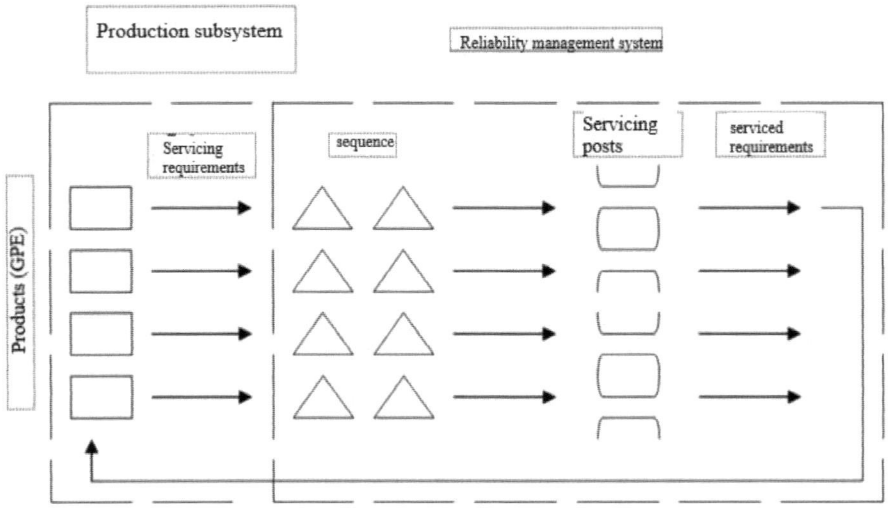

Figure 1. The structure of the grain processing equipment reliability management subsystem

The production system is a source of requests for maintenance and repairs. Requirements, or requests, for service enter the service subsystem, if it is busy with service, then the requests wait in the queue. At the service stations, they are satisfied and the serviced requests leave the maintenance and repair system.

The parallel system principle provides for the creation of a parallel subsystem to ensure the operability of the technological system.

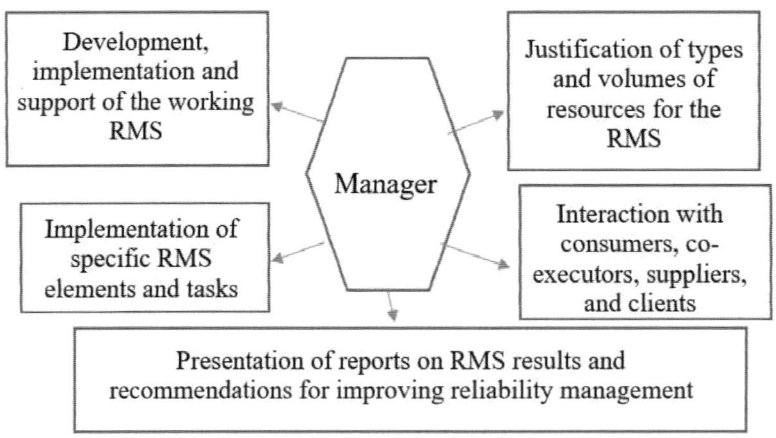

Figure 2. The main functions of a RMS manager

The first principle is organization of a specialized subsystem that is engaged in restoring the operability of the technological system: the reliability management system (RMS). In accordance with the state standard, RMS is developed for restoring the operability of the technological system. Its main task is to manage the reliability and quality of the enterprise. Therefore, it should be part of the management structure. The direct management of RMS is carried out by a company manager [3]. These responsibilities can be assigned to the head of the quality department (Fig. 2).

The second principle is the organization of service of the technological system. It provides for the entry of reliability management into the quality management system, risk management and the general structure of enterprise management (Table 1).

Table 1. Strategic, tactical and operational decisions of risk management of a technological system [4]

Strategic	Tactic	Operational
Top manager. Representative of the owner. Manager	Middle-level executive. Manager. Project manager	Project manager. Project group
Identification of objectives, means, risks, participants, and creation of a basis for making tactic and operative decisions	Identification of means ensuring implementation of objectives and management of tactic goals within the strategic level	Implementation of decisions taken at the tactic level, and operative risk management

The third principle is an analysis of the types and consequences of failures of a technological system, due to the application of the national risk management standard. The principle gives recommendations to achieve the set goals. Failure Modes and Effects Analysis (FMEA) is a method of systematic system analysis to identify potential failures, their causes and consequences, as well as the impact of failures on the system (the system as a whole or its components and processes) [5] .

The principle of redundancy is a special case of using a parallel system for a system element. It involves the creation of redundancy due to a loaded and cold reserve and additional connections. These can be such backup methods as duplication, diagnostic methods, information backup.

The principle of strategic management is a planned preventive system (by the operating time - motorcycle hours, consumption of a certain amount of fuel or performance of a certain volume of works in conditional reference hectares), by the state (the list and frequency of maintenance is determined by the state of machines), by the event, and for imported machinery and equipment - "service for reliability" [6, p. 8].

Optimization of the organization structure of the technical service subsystem is based on the model of its functions and the mathematical apparatus of the theory of probability and the theory of Markov processes. It is necessary to develop tools for technical service on a different basis [7, p. 3807-3815]. Optimization of the technical service subsystem by the autonomous, centralized, and combined (hybrid) methods.

The principle of forming the information redundancy of a technological system is based on predictive modeling and mathematical modeling [8, p. 715-719].

4. Conclusion

The reliability of a technological system can be achieved by providing technical services and relying on the following principles:

– creation of a parallel system (creation of a parallel subsystem to ensure the operability of the technological system);

– organization of a specialized subsystem that is engaged in restoring the operability of a technological system: a reliability management system (due to the application of the National Standard of the Russian Federation);

– the entry of reliability management into the quality management system, risk management and the general structure of enterprise management;

– analysis of the types and consequences of failures of the technological system (due to the application of the national standard for risk management);

– redundancy is a special case of a parallel system, formation of redundancy due to a loaded and cold reserve and additional connections (duplication, diagnostic methods, informational, etc.); formation of information redundancy on the basis of predictive modeling and mathematical modeling;

– strategic management: planned and preventive systems, states, events, reliability;

– optimization of the technical service subsystem by modeling its functions;

– development of tools for providing technical services and using them.

References

[1] Reliability in technology. Technological systems: terms and definitions. GOST 27.004-85. Industrial product dependability. Technological systems. Terms and definitions, available at: http://client.consultant.ru/site/list/?id=1009757373

[2] Reliability in technology: terms and definitions. GOST 27.002-2015, available at: http://client.consultant.ru/site/list/?id=1009757595

[3] Reliability in technology. Reliability management system. Basic provisions: National standard of the Russian Federation GOST R 27.001-2009. Dependability in technics. Dependability management system. Basic principles, available at: http://client.consultant.ru/site/list/?id=1009757509

[4] Project Risk Management: Basic Provisions: National Standard of the Russian Federation. GOST R 52806-2007. Project risk management. General, available at: https://www.russiangost.com/p-68762-gost-r-52806-2007.aspx

[5] Risk management. A method for analyzing the types and consequences of failures. National standard of the Russian Federation. GOST R 51901.12-2007 (IEC 60812: 2006) Risk management. Procedure for failure mode and effects analysis

[6] Arzhenovskiy A G 2020 *Improving the optimization of technical service, methods and means of functional operational diagnostics of mobile energy facilities in the agro-industrial complex*, thesis' abstract (Zernograd: Azov-Chernomorsky Institute-branch of Don SAU) 44 p

[7] Seregin A A, Nikitchenko S L, Valuev N V, Kurochkin V N and Smykov S V 2018 Hinged aggregate for technical maintenance of machines: modeling, testing and conditions of application *J. of Mechanical Science and Technology* **32(8)** 3807-3815

[8] Kurochkin V N, Lebedev A T, Rudenko N E, Zhevora Y I and Doronina N P 2018 Mathematical modeling of the technological system of the harvesting and transport process *India Res. J. of Pharmaceutical, Biological and Chemical Sci.* **9(6)** 715-719

Determination of stripping zones and axle spacing between the cylinders in two-cylinder stripping reaper (header)

V A Sakharov and A A Kuvshinov

Federal State Budget Scientific Institution Federal Research Center All-Russian Scientific Research Institute of Soybean, 19, Ignatievskoe highway, Blagoveshchensk, Amur Region, 675027, Russia

E-mail: info@vniisoi.ru

Abstract. Stripping devices become an alternative to harvester-threshers in the harvesting season. The research objective is improvement of harvesting operations, including the introduction of the method of soya stripping. The research showed that the use of a single-cylinder stripping device was not optimal, for harvesting losses exceeded agricultural norms. It was decided to use a two-cylinder scheme. The implementation of this design solution reduced the grain loss rate, but still it was not enough to achieve compliance with the requirements of classical soybean harvesting. For high-quality soybean stripping, it is necessary to calculate the zones of stripping and axle spacing between the first and second stripping cylinders, as well as the profile of the stripping combs. Schemes presented are calculation of the stripping zones and determination of the axle spacing between the cylinders, taking into account their positional relationship in the stripping reaper.

1. Introduction

When studying the possibility of stripping of agricultural plants, morphological features of the structure of stems and productive parts are of particular importance. The research [1] considers eccentric swinging stripping device for harvesting the ground cherry C. humilis. The results of the experiment with 3 factors showed that the optimal values of the selected parameters are: rotation speed of 11.26 rpm, feed rate of 0.227 m / s and comb turning angle of 5°. The rate of fruit gathering and damage amounted to 95.21% and 4.56%, respectively.

The research is carried out into improvement of the efficiency and reduction in damage during harvesting of L. barbarum [2] by developing a portable vibro-comber type brush harvester. The analyses of the findings showed the best ratio of the selected parameters: the speed of rotation of the comber type brush – 64.52 rpm, the speed of rotation of the cylnder-29.68 rpm, the style of the comber type brush-II. The results of field experiments showed that the completeness of the gathering amounted to 89.12%, and the damage rate - 6.24%.

The research paper presents the findings of investigations on the development and substantiation of the parameters and modes of operation of the working bodies of the machine for harvesting seeds of forage grasses [3]. Analytical dependences that characterize the capture conditions and the length of the ridges were found, and the coefficient of influence of ridges on the ears of plant stems was determined.

The article [4] provides the theoretical justification for the optimal layout of the stripping device for harvesting grain crops; the parameters of the stripping combs; substantiation of their height, width of the gap between the combs and optimal slope of the combs.

When analyzing foreign sources, special attention should be paid to the fact that the problem of crops stripping (in particular legumes) is poorly studied. Basically, the improvement of soybean harvesting consists in the following: creating a mathematical model for estimating the loss of soybean seed yield depending on the changing height of the cut, used for harvesting, and sowing density [5]. Series of field tests were conducted to study the influence of various key operating parameters on the quality of soybean harvesting, and an attempt was made to systematically determine the optimal combination of parameters [6].

The comparison of soybean harvesting was carried out between different cultivation systems that depend on the height of the lower bean attachment and the length of the plant [7]; to improve the quality of the harvesting process, the quality of soybean harvesting was assessed using combine harvesters with different reapers and rates of movement [8]. The main mechanical properties of green soybean seeds and beans were studied (elastic modulus of various parts of green soybean seeds, bean envelope, bean, Poisson's ratio of seeds) [9]; the physical and physiological properties of soybean seeds were assessed at three rates of movement of the combine harvester using statistical control over the technological process [10].

The shape of the comb used for grain crops stripping is not suitable for soybean stripping. A special feature of soybean plants is that the beans are located along the entire height of the plant, starting from 8 to 12 cm from the field surface. In plants of grain crops (wheat), the ear is weakly attached to the stem at the upper point and is easily stripped with a comb and carried away by the air flow created by the stripping cylinder. Losses during harvesting of soybeans by the stripping method amount to 12 - 18 % on average. The main problem is the impact of the stripping comb on the soybean, which leads to its opening and flight of the lower beans and grains of the plant forward in the direction of movement.

The purpose of the research is to determine the stripping zones for soybean stripping with the help of cylinders, to determine axle spacing between cylinders in the two-cylinder reaper (header).

2. Materials and methods

The goal was achieved using a mathematical tool, using trigonometric formulas, which, in comparison with the use of the graphical method, eliminates calculation errors.

3. Results

It is necessary to determine the lower level of stripping of soybean stalk with cylinders of the stripping header (reaper).

When designing two-cylinder reaper, it is necessary to determine the diameter and positional relationship of the cylinders. One of the criteria is the height of stripping with the first (upper) cylinder, which owing to ΔH does not coincide with the lower point of the radius of the stripping combs along the surface normal (figure 1).

To determine the parameter ΔH, we use figure 2, where CD is the plant stalk, r is the radius of the cylinder (the minimum radius for the separation of soy beans), and R is the radius of the stripping cylinder at the ends of the stripping combs.

Forward movement of the stripping cylinder: from right to left, rotation – clockwise. The position of the CD stem is the extreme vertical; in this position the section A_2A_3 shall be combed. The stem passes through the lower point B_2 of the stripping cylinder in the position C_1D_1 at the angle β.

Let us compare the vertical position of points A_2 and B_2:

$$B_1B_2 = R - R \cdot \sin\alpha = R(1 - \sin\alpha) \tag{1}$$

where $\alpha = \arccos(r/R)$;

when the ratio is $r/R = 0.75$; $\alpha = 41.4^{\circ}$; $B_1B_2 = 0.339R$;

when the ratio is $r/R = 0.5$; $\alpha = 60^\circ$; $B_1B_2 = 0.134R$;
when the ratio is $r/R = 0.25$; $\alpha = 75.5^\circ$; $B_1B_2 = 0.032R$

We find the lower point of combing with the help of the first (upper) cylinder using the analytical method (figure 2).

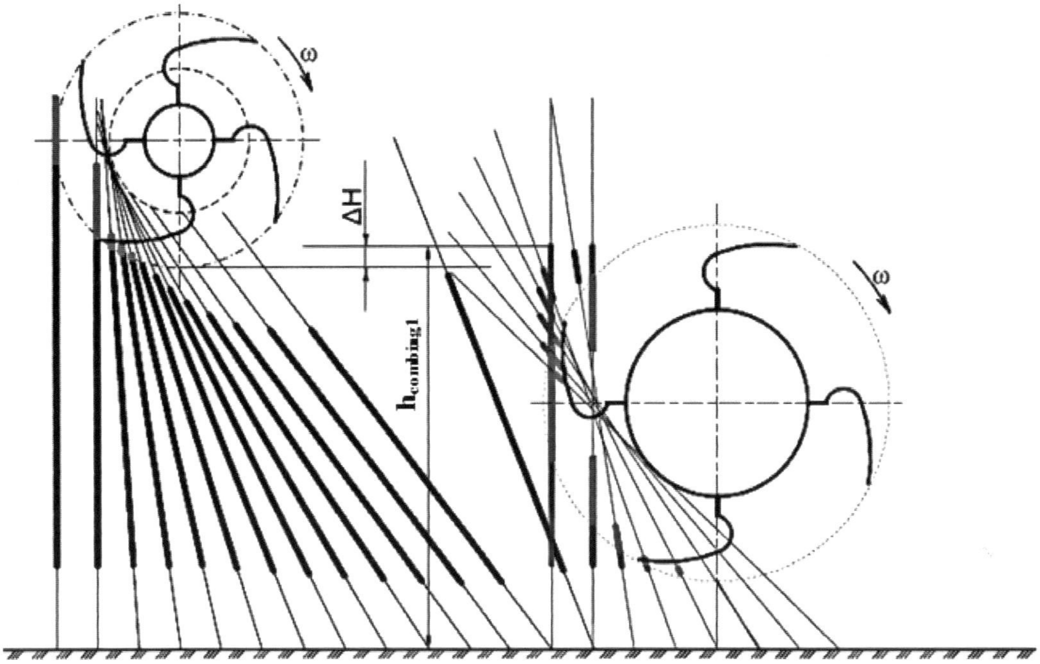

Figure 1. Scheme for determining the zones of the upper and lower cylinders

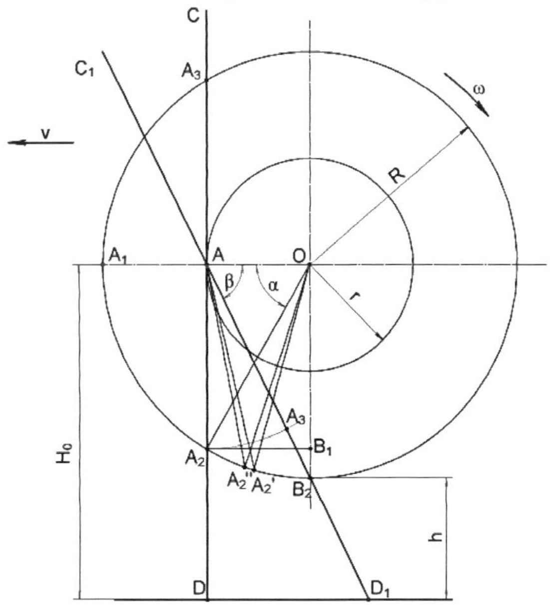

Figure 2. Diagram for the determination of ΔN

We consider the triangle AOA_2: OA = r, OA_2 = R. By increasing the angle α with a certain step up to 90°, the triangle AOA_2 will be transformed and at α=90° it will take the position of the triangle AOB_2.

We consider two intermediate arbitrary positions of the triangle AOA'_2 and AOA''_2. Using the cosine theorem, we find the distances AA'_2 and AOA''_2:

$$AA'_2 = \sqrt{r^2 + R^2 - 2rR \cdot \cos\alpha'} \quad (2)$$

$$AA''_2 = \sqrt{r^2 + R^2 - 2rR \cdot \cos\alpha''}, \quad (3)$$

We consider the triangle $AA'_2A''_2$.

Considering the arc $A'_2A''_2$ to be a line segment with a certain assumption, its magnitude is determine by the formula:

$$A'_2A''_2 = \frac{\pi R}{180} \cdot (\alpha' - \alpha''). \quad (4)$$

Using the theorem of cosines, we find angle $\angle A'_2AA''_2$:

$$\angle A'_2AA''_2 = \arccos\{[(AA'_2)^2 + (AA''_2)^2 - (A'_2A''_2)^2]/2(AA'_2) \cdot (AA''_2)\}. \quad (5)$$

Using the above formulas, changing the angle α and taking into account that $\angle A'_2AA''_2$ is an increment of the angle β, we find that for the variant R = 150 mm, r = 75 mm, the angle β changes from 63,44°, AA_2 – from 167,71 to 129,9 мм, and ΔAA_2 respectively 37,81÷0 mm.

Let's determine where the point A_2 (the lower point of combing of the stem in the CD position) will move when the stem position is C_1D_1. The movement of the stem relative to the original position can be described by two movements: rotation relative to point A and linear movement of point A_2 along the line C_1D_1.

The diagonal AB_2 is equal to:

$$AB_2 = \sqrt{r^2 + R^2} \quad (6)$$

$$\beta = \mathrm{arctg}\frac{R}{r}. \quad (7)$$

The distance AA'_2 is:

$$AA'_2 = R \cdot \sin\alpha + H_0\left(\frac{1}{\sin\beta} - 1\right) \quad (8)$$

where AA'_2 is the position of point A_2 on C_1D_1,

$H_0(\frac{1}{sin\beta} - 1)$ – travel x (figure 3)

$$x = \frac{H_0}{sin\beta} - H_0 = H_0(\frac{1}{sin\beta} - 1) \quad (9)$$

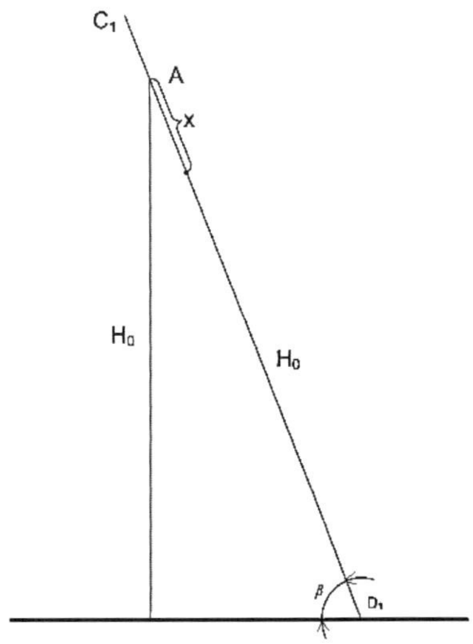

Figure 3. Diagram of stem movement

We compare AB_2 and AA'_2. Set R = 150 mm, r = 75 mm, H_0 =550 mm: r/R = 0.5; β=63.43°; AB2 = 167.7 mm; AA'_2 = 194.4 mm; $AA'_2 > AB_2$, i.e. the combed part of the stem is below the point B_2.

Thus, when the stripping cylinder approaches the field surface, the maximum combing will tend to occur at the lower point of the cylinder B_2, and when this distance increases - at the point A_2 of the vertically standing stem (figure 2). The figure shows that due to the inclination of the stem, the lower level of the main cylinder and the height h of the combs above the field surface do not coincide.

The lower level of combing h with the help of the main cylinder is determined by the formula (10):

$$h_{comb} = \frac{h}{\sin\beta} = \frac{h}{\sin(\mathrm{arctg}\frac{R}{r})}. \tag{10}$$

To determine the position of the main cylinder above the surface of the field in accordance with a given lower level of the combing associated with the location of the lower beans on the plant stem, use the formula (11):

$$h = h_{comb}\sin\beta = h_{comb}\sin\left(\mathrm{arctg}\left(\frac{R}{r}\right)\right). \tag{11}$$

Using formulas (5) and (9), setting the step of changing the angle α in 2°, for H_0=550 mm, we make table 1.

The numbers in the first column of table 1 show the increase in amount of combing, by turning the combed plant on the angle β and the second column is the reduction in the combing due to the "pulling" of the stem. The value of their difference will show the value of the additional stem combing against the stem combing in the vertical position.

Numbers with negative values indicate that the part of the stem that is combed in a vertical position, at these angles of inclination of the stem, is outside the outer radius of the combs of the combing drum. The maximum difference adds the largest amount of combing to the combing of the stem in an upright position. In this case, the maximum combing will be 129.90 + 5.49 = 135.49 mm with the angle of inclination of the stem of about 82° (table 1).

IOP Conf. Series: Materials Science and Engineering 1111 (2021) 012032 doi:10.1088/1757-899X/1111/1/012032

Table 1. The value of additional combing.

Angle β, °	Increasing the combing by turning ΔAA_2, mm	Reduction of combing due to «pulling» x, mm	Difference, mm
63,44	37,81	64,90	-27,09
65,05	35,45	56,65	-21,20
66,68	33,06	48,95	-15,83
68,32	30,64	41,80	-11,16
69,98	28,19	35,20	-7,01
71,67	25,72	29,15	-3,43
73,38	23,22	24,20	-0,98
75,11	20,70	19,25	1,45
76,87	18,16	14,85	3,31
78,65	15,61	11,00	4,61
80,46	13,03	7,70	5.33
82,30	10,44	4,95	5,49
84,17	7,84	2,75	5,09
86,08	5,24	1,10	4,14
88,02	2,62	0,55	2,07
90	0	0	0

In further calculations, it is necessary to determine the axle spacing in the two-cylinder header.

The presented scheme allows you to calculate the distance of installation of stripping cylinders for high-quality soybean combing. For reasons of minimizing losses and improving the composition of the combed soy grain heap, the upper cylinder combs the tops of plants, about 1/3 of the stem (the height from the field surface to the stripping combs is 40 – 60 cm, depending on the height of the stem), the lower cylinder - 2/3 of the stem.

In this case, the angle of inclination of the stem will be $\alpha = \arcsin\left(\frac{2}{3}\right)$, or about 42°, which does not exceed the angle of bent fracture of the stem, equal to about 20°, obtained experimentally. We accept the condition that the stalk after combing with the help of the first cylinder should take a vertical position before combing with the second cylinder. On the basis of this, we consider the diagram (figure 4).

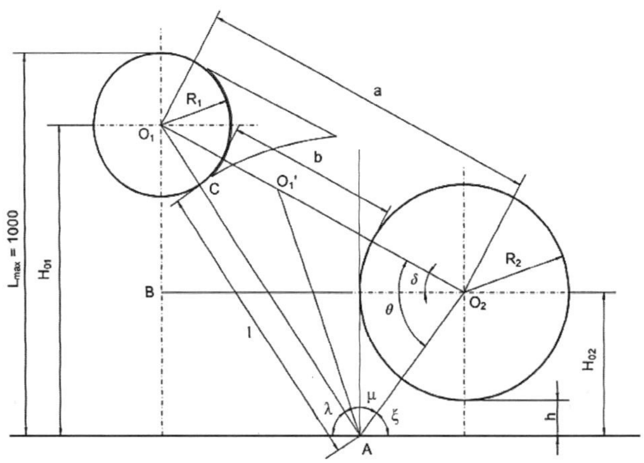

Figure 4. Diagram for determining the axle spacing between the stripping cylinders: l – height of the soybean stalk; H_{01} – height of the first stripping cylinder; H_{02} – height of the second stripping cylinder; O_1 – center of the first stripping cylinder; O_2 – center of the second stripping cylinder; R_1 – radius of the first stripping cylinder; R_2 – radius of the second stripping cylinder; a – axle spacing; h – distance from the combs of the second stripping cylinder to the field surface

IOP Conf. Series: Materials Science and Engineering 1111 (2021) 012032 doi:10.1088/1757-899X/1111/1/012032

We consider the triangle AO_1O_2, the angle and adjacent angles.
We determine the angles:

$$\xi = \mathrm{arctg}\frac{H_{02}}{R_2} = \mathrm{arctg}\left(\frac{h}{R_2} + 1\right) \tag{12}$$

$$\lambda = \arcsin\frac{H_{01}}{l+R_1} = \arcsin\left(\frac{l-R_1}{l+R_1}\right) \tag{13}$$

$$\mu = 180^\circ - \xi - \lambda\,. \tag{14}$$

We determine the angle of inclination of the δ plane passing through the axes of the cylinders to the field surface at the maximum value of the stem height l_{max}:

$$\delta = \arcsin\frac{O_1B}{a} = \arcsin\frac{H_{01}-H_{02}}{a}\,. \tag{15}$$

The angle $\theta = \xi + \delta$
We consider an arbitrary triangle $AO_1'O_2$, where the point O_1' lies on the line O_1O_2.
We determine the dependence of the axle spacing with the help of the cosine theorem:

$$O_1'A = \sqrt{AO_2^2 + O_1'O_2^2 - 2(AO_2)(O_1'O_2)\cos\theta.} \tag{16}$$

Denoting $O_1'A = l + R_1$, $AO_2 = \frac{R_2}{\cos\xi} = C$, $O_1'O_2 = a = y$, we square both parts of the equation and get the reduced form of a quadratic equation.
The real root of the equation or substituting values:

$$y^2 - Dy + E = 0$$
$$\text{where } 2C\cdot\cos\theta = D,$$

$$[C^2 - (l + R_1)^2] = E.$$

Real root of the equation:

$$y_1 = \frac{D}{2} + \sqrt{\left(\frac{D}{2}\right)^2 - E}. \tag{17}$$

Or substituting the value:

$$a = C\cdot\cos\theta + \sqrt{[C^2\cdot\cos^2\theta - C^2 + (l + R_1)^2]}. \tag{18}$$

4. Discussion

Taking the values R_1 = 175 mm (at the ends of the combs), R_2 = 330 mm, H_{01} = 850 mm, H_{02} = 350 mm, l_{max} = 1000 mm, l_{min} = 600 mm, we get the maximum and minimum axle spacing y_{max} = 1400 mm, y_{min} = 800 mm.

Thus, in a two-cylinder reaper, when adjusting the axle spacing along the O_1O_2 line for the radii of the stripping cylinders and plant heights set by us, the adjustment range should be at least 600 mm.

When calculating the diameter of the main (second) stripping cylinder, initially taken as 2/3 of the length of the stem, it is necessary to make a clarifying calculation to comply with the following inequality:

$$H_{02} + R_2\sin\alpha \geq H_{01} - R_1 + \Delta H \tag{19}$$

where H_{01} and H_{02} – heights of the axes of the corresponding cylinders relative to the field surface, R_1 and R_2 – radii of the stripping cylinders at the ends of the combs, α is from the diagram in figure 2, where $\alpha = arcos\left(\frac{r}{R}\right)$, ΔH - difference between the lower point of the upper (first) cylinder and the level of the stem combing of the upper cylinder (figure 1), determined by the above method. If there is nonfulfillment of this inequation, the cylinder diameter must be adjusted.

5. Conclusion

Using the presented formulas and calculation methods, it is possible to determine the diameters and axle spacing of two-cylinder reapers for stripping soya of various varieties having certain morphological features.

Acknowledgments
The work was carried out within the framework of the FR program of the State Academies of Sciences for years 2013-2020. The direction of the research is 162: Fundamental problems and principles of development of intensive machine technologies and energy-saturated equipment of a new generation for the production of major food groups.

References

[1] Du X, He J, He Y and Fang D 2020 Design and experiment of eccentric swing combing device for Cerasus humilis *INMATEH - Agricultural Engineering* **60** 89-98
[2] Chen J, Zhao J, Chen Y, Bu L, Hu G and Zhang E 2019 Design and Experiment on Vibrating and Comb Brushing Harvester for Lycium barbarum, Nongye Jixie Xuebao *Transactions of the Chinese Society for Agricultural Machinery* **50** 152-161
[3] Zhortuylov O, Soldatov V, Zhumatay G, Bekenov U, Zhortuylov A and Zhumabaev S 2018 Parameters and operating modes of the working organs of the machine for harvesting forage grass seeds by comb on the root *Acta Technica CSAV* **63** 10
[4] Dzhambursin A S and Turymbetova G D 2017 Substantiation of expedient parameters and operating modes of the stripping device for harvesting grain crops in Kazakhstan *Int. J. of Mechanical Engineering and Technology* **8** 691-698
[5] Rębilas K, Klimek-Kopyra A, Bacior M and Zając T 2020 A model for the yield losses estimation in an early soybean (Glycine max (L.) Merr.) cultivar depending on the cutting height at harvest *Field Crops Research* 254
[6] Jin C, Guo F, Xu J, Li J, Yin X et al 2019 Optimization of working parameters of soybean combine harvester (Nongye Gongcheng Xuebao) *Transactions of the Chinese Society of Agricultural Engineering* **35(13)** 10-22
[7] Beiküfner M, Hüsing B, Trautz D and Kühling I 2019 Comparative harvest efficiency of soybeans between cropping systems affected by first pod height and plant length *Organic Farming* **5(1)** 3-13
[8] de Menezes P C, da Silva R P, Carneiro F M, de Oliveira M F, Voltarell M A et al 2018 Can combine headers and travel speeds affect the quality of soybean harvesting operations? *Revista Brasileira de Engenharia Agricola e Ambiental* **22(10)** 732-738
[9] Zhao Y, Liu S, Xiao H, Mao E, Liu D et al 2018 Finite element analysis and experimental validation of the compression characteristics to green soybean *Int. Agricultural Engineering J.* **27(3)** 290-300
[10] Paixão C S S, Chrispin C P, Silva R P, Girio L A S and Voltarelli M A 2017 Physical and physiological quality of soybean seeds at three speeds of the harvester *Revista Brasileira de Engenharia Agricola e Ambiental* **21(3)** 214-218

International Conference: Actual Issues of Mechanical Engineering (AIME 2020) IOP Publishing
IOP Conf. Series: Materials Science and Engineering 1111 (2021) 012033 doi:10.1088/1757-899X/1111/1/012033

Problem solution of optimal pathfinding for the movement of vehicles over rough mountainous areas

K Zhigalov[1], D K-S Bataev[2], E Klochkova[3], O A Svirbutovich[4] and G A Ivashchenko[5]

[1] V.A. Trapeznikov Institute of Control Sciences of Russian Academy of Sciences, Moscow, Russia
[2] Integrated Research Institute named after Kh.I. Ibragimov, Russian Academy of Sciences of the Chechen Republic, 21A, Staropromyslovsky ave., Grozny, 364051, Russia
[3] Moscow University of the Ministry of Internal Affairs of the Russian Federation n.a. V. Ya. Kikotya, Moscow, Russia
[4] Irkutsk National Research Technical University, 83, Lermontova str., Irkutsk, 664074, Russia
[5] Bratsk State University, 40, Makarenko str., Bratsk, 665709 Russia

E-mail: kshakalov@mail.ru

Abstract. In recent years, a large number of studies were carried out in the field of finding and constructing the optimal path of movement. However, the task remains relevant due to the emergence of new aspects in this problem. The paper discusses and describes the existing algorithms for finding the shortest path on the graph. The material can be useful in building navigation systems and automated control in terms of the choice of optimal path.

1. Introduction

Solving the problem of finding the optimal route of movement has become more and more relevant in recent years [1]. The solution to the problem of finding the optimal path of movement has become increasingly popular in recent years [2, 3]. A sufficient number of navigation software has already been developed. The analysis showed that although a sufficient number of both international and national software products are presented on the market, they all a particular disadvantage. Most of the systems are designed to build a path only taking into account the extensive asphalt road network and without regard of the presence of unpaved, country roads, as well as the technical characteristics of the vehicle itself during the pathfinding.

The initial data are the coordinates of the initial and final points of movement, while the path of movement passes through mountainous area, as well as a description of the area, which includes its altitude, steepness, description of soil and rocks. Possessing information about the environment, as well as about the existing road network, it is necessary to build a possible path of movement, taking into account the technical characteristics of vehicles. This formulation of the problem may be especially relevant in the conditions of mining operations [4-6].

The criterion for the choice of a path will be the shortest time to overcome the path from the starting point to the destination. It is necessary to note that the shortest distance does not always mean

IOP Conf. Series: Materials Science and Engineering 1111 (2021) 012033 doi:10.1088/1757-899X/1111/1/012033

that this is the fastest way. When building a path, it is necessary to take into account that it is safe for the movement of vehicles apart from all its other characteristics.

2. Pathfinding

To describe the area, we will use the graph model. The road network will be represented as a oriented graph.

$G: = (V,E)$, where

V - it is a set of vertices or nodes that represent intersections of roads;

E - is a set of graph edges - road sections that connect their intersections.

Each edge of the graph will have its own "value", i.e. some costs to overcome this section of the road. Thus, we reduced the solution of the problem to find a path on the graph.

Our analysis of the existing algorithms to find the shortest path on a graph showed that all algorithms have disadvantages.

Table 1. Description of algorithms for finding the shortest path on a graph

Algorithm	Brief description
Breadth First Search	the path found as a result of the algorithm is not always the shortest; the search is carried out evenly in all directions and therefore is not effective
Dijkstra algorithm	the path found as a result of the algorithm is the shortest; the search is carried out evenly in all directions, and therefore is not effective
Depth first search algorithm	as a result of the algorithm, the optimal path is not always found; the speed of the algorithm largely depends on the correct choice of the stop depth
Best-First Algorithm	the found solution is not optimal; facing an obstacle, the path bends around it
Divide-and-conquer algorithm	the found solution is not optimal, it does not allow overcoming major obstacles
Crash algorithm	a solution is not always found, obstacles are overcome, but the search is not effective
Algorithm A*	the optimal solution is always found, overcoming the obstacle, if there is such a possibility

As an algorithm suitable to solve the problem, the authors proposed to use Algorithm A*.

The principle of the algorithm is based on a heuristic search, as a result of which all nodes are sorted according to the approximation of the best path going through a given node.

In general, the heuristic function can be represented as follows:

$$f(S) = g(S) + h(S), \text{ where} \quad (1)$$

$g(S)$ — real current value of the node S,

$h(S)$ — a heuristic function that estimates the cost of the best sequence of actions starting at a node S and ending with a solution.

At each stage of the algorithm, a set of decisions is assessed, the result of which is the creation of the best path that will have the minimum cost. In order to solve the set task of pathfinding of a vehicle in mountainous area, it is necessary to compile a calculation function that takes into account the area and the capabilities of the existing vehicle.

Figure 1 shows a block diagram of the search for the optimal path using Algorithm A*.

*** Formatting Issue - Best Available Paper/Graphic ***

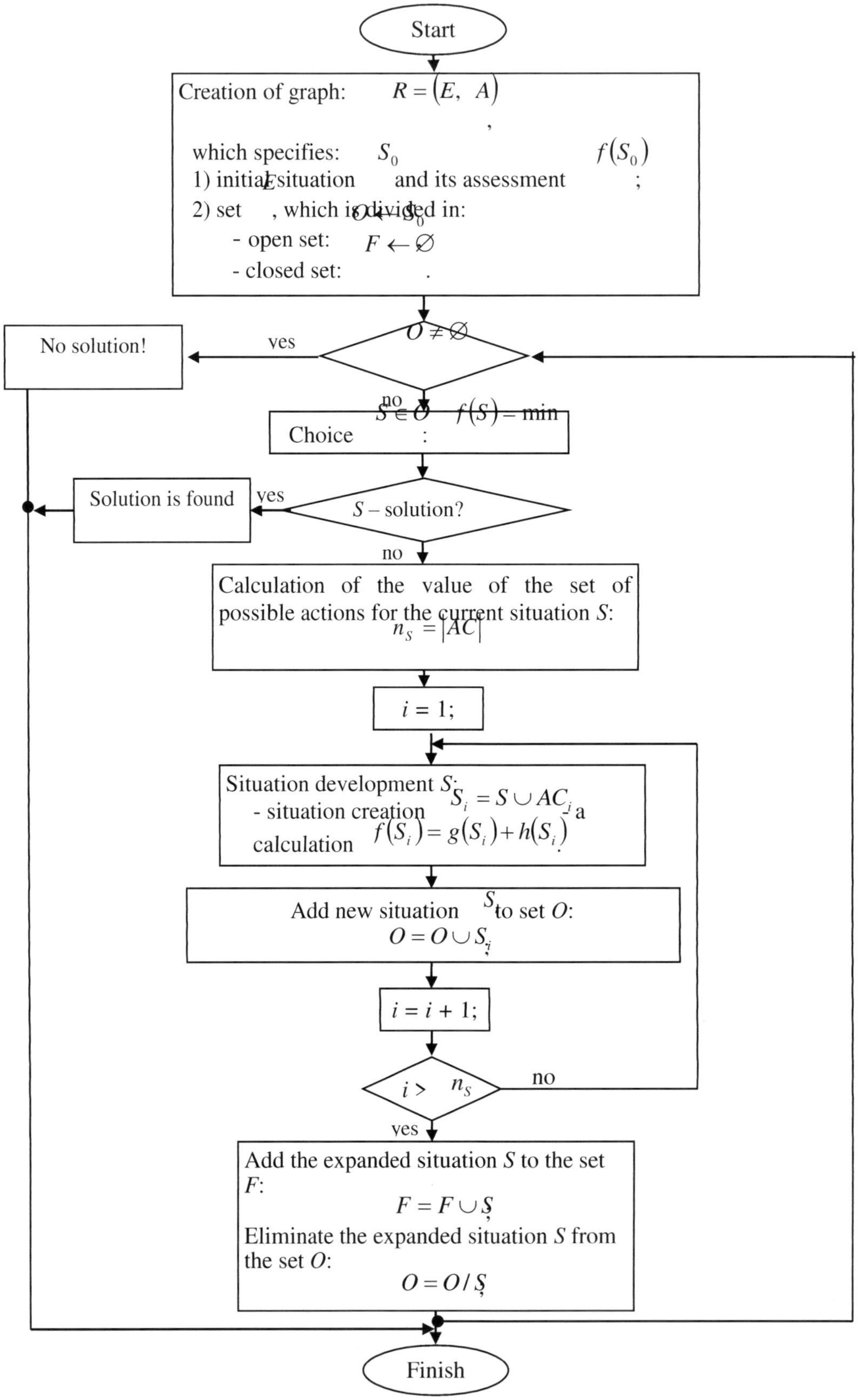

Figure 1. Block diagram of Algorithm A*

3. Creation of calculation function

Let us compose a calculation function for a typical car, taking into account the general technical characteristics. Obviously, it is not possible to give an accurate assessment of the time of movement. Therefore, as an approximate assessment we will use the Naismith rule.

Vehicle speed in mountainous areas is selected depending on the condition of the road. On straight-line sections in conditions of good visibility and good road condition, it should not exceed an average of 45 km / h. It is recommended to drive at the same speed on sharp and closed turns. During the ascent of 6-8 degrees it is recommended to drive at 10 km / h, and at no more than 25 km / h on descents.

The most logical step towards the solution of this problem will be to compose a certain function of the assessment of time to overcome various parts of the area, which will satisfy the following requirements:

1) continuity - this will allow assessing the travel time of the sections for all possible descent angles
2) a certain level of sensitivity of the function to the factor - this will allow diversifying the areas of the area with slightly different descent angles;
3) the closeness of the values of the calculated assessments to the empirical results;
4) calculation ease.

A simple solution would be to create an interpolation function with nodes at the following points:

Road descent	The time a car covers 1 km
$\alpha = 0^0$	T=1,2 (min)
$\alpha = 10^0$	T=6 (min)
$\alpha = 20^0$	T=6 (min)
$\alpha = -10^0$	T=2,4 (min)

We use the Lagrange interpolation formula:

$$L_n(x) = \sum_{i=0}^{n} y_i \times \frac{(x - x_0)(x - x_1)...(x - x_{i-1})(x - x_{i+1})...(x - x_n)}{(x_i - x_0)(x_i - x_1)...(x_i - x_{i-1})(x_i - x_{i+1})...(x_i - x_n)},$$

where

$L_n(x)$ – interpolating polynomial;

n – number of interpolation nodes;

x_i – the value of the argument at the interpolation node;

y_i – function value at interpolation node.

The resulting interpolation function is a heuristic assessment that allows assessing the complexity of the path.

4. Calculation scheme of the assessment of path time

During the description of the area, we will use a digital area map based on the use of a raster model of spatial data.

In this case, the calculation scheme for the assessment of travel time of a given path can be presented in the following form (Figure 2):

1) to choose the calculated path;
2) to divide the path into elementary components with reference vertices in accordance with the scale of the digital map;
3) to compose a "relief profile" of the path.

4) When compiling a profile, the height at each reference vertex is determined;
5) to calculate the descent value α_i for each segment:

$$\alpha_i = arctg\left(\frac{h_{i+1} - h_i}{d_i}\right)$$, where

h_i and h_{i+1} – height of ith and $(i+1)^{th}$ heights respectively;

d_i – distance on the map between vertices, calculated by the formula:

$d_i = k_i \times t$,

$$k_i = \begin{cases} 1, \text{ if movement is horisontal or vertical;} \\ \sqrt{2}, \text{ if movement is diagonal;} \end{cases}$$

t – scale accuracy

5) to calculate the total travel time:

$$T = \sum_{i=1}^{n} T(\alpha_i) \cdot \frac{d_i}{1000\ м}. \quad (2)$$

The proposed calculation scheme does not take into account all the aspects related to the characteristics of the area. However, to solve the problem in the first approximation, this is quite enough, since it will be quite simple to include them in the calculation system as additional parameters in the future and the obtained results are quite close to real ones.

In order to solve the problem, it is necessary to optimize the algorithm A*, taking into account the proposed principle of the calculation of objective function.

The optimization of the algorithm A* is in the fact that in order to create a travel path, it is necessary to set not only the values of the costs of moving along the corresponding sections of the path, but also the description of these sections for the possibility of overcoming them by vehicles with different technical characteristics. According to the available information, the algorithm will calculate the function and select the optimal path.

Our system is developed exclusively for mountainous areas, without taking into account vegetation, type of soil and the passability of mountains depending on the location of the system of ridges, spurs, valleys and gorges. During the travel in mountainous area, the main obstacles that arise in front of the vehicles are descents (more than 20°) and ascents (more than 30°), as well as the angle of lateral static stability. Possessing information about the area, the algorithm will be able to determine the passability of a specific obstacle for a given technical means.

Thus now let us determine exactly how the algorithm will determine the possibility of overcoming an obstacle using the area data and taking into account the characteristics of a particular car.

The calculation of the ascent / descent angle (influencing the choice of the path, determining whether the obstacle is overcome) depends on the map scale. The calculation of it is more accurate, when the scale is larger. It follows that when using small-scale maps, the object (car) will be represented as a point object, and in this case, during the movement, only the possibility of overcoming ascents/descents will be taken into account, and the influence of the angle of inclination (the angle of cross-passability) will be ignored. In order to solve this problem, we propose to transform the car from a point object view into a larger object, that is, to use a larger scale.

International Conference: Actual Issues of Mechanical Engineering (AIME 2020) IOP Publishing
IOP Conf. Series: Materials Science and Engineering 1111 (2021) 012033 doi:10.1088/1757-899X/1111/1/012033

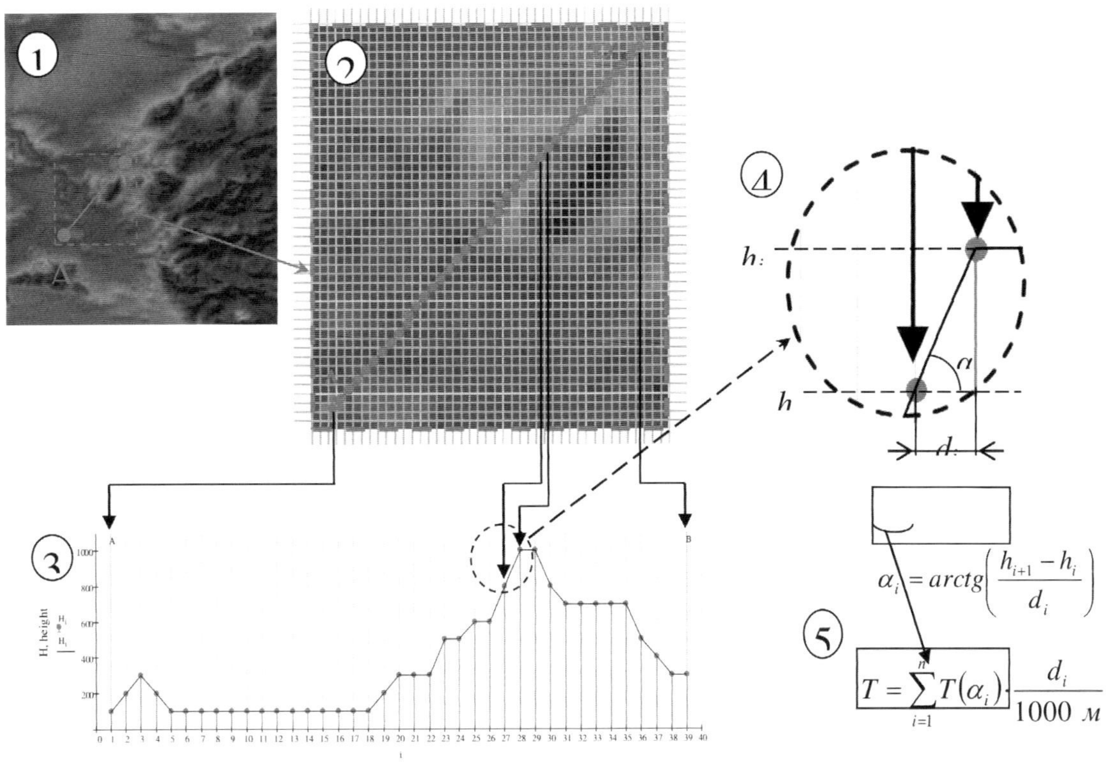

Figure 2. Calculation scheme of the assessment of path time

5. Determination of the angle of lateral static stability

Let us consider the type of obstacle that depends on the angle of lateral static stability.

The angle of lateral static stability is included in the technical characteristics of a car and is defined as the angle by which a car must be bended around the longitudinal axis in order to tip it over (Fig. 3). At the same time, various lateral forces are constantly acting on any vehicle, contributing to its deviation from the rectilinear direction.

Figure 3. Lateral static stability angle

To find this angle, we can use the methods of analogy and similarity. To simplify the task, a car can be represented as a parallelepiped

*** Formatting Issue - Best Available Paper/Graphic ***

A B C D_1

ABCD , in this case, the coordinates of the point of intersection of the
B *D*

diagonals and parallelogram are known (Figure 4a).

It is at this point that the coordinates of the current position of a car will be calculated, which are necessary for next calculations.

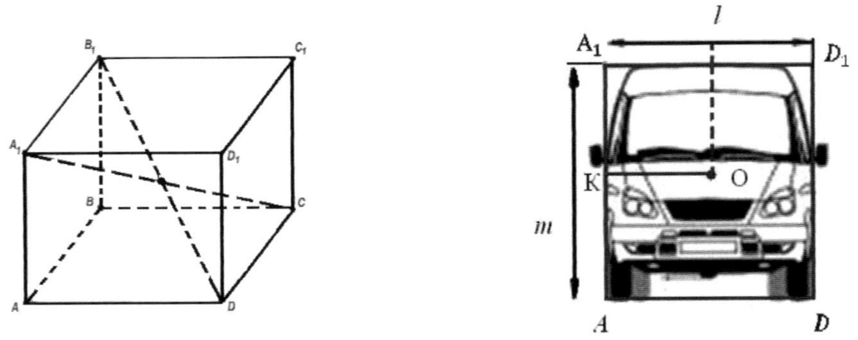

a) b)

Figure 4. Car projection

Accordingly, when a car is bended at an angle β, the position of the KO, and, consequently, the coordinates and the value of heights at point K and point O will change (Figure 5).

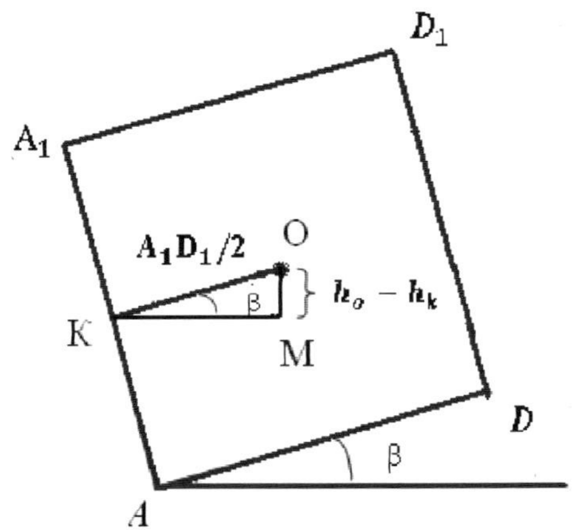

Figure 5. Lateral bend angle of a vehicle

In this case, the intersection of the straight lines KO and KM (KM is parallel to the line of the Earth's plane, from which the angles of ascent and descent are calculated) will create an angle equal to the angle to which the car was bended.

As a result, all the calculations are reduced to find the value of the angle β in the triangle OKM, in which OK is calculated as ½ of the width of a car that is ½ *l*; the values of heights at the point O and K of a car are known (using a height map of large scale). Therefore, the angle β will be calculated as:

$$\beta = \arcsin$$

It is necessary to make a note about the used scale once again. The larger the scale, the more reliable the angle β will be calculated. Since in its calculations, we proceeded from the idea that a car is not represented as a point object, but as a parallelogram, which takes into account the characteristics of a car, one of which (namely the width of a car), we used in the calculations. Otherwise, the algorithm will not see the height difference between point O (a fixed point for pathfinding) and point K (the intersection of the perpendicular dropped from point O to the side plane of a car, the length of which is $1/2l$).

6. Determination of the angle of ascent / descent of a car

The same method will be used to determine the ascent / descent angle of a vehicle. Then, we present the side surface of a car as a rectangle DD_1C_1C (Fig. 3c) with the point O (the point of intersection of the diagonals) and the perpendicular ON (equal to ½ D_1C_1, where D_1C_1 is the length of a car).

Then, all the calculations for the determination of the ascent / descent angle will be reduced to the determination of the angle α from the triangle NOP, in which NP is the height difference, and NO is ½ D_1C_1 (Fig. 6). Therefore, the angle α will be calculated as follows:

$$\alpha = \text{arctg}$$

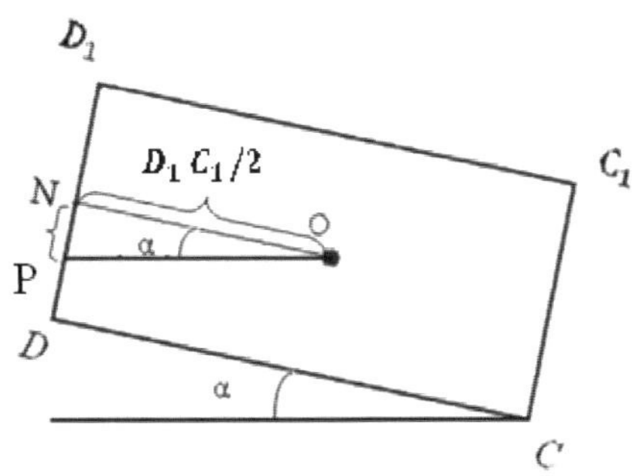

Figure 6. Determination of the angle of ascent/descent of a car

Thus, at the moment we know all the parameters and calculation methods by which the best path will be chosen by the algorithm A*.

At the same time, the support points are rigid, not elastic, due to the use of pneumatic tires, in order to exclude factors that change the stability of the machine with different conditions of the tires during operation [7, 8].

7. Conclusion

The problem of optimal pathfinding for the movement of vehicles over rough mountainous area is relevant. To solve this problem it is proposed to use the algorithm for finding the shortest path A*, which has proven itself during pathfinding on rough area with possible obstacles in the movement path. In order to improve the efficiency of the algorithm, the authors proposed optimization, which allows taking into account not only the distinctive features of the area, but also the technical characteristics of vehicles, on which their passability depends during pathfinding. To assess the obtained paths, a computational function is proposed, which depends on the area.

The proposed solution to the problem of pathfinding over rugged mountainous area may be more effective than the use of other algorithms.

References

[1] Mazakov E B, Matrokhina K V and Trofimets V Y 2021 *Traffic management at the enterprises of the mineral industry. Advances in raw material industries for sustainable development goals* (London: CRC Press) p 528

[2] Safiullin R N, Afanasyev A S and Rezchenko V V 2019 The Concept of Development of Monitoring Systems and Management of Intelligent Technical Complexes *J. of Mining Institute* **237** 322-330. DOI: 10.31897/PMI.2019.3.322

[3] Afanasyev A S, Egoshin A M and Alekseev S V 2018 Justification of application of the logistic approach in the models of road safety management *IOP Conf. Ser.: Earth Env.* **194** 218

[4] Klyuev R, Bosikov I, Gavrina O, Madaeva M and Sokolov A 2021 Improving the energy efficiency of technological equipment at mining enterprises *Advances in Intelligent Systems and Computing* **1258** 262-271. DOI: 10.1007/978-3-030-57450-5_24

[5] Klyuev R, Fomenko O, Gavrina O, Turluev R and Marzoev S 2021 Energy indicators of drilling machines and excavators in mountain territories *Advances in Intelligent Systems and Computing* **1258** 272-281. DOI: 10.1007/978-3-030-57450-5_25

[6] Kosarev O V, Tcvetkov P S, Makhovikov A B, Vodkaylo E G, Zulin V A and Bykasov D A 2019 Modeling of industrial IOT complex for underground space scanning on the base of Arduino platform *Topical Issues of Rational Use of Natural Resources – Proc. Of The Int. Forum-Contest of Young Researchers* pp 407-412

[7] Klyuev R V, Bosikov I I, Mayer A V and Gavrina O A 2020 Comprehensive analysis of the effective technologies application to increase sustainable development of the natural-technical system *Sustainable Development of Mountain Territories* **2** 283-290. DOI: 10.21177/1998-4502-2020-12-2-283-290

[8] Abdulaev E K, Makharatkin P N and Pumpur E V 2019 A priori ranking and an analysis of factors affecting tire wear *IOP Conf. Ser.: Earth Env.* **378(1)** 012001. DOI: 10.1088/1755-1315/378/1/012001

Planning model of a group UAV flight in three-dimensional space

N A Maksimov[1], K Zhigalov[2], A V Gorban[3] and I V Ignatev[4]

[1] Moscow Aviation Institute, Moscow, Russia
[2]V.A. Trapeznikov Institute of Control Sciences of Russian Academy of Sciences, Moscow, Russia
[3] Irkutsk National Research Technical University, 83, Lermontova str., Irkutsk, 664074, Russia
[4] Bratsk State University, 40, Makarenko str., Bratsk, 665709 Russia

E-mail: n-a-maximov47@yandex.ru

Abstract. The paper describes the algorithm and model for planning group actions of unmanned aerial vehicles (UAVs) to monitor a given set of objects. A distinctive feature of the task is the shooting of objects from a predetermined height. This makes it necessary to move from considering a "flat" model to planning a flight in three-dimensional space. In addition, the authors take into account the possibility of destroying individual UAVs and the need to reschedule the flight route for the remaining UAVs in order to accomplish the assigned task.

1. Introduction

Nowadays, the use of unmanned aerial vehicles for the solution of observation problems is very important [1]. More and more often there are situations when there is a need for prompt receipt of information, while the information about situation necessary to obtain can be located both in hard-to-reach and crowded places.

There is a wide range of available unmanned aerial vehicles on the Russian market and modern experience of their practical use has shown that already now they have sufficient technical characteristics not only to record the information received, but also to process it.

The issues of targeted flight (mission) planning, taking into account the air situation and possible enemy countermeasures have been resolved for a single UAV [2, 3]. However, there is a need to resolve these issues for the group of UAVs. The method of sequential flight planning of a group of several UAVs in the general case is not able to ensure the optimal use of the existing group. Therefore, it becomes necessary to develop algorithms that allow simultaneous control of several UAVs.

2. Problem statement

The purpose of this paper is to solve the problem of planning a fly-by of many objects by a group of unmanned aerial vehicles and building its three-dimensional model.

The following is given:

- the characteristics of each UAV model, which can be taken from the flight manual or from other documents of the chief designer of UAV;
- the coordinates of the target and the specified height of their shooting;

- the map of the area;

It is required:

- to explore the principles, ways of creation, algorithms and special software (SS) for planning group use of UAVs;
- to develop a layout of the open source software for automated routing and conduct experimental research.

3. Development of a model for automated routing. Algorithm for the distribution of routing points for UAVs

The problem of the distribution of routing points for UAVs is reduced to the solution of the classical problem of transport routing. The solution of the problem makes it possible to find the routes for several vehicles starting their route from one point (depot) [6]. In order to solve the problem, the authors propose to use a modified "greedy" algorithm [7]. Among the main factors that influenced this choice, in addition to the general restrictions imposed on the possibility of using UAVs, the authors used the following:

- when planning to fly over a certain area, a significant number of unmanned aerial vehicles can be used, as a result, the number of points required to create a route can be huge;
- each considered UAV has unique technical characteristics that can change over time directly in the process of movement;
- when using UAVs, there is a possibility of their failure, one of the reasons for such a situation may be a sharp change in weather conditions, as a result of which there is the increase in the consumption of energy resources required to reach the routing point.

Taking into account the above mentioned aspects, we can conclude that any pre-created route may turn out to be incorrect even with a slight error in the calculations, and making amendments to the already formed route leads to its complete recalculation.

Let us consider the UAV group flight algorithm (Figure 1):

1) At the initial stage, a search is performed for available route points that are at the greatest distance from the starting point and from each other. The criterion of proximity in this case is the technical characteristics of the UAV (the availability of the necessary energy resources to reach the found route point and return movement, as well as the ability to climb to the height specified by the route point).

2) After the drone reaches the first route point, it is necessary to search for the nearest reachable route points. One of which is defined as a new route point for the UAV.

1) In the case that the search gave negative results and there are no more reachable waypoints, the UAV returns to the starting point.

Let us introduce some clarifications into the operation of the algorithm:

– each UAV has information on the location of all points requiring a visit, when a point is included in the visiting route by one of the UAVs, others no longer visit this point, it becomes inaccessible to them;
– in the case of emergency, functional loss of UAV, its unvisited points are opened for access to other unmanned aerial vehicles;
– before the departure, the analysis of the flown territory is carried out for the presence of "restricted" zones, which are marked on the map and this terrain becomes inaccessible to the UAV. Such zones can be zones closed for UAV flights or those where there are threats of UAV functional loss. Terrain analysis is also carried out to determine possible obstacles to the movement of UAV. The UAV route should be built taking into account the bypass of the marked obstacle in order to avoid collision
– the flight range of unmanned aerial vehicles is significantly influenced by weather conditions (mainly wind).

In order to solve the problem, we will take into account the following characteristics of UAV:

- possible speed of movement;

IOP Conf. Series: Materials Science and Engineering 1111 (2021) 012034 doi:10.1088/1757-899X/1111/1/012034

- maximum lifting height;
- lifting speed;
- angular velocity;
- maximum flight time.

In order to describe the route point, the following characteristics are entered: coordinates, visit height.

Any obstacle is described by its coordinates, height and radius.

Dangerous zones are characterized by location coordinates and radius.

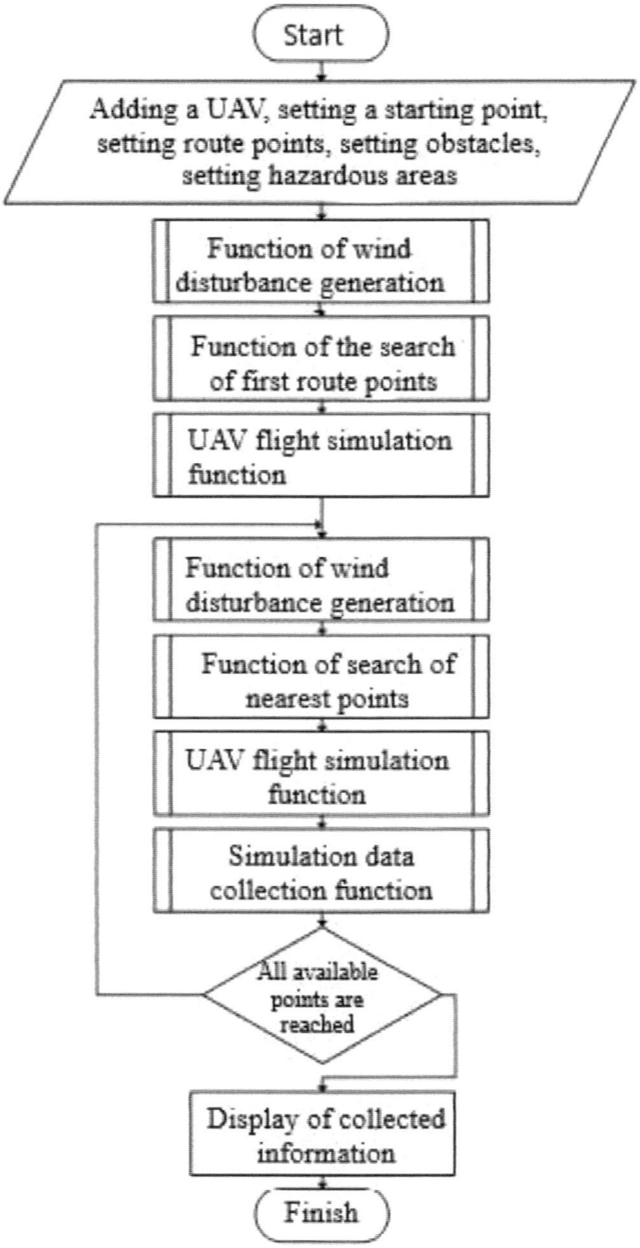

Figure 1. Block diagram of simulation algorithm

4. Development of software for modeling automated routing

Software development was carried out using Qt Framework in the high-level C ++ language using the Qt classes. The mingw32 compiler was used to build the project on the Windows platform and linuxg ++ for Ubuntu.

5. UAV movement

In order to calculate the motion of UAV in the simulation, a discrete constant step (h=19729 m/sec) was used. Before the next shift of the unmanned aerial vehicle, the speed and direction of its movement are calculated, taking into account the influence of the environment.

The UAV can change the direction of movement as needed. The direction of turn depends on the current position of a drone, angular speed and distance to the route point.

The angle of rotation to which UAV must be deployed can be calculated as follows:

$$\theta_{req} = \operatorname{atan2}\left(\left(y_{drone} - y_{point}\right), \left(x_{drone} - x_{point}\right)\right)$$

where x_{drone} and y_{drone} - UAV coordinates, x_{point}, y_{point} - route point coordinates.

To make a decision on the need to UAV rotation, the current angle of the direction of flight and the required (calculated) angle are compared.

The rotation of UAV does not occur instantly, for its implementation, a certain spatial volume is required, in order to calculate the required distance to complete the rotation, we use the following formula:

$$L = sin\left(\frac{\theta_{req}}{2}\right) \cdot \left(\frac{360}{V_{rotate}} \cdot V_{sum}\right)$$

where L – chord length; V_{rotate} – rotation speed; V_{sum} - UAV speed with wind influence.

Depending on the obtained result, the radius of the rotation is determined, if the rotation requires a distance greater than to the road point, then the rotation will be carried out along a large radius, if less - along a small one. The example of the choice of the direction of rotation is shown in Figure 2.

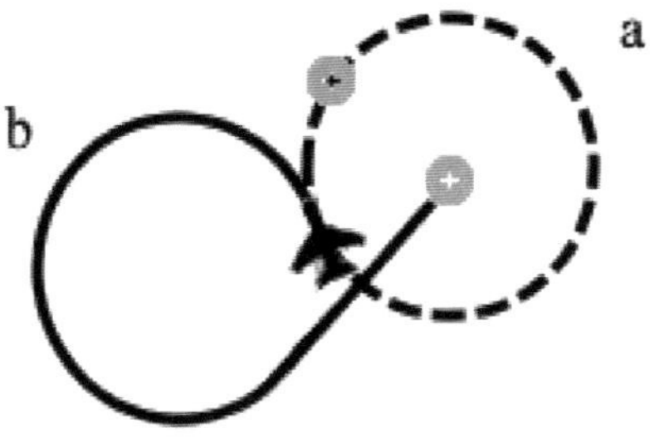

Figure 2. Choice of rotation direction based on route point location

6. Height gain

One of the characteristics of a route point is its height. This characteristic was introduced in order to be able to set the height from which to shoot the object located in it. This is explained by the rugged (mountainous) terrain over which UAV flies, or urban or industrial buildings, the height of which can reach several hundred meters. If the distance to the nearest route point is less than the distance necessary to lift UAV (taking into account its maximum vertical speed), an UAV will move around the route point until it reaches the required height. The example of such a movement is shown in Figure 3.

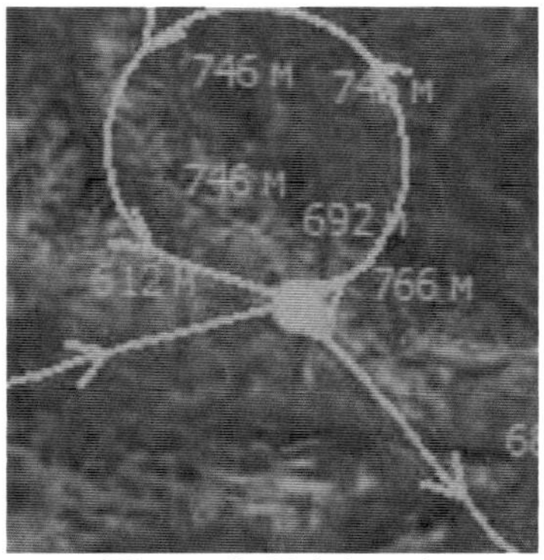

Figure 3. Height gain

7. Obstacle avoidance

One of the problems that an unmanned aerial vehicle may face is an obstacle, such as a mountain, a high building, etc. When it faces an obstacle, an UAV can either change its direction of movement, or gain height and fly around it vertically. Bypassing an obstacle in the horizontal plane is almost always possible, however flying vertically is possible only if the technical characteristics of UAV allow it (it is necessary to take into account the maximum height of UAV and the height of an obstacle).

There are several algorithms to find the ways to avoid obstacles during the movement from one point to another [8, 9]. The most famous ones are the wave tracing algorithm and the A* algorithm. In most cases, these algorithms accurately find a route among several obstacles of a complex form. However, they are not quite suitable for our task. The main reason for this is resource consumption. When using these algorithms, the terrain map is divided into cells and taking into account the speed with which the UAV moves and the model of its movement, it will be necessary to consider areas that are large in size in order to solve the problem the use of these algorithms. Another reason is the virtually complete absence of complex obstacles on the flight route as almost all industrial facilities are located on flat terrain. Therefore, most of the objects that can be met on the path of UAVs have an extremely simple form and can be represented in the form of circles described around the object. In the same way it is possible to imagine the areas initially prohibited for UAVs.

First, the stage of checking the route for the presence of obstacles includes the search for the first obstacle on the way of UAV and evaluation of its length. If the analysis shows the presence of an obstacle, then the distance from UAV to the obstacle is calculated, as well as the distance required for height gain. In the case when there is no opportunity to gain height (there is not enough available distance, technical characteristics of UAV are insufficient), the search is carried out for the nearest point to bypass an obstacle, which will become the route point.

The operation to find obstacles is repeated at each new route point, until all obstacles are overcome.

8. Wind generation algorithm

As we have already mentioned above, weather conditions and wind in particular, have a significant effect on the duration of UAV's stay in the air, its speed and fuel consumption. To assess this influence, we have proposed the algorithm that simulates the behavior of wind.

At the initial stage of modeling, the wind direction and speed were set randomly and its further behavior was subject to several conditions:

- the calculation of changes in wind characteristics occurs at each step of the simulation;
- there was the assumption that there was a direct relationship between the wind speed and the probability of its change (the increase in wind speed leads to the increase in the probability of its decrease and vice versa);
- every change in wind characteristics can be cut off randomly;

- there is a possibility that the wind strength and speed will remain unchanged in the next step of the simulation;
- the direction and speed of wind change randomly without sharp jumps.

9. Interface description

During the development of the interface, the main task was to provide a user with only the information he needed at the moment. In this regard, all settings were moved to separate windows.

The main window of the program can be conditionally divided into two areas:

1) The map with objects displayed on it. This area occupies about 80% of the window. In this area, the movement of the UAV is drawn and all routes are displayed at the end of the movement.

2) The area with buttons and current simulation information. Here the information about the state of UAV is reflected and the simulation time and the number of visited points are displayed at the end of the simulation.

The window of map selection allows opening the required image using the explorer. The map preview function is available to the user only until the selection is confirmed. The displayed map is scaled to fit the actual size of the window.

After the final selection of the map, the “Modeling” and “Set parameters” icons become available. The main windows used in the simulation are as follows.

10. Experimental verification of developed software

In order to check the performance of the software, it was launched with a set of predefined parameters, most of which are presented below.

When modeling, a group of four UAVs with the following characteristics was considered:

1) UAV # 1 - flight speed 67 km / h, angular speed 25 rpm, gain 10 m / min, maximum height 2000 m, maximum flight time 370 min;

2) UAV No. 2 - flight speed 67 km / h, angular speed 17 rpm, gain 14 m / min, maximum height 933 m, maximum flight time 262 min;

3) UAV No. 3 - flight speed 52 km / h, angular speed 17 rpm, gain 16 m / min, maximum height 1167 m, maximum flight time 393 min;

4) UAV No. 4 - flight speed 40 km / h, angular speed 11 rpm, gain 9 m / min, maximum height 907 m, maximum flight time 239 min.

In addition to the number and characteristics of UAVs, 13 route points and a starting point are manually set.

Starting point parameters: (122, 410), height 275 m.

Route point parameters:

Route point number (RPN)	X	Y	Height	Route point number (RPN)	X	Y	Height
1	520	50	365	8	560	285	365
2	586	150	644	9	432	225	233
3	816	351	644	10	746	432	644
4	76	36	260	11	43	200	540
5	206	124	393	12	181	236	672
6	374	63	609	13	326	267	602
7	286	364	302	-	-	-	-

At the end of the simulation, a simplified version of the navigator calculation will be shown for each UAV that participated in the simulation. The navigator calculation contains information on the route of movement, rotation points and flight altitude. The right side contains information about the number of unvisited points, the simulation time and navigational instructions. The simulation result is shown in Figure 4.

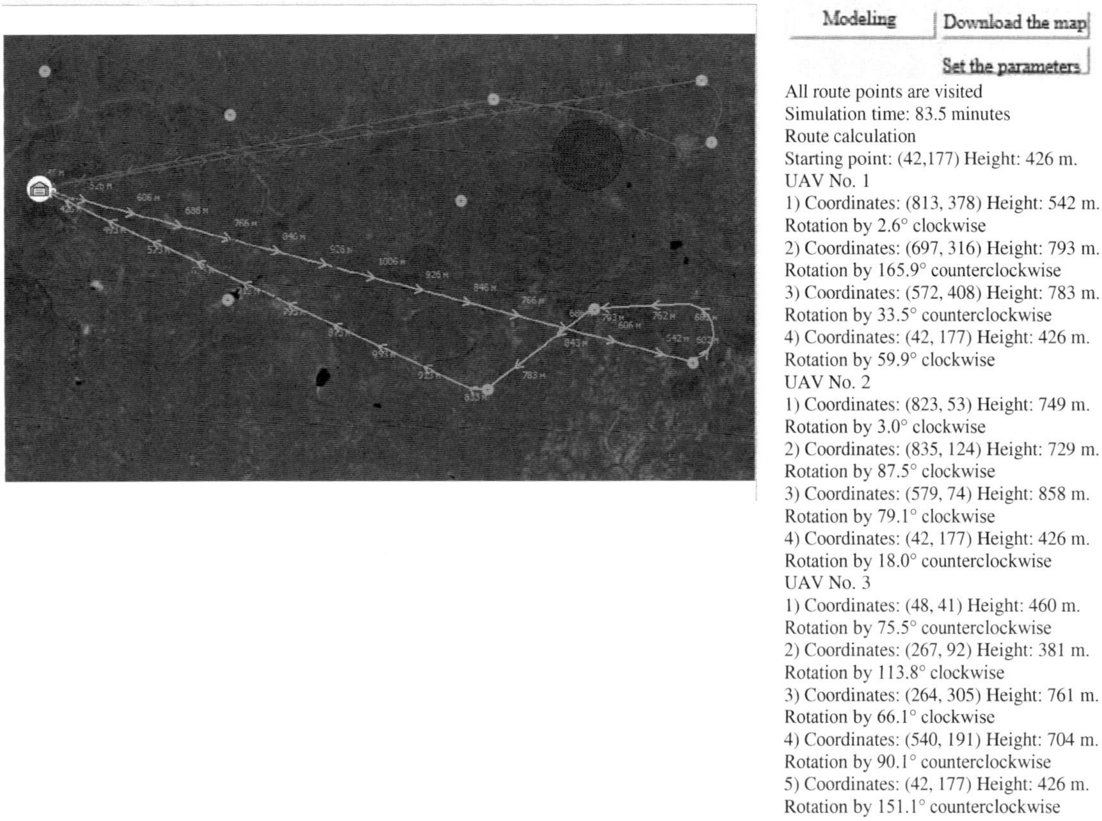

Figure 4. Presentation of the simulation result on the map

According to the simulation results, it can be concluded that all route points have been visited, i.e. the planning algorithm was correct. The simulation time on a standard office machine was 103.6 minutes.

The results of the engineering and navigational calculation are given below.

The starting point for all UAVs is common with coordinates: x=122, y=410, height 275 m.

UAV №1: RPN1, rotation by 17.1 degrees counterclockwise;

RPN2, rotation by 96 degrees clockwise;

RPN3, rotation by 51.3 degrees clockwise;

RPN (starting) coordinates 122, 410, height 275 m rotation by133.6 degrees clockwise

UAV №2: RPN4, rotation by80 degrees clockwise;

RPN5, rotation by132 degrees clockwise:

RPN6, rotation by63.5 degrees counterclockwise;

RPN7, rotation by125,4 degrees clockwise;

RPN (starting) coordinates 122, 410, height 275 m rotation by 34.3 degrees clockwise

UAV №3: RPN8 rotation by1 degrees clockwise;

RPN9, rotation by137.7 counterclockwise

RPN10, rotation by160, 37 counterclockwise;

RPN (starting) coordinates 122, 410, height 275 m rotation by152.58 degrees clockwise

UAV №4: RPN11, rotation by99.6 degrees counterclockwise;

RPN12, rotation by130.8 degrees clockwise;
RPN13, rotation by15.95 degrees counterclockwise;
RPN (starting) coordinates122, 410, height 275 m rotation by132.4 degrees clockwise

11. Conclusion

As a result of this research, we can single out the developed algorithm for planning UAV flight routes in a group flight, taking into account the effect of the wind, the need to shoot from a given height and fly around obstacles of the route. The experimental software was created on the basis of the developed algorithm. The testing the software on a sufficiently large number of test examples showed its performance in a wide range of initial data (the number of UAVs is -2-10, the number of route points is up to 150). The methods used in the work have been applied in other industries [2-4].

References

[1] Pham D T 2009 The Bees Algorithm: Modelling Foraging Behaviour to Solve Continuous Optimisation Problems *Proc. of the Institution of Mechanical Engineer, Part C* **223(12)** 2919-2938

[2] Klyuev R V, Bosikov I I, Revazov V Ch, Krysanov K S and Dzeranov B V 2020 A comprehensive analysis of the properties of reservoir rocks to determine the productive oil horizons *IOP Conf. Ser.: Mater. Sci. Eng.* **952** 012017. DOI: 10.1088/1757-899X/952/1/012017

[3] Klyuev R V, Bosikov I I, Mayer A V and Gavrina O A 2020 Comprehensive analysis of the effective technologies application to increase sustainable development of the natural-technical system *Sustainable Development of Mountain Territories* **2** 283-290. DOI: 10.21177/1998-4502-2020-12-2-283-290

[4] Klyuev R V, Bosikov I I, Egorova E V and Gavrina O A 2020 Assessment of mining-geological and mining technical conditions of the Severny pit with the use of mathematical models *Sustainable Development of Mountain Territories* **3** 418-427. DOI: 10.21177/1998-4502-2020-12-3-418-427

Economic efficiency of increasing the reliability of electric transport in urban transport systems

M K Chernyakov[1], K Ch Akberov[2], B V Malozyomov[1], V Yu Listkov[3], I Yu Kharlamova[2], E N Saricheva[2], I A Shuraev[4]

[1]Novosibirsk State Technical University, 20, Karla Marksa Av., Novosibirsk, 630073, Russia
[2]Novosibirsk State University of Economics and Management, 56, Kamenskaya st., Novosibirsk, 630087, Russia
[3]Siberian University of Consumer Cooperation, 26, Karla Marksa Av., Novosibirsk, 630087, Russia
[4]Siberian Institute of Management - RANEPA branch, 6, Nizhegorodskaya st., Novosibirsk, 630102, Russia

E-mail: mbv5@mail.ru

Abstract. In this article, the authors propose a new approach to determining the efficiency of increasing the reliability of electric transport, based on the symbiosis of the effective implementation of a diagnostic technical set of measures and minimizing the cost of operating electric transport. As a result of its use, it was determined that increasing the reliability of electric transport systems can lead to a decrease in both specific operating costs and specific capital investments as a result of an increase in the annual productivity of electric transport. This is due to the reduction in the time it takes to take it out of operation to eliminate the consequences of failures and to carry out maintenance and repair.

1. Introduction

Currently, there are different views on determining the technical reliability of complex technical systems, including electric transport [1, 2]. One of the important approaches is the economic assessment of increasing the reliability of electrical transport [2, 3]. In this article, the authors propose a new approach to determining the efficiency of increasing the reliability of electric transport, based on the symbiosis of the effective implementation of a diagnostic technical set of measures and minimizing the cost of operating electric transport. As a result of its use, it was determined that increasing the reliability of electric transport systems can lead to a decrease in both specific operating costs and specific capital investments as a result of an increase in the annual productivity of electric transport. This is due to the reduction in the time it takes to take it out of operation to eliminate the consequences of failures and to carry out maintenance and repair [4].

As mentioned earlier, one of the important problems in the operation of trolleybuses, which is a type of electric transport, at the present stage is to improve the quality of services for the transportation of passengers. Along with this, the task is to reduce the cost of operation and repair. These requirements are contradictory to a certain extent; therefore, it is more correct to speak about the joint solution of two problems by optimizing the level of quality of services in the transportation of passengers [5]. At the same time, the minimum of total transportation costs was taken as the criterion of optimality [6].

The reliability of electric transport is one of the important components of quality, and the above tasks fully relate to this characteristic. To solve them in relation to the level of reliability of a trolleybus, mathematical models were developed that describe the dependence of the costs of creation E_c, operation E_e and repair E_p on the reliability indicators of units and parts of trolleybuses [7]. Two approaches to solving the problem were proposed: optimization of the reliability level of electric transport as a whole and optimization of the reliability of its elements with the subsequent determination of the structural reliability of the trolleybus as a whole.

2. Results

Let us consider an approach to determining the optimal level of trolleybus reliability. In the general case, the cost of a trolleybus E_{c1} with increased reliability can be expressed by the function

$$E_{c1} = E\left(E_{c0}, H_0, H_1\right), \tag{1}$$

where E_{c0} is the cost of a trolleybus of initial (basic) reliability; H_0, H_1 - trolleybus reliability levels, measured by some quantitative indicators, respectively, before and after the reliability improvement.

Since the cost of building a trolleybus also depends on its capacity, in equation (1), E_c we mean the unit cost per unit of capacity. The function $E\left(E_{c0}, H_0, H_1\right)$ must have the following properties:

1. $E\left(E_{c0}, H_0, H_1\right) > 0$, since the costs are non-negative;
2. $E\left(E_{c0}, H_0, H_1\right) > E_{c0}$;
3. $\lim\limits_{H_1 \to 1} E\left(E_{c0}, H_0, H_1\right) = \infty$, since it is impossible to achieve absolute reliability in real design and production conditions;
4. $E\left(E_{c0}, H_0, H_1\right) \leq E\left(E_{c0}, H_0, H_2\right)$ if $H_0 < H_0'$ and, if $E\left(E_{c0}, H_0, H_1\right)$, that is, $E\left(E_{c0}, H_0, H_1\right)$ it does not decrease with respect to a fixed value and does not increase with H_0 respect to a fixed value H_1, because an increase in reliability is almost always associated with additional costs.

To quantify the level of reliability H in expression (1), we use a complex indicator reflecting, if possible, all the components of reliability: reliability, durability, maintainability. Utilization and downtime ratios are often used as such an indicator under operating conditions K_{pr}, which depend on both uptime and recovery time.

The properties 1- 4 of the function (1) are satisfied by an exponential function of the form $f(x) = k \cdot x^a$. Then we have

$$E_{c1} = E_{c0} \cdot \left(\frac{1 - K_{ti0}}{1 - K_{ti1}}\right)^a = E_{co}\left(\frac{K_{pr0}}{K_{pr1}}\right)^a, \tag{2}$$

where K_{ti0}, K_{pr0} are the coefficients of technical utilization and downtime of the trolleybus by the time of the start of measures to improve reliability; K_{ti1}, K_{pr1} - coefficients of technical utilization and downtime of the trolleybus after implementation of measures to improve reliability; α - statistical coefficient, reflecting the degree of increase in reliability indicators from the funds invested in its increase. The statistical coefficient α is determined first as the mathematical expectation of the known distribution laws of a random variable for the investigated trolleybus node, and at subsequent steps - according to the statistical data of observations after the implementation of measures [8, 9].

The costs of maintenance and repair of the trolleybus during operation will be

$$E_{top} = E_{to} + E_p \ , \tag{3}$$

where E_{to} is the cost of maintaining a serviceable trolleybus; E_p - costs of repairing a trolleybus.

Expressing E_{to} and E_p in terms of the unit costs e_{to}, e_n, respectively, per unit time of the trolleybus being in good and bad condition

$$E_{to} = e_{to} \cdot K_{ti} \cdot t = e_{to} \cdot \left(1 - K_{pr}\right) \cdot t \ , \tag{4}$$

$$E_p = e_n \cdot \left(1 - K_{ti}\right) \cdot t = e_n \cdot K_{np} \cdot t \ . \tag{5}$$

Considering that the costs of repairs during use are carried out at the moment t [10], they must be brought to the initial moment, that is, to the time of creation. Instead of the discrete reduction method by using the standard factor e_n, the continuous reduction method is applied. Then the maintenance costs can be represented as

$$E_{to}^{np} = e_{to} \cdot \left(1 - K_{np}\right) \cdot \frac{1}{\gamma} \cdot \left[1 - \exp\left(1 - \gamma \cdot t\right)\right] = e_{to} \cdot \left(1 - K_{np}\right) \cdot t_{np} \ , \tag{6}$$

where $\gamma = \frac{1}{8760} \cdot \ln\left(1 + E_n\right)$; t_{np} - conditional time, taking into account the reduction of costs to the initial moment, that is

$$t_{np} = \frac{1}{\gamma} \cdot \left[1 - \exp\left(-\gamma \cdot t\right)\right] \tag{7}$$

Similarly, for the second component of formula (8), taking into account (2)

$$E_p^{np} = e_n \cdot K_{np} \cdot t_{np} \tag{8}$$

The total costs in the field of creating and using a trolleybus of increased reliability, taking into account the diagnostic complex according to equations (2) (5), will be

$$E_{\Sigma 1} = E_{co} \cdot \left(\frac{K_{npo}}{K_{np1}}\right)^{\alpha} + E_{mo}^{np} + E_p^{np} = E_{co} \cdot \left(\frac{K_{npo}}{K_{np1}}\right)^{\alpha} + e_{mo} \cdot \left(1 - K_{np1}\right) \cdot t_{np} + e_n \cdot K_{np1} \cdot t_{np} \tag{9}$$

From (9) it can be determined that at the minimum value $\frac{dE_{\Sigma 1}}{dK_{np1}} = 0$ the optimal level of reliability corresponds

$$K_{np}^{opt} = \left[\frac{\alpha \cdot E_{co}}{\left(e_n - e_{to}\right) \cdot t_{np}}\right]^{\frac{1}{\alpha+1}} \cdot K_{npo}^{\frac{\alpha}{\alpha+1}} \tag{10}$$

According to equation (10), it is possible to determine, per one trolleybus, the optimal level of reliability, characterized by the downtime coefficient K_{np1}^{opt} or the coefficient of technical utilization $K_{ti1}^{opt} = 1 - K_{np1}^{opt}$, and also determine the corresponding total costs $E_{\Sigma 1}^{\min}$ by expression (9), as well as the magnitude of the effect per one trolleybus with the optimal level of reliability, equal

$$E_{nb} = E_{\Sigma 0} - E_{\Sigma 1}^{\min} \tag{11}$$

Under operating conditions, the reliability of the equipment is characterized by its reliability, durability and maintainability. The economic effect of increasing the level of reliability can be defined as the sum of the economic effects of increasing reliability, durability and maintainability, excluding additional costs that were required to increase the reliability of the trolleybus, that is

$$E_{ne} = \left[\left(E_b + E_d + E_p \right) - E_n \cdot K \right] \cdot N \text{, rub / year,} \quad (12)$$

where E_n is the standard payback ratio; K - one-time costs to improve the reliability of the trolleybus; N - the number of operated trolleybuses.

The annual economic effect of increasing the reliability of the trolleybus can be defined as

$$E_b = \frac{L_{g1}}{L_{cp1}} \cdot U_1 - \frac{L_{g2}}{L_{cp2}} \cdot U_2 + \frac{\Delta t_{tb} \cdot E_n \cdot C_t}{8760} \text{, rub / year,} \quad (13)$$

where L_{g1}, L_{g2} - the annual runs of the trolleybus before and after the increase in reliability, km; L_{cp1}, L_{cp2} are the reliability of the trolleybus, characterized by the mean time between failures of the trolleybus before and after the increase in reliability, km / failure; U_1, U_2 - damage from trolleybus failure before and after the safety increase, rubles / refusal; C_t - trolleybus price; 8760 - annual working time fund, h; Δt_{tb} - saving time as a result of the reliability of the trolleybus electrical equipment. Time saving is defined as

$$\Delta t_{tb} = \frac{L_{g1}}{L_{cp1}} \cdot \tau_{v1} - \frac{L_{g2}}{L_{cp2}} \cdot \tau_{v2} \text{, rub / year,} \quad (14)$$

where τ_{v1}, τ_{v2} are the average downtime of the trolleybus under repair due to failure before and after the increase in reliability, h.

The annual economic effect from increasing the durability of the trolleybus is defined as

$$E_d = \sum_{i=1}^{k_p} \left(\frac{L_{g1}}{L_{pi1}} \cdot c_{i1} - \frac{L_{g2}}{L_{pi2}} \cdot c_{i2} \right) + \sum_{i=1}^{k_p} \left(\frac{L_{g1}}{L_{pi1}} \cdot \tau_{i1} \cdot c_{tm} - \frac{L_{z2}}{L_{pi2}} \cdot \tau_{i2} \cdot c_{tm} \right) + \sum_{i=1}^{k_p} \frac{\Delta \tau_{mi}}{8760} \cdot E_n \cdot C_t \text{,} \quad (15)$$

where L_{pi1}, L_{pi2} - trolleybus runs between types of repairs before and after increasing durability, km; , c_{i1}, c_{i2} - the cost of the i-th type of repair before and after increasing the durability of the trolleybus, rubles; , τ_{i1}, τ_{i2} - the duration of the downtime of the trolleybus in the i type of repair before and after increasing the durability, h; c_{tm} - cost of 1 hour of trolleybus downtime for repair, rubles; $\Delta\tau_{mi}$ - saving trolleybus hours during repairs as a result of increasing the durability of the trolleybus [11, 12].

The calculation according to expression (15) is carried out only for those types of repairs, the runs between which, as a result of increased durability, have changed taking into account the cyclicity coefficient [13, 14].

The annual economic effect from improving the maintainability of trolleybus equipment will be

$$E_p = \frac{L_{g1}}{L_{cp1}} \cdot \tau_{v1} \cdot c_{tm} - \frac{L_{g2}}{L_{cp2}} \cdot \tau_{v2} \cdot c_{tm} \quad (16)$$

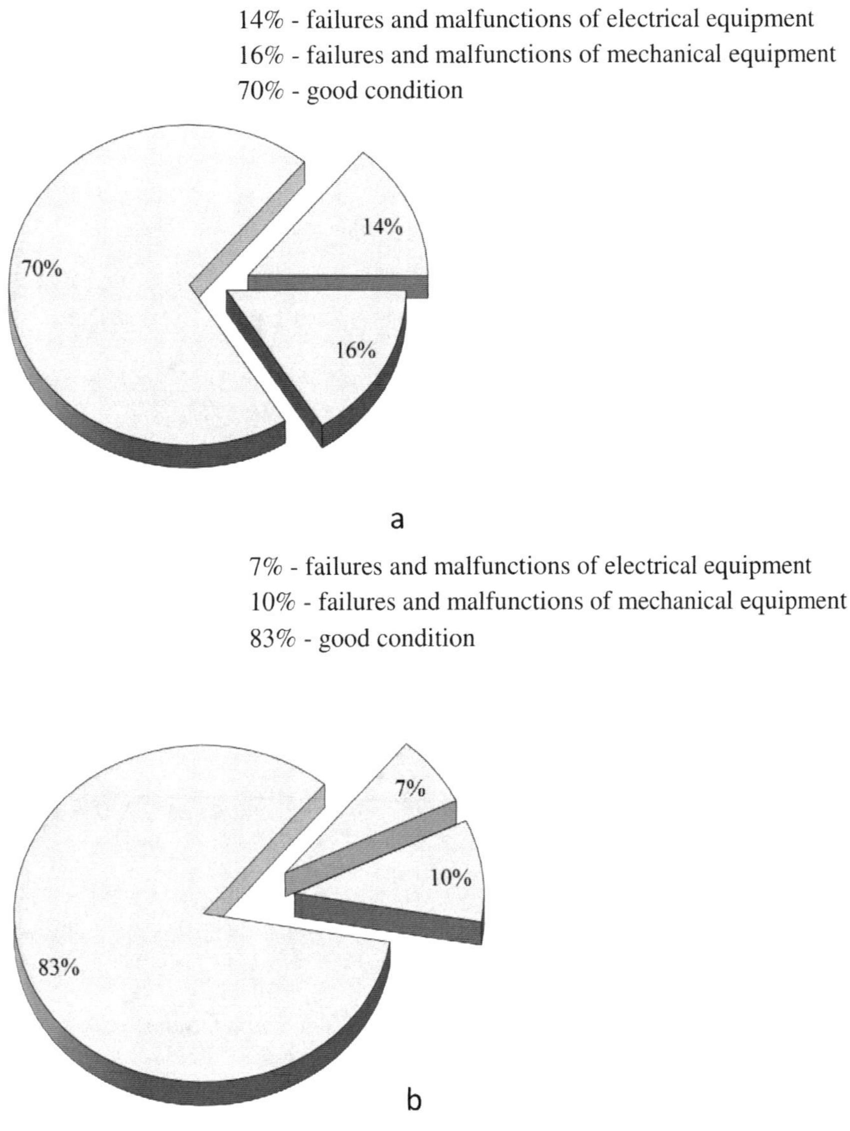

Figure 1. Change in serviceable and faulty technical conditions of the electrical complex of the trolleybus (SPTS): a - before the implementation of the SPTS; b - after the implementation of the SPTS.

The payback period of the one-time costs K for improving the reliability of the electrical equipment of the trolleybus is defined as

$$T_{ok} = \frac{K}{E_b + E_d + E_p} = \frac{3200000}{2000000} = 1.6 \ years. \quad (17)$$

Thus, according to the presented method, the payback period of the proposed diagnostic system is $T_{ok} = 1.6$ years.

3. Conclusion

As a result of the analysis of the collected statistics on trolleybus equipment failures, it was determined that the number of failures decreased over the year from 30% to 17% on average, that is, almost 2 times (Figure 1). Of these, equipment failures decreased in 2 times, and failures of mechanical equipment decreased by 60%, which is a confirmation of the dependence of the serviceability of mechanical equipment on the serviceability of equipment. The remaining 7% of equipment failures were attributed to random, which depend on external operational factors. Further studies should be devoted to the determination of these factors.

References

[1] Nazarova M N and Palaev A G 2017 Diagnostics and repair of centrifugal oil transfer pump rotor shaft *IOP Conf. Ser.: Earth Env.* **87** 092016

[2] Gabov V V and Romanova V S 2017 Investigation of Layer-by-layer Destruction of Rocks in High-frequency Cone Crusher *IOP Conf. Ser.: Earth Env.* **87** 022006. DOI:10.1088/1755-1315/87/2/022006

[3] Strizhenok A V and Korelskiy D S 2016 Assessment of the state of soil-vegetation complexes exposed to powder-gas emissions of nonferrous metallurgy enterprises *J. of Ecological Engineering* **17(4)** 25-29

[4] Grosse K U 2012 Non-destructive testing and technology for monitoring the technical condition of structures for quality control and supervision of construction sites *ALITinform: Cement. Concrete. Dry mixes* **6** 62-77

[5] Ardashkin I B, Yakovlev A N and Martyushev N V 2014 Evaluation of the resource efficiency of foundry technologies: Methodological aspect *Advanced Materials Research* **1040** 912-916. DOI: 10.4028/www.scientific.net/AMR.1040.912

[6] Loshkarev I Yu and Chernyshov A S 2013 Unbrakable control. Features of methods of nondestructive testing, in: *Actual problems of power engineering of agrarian and industrial complex: Mater. of IV Int. sci. and pract. conf.* (Ed A V Pavlova) pp 184-186

[7] Malozyomov B V, Babaeva O V and Andreev A I 2014 Posteriori analysis of the reliability of transport systems *Scientific problems of transport in Siberia and the Far East* **1-2** 93-95

[8] Vorfolomeev G N, Evdokimov S A, Malozyomov B V, Schurov N I and Shalnev V O 2004 Reliable sources of a feed of the direct current, in: *8th Korea-Russia Int. Symp. on Science and Technology Proc.: KORUS 2004. sponsors: Tomsk Polytechnic University, University of Ulsan, Novosibirsk State Technical University* pp 316-320

[9] Daimer J, *Graphite electrode for electrothermal reduction furnaces, electrode column and method for making graphite electrodes* Patent for invention RUS No. 2374342 (12 May 2005)

[10] Shabalina A V, Lapin I N and Belova K A 2015 Graphite electrodes for electric arc furnaces *Black metals* **12(1008)** 20-21

[11] Malozyomov B V, Wilberger M E and Kulekina A V 2015 The most typical damage and methods of diagnosing the traction motor *Transport: science, technology, management* **10** 60-65

[12] Nikolaev A A, Nikolaev A V, Kirpichev D E and Tsvetkov Yu V 2008 Formation of a diffuse cathode spot on a graphite electrode with an arc discharge *Physics and chemistry of material processing* 43-48

[13] Shchurov N I, Porsev E G and Vil'berger M E 2009 Asymmetrical and nonsinusoidal operation modes of multipulse rectifiers *Russian Electrical Engineering* **80(12)** 680-684

[14] Borisenkov S, Votintsev A and Roth H 2003 Quality control: non-destructive testing of brazed joints using X-ray radiation *Components and technologies* **28** 168-170

Classification of heavy high-viscosity oils for improving the efficiency of reservoir development

I Sh Mingulov, V V Mukhametshin, Sh G Mingulov and L S Kuleshova

Ufa State Petroleum Technological University, Branch of the University in the City of Oktyabrsky, 54a, Devonskaya St., Oktyabrsky, Republic of Bashkortostan, 452607, Russian Federation

E-mail: vv@of.ugntu.ru

Abstract. The article aims to study the influence of oil viscosity on its density in order to obtain a statistical dependence. The influence of oil density on the rheological behavior of the liquid was established, the boundaries of the dependences were determined and the reasons for its growth anomalies were revealed. The results were obtained by generalizing the measurement of viscosity of degassed oils from various oil and gas producing provinces of Russia. They make it possible to study the physicochemical characteristics of high viscosity oils for their effective production.

1. Introduction

In recent decades, in many countries, the largest reserves of heavy high-viscosity oils were discovered. They amount to tens of billions of tons. The largest commercial reserves of such hydrocarbons have been discovered in Russia, Kazakhstan, Canada, China, Venezuela, Argentina, etc.

In Russia, high-viscosity oil fields include Russkoye, Van-Yeganskoye, Ai-Yagunskoye (Western Siberia), Shafranovskoye, Arlanskoye, Voyadinskoye, Khmelevskoye, Baysarovskoye (Republic of Bashkortostan), Nurlatskoye (Republic of Tatarstan), Gremikhinskoye (Republic of Udmurtia), (Komi Republic), Varandey, Taravei, Kharyaginskoye, etc. [1].

In [2], it has been proposed to classify high-viscosity oils into three groups. The first group is composed of oils with a viscosity of 30–100 mPa·s, the second one – of oils with a viscosity of 100–500 mPa·s, and the third one – with a viscosity above 500 mPa·s. At the same time, the intervals of variation in oil density for these groups correspond to 834-929, 882-955 and above 924 kg/m^3. According to the classification adopted by the Second International Conference in Venezuela with the participation of Unitar (educational and research institute of the UN), liquids with a viscosity of more than 10,000 mPa·s are conventionally referred to bitumen. Most of the well-developed high-viscosity oil fields belong to the first and partially second groups. In [3], oils have been classified as viscous, high-viscosity and heavy with viscosity values less than 0.04; 0.04–0.72 and more than 0.74 Pa·s.

In [4], oils have been classified into ordinary and heavy with viscosity values of 0.05 Pa·s. As can be seen, the boundary of separation of viscous and high-viscosity oils in [4] compared with [3] has significantly shifted towards higher viscosity values.

In Venezuela, Canada and the United States, high-viscosity oils and bitumen are classified only by density. Oil with a density of less than 876 kg/m^3 is light, 876–928 kg / m^3 - medium, 928–1000 kg /

IOP Conf. Series: Materials Science and Engineering 1111 (2021) 012036 doi:10.1088/1757-899X/1111/1/012036

m^3 - heavy, and 1000–1076 kg / m^3 - superheavy [2].

The value of oil viscosity is of particular importance in reservoir conditions when obtaining the final oil recovery factor of the reservoir [6–11].

2. Materials and methods

In [3], a correlation between the viscosity of hydrocarbons and their density has been established. The data of more than 300 measurements of parameters of high-viscosity oil fields were used. The measurements were divided into density intervals in which the averaged values of viscosity were calculated. The test points were located in four sections, with different slope angles and clear boundaries.

Table 1 shows the numerical values of oil viscosity, density, mass fractions of resins, asphaltenes and coke in hydrocarbons of each class. The mass fractions of these components increase from class to class [3].

Table 1. Oil parameters

Parameters	Limits of variation in parameters of hydrocarbons by classes			
	I	II	III	IV
Viscosity, Pa · s	< 0.04	0.04–0.74	0.74–720	> 720
Density, kg/m^3	< 875	875–955	955–995	> 995
Correlation coefficient	0.47	0.52	0.38	0.37
Standard deviation of correlation coefficients	0.22	0.27	0.23	0.20–0.52
Mass fraction of resins, %:				
limits of changes	0–7	1.00–9.00	4.60–17.20	11.70–75
average value	2.23	4.12	9.95	21.76
Student's test of the compared classes	4.23	9.07	7.62	
Mass fraction of resins, %:				
limits of change	0.3–20.0	4.2	11.7–36.0	21.4–57.3
average value	10.9	16.8	20.6	26.1
Student's test of the compared classes	7.36	3.02	2.36	
Classes of hydrocarbons	light and viscous oils	high-viscosity oils	heavy oils	natural bitumens

To carry out hydrodynamic calculations of pipelines, it is advisable to consider the dependence of oil viscosity on its density; in laboratory conditions, it is sufficient to measure the density of oil, and from the statistical dependence it is possible to calculate the oil viscosity.

In Table 2, for a number of fields in Russia and the CIS countries, a sample of high-viscosity oil was used [5, 12]. For each oil, measurements of kinematic viscosity were used at a temperature of 20°C (ν20). These data were plotted on a graph (Figure 1) in the "oil viscosity - oil density" coordinates.

Table 2. Density and effective viscosity of oils and emulsions of a number of high-viscosity oil fields at t = 20 ° C [6]

Oilfield	Density degassed - bath oil, kg / m ³	Viscosity of degassed oil, mPa · s	Oil water cut , unit fraction	Viscosity of the emulsion, mPa · s
Arlanskoe	892	30	0.59	300
Karacha-Elginskoe	910	32	0.60	265
Taimurzinskoe	912	35	0.60	280
Chekmagushevskoe	901	57	0.58	210
Shafranovskoe	920	120	0.55	1700
Vyatskaya square	899	45	0.60	210
Gareevskoe	832	27.3	0.30	190
Kushkul	895	36	0.50	210
Pavlovskoe	905	230	0.40	950
Naratovkoe	900	64.7	0.45	320
Voyadinskoe	908	101	0.55	528
Baysarovskoe	900	104	0.52	560
Khmelevskoe	899	98	0.65	510
Sauzbashevskoe	902	63.2	0.55	410
West- Surgutskoe	885	46	0.64	250
Russian	910	150	0.30	1600
Usinskoe	962	3000	0.20	3500
Gremikhinskoe	919	107.6	0.60	1200
Mishkinskoe	904	89	0.55	2890
Kiengop	912	67.0	0.47	750
Nurlatskoe	931	228	0.12	1000
Kenkiyak	910	250	0.56	4000
Kotur- Tepe	905	30	0.58	330

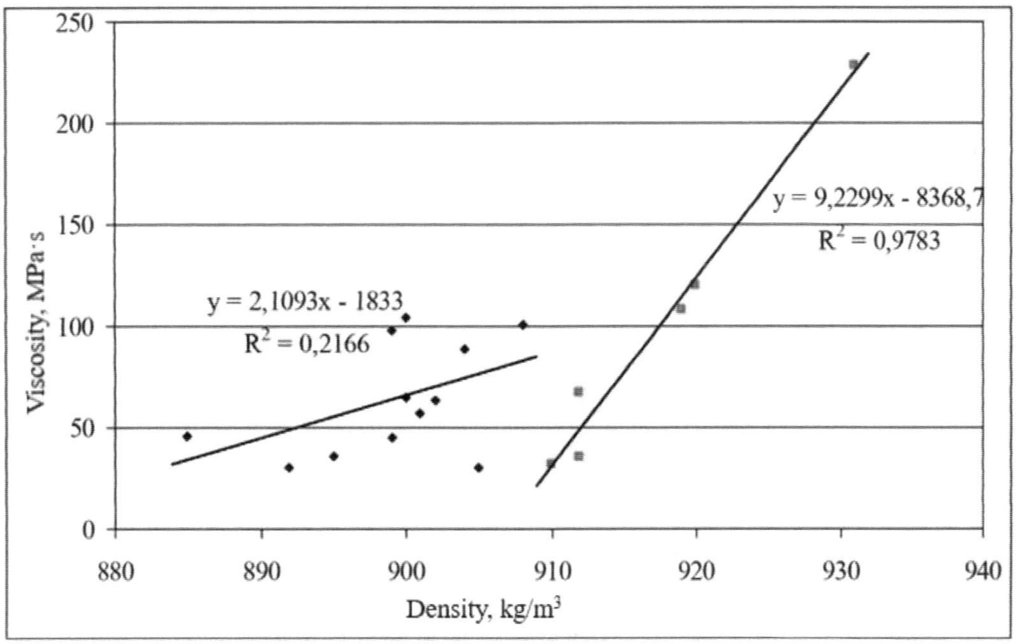

Figure 1. Dependence of dynamic viscosity of oils on density

3. Results and Discussion

Figure 1 shows the existence of two groups of oils, forming relationships with different tilt angles. The authors analyzed reasons for the presence of these two characteristic areas. In [5], a similar attempt was made to obtain a dependence without dividing oils into any groups.

As can be seen from Figure 1, there are two groups of oils that are different from each other. The first group includes oils with a density of up to about 910 kg/m^3. The dependence is described by the formula ($R^2 = 0.22$):

$$\mu = 2.1\,\rho - 1833, \tag{1}$$

where μ – oil viscosity, mPa · s; ρ – oil density, kg/m ^{3}.

The second group is described by the dependence ($R^2 = 0.98$):

$$\mu = 9.23\,\rho - 8368. \tag{2}$$

Figure 1 shows the qualitative differences in the rheological behavior of oils of both groups. It is quite obvious that this is due to the composition of oils and the content of high-molecular-weight asphalt-resinous and paraffinic substances (ASPV). These substances determine the nature of structure formation and anomalies in rheological properties.

The boundary corresponding to the oil density of 910 kg/m^3 separates different classes of oils.

Obviously, oils with a lower density should be referred to conventional high-viscosity oils. Oils with a higher density should be classified as heavy oils.

The proportionality coefficient of the first section is small and amounts to about 2.02 m^2/s.

The second section of the dependence has a coefficient of 8.86 m^2/s.

The reason for an intensive increase in viscosity in the second section is a significant increase in the content of molecules of heavy hydrocarbons - resins and asphaltenes in oils, forming associates of molecules and a certain spatial structure that resists flow; on the other hand, during the flow of this liquid, the interaction of associates of molecules with each other is possible, similar to the flow of concentrated water-oil emulsions with a water cut of more than 0.5 by volume, in which an increase in viscosity occurs due to the inter-droplet interaction in the flow.

In [13], it was shown that upon complete dissolution of paraffin, the oil viscosity depends on the content of resins and asphaltenes. When oil is cooled, paraffin crystallizes. The combination of these factors causes the structure formation and anomalies in the rheological behavior.

The presence of resins and asphaltenes reduces the saturation temperature of oil with paraffin due to their adsorption on aggregates of paraffin molecules.

Resins are components with a density of 1000 kg/m^3 and a molecular weight of 1200. Their content in oil can reach 30% or more.

Asphaltenes are solids with a density of 1000–1200 kg/m^3 and a molecular weight of up to 10,000. Their amount reaches 15%. Asphaltenes form associates of molecules, stabilized by solvation layers of resins and naphthenoaromatic hydrocarbons. Such associates form a spatial coagulation structure which has tixotropic properties [14].

The anomaly of the rheological behavior depends on the shear stress. With the growth of the latter, the destruction of structural bonds and the transition of the fluid model to the Newtonian one occur [1, 15]. The relationship between the content of resins, asphaltenes and the oil viscosity can be illustrated by light absorption coefficient K_{la}. With an increase in K_{la}, the viscosity increases, which indicates the degree of influence of the content of asphalt-resinous substances on the rheological behavior of oil.

The relationship between the content of resins, asphaltenes and oil viscosity is illustrated in Figure 2, obtained by measuring the light absorption coefficient K_{la} using the FEK-56 photocalorimeter at the Arlan oil field of the Republic of Bashkortostan. The experimental points plotted on the graph show a strong relationship between these parameters, which indicates the degree of influence of the content of asphalt-resinous substances on the rheological behavior of oil.

The presence of paraffin infleucnes the viscosity of oil at temperatures below the crystallization temperature. The crystals of paraffin in the liquid, as well as solid particles of asphaltenes, cause an increase in the coefficient of friction in the liquid.

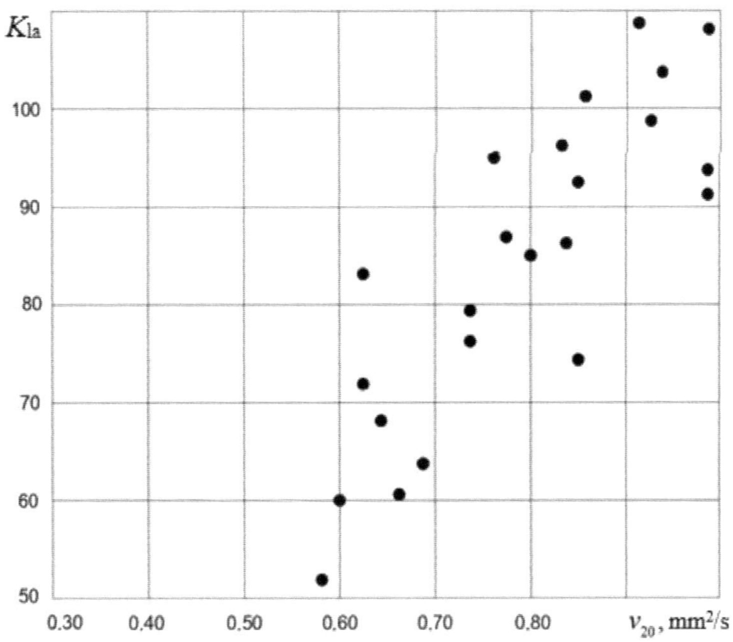

Figure 2. The relationship between the kinematic viscosity of oil and the coefficient of light absorption for the Arlan oil field

At rest, these crystals form a structure in a liquid with an initial (static) shear stress. When the liquid moves away, this structure is destroyed rather quickly with a decrease in the effective viscosity.

4. Conclusion

1. It has been established that the dependence of oil viscosity at its values in the range from 30 up to 250 mPa·s on its density in the range from 880 to 930 kg/m^3 has two linear sections, separated by a density value of about 910 kg/m^3.

2. The coefficient of proportionality of the first section is small and amounts to about 2.02 m^2/s. The second section has a coefficient of 8.86 m^2/s.

3. The reason for an intensive increase in viscosity in the second section is a significant increase in the content of molecules of heavy hydrocarbons in oils - resins and asphaltenes, which form associates of molecules and a certain spatial structure that resists the flow. On the other hand, during the flow, the interaction of associates of molecules with each other is possible. It is similar to the flow of concentrated water-oil emulsions with a water cut of more than 0.5, in which an increase in viscosity occurs due to the inter-droplet interaction in the flow.

References

[1] Mamonov F A 2002 *Field Transportation of High-Viscosity Oil with Preliminary Discharge of Produced Water* (Ufa: USPTU) 248 p

[2] Khalimov E M, Klimushin I M and Ferdman L I 1987 *Geology of High-Viscosity Oil Fields in the USSR* (Moscow: Nedra) 172 p

[3] Verevkin K I and Diyashev R N 1982 Classification of Hydrocarbons when Choosing Methods for their Production *Oil Industry* **3** 31-34

[4] Aigistova S Kh, Muslimov R Kh and Kasimov R S 1980 On the Classification and Rational use of High-Viscosity Oil in Tatarstan *Oilfield Business* **2** 13-14

[5] Ishmukhametov A V, Mingulov Sh G and Bulchaev N D 2017 *Operation of Wells with Viscous*

Paraffinic Oils (Saint Petersburg: Nedra) 311 p

[6] Kuleshova L S, Mukhametshin V V and Safiullina A R 2019 Applying information technologies in identifying the features of deposit identification under conditions of different oil-and gas provinces *J. of Phys.: Conf. Ser.* **1333(7)** 072012. DOI: 10.1088/1742-6596/1333/7/072012

[7] Soloviev N N, Mukhametshin V Sh and Safiullina A R 2020 Developing the efficiency of low-productivity oil deposits via internal flooding *IOP Conf. Ser.: Mater. Sci. Eng.* **952(1)** 012064. DOI: 10.1088/1757-899X/952/1/012064

[8] Mukhametshin V Sh 2020 Rationale for the production of hard-to-recover deposits in carbonate reservoirs *IOP Conf. Ser.: Earth Env.* **579(1)** 012012. DOI: 10.1088/1755-1315/579/1/012012

[9] Akhmetov R T, Mukhametshin V V and Kuleshova L S 2019 Simulation of the absolute permeability based on the capillary pressure curves using the dumbbell model *J. of Phys.: Conf. Ser.* **1333(3)** 032001. DOI: 10.1088/1742-6596/1333/3/032001

[10] Mukhametshin V Sh, Tyncherov K T and Rakhimov N R 2020 Geological and field control and diagnostics of treatment efficiency of a wellbore zone within carbonate reservoirs *IOP Conf. Ser.: Earth Env.* **579(1)** 012021. DOI: 10.1088/1755-1315/579/1/012021

[11] Mukhametshin V Sh 2020 Justification for increasing the performance of hydrochloric acid treatment in wells of fields with carbonate reservoir *IOP Conf. Ser.: Mater. Sci. Eng.* **905(1)** 012083. DOI: 10.1088/1757-899X/905/1/012083

[12] Valeev M D, Leontev S. A, Maier A V and Mokhov M A 2018 *Theory and Practice of High-Viscosity Oil Production from Water-Flooded Wells* (Moscow: National University of Oil and Gas "Gubkin University") 268 p

[13] Khromovichev M N 1984 Rheology of Heavy Oils and their Emulsions, in: *Domestic Experience: Express Information "Oilfield Business"* (Moscow: VNIIOENG) pp 14–18

[14] Devlikamov V V, Khabibulin Z A and Kabirov M M 1984 *Abnormally Viscous Oils* (Moscow: Nedra) 240 p

[15] Steinborn R and Flock D L 1983 The Rheology of Heavy Crude Oils and their Emulsions *J. of Canadian Petroleum Technology* **22(5)** 38–52. DOI: 10.2118/83-05-03

IOP Conf. Series: Materials Science and Engineering 1111 (2021) 012037 doi:10.1088/1757-899X/1111/1/012037

Techniques for optimization of gas extraction from production wells annulus

Sh G Mingulov and I Sh Mingulov

Ufa State Petroleum Technological University, Branch of the University in the City of Oktyabrsky, 54a, Devonskaya St., Oktyabrsky, Republic of Bashkortostan, 452607, Russian Federation

E-mail: task3@mail.ru, info@of.ugntu.ru

Abstract. The article experimentally establishes the regularities of the separated gas pressure growth in the annulus depending on geological and technical characteristics of producing wells; it obtains an experimental dependence for calculating the period of stabilization of the gas pressure in the annulus when the pressure is equal to the pressure in the discharge manifold of the well.

1. Introduction

One of the ways to increase the efficiency of oil field development is to intensify oil production from wells [1–8]. It is known that to increase the flow rate of wells operated by traditional mechanized methods, various technologies and technical means are used to reduce the gas pressure in the annulus [9]. Ejectors that are installed in the tubing string above the liquid level or directly at the wellhead are used for forced pumping of gas from the annular space of the well into the flow header [10]. There are also technologies for pumping gas with the help of additional plungers installed on the rod string, as well as an overhead compressor driven by the balance beam of the pumping unit [11].

The experience of using the technology of forced pumping of gas from the annulus showed that the flow rate of one well increases on average by more than 2.7 t / day [12].

To calculate the parameters of gas compression from the annulus to the reservoir, it is necessary to have data on the volumes of gas supplied from the pump intake to the annulus of the well per unit of time.

For the experimental determination of these volumes, it is necessary to measure the dynamics of the gas pressure in the annulus after the pressure is reduced to atmospheric value. In this case, the pumping unit must continue to operate continuously in order to avoid violations of the conditions for gas extraction from oil and its separation at the pump intake.

The dynamics of changes in the gas pressure in the annulus ultimately makes it possible to determine the amount of gas separated at the pump intake and supplied to the annulus, which will be described below.

2. Materials and methods

The essence of the experiment is to pump out gas from the annular space to a pressure corresponding to atmospheric pressure, followed by turning off this compressor and periodically recording the pressure in the annular space for a period sufficient for its stabilization.

The simplest and most accessible method for constructing a pressure dynamics curve is its periodic registration after the pump is started up and brought into operation. In the initial period after killing, the downhole pumping unit pumps out fluid from the annulus. Further on, the dynamic fluid level is stabilized. During the period before and after the stabilization of the pumping mode, periodic beating of the dynamic level and recording of the change in gas pressure in the annular space over time are performed.

3. Results and Discussion

Figure 1 shows the obtained curves of gas pressure dynamics in the annular space of wells No. 740 and No. 716 of the Chutyrsko-Kiengopskoye field. In well No. 740 (curve 1), already after 12 hours, the gas pressure stabilized, which equaled the pressure in the reservoir (about 1.45 MPa).

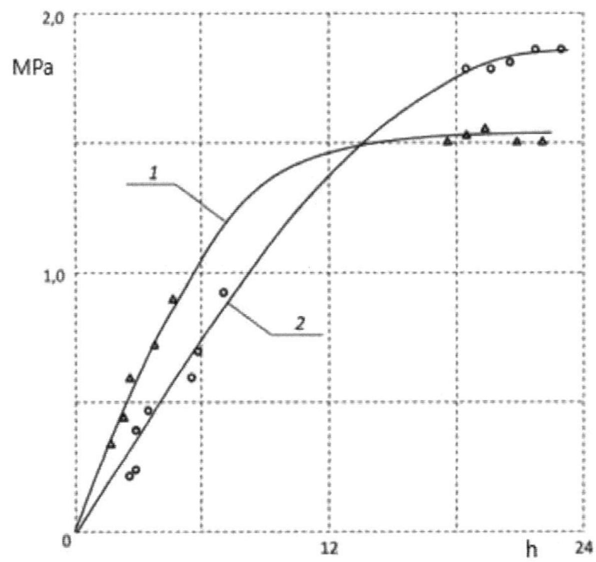

Figure 1. Change in gas pressure in the annulus after pressure relief to 0.1 MPa in wells No. 740 (1) and No. 716 (2) of the Chutyrsko-Kiengopskoye field

At the same time, full stabilization of gas pressure in well No. 716 (curve 2) did not occur.

The reason for this was the higher dynamic fluid level (852 m) in well No. 716 compared to well No. 740 (462 m), despite approximately the same gas-oil ratio (about 6.5 m3 / t) and oil flow rates of wells (6.0 and 5.2 m^3 / day).

Along with the indicated volume of the well filled with gas, the pressure dynamics are influenced by the gas-oil ratio, the pressure at the pump intake and the conditions for the separation of the free gas phase.

Therefore, the gas flow rate entering the annulus above the pump will be proportional to the value:

$$Q_g = (G_f - G_{left}) \frac{Z \cdot P_O T_{well}}{P_{pr} T_O} Q_{oil} б, \quad m^3/day, \qquad (1)$$

where G_f is the gas factor, m^3/m^3; Z is the gas constant; P_O, T_O--pressure and temperature, equal to 0.1 MPa and 273 K, respectively; T_{well} – fluid temperature at the pump intake, K; Q_{oil} – well oil production rate, m^3 / day; $б$ – gas separation coefficient.

The G_{left} value in (1) is determined from the curve of one-time oil degassing.

As an example, Figure 2 shows the curves obtained for fields in Western Siberia. In this figure, at a given pressure value, the amount of gas released from oil is determined, and the residual amount of G_{left} gas is determined from it.

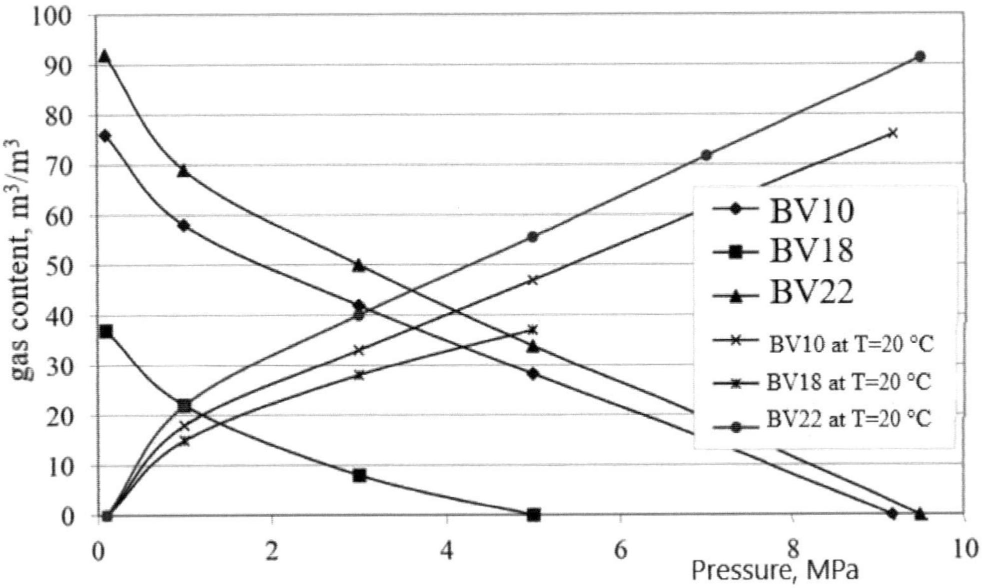

Figure 2. Differential degassing curves for oils: gas content, m^3/m^3, pressure MPa

Considering that during the period of reaching the gas pressure, some changes occur at the pump intake, it is permissible to take its average value as the calculated value of the G_{left} coefficient. The G_{left} value can also be calculated by the ratio:

$$G_{left} = G_f\left(\frac{P_{pr} - 0.1)}{P_{sat} - 0.1}\right). \quad (2)$$

If we take V_a – as the volume of the annular space above the dynamic level, then the change in the ratio V_g/V_a in time will characterize the rate of filling this space with gas and increasing pressure in it.

Similarly, if we denote P_a - the current pressure in the annulus, and P_r - pressure in the reservoir, then the change in the P_a / P_r ratio per unit time will show the intensity of the gas pressure growth above the dynamic level.

In this case, the P_a/P_r ratio will depend on the volume of gas V_g entering the annulus.

Thus, the parameters P_a/P_r and V/V_a are related in a certain way.

The amount of gas entering the annulus per day is calculated:

$$V_g = Q_g \cdot t, \text{ m}^3, \quad (3)$$

where t is time, days.

In order to establish such a relationship in several wells in fields with a large scatter of GOR values, using the measurements described above, the curves of the dynamics of dimensionless pressures were plotted as a function of time (Figure 3).

Figure 3 shows that the nature of change of gas pressure in the annulus is approximately the same. With an increase in the value of V_g/V_a, the period after which the pressure in the annulus reaches the pressure in the reservoir ($P_a/P_r = 1$), decreases.

From a practical point of view, the period of reaching the gas pressure in the annular space of the reservoir pressure is especially interesting. The inverse value of this period($1/t_{set}$), characterizes the intensity of gas inflow and growth of its pressure in the annulus. The value of the pressure in the annulus during its stabilization will correspond to the pressure in the discharge manifold of the well due to the presence of a check valve at its wellhead.

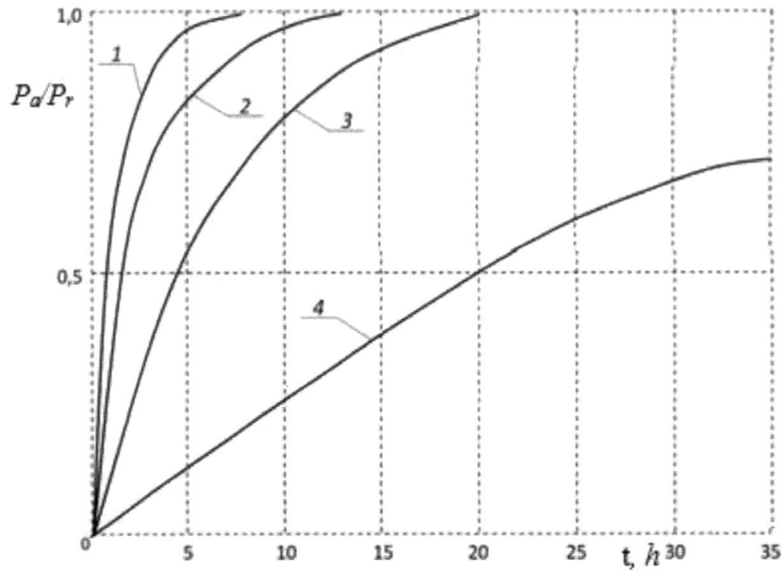

Figure 3. Dependence of the relative value of P_a/P_r on the period of operation of the installation of the downhole sucker rod pump: 1 - well. No. 279; 2 - well. No. 336; 3 - well. No. 740; 4 - well. No. 598

Figure 4 shows the dependence of $1/t_{set}$ on the parameter V_g/V_a. This dependence is described by the formula:

$$1/t_{set} = 3.63 \cdot 10^{-3}\ (V_g/V_a)^{1,75} \quad (4)$$

or

$$t_{set} = 3.63 \cdot 10^{3} (KG_f Q_{OIL} б/V_a)^{-1,75}. \quad (5)$$

Formula (5) makes it possible to calculate the period t_{set}, during which the gas pressure in the annular space will reach the pressure in the discharge manifold of the well.

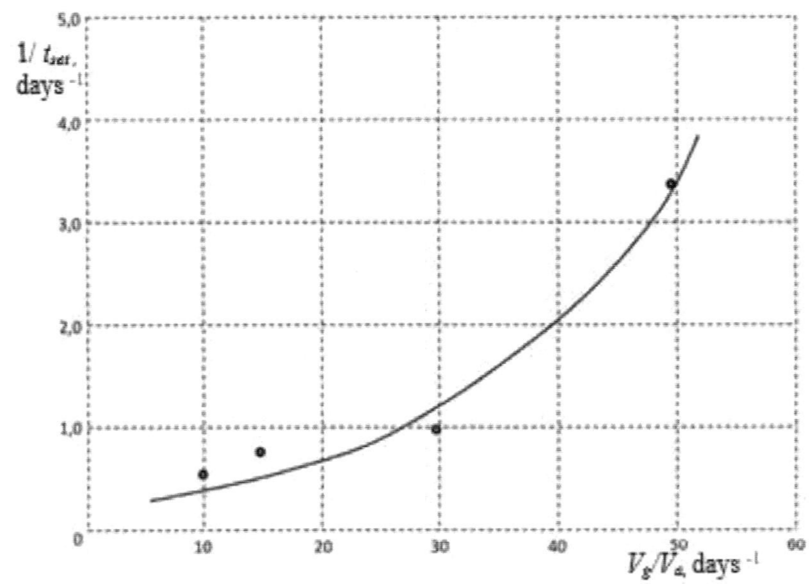

Figure 4. Dependence of the parameter $1/t_{set}$ on the value V_g/V_a

4. Conclusion

1. The regularities of the pressure increase of the separated gas in the annulus were experimentally established depending on the geological and technical characteristics of the wells.

2. The study obtains an experimental dependence for calculating the period of stabilization of the gas pressure in the annulus when the pressure is equal to the pressure in the discharge manifold of the well.

References

[1] Briones M, Zambrano J A and Zerpa C 2002 Study of Gas-Condensate Well Productivity in Santa Barbara Field, Venezuela, by Well Test Analysis *SPE Annual Technical Conf. and Exhibition (San Antonio, Texas, 29 September-2 October 2002)* p 9. DOI: 10.2118/77538-MS

[2] Mukhametshin V V and Kuleshova L S 2020 On uncertainty level reduction in managing waterflooding of the deposits with hard to extract reserves *Bull. of the Tomsk Polytechnic University. Geo Assets Engineering* **331(5)** 140–146. DOI 10.18799/24131830/2020/5/2644

[3] Yakupov R F and Mukhametshin V Sh 2013 Problem of efficiency of low-productivity carbonate reservoir development on example of Turnaisian stage of Tuymazinskoye field *Oil Industry* **12** 106–110

[4] Kuleshova L S, Mukhametshin V V and Safiullina A R 2019 Applying information technologies in identifying the features of deposit identification under conditions of different oil-and gas provinces. *J. of Phys.: Conf. Ser.* **1333(7)** 072012. DOI: 10.1088/1742-6596/1333/7/072012

[5] Economides J M and Nolte K I 2000 *Reservoir stimulation* (West Sussex, England: John Wiley and Sons) 856 p

[6] Khuzina L B, Mukhametshin V Sh, Tyncherov K T and Shaikhutdinova A F 2018 On the choice of the oscillators' installation site *Int. J. of Civil Engineering and Technology* **9(9)** 1952-1959

[7] Akhmetov R T, Mukhametshin V V and Kuleshova L S 2019 Simulation of the absolute permeability based on the capillary pressure curves using the dumbbell model *J. of Phys.: Conf. Ser.* **1333(3)** 032001. DOI: 10.1088/1742-6596/1333/3/032001

[8] Chudinova D Yu, Kotenev Yu A, Sultanov Sh Kh and Mukhametshin V Sh 2018 The neural network for grouping wells of different facies zones of productive layer *IOP Conf. Ser.: Earth Env.* **194(8)** 082008. DOI: 10.1088/1755-1315/194/8/082008

[9] Sevastyanov A, Ivanov A A and Fatkullin A S 2014 Technology of gas removal from oil wells annular area *Oilfield Engineering* **9** 30-31

[10] Topol'nikov A S, Urazakov K R and Molchanova V A 2009 Modeling of Gas Bypass from the Annulus During Operation of Wells with Low Production Rate Complicated by High GOR *Scientific-Technical Herald "NK Rosneft"* **2** 33–37

[11] Mak-Koi C 2004 Gas Compressor Activated by the Walking Beam of a Jack May Be Useful in Various Field Operations *Oil and Gas Technologies* **3** 44–46

[12] Olarte J D, Haldar S, Said R, Ahmed M, Burov A, Stuker J, Kharrat W and Wortmann H 2011 New Approach of Water Shut off Techniques in Open Holes - and World First Applications of Using Fiber Optic Services With Tension-Compression Sub *SPE/DGS Saudi Arabia Section Technical Symp. and Exhibition (Al-Khobar, Saudi Arabia, 15-18 May 2011)* p 10. DOI: 10.2118/149116-MS

Developing a drilling fluid compensator for increasing the technological condition of drilling equipment

T N Minnivaleev and A N Minnivaleev

Ufa State Petroleum Technological University, Branch of the University in the City of Oktyabrsky, 54a, Devonskaya St., Oktyabrsky, Republic of Bashkortostan, 452607, Russian Federation

E-mail: timxn@yandex.ru, info@of.ugntu.ru

Abstract. The phenomenon of fluctuations in pressure and fluid flow, as well as hydraulic shocks in most cases lead to negative consequences that affect the technical condition and operation of various equipment. The article discusses the causes of pressure fluctuations of the flushing liquid in the drill string and their consequences, and provides a developed compensator for the flushing liquid, which allows to smooth the flow pulsation.

1. Introduction

Due to the increasing requirements for the economic efficiency of the drilling process, it is important and relevant to improve the layout of the bottom of the drill string in order to increase the durability of drilling equipment, reduce the cost of carrying out down-lifting operations and improve the technical and economic indicators of drilling in General. One of the ways to improve drilling efficiency is to reduce the dynamism of the drill tool when drilling wells and thereby increase the mechanical speed of drilling and the durability of the drill tool [1–8].

2. Materials and methods

In solving the problem of hydrodynamics, the basic laws of theoretical mechanics and hydromechanics were used in relation to the accepted technological scheme for installing a compensator in the layout of a drilling tool, as well as empirical methods.

3. Content of the method and evaluation of its effectiveness

Aznakaevsky PBR LLC "Drilling" drilled a large number of wells. Currently, oil drilling is conducted mainly by turbodrills and downhole motors. Direction is drilled with drilling bits with diameter of 393 mm and then the well is being built with casing pipes with a diameter of 324 mm, the conductor is drilled with drilling bits with diameter of 293,3 mm and then the well is being built with casing pipes with a diameter of 215,9 mm.

The following drill pipes are used: drill pipe 127x9, 19 Z-147 TBPN and TBPC, weighted drill pipe 178x90 Z-147, light-alloy drill pipe 147x11 Z-147.

Piston drilling pumps BRN-1 and NBT-600, which are slow-moving, are widely used for drilling wells. They are a source of uneven flow and pressure of the drilling mud, which, in turn, leads to vibrations, alternating loads in the drill string, uneven operation of turbo drills and screw downhole engines, and, therefore, uneven rotation of the bit and drilling of the rock. Hydraulic shocks occur both above the turbo drill and on any part of the drilling tool. When drilling deep wells, the start and stop of drilling pumps can be accompanied by short-term, sharp increases in pressure in the circulation system, which also negatively affects the operation of the drilling rig. This is evidenced by the periodic operation of safety valves and ruptures in the diaphragms of compensators for drilling pumps.

The «GEOTEST» computerized station allows monitoring a wide range of parameters of the well drilling process.

The purpose of the station "GEOTEST»:

1) Control of the main drilling parameters, such as the axial load on the bit, pressure and flow of the flushing liquid, mechanical drilling speed, etc. Assessment of the situation and prevention of accidents and complications.

2) Automated collection of geological, geochemical and technological information during drilling.

3) Lithological dissection of the section. Isolation of reservoirs and estimation of saturation character.

4) Documentation of the drilling process. Data transfer from the drilling rig.

The pressure sensor of the washing liquid of the GEOTEST station is a strain gauge pressure Converter. Installation of the pressure sensor to the discharge line is carried out through the medium separator of the standard pressure gauge by means of a tee. The pressure sensor allows the continuous measurement of the oscillations of the washing fluid in the process of drilling a well.

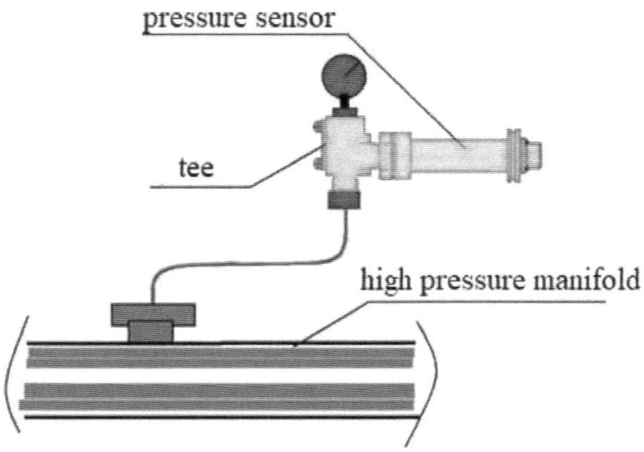

Figure 1. Mounting the flushing liquid pressure sensor

The results of measurements of drilling parameters for oil well No. 7063 on Kholmovskaya square are presented below. Geological and technical studies were carried out directly in the process of drilling a well in order to solve a set of geological and technological problems aimed at rapid identification of oil-promising reservoir formations in the section of a drilling well, lithological dissection of the well section, and optimization of the drilling mode. The research was conducted in the range of 355-1410 meters. Technical parameters of drilling (rvx – pressure of the flushing liquid, Wd – load on the bit, Vmex – mechanical drilling speed, Vreys – trip drilling speed) of well No. 7063 and the bits used in this case are shown in table 1.

Up to a depth of 1236 meters, a flushing liquid with a density of 1000 was used, then drilling was carried out with a drilling fluid with a density of 1120. The values of the flushing fluid pressure at the inlet (figure 1), the bit load (figure 2), and the sinking speed (figure 3) registered during drilling are shown in Fig.2 in the intervals: a) 355–570 meters; b) 570–850 meters; c) 850–1120 meters; d) 1120–1410 meters.

Geological description of the sludge from well No. 7063:

- 355-427 m – artinsky tier (anhydrites, gypsum);
- 427-610 m – Upper Carboniferous (weakly known Dolomites, light gray limestones);

- 610-700 m – Myachkovsky horizon (grey limestone with dolomite interlayers;
- 700-770 m – Podolsky horizon (limestone light gray, gray);
- 770-835 m – Kashirsky horizon (limestones, mudstones);
- 835-880 m – verey horizon (mudstones, gray limestones);
- 880-915 m – Bashkir tier (light gray limestone);
- 915-980 m – protvinsky horizon (limestones, Dolomites);
- 980-1160 m – Serpukhovsko-Oka horizon (the Dolomites are light brown);
- 1160-1220 m – Carbonate rocks of the Aleksinsko-Tula horizon (limestones, mudstones, dark gray, black);
- 1220-1410 m – Terrigenous rocks of the Tula-Bobrikov horizon (dark grey mudstones, siltstones, brown sandstones).

Table 1. Technical parameters of drilling and used bits

Interval of depth, m		Mud Pressure, MPa	Wd, t	Drilling bit		ROP, m/h	Run Speed, m/h
beginning	end			Type/Size	№		
354	616	100	18	215.9 ТЗГН	442	29.1	15.2
616	792	105	17	R-15 215.9	1	44	16.6
792	980	90	16	R-15 215.9	43	20.4	9.1
980	1069	89	10	R-05 215.9 ТЗГНУ	19	8.9	3.7
1069	1081	90	12	R-05 215.9 ТЗГНУ	19	4.8	1.6
1081	1136	95	15	R-05 215.9 ТЗГНУ	19	13.7	3.3
1136	1164	85	20	R-05 215.9 ТЗГНУ	49	8.8	1.9
1164	1236	110	10	R-05 215.9 ТЗГНУ	49	5.5	3.6
1236	1267	88	15	R-05 215.9 ТЗГНУ	55	2.5	0.7
1267	1337	90	19	R-05 215.9 ТЗГНУ	55	3.1	2.1
1337	1357	81	14	R203M144		4.9	1
1357	1391	89	13	R203M120		3	1.4
1391	1420	92	22	REED/120.7	120	1.3	0.7

Analyzing Fig. 2, graph 1 of which represents the change in the pressure of the flushing liquid during drilling, it becomes obvious the amount of unevenness of the pressure of the flushing liquid during drilling. At an average operating pressure of 9 MPa, the difference between the highest and lowest pressure values reaches 5 MPa or more.

During the drilling process, the bit is also a source of pressure fluctuations and vibrations. When drilling wells in the Aznakaevsky PBR, it was noted that on average, when drilling an oil well, 1 diaphragm of the compensator of the drilling pump fails, the pressure in which is set equal to 2/3 of the working pressure. However, when drilling in the Serpukhov-Oka horizon, which has a fractured nature of the rock, causing the collapse of the rock, clamping the turbodrill body and plugging the holes of the bit and jamming it, as a result of the rupture, it is necessary to replace up to 3-4 compensator diaphragms, which causes a simple drilling for more than 1 hour for each burst of the diaphragm. Along with the numerous costs of drilling wells, a company also has to bear the cost of replacing the orifices of compensators for drilling pumps, with the average cost of the diaphragm D-40 4000 rubles.

Based on this, it can be concluded that the diaphragm compensators used in drilling do not allow to properly extinguish pressure fluctuations in the drill string, and, consequently, additional compensators must be installed in the layout.

We have developed a flushing fluid compensator [2], which is recommended to be installed to dampen pressure and flow fluctuations in wells drilled with bits up to 215.9 mm in diameter. For wells drilled with larger diameter bits, to solve the problem of smoothing the uneven pressure of the drilling mud, it is recommended to use an upgraded flushing fluid compensator equipped with a housing 1, a spring 2, a closed chamber 3 formed by a spring-loaded piston 4 and a partition 5. the Closed chamber is connected by channels 6 to the annulus space (Fig. 3). This compensator can be installed both above the bit and on any part of the drilling tool by means of locking threads, which are attached to the compensator body, and adapters.

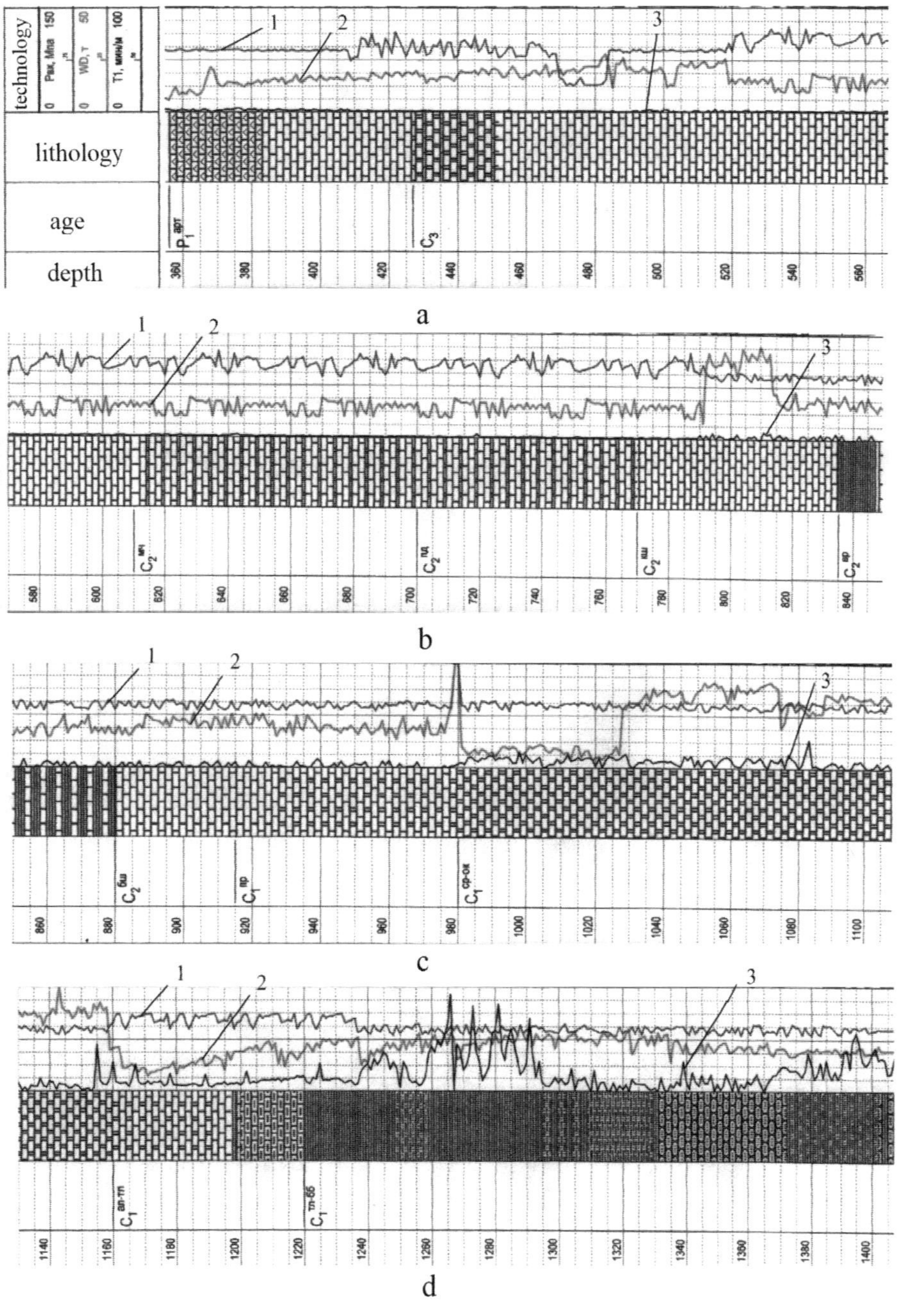

Figure 2. Recorded drilling parameters of well No. 7063

We solved the problem of the dynamics of the mathematical model of this compensator to determine the main dimensions of the upgraded version of the compensator and to determine the degree of damping of pressure pulsations.

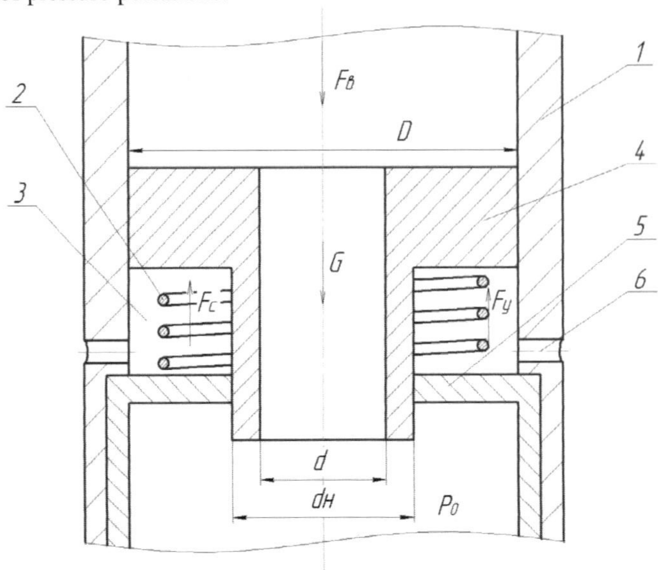

Figure 3. Schematic diagram of the piston compensator stage

The sequence of calculating the compensator is shown below.

Initial data: D, d, $d_н$, G - weight of the piston; $P_u = P_u(t)$ – pressure in the upper part of the flow; P_c- pressure in the annular space (for simplification of calculations, it is assumed constant); C - spring stiffness coefficient.

In the future, the General law of change is adopted P_u.

$$P_u = A\sin\omega t + B, \tag{1}$$

where A is the oscillation amplitude; B is the average pressure; ώ – circular frequency; t – time.

It is necessary to determine the law of pressure change in the flow area located below the piston compensator P_0.

Let us make a differential equation of the piston movement. Forces acting on the piston (main forces) F_u – the force of the flow pressure on the upper part of the piston:

$$F_u = P_u S_p,$$

where $S_p = \frac{\pi(D^2-d^2)}{4}$ is the surface area of the piston.

F_e – the spring's elastic force, which is determined by the formula:

$$F_e = c\Delta l,$$

where Δl is the deformation of the spring; C is the coefficient of stiffness of the spring.

F_c – the force of pressure on the piston from below (from the chamber under the piston, where the pressure is assumed to be equal to the annulus):

$$F_c = P_c S_b{}',$$

where $S_b' = \frac{\pi(D^2-d_b^2)}{4}$ is the surface area of the piston pressure from below G – piston gravity; $m = \frac{G}{g}$ – weight of the piston

The differential equation of the piston movement will have the form:

$$\frac{G}{g}\frac{d^2x}{dt^2} = P_u S_u + G - c\Delta l - P_c S_b{}'. \tag{2}$$

Entering symbols $\frac{c}{m} = k^2$; $\frac{S_u}{m} = a$; $\frac{S_b{}'}{m} = u$; $D = P_c u + Ba$.

Solving equation (2), we get:

$$X(t) = -\frac{Aa\omega}{k(k^2-\omega^2)}\sin kt - \frac{D}{k^2}\cos kt + \frac{Aa}{k^2-\omega^2}\sin\omega t + \frac{D}{k^2}. \tag{3}$$

In this case, the speed of the piston will be:

$$V_p = \frac{dx}{dt} = \frac{-A a\omega}{k^2-\omega^2}\cos kt + \frac{D}{k}\sin kt + \frac{Aaw}{k^2-\omega^2}\cos \omega t. \tag{4}$$

To establish the dependency between the average velocities of fluid in different stream sections, select certain amount of liquid and apply the theorem on change of momentum during an infinitely small interval of time dt, the result is the desired equation that determines the pressure of the flushing fluid at the exit of the compensator:

$$P_0 = P_u(t) - \frac{\rho}{S_u} V_p^2 (S_p - S_p{}'), \tag{5}$$

Where ρ is the fluid density.

Using equation (5), given the geometric parameters of the compensator and knowing the law of change in the pressure of the washing liquid in the section located at the inlet to the compensator, you can determine the pressure of the washing liquid at the outlet of the compensator.

Solving the inverse problem, we determined the optimal size of the developed compensator.

Thus, the unevenness of pressure and fluid flow in the drill string reaches significant values and negatively affects the drilling process as a whole. Therefore, the development of compensators that reduce the uneven pressure of the drilling fluid in the drill string is an urgent task. The compensator developed by us is one of the ways to solve this problem.

4. Conclusion

Despite the presence of the compensator, installed on the discharge line of the mud pump, the pressure fluctuation of drilling fluid in the drill string, emerging from the pump and from the work of the entire drilling tool as a whole, reaches a considerable value. Therefore, the development of compensators that reduce the uneven pressure of the drilling fluid in the drill string is an urgent task. The compensator developed by us is one of the ways to solve this problem.

References

[1] Minnivaleev T N 2018 Efficiency determination of amplitude decay of pressure fluctuation of circulating fluid with the help of a multi-stage piston-type downhole compensator *Advances in Engineering Research* **157** 397-400

[2] Polyakov V N, Chizhov A P, Kotenev Yu A and Mukhametshin V Sh 2019 Results of System Drilling Techniques and Completion of Oil and Gas Wells *IOP Conf. Ser.: Earth Env.* **378(1)** 012119. DOI: 10.1088/1755-1315/378/1/012119

[3] Mukhametshin V Sh, Kotenev Yu A and Sultanov Sh Kh 2018 Assessment of the Need to Stimulate the Development of Hard-to-Recover Reserves in Carbonate Reservoirs *IOP Conf. Ser.: Earth Env.* **194(8)** 082027. DOI: 10.1088/1755-1315/194/8/082027

[4] Mukhametshin V V 2018 Bottomhole formation zone treatment process modelling with the use geological and geophysical information *IOP Conf. Ser.: Earth Env.* **194(2)** 022024. DOI: 10.1088/1755-1315/194/2/022024

[5] Kuleshova L S and Mukhametshin V V 2019 Elimination of uncertainties in predicting well interaction using indirect geological field information *IOP Conf. Ser.: Earth Env.* **378(1)** 012115. DOI: 10.1088/1755-1315/378/1/012115

[6] Minnivaleev A N, Kuzeev I R, Minnivaleev T N, Minnivaleeva A N and Gainanshin M F 2020 Hydromechanical device to clean the inside of pipes from paraffin deposits *Int. J. of Advanced Trends in Computer Science and Engineering* **9(1)** 124-127. DOI: https://doi.org/10.30534/ijatcse/2020/19912020.

[7] Mukhametshin V V and Kuleshova L S 2019 Prediction of production well flow rates using survey data *IOP Conf. Ser.: Earth Env.* **378(1)** 012116. DOI: 10.1088/1755-1315/378/1/012116

[8] Khisamov R S, Abdrakhmanov G S, Kadyrov R R and Mukhametshin V V 2017 New technology of bottom water shut-off *Oil Industry* **11** 126–128. DOI: 10.24887/0028-2448-2017-11-126-128

IOP Conf. Series: Materials Science and Engineering 1111 (2021) 012039 doi:10.1088/1757-899X/1111/1/012039

Methods for increasing the efficiency of hydrochloric-acid treatments of wells

V V Mukhametshin[1] and L S Kuleshova[2]

[1] Ufa State Petroleum Technological University, 1, Kosmonavtov St., Ufa, Republic of Bashkortostan, 450062, Russian Federation
[2] Ufa State Petroleum Technological University, Branch of the University in the City of Oktyabrsky, 54a, Devonskaya St., Oktyabrsky, Republic of Bashkortostan, 452607, Russian Federation

E-mail: vv@of.ugntu.ru, markl212@mail.ru

Abstract. The authors generalize the experience of conducting hydrochloric-acid treatments in the wells of high-viscosity oil deposits with hard-to-recover reserves in carbonate reservoirs. Intervals of changes in the values of geological and technological parameters with the efficiency of hydrochloric-acid treatment over 50% have been estimated. The impact parameters have been determined in various geological-field conditions. The causes of catastrophic water cut of wells as a result of impact have been established.

1. Introduction

Increasing the efficiency of impact on the bottom-hole zone of wells using hydrochloric-acid solutions is one of the factors of effective development of oil deposits in carbonate reservoirs. To solve this problem, it is necessary to provide a geological and technological justification for the selection of wells and technological parameters of impact in specific geological conditions of deposits [1-8].

A significant number of technologies proposed to date for stimulating the bottomhole zone using hydrochloric acid indicates a wide variety of productive formations in terms of bedding conditions, geological-physical and physicochemical properties of reservoir rocks and fluids saturating them, differences in the development technology, which must be taken into account when impacting to increase its effectiveness.

The experience of carrying out hydrochloric acid treatments (HAT), highlighted in the literature by various researchers, shows that in different geological and field conditions, this method of influencing the bottomhole formation zone has different efficiency and its value is determined by the influence of a wide variety of factors, among which the main ones are:

- type of HAT (foam acid, thermoacid, regular HAT, etc.);
- technological parameters of HAT (acid volume, injection pressure, acid concentration, etc.);
- geological features of formations (porosity, permeability, fracturing, etc.);
- technological performance of wells and deposits (current reservoir pressure, water cut, frequency of HAT, etc.).

The effectiveness of treatments is assessed in different ways: by the change in the productivity and flow rate of wells before and after the treatment; by the relative increase in flow rate; by the total additional oil production; by the change in inflow profiles; by success rate (the ratio of the number of effective operations to the total number of treatments). At the same time, along with the intensification of oil production, in some cases there is also a change in oil recovery. For example, under the conditions of the Yablonevskoye field, when developing it in the mode of dissolved gas, the radius of the zone of the carbonate formation, which was actively affected by hydrochloric acid, which increased the natural pore permeability, did not exceed several tens of meters. Active hydrochloric acid penetrates much deeper into the formation along some of the largest fractures. Vertical layer-by-layer macrocracks have a small specific surface area, and therefore concentrated acid moves along them at high speed over long distances, commensurate with the distance between the wells. Along the way, it corrodes fracture taper points and thereby increases fracture permeability. In this case, the speed of movement of the acid can reach several hundred meters per day. Calculations showed that oil recovery due to fracturing and acid treatments increased 1.62 times.

At the same time, during waterflooding of productive formations, unsystematic hydrochloric acid treatments may lead to a decrease in waterflooding and a decrease in oil recovery. For example, in the conditions of carbonate deposits of the Kuibyshev region, when acid is pumped under pressure, productive formation thickness decreases by 21 - 75%.

There have been cases where the injected acid appears in the production of adjacent production wells. As a result of deep penetration of acid into natural fractures of a carbonate reservoir and a multiple increase in the permeability of narrow fractured intervals, the heterogeneity of the carbonate reservoir in permeability may increase and the coverage of the pore volume of the reservoir by displacement due to the predominant movement of water along the fractures may decrease. This is the negative effect of haphazard acid treatments on the porous-fractured carbonate reservoir. A decrease in the reservoir sweep efficiency is also noted in the fields of Tatarstan, Udmurtia, and the Perm region. After the wells are commissioned from drilling, the sweep coefficient in thickness does not exceed 0.3; after the first HAT, it rises to 0.5; after the second HAT, it decreases again to 0.3; and after the third and fourth HATs, it does not exceed 0.2.

On the other hand, in low-permeability carbonate reservoirs (less than $100 \cdot 10^{-3}$ μm^2) with a large specific surface, hydrochloric acid is neutralized directly at the wellbore walls, and only the products of its interaction with acid penetrate deep into the formation, which also reduces the efficiency of treatments with an increase in their frequency. Moreover, in this case, the depth of penetration of the active acid into the formation does not exceed several meters.

In this regard, in order to increase the effectiveness of the impact and eliminate the adverse consequences of hydrochloric acid treatments on the oil recovery process, various varieties of them have been proposed and used.

2. Materials and methods

The geological-field analysis of impact of different geological and technological parameters on the efficiency of conducting usual and pressurized hydrochloric-acid treatments has been carried out using the deposits of high-viscosity oil in carbonate reservoirs of the Turnaisian stage of the North-Western part of Bashkortostan.

The parameters are as follows: effective oil-saturated thickness (N_e), average thickness of oil-saturated interlayers (N_p) and their number (n); coefficient of layer's porosity (M_g) in the well; share of reservoir rocks in the total thickness of the layer (K_p); time period from the beginning of operation of wells to the hydrochloric-acid treatment (t); maximum oil flowrate in the well before conducting the treatment (Q_{nMax}); flow rate (Q_{n1}); water cut (f_1); cumulative oil production (Q_{cum}) at the time of hydrochloric-acid treatment; volume (V_K) and maximum pressure (P_{Max}) of pumping acid into the layer.

The selection of geological and technological parameters that have the greatest impact on the effectiveness of hydrochloric-acid treatment has been carried out using a sequential Wald procedure. Informational value has been determined according to the Kulbak criterion.

3. Results and Discussion

Distribution of wells depending on changes in the values of informative geological and technological parameters with the effectiveness criterion being represented with an increase in oil flowrate, shows that the efficiency rate increases significantly for wells with worse reservoir properties and greater geological heterogeneity. At N_p<4.2 m, n >3, M_g < 13.3 %, K_p< 0.66, the efficiency makes more than 50%. Low reservoir properties and increased geological heterogeneity cause weak layer development in the drainage zone of the well and condition the presence of a reserve of oil production, on the one hand, and the presence of significant filtration resistances in the bottom-hole zone of the reservoir, on the other hand.

The need for a reserve of oil production is conditioned by the fact that it is advisable to impact on the wells with a service life of less than 160 months and with accumulated oil production of less than 24.6 thousand tons. At the same time, the probability increases for wells with a higher $Q_{\text{пмax}}$ value. This fact is explained by the clogging of the bottom-hole zone due to the deposition of resins, asphaltenes and paraffins as a result of intensive oil production and a significant decrease in layer pressure in the drainage zone.

Determining the volume of acid injected into the layer is an important point in conducting the hydrochloric-acid treatment [9-14]. The volume depends on the geological features of the layer at the site of its opening in the well and the technological features of wells and deposits. The use of relevant dependencies allows us to obtain inequalities for determining the volume of injected acid and injection pressures: V_k>0.71 Ne; V_k >1.49 N_p; V_k >0.059 Q_{n1}; V_k > 0.175 f_1; V_k >4.4·10^{-4} Q_{cum}; P_{max}<0.98 V_k.

Inequalities (1) allow diagnosing the required volume of acid by the maximum value of the parameter.

However, the results obtained are probabilistic. To get an unambiguous answer to the question about efficiency, the values of total diagnostic coefficients (TDCs) have been calculated, and the interval with the proved positive effect have been estimated. This interval changes from 8.9 to 66.8.

The use of TDC distributions allows changing the technological parameters of the impact and transferring wells from zones with a negative effect and zones of uncertainty to zones with a positive effect.

The distribution of wells depending on changes in the values of informative geological and technological parameters with the efficiency criterion being represented with a decrease in the water cut of the extracted products, shows that at N_e>11.5 m; N_p>4.3 m; n<3; K_P< 0.73; M_g<13 % , the efficiency rate of the impact makes more than 50%. It is advisable that the values of technological parameters are as follows: t<118 months; Q_{incl}<17.8 thousand tons; Q_{nmax}<349 t/month; Q_{n1}>77 t/month; V_k<7.5 m^3; P_{zak}<7.6 MPa.

The obtained inequalities also allow us to diagnose the efficiency of treatment operations at a qualitative level, both within the analyzed fields and similar ones in terms of geological and field characteristics, by searching wells for appropriate informative features, as well as to approximate the impact parameters. However, these results, as well as in the case when the efficiency criterion is represented with an increase in oil flowrate, allow assessing only probable impact and cannot give a clear-cut answer. The TDC values have been calculated to resolve this issue. It has been revealed that the TDC change from 14.4 to 40.9 provides for unambiguously positive efficiency of hydrochloric-acid treatment.

Comparison of research results for both variants shows that a significant number of wells in the first variant make part of the group with positive effect, while the second variant represents the group with negative effect, and vice versa, i.e. the higher the oil flowrate, the higher the water cut.

Comparison of the intervals of changes in the values of relevant parameters with the efficiency of hydrochloric-acid treatment over 50% indicates the presence of common intervals in both variants. The values of these intervals are shown in Table 1.

It is the total intervals that explain the fact that some wells are characterized by a decrease in the water cut of oil products after exerting the impact, along with an increase in oil flowrate.

Table 1. Intervals of changes in the values of relevant geological and technological parameters with the efficiency of hydrochloric-acid treatment over 50%

Interval by variants		Total interval for variants 1 and 2
1	2	
N_e>7.7	N_e>11.5	N_e>11.5
M_g<13.3	M_g<13.0	M_g<13.0
K_p<0.66	K_p<0.76	K_p<0.66
t<160	t<118	t<118
$Q_{\text{пмах}}$>180	$Q_{\text{пмах}}$<349	280<$Q_{\text{пмах}}$<349
$Q_{\text{п1}}$<125	$Q_{\text{п1}}$>77	77<$Q_{\text{п1}}$<125
Q_{cum}<24.6	Q_{cum}<17.8	Q_{cum}<17.8
7.5<$P_{\text{мах}}$<9.0	$P_{\text{мах}}$>7.6	7.6<$P_{\text{мах}}$<9.0

Besides, simultaneous increase in oil flowrate and water cut can be explained by the absence of common intervals for certain parameters (Table 2).

It should be particularly noted that there is no total interval of change in volumes and specific volumes of the injected acid. The results show that additional oil production may provoke a significant increase in water cut of the products.

Table 2. Intervals of changes in the values of geological and technological parameters with the efficiency of hydrochloric-acid treatment over 50%

Interval by variants	
N_p<4.6	N_p>4.2
n>3	n<3
$V_к$>7.8	$V_к$<7.5
$V_к/N_e$>0.71	$V_к/N_e$<0.65
$V_к$>0.71N_e	$V_к$<0.65N_e

4. Conclusion

Thus, the obtained results provide for the geological and technological justification of increasing the efficiency of hydrochloric-acid treatment of wells for high-viscosity oil deposits in the carbonate reservoirs of the Turnaisian stage.

References

[1] Gomari K A R, Karoussi O and Hamouda A A 2006 Mechanistic Study of Interaction between Water and Carbonate Rocks for Enhancing Oil Recovery *SPE Europec/EAGE Annual Conf. and Exhibition (Vienna, Austria, 12-15 June 2006)* p 8. DOI: 10.2118/99628-MS

[2] Kuleshova L S, Mukhametshin V V and Safiullina A R 2019 Applying information technologies in identifying the features of deposit identification under conditions of different oil-and gas provinces *J. of Phys.: Conf. Ser.* **1333(7)** 072012. DOI: 10.1088/1742-6596/1333/7/072012

[3] Yakupov R F and Mukhametshin V Sh 2013 Problem of efficiency of low-productivity carbonate reservoir development on example of Turnaisian stage of Tuymazinskoye field *Oil Industry* **12** 106–110

[4] Zeigman Yu V, Mukhametshin V Sh, Khafizov A R, Kharina S B, Abutalipova E M and Avrenyuk A N 2017 Peculiarities of selecting well-killng fluids composition for difficult conditions *Oil Industry* **1** 66–69

[5] Hognesen E J S and Strand T A 2005 Waterflooding of preferential oil-wet carbonates: Oil recovery related to reservoir temperature and brine composition *SPE Europec/EAGE Annual Conf. (Madrid, Spain, 13-16 June 2005)* p 9. DOI: 10.2118/94166-MS

[6] Shaidullina R M, Amirov A F, Muhametshin V Sh and Tyncherov K T 2017 Designing Economic Socialization System in the Educational Process of Technological University *European J. of Contemporary Education* **6(1)** 149–158. DOI: 10.13187/ejced.2017.1.149

[7] Yakupov R F, Mukhametshin V Sh, Zeigman Yu V, Chervyakova A N and Valeev M D 2017 Metamorphic aureole development technique in terms of Tuymazinskoye oil field *Oil Industry* **10** 36–40. DOI: 10.24887/0028-2448-2017-10-36-40

[8] Yakupov R F, Gimazov A A, Mukhametshin V Sh and Makaev R I 2018 Analytical method for estimating efficiency of oil recovery technology in case of bottom water-drive reservoir, verified on the hydrodynamic model *Oil Industry* **6** 66–69. DOI: 10.24887/0028-2448-2018-6-66-69

[9] Rabie A I and Nasr-El-Din H A 2015 Effect of Acid Additives on the Reaction of Stimulating Fluids During Acidizing Treatments *SPE North Africa Technical Conf. and Exhibition (Cairo, Egypt, 14-16 September 2015)* p 27. DOI: 10.2118/175827-MS

[10] Akhmetov R T, Mukhametshin V V and Kuleshova L S 2019 Simulation of the absolute permeability based on the capillary pressure curves using the dumbbell model *J. of Phys.: Conf. Ser.* **1333(3)** 032001. DOI: 10.1088/1742-6596/1333/3/032001

[11] Khuzina L B, Mukhametshin V Sh, Tyncherov K T and Shaikhutdinova A F 2018 On the choice of the oscillators' installation site *Int. J. of Civil Engineering and Technology* **9(9)** 1952-1959

[12] Imqam A, Bai B, Wei M, Elue H and Muhammed F A 2016 Use of Hydrochloric Acid To Remove Filter-Cake Damage From Preformed Particle Gel During Conformance-Control Treatments *SPE Production & Operations* **31(3)** p 11. DOI: 10.2118/172352-PA

[13] Kotenev Yu A, Mukhametshin V Sh and Sultanov Sh Kh 2018 Energy-efficient technology for recovery of oil reserves with gas injection *IOP Conf. Ser.: Earth Env. Sci.* **194(8)** 082019. DOI: 10.1088/1755-1315/194/8/082019

[14] Mukhametshin V Sh, Kotenev Yu A and Sultanov Sh Kh 2018 Assessment of the Need to Stimulate the Development of Hard-to-Recover Reserves in Carbonate Reservoirs *IOP Conf. Ser.: Earth Env. Sci.* **194(8)** 082027. DOI: 10.1088/1755-1315/194/8/082027

Use of criticality principle in information protection systems at machine-building enterprises

K Yu Zhigalov[1], R R Turluyev[2] and O V Dudareva[3]

[1]V.A. Trapeznikov Institute of Control Sciences of Russian Academy of Sciences
[2]Chechen State University, 17a, Dudaev Boulevard, Groznyy, 364015, Russia
[3]Irkutsk National Research Technical University, 83, Lermontov str., Irkutsk, 664074, Russian Federation

E-mail: kshakalov@mail.ru

Abstract. At present, the security of information systems plays a key role not only for large enterprises, but also for individuals. The information presented in the paper considers an innovative approach to the organization of multi-stage protection of information networks based on access matrices created and specified by data on the degree of criticality of protected files against illegal actions of third parties. The authors proposed to use several levels of criticality with three degrees: critical for program running; important for programs; not critical to programs. Besides, an algorithm for determining the criticality level by file extension as universal is proposed as the basic one. Timely response to changes in critical files not only of the operating system, but also of all programs significantly increases the security of enterprise systems. Moreover, the proposed method can be used to minimize the loss of important data.

1. Introduction

The main feature of modern enterprise information systems is that the system components are distributed in space and physical communication between them is carried out both through network connections and software [1, 2].

When users work remotely (as it happens now), remote attacks on the enterprise information infrastructure take the leading position in the number of attempts and the success of their application. In this regard, the control over the infrastructure software at the file level is becoming ever more relevant [3].

The research has shown that companies around the world lose about 6% of their revenues due to incidents related to various ways of information fraud and theft. The number of crimes in this area by country is as follows:

- USA 80%;
- Great Britain 85%;
- Germany 75%;
- France 80%;
- Russian Federation 90%.

The major disadvantages of the current information protection systems include the following:

- visibility of security programs to end users;
- large amounts of stored backup information.

It is known that in this particular case, forewarned is forearmed.

Only backup techniques can protect against accidental or deliberate destruction or modification of files by users with full access to them [4].

However, there are several disadvantages to existing backup methods, including:

• need for large amounts of physical storage to accommodate backup files. As practice shows, the amount of backup storage often exceeds the amount of data used by the company by 2-3 times even in the archived form;

• significant time required to recover data from backups. This is caused by the search for the necessary archive, unpacking the files and forwarding them over the network to the server or workstation on which they were used before their loss;

• no automated file integrity monitoring system. This aspect significantly complicates the search for lost information, since the emergency case can be detected by the user only after some time;

• significant network loading when performing backup procedures from several computers (therefore, this type of work is carried out mainly at night);

• lack of technical capability to verify the correctness of backup procedures.

2. Access matrices

To simultaneously minimize all factors, it is advisable to use the protection systems of a fundamentally different level based on access matrices created on the basis of criticality levels of certain files.

The Access Matrix is a table that displays access control rules [5].

File criticality (in this paper) is the ability of the software, to which the file belongs, to work if it is lost.

For several years Microsoft in its operating system (hereinafter referred to as the OS) based on Windows has been using the criticality criteria of some files in backup methods, which serve to:

• restore the previous OS;

• restore the OS operability.

In the first case, the operating system logs all configuration file changes.

In the second case, Microsoft proposes to rewrite important system files from the original storage (operating system installation).

However, these methods do not affect additional software installed by the user independently and the software not related to the operating system.

According to the studies, not all files installed with programs are critical to their work [6].

Thus, criticality criteria can be generically divided into the following levels:

- critical to the work of the program as a whole. If these files are lost, the program does not start or perform its main functions;

- critical for program operation (usually files of different program modules). If this group of files is lost, the program starts, but does not perform some secondary functions;

- not critical for operation (text files – instructions, backup files for recovery of software performance, files of previous updates, files intended for different configurations, log files). If this group of files is lost, the program can start and perform its main functions.

All software manufacturers create it individually, and therefore there is no single list of file extensions related to the above groups, either directly or indirectly. This fact forces to test each program and each of its files manually and put the results in a table (Table 1). To simplify the process, it is useful to initially group files by extension and scan all files that have this extension at the same time.

3. Criticality check algorithm

The verification algorithm is as follows:

1. Copy the directory containing the program into the archive;
2. Find files with one common extension and delete them;
3. Try to run the program, and, if possible, test the program. If the program does not start, the files fall into the category (1). If the program is running and performs only basic functions, the files fall into

the category (2). If the program is running and performing its functions, the files fall into the category (3);

4. Tabulate the result;
5. Return the deleted files from the archive;
6. Repeat with a different file type.

Table 1. File extension criticality

Operating system	Software	Software version	File extension	c r
W	Avast		exe	
W	Office	2		2

The program is tested using the black box method. The method is based on the fact that the tests are performed in accordance with software specifications or other documents describing the requirements for the system.

Not all software files with the same extension are equally critical to the program. If the number of files is small, there is a need to check all these files and assign the criticality criterion individually to each file. If the number of files is quite large, there is a need to limit to a quite large sample to determine the percentage of files critical for the program among the "not critical":

$$C = \frac{\text{Number of critical files}}{\text{Number of checked files}} \times 100\% \quad (1)$$

where C – coefficient of critical files among non-critical files. If a very small number of critical files were found in the sample, it makes sense to find these files to minimize the archive.

To unify file criteria matching, it is possible to use cloud technologies to increase the criticality table at the expense of future system users.

Upon completion of file criticality tables, a file system access matrix is created based on them (Table 2).

The criticality matrix adds system action options for each group of files and users.

According to the studies, the following possible program operation modes were integrated into each group:

• read lock (in fact, this is blocking the user's access to these files);

• record lock (the user can read but not write files);

• monitoring the user actions (in this case, the Program can log user actions with these files (deleted/read/edited/copied/started);

• system-safe monitoring (the program logs the user's actions and creates the following folder on the backup storage server: \date\user login\... Thus preserving the file structure from the root, we copy files deleted or modified by the user).

• safe optional monitoring (the program logs the user's actions and creates the following folder: \date\user login\... It saves only deleted or modified files based on the file criticality table).

Table 2. Access matrix (example)

General data	Group data	File extension	File route	File cri	Software operation mode					Individual keys according to the user's name or group	
					Read lock	Record lock	Action monitoring	System-safe monitoring	safe optional monitoring	Semenov	Sale Support
					xx	yy	zz	vv	Uu		
Software files	OS files	dll		1		yy				1	
		sys		1							
	Software setup files	exe		1						1	
		bat		1		yy					
	Temporary files	tmp			-		zz			1	1
	System and software login	log		3			zz			1	1
Data files	Text documents	doc								3	
		docx					z			2	
		odt									
		txt									
	DB	1cd								4	
		lgp		1					Uu		2
		sql									
		mdf									
		rdf									
	s	jpeg								3	
		raw						vv			
		bmp									
		gif									
				2	xx	yy yy				1	

4. Conclusion

The positive aspects of the access matrix include the following:

1. It can be used in various kinds of data protection software by differentiating access to them; protects critical data with backup.

2. The matrix makes it possible to choose the necessary level of protection and thereby save resources of the company's internal network.

3. The file criticality matrix significantly reduces the time to configure the information protection system.

The negative aspects of the access matrix include the following:

1. The need to test all programs manually during the initial phase of technology implementation.

2. A matrix in the hands of an attacker will allow quickly copying important data.

The research has shown that an average from 25% (Windows) to 45% (Linux/Unix) of non-critical files can be contained on a computer.

The proposed approach to information security has a certain potential. This method allows responding to all changes in critical files in a timely manner (deleting a file), which significantly improves security and essentially eliminates the possibility of losing important data due to extremely long period since their change/deletion.

References

[1] Klyuev R V, Bosikov I I, Mayer A V, Gavrina O A 2020 *Sustainable Development of Mountain Territories* **V.12 2(44)** 283-290

[2] Bosikov I I, Klyuev R V, Tavasiev V Kh, Gobeev M A 2020 *IOP Conference Series: Materials Science and Engineering* **918** 012223

[3] Klyuev R, Turluev R, Khadzhiev A 2020 *International Ural Conference on Electrical Power Engineering (UralCon)* INSPEC Accession Number: **20034001** 324-329

[4] Bovkun A, Korodyuk I 2019 Analysis of the development of small innovative enterprises in the construction industry. *IOP Conf. Ser.: Mater. Sci. Eng.* **667**

[5] Skorobogatova Y A, Bovkun A S, Ivanov M Yu, Shilova O S 2020 The role and place of small innovative enterprises in the construction industry in the modern economy of single industry towns. *IOP Conf. Ser.: Mater. Sci. Eng.* **880**

[6] Repinskiy O D, Konyukhov V Yu, Bovkun A S, Schupletsov A F 2021 Improving the competitiveness of Russian industry in the production of measuring and analytical equipment. *Journal of Physics: Conference Series* **1728**

The effect of the grain size on the cycle life of spring steel 50 CrMnV

V G Pachurin[1], V V Galkin[1], A A Filippov[1], G V Pachurin[1], M V Mukhina[2] and Zh V Smirnova[2]

[1] Nizhny Novgorod State Technical University named after R. A. Alekseev, 24, Minina str., Nizhny Novgorod, 603950, Russia
[2] Minin Nizhny Novgorod State Pedagogical University, 1, Ulyanova str., Nizhny Novgorod, 603002, Russia

E-mail: mukhina_mv@mininuniver.ru

Abstract. Most of the breakdown of machine parts and metal products during operation is fatigue-related due to the impact of load cycling during operation, which often leads to serious financial losses, and sometimes even fatalities. Therefore, the issues of increasing cycle life are relevant. In this work, we propose a method for predicting the cycle life of spring-type steel: the expected cycle life of steels 50CrMnV and 60SiCr in the range of 105-106 cycles can be preliminarily estimated using the equation depending on the grain size. The method allows predicting the cycle life and significantly reducing the time of its evaluation for steels such as 50CrMnV and 60SiCr by eliminating long-term fatigue tests. In addition, it also allows choosing the best performance from process treatments for the cycle life parameter without long labor-intensive fatigue tests.

1. Introduction

One of the most important tasks of the industry is to accelerate the development of a process database to ensure competitive products in the domestic and international markets [1-4]. The quality of machine-building products is characterized by their performance properties [5-9].

During operation, most machines and equipment are exposed to load cycling, thus, the most common type of fractures is fatigue failure, which often leads to serious financial losses and sometimes fatalities [10-12]. Therefore, the issues of fatigue life of parts, components, machines and devices in general are priority areas of modern science and the most important task of industry.

In industry, metal products are widely used, which are made by various processing methods. As a rule, all strength metal parts of machines are made from blanks obtained by hot or cold forming [13-15]. Various operating conditions differ in size, nature of the load, temperature conditions, and environment, which have different effects on the processes of metal hardening [16, 17] and its resistance to fatigue failure. Many parts of machines and structures made from such blanks must withstand a huge number of loads and stresses.

Depending on a number of factors, changes in loads can occur either within the deformation or stress limits. Constant interest in the problem of fatigue failure of metals and alloys is associated with the problems of fatigue failure of critical parts of metal structures, since the brittle destruction of such parts is often preceded by the growth of a fatigue crack, which reduces the load-bearing capacity. The use of fracture mechanics approaches made it possible to evaluate and predict the crack resistance and cycle life of metal structures. It is necessary that the methods of testing metal materials for fatigue and fatigue crack growth are sensitive to the structural condition of the material. In addition, during fatigue tests, the kinetics of damage accumulation can be traced.

To date, in the process of designing metal products, there is no single technical plan of pressure treatment, taking into account the prediction of cycle life and the relation between the deformable state of the metal and the structure formed by it. While there is a lack of effective means of information support of engineering and design, which concerns forming processes and cycle life of the material obtained as a result of conversion technologies.

The number of products in the form of forged products and sheet blanks are formed of different microstructures, which is characterized by grain size, structure and residual stresses of different levels. The main thing in fatigue failure is not the average characteristics of strain resistance, but the measurements in specific places and elements of products that are primarily responsible for their service life. For the relation of local deformed measurement with their structural changes, their quantitative assessment is necessary.

It is important to offer information support for the process engineering of manufacturing plastically deformed metal products with specified fatigue characteristics. The current practice of process engineering does not allow us to develop technologies for metal forming, based on the requirements for their service properties, in particular, cycle life.

The factors that affect the structural condition of the material including grain size, type of microstructure and heat treatment, and residual stresses at the micro and substructure levels. At the same time, many domestic and foreign researchers note that the most important structural parameter of polycrystalline metal materials that affects the origin and propagation of cracks is the grain size, since grain junction lines can be effective barriers to sliding processes.

In high-strength metal materials, the determining structure factor may be the size of the sub-grain or one of the structure components. Most often, as the grain size decreases, the fatigue strength of samples increases, although grinding the structure does not always lead to a change in service life.

Plastic deformation in the cold or hot state affects the structure of the metal at all its levels, and it can simultaneously be combined with heat forming. At the level of fine structure, it changes the density and structure of defects in the crystal lattice, at the micro level - the grain size, morphology, and the value of residual stresses, at the macro level it causes residual macro stresses due to uneven deformation in the measurement and texture of the workpiece. The main technological factors include: the degree of deformation, its mechanical schematic, and temperature-speed processing conditions. For the entire product as a whole, the service properties are also affected by the inhomogeneity of the structure in the entire part. It is determined by the size of the part, as well as the geometry of structure elements or individual measurement of the part (blank), which are the most loaded and responsible for the service life of the product. The latter factor, in particular, has a great influence, since under real operating conditions, as a rule, the main load is taken by specific elements: in fasteners (nuts, bolts) - thread; in elastic elements (springs) - internal flat surface; in sheet stiffness - rifts; in cutting products (axe-knife group) - blade, etc.

A fairly large number of both foreign and domestic studies are devoted to the fatigue of structured materials, a review of which is presented in the works. They provide information on the fatigue resistance of metals and alloys deformed at room temperature and uniform loading.

During conversion technology processes of the source material by pressure processes, the mechanical properties of metals and alloys depend on the structure determined by the conditions of their processing. There is a clear relation between the structure parameters, in particular the grain size, and the mechanical and service properties of the metal.

The grain size represents the structure parameter that is most easily distinguished and that can establish analytical relation between structure and properties. This allows to indirectly evaluate the mechanical characteristics and service properties of the material, and as a result, the technical specifications of major products set the grain score.

Hot forming of metals and alloys has a huge application in production in all its forms. Under conditions of hot deformation at temperatures above 0.4 melting temperature, the process of recrystallization occurs, which is reduced to the formation of new grains to replace the deformed ones.

As the main parameter affecting the fatigue resistance of plastically deformed metals and alloys under hot processing conditions, it is proposed to consider the size of the recrystallization grain, since the cycle life of the material decreases with an increase in the degree of recrystallizing.

2. Materials and methods

This paper presents the results of studying the relation between the structure formed by the manufacturing technology and the fatigue resistance of metal products.

Micrographic investigation was carried out with 50CrMnV spring steel in order to determine the grain size for samples with different degrees of process induced distortion (degree of reduction). For this purpose, flat samples were cut by electrical discharge machining from rolled into wedges and heat-treated metals.

The grain size was determined by the length of the rolled sheet. For this purpose, based on the use of optic digital microscopy, a specialized software package was developed using the NI Vision software and the LabVIEW development environment of National Instruments.

The plan includes two programs. The first one provides quantitative measurements and calculation of the metal microstructure. To do this, digital photos of the microstructure are binarized, as a result of which the image elements are divided into objects – grains and background.

The samples were tested for fatigue (symmetric bending) at the same stress, providing 10^5-10^6 cycles.

3. Results and discussions

The microstructure along the entire length of spring steel sheet 50CrMnV consists of sorbite and ferrite (Figure 1).

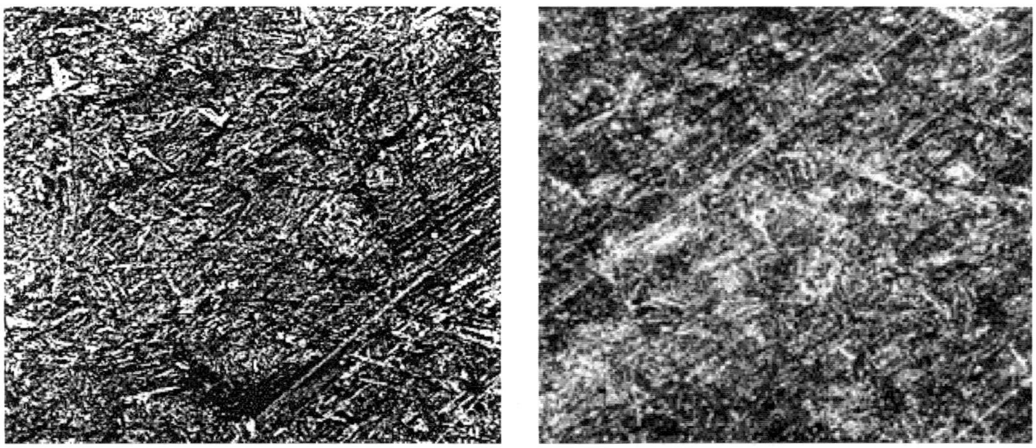

a b

Figure 1. Microstructure of the studied zones of the spring leaf: a-central hole; b-cross-section with the reduction degree of 40%

The analysis of the experimental data as a result of fatigue tests with different degree of reduction has shown that when removing layers of rolled into wedges metal, sheet fatigue life of the samples increases with the reduction degree of 23%, but the fatigue life of the samples reduces with the reduction degree of 40%.

Figure 2 shows a comparison of the results of the fatigue test with the grain sizes of the samples.

IOP Conf. Series: Materials Science and Engineering 1111 (2021) 012041 doi:10.1088/1757-899X/1111/1/012041

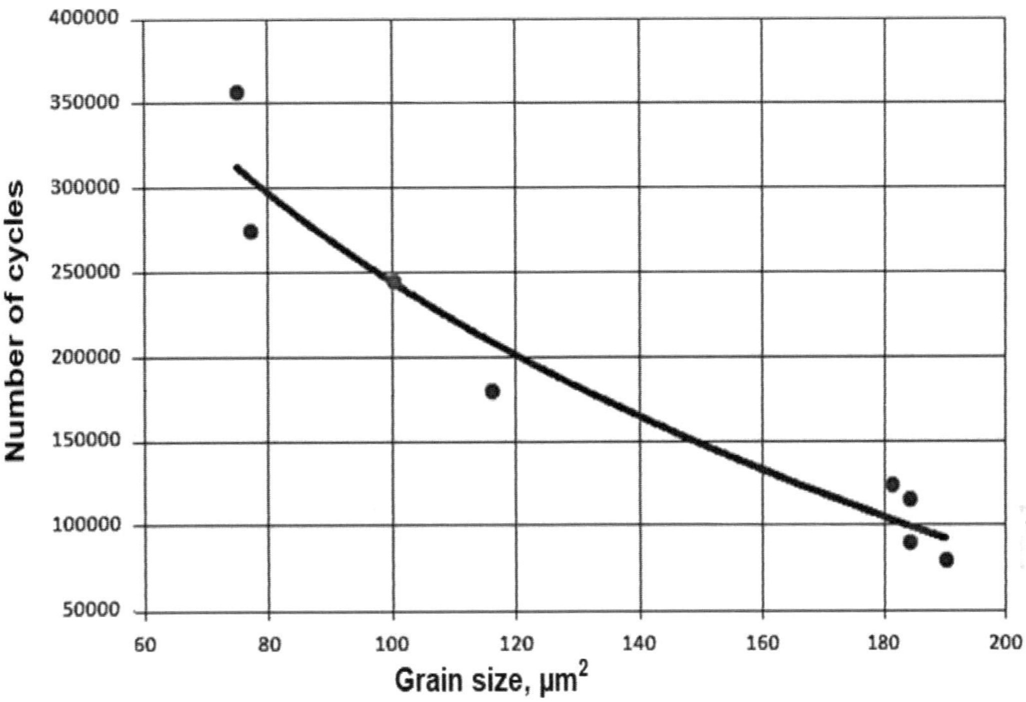

Figure 2. Dependence of the number of cycles to the failure of steel 50CrMnV on the grain size

According to the mathematical processing of experimental data, the equation of the dependence of the number of cycles to failure (cycle life) on the grain size of steel 50CrMnV is obtained:

$$Y = -237125.08 \, \mathrm{Ln}\,(x) + 1336448.22 \qquad R^2 = 0.93, \qquad (1)$$

where *Y* is the number of cycles to failure; *x* is the grain size, microns; *R* is the correlation coefficient.

To implement the method, it is sufficient to determine the grain size of steel and calculate the expected cycle life using equation 1.

The method allows one to predict the cycle life and significantly reduce the evaluation time for steel 50CrMnV, for which it is enough to determine the grain size and calculate the expected cycle life using equation 1. In addition, this method allows choosing the optimal mode from competing processing treatments for the cycle life parameter by comparing the grain size: the grain is smaller, the cycle life is higher.

Experimental verification of the method was also carried out on steel 50CrMnV and 60SiCr. The results confirmed the reliability of this method. Deviations of cycle life in experiments from the data obtained by equation 1 did not exceed 6%. This gives reason to believe that the grain size is the main control parameter that characterizes fatigue resistance.

4. Conclusion

The proposed method allows predicting the cycle life and significantly reducing the time of its evaluation for spring steel 50CrMnV, by eliminating long, expensive, energy - and time-consuming fatigue tests.

In addition, it also allows choosing the optimal performance from competing processing treatments for the cycle life parameter without conducting long-term labor-intensive fatigue tests by simply comparing the grain size of steels: when the grain size is smaller, the cycle life is higher.

References

[1] Pachurin G V 2012 *Russian Engineering Research* **32(9–10)** 661–664
[2] Terentyev V F and Petukhov A N 2013 *Fatigue of high strength materials* (Moscow: IMET RAS – CIAM) 515 p
[3] Gurov V D and Vladimirov A G 2005 *Steel* **12** 52-54
[4] Sokolov A A and Artiukhin V I 2008 *Metalware* **2(18)** 50-54
[5] Amirov M G and Lavrinenko Yu A 1992 *Basis of technology for automated cold upsetting. Study Guide* (Ufa: UAI) 142 p
[6] Moiseichick E A 2014 *Bulletin of MGSU* **11** 58-67
[7] Bunatyan G V 2010 *Metalware* **1(22)** 12-15
[8] Zinuti A and Saro J 2003 *Metalware* **2(03)** 41-47
[9] Kulesha V A 2000 Distinctive features of steel production for high quality metalware *The works of the third congress of rolling mill operator* (Moscow) pp 543-546
[10] Galkin V V, Mikhalenko F P and Ulanov P V 2001 Analysis of the performance of the leaf springs sheets obtained by treatment of rolled strip *KSHP OMD* **8** 3-7 (in Russian)
[11] Mishin A F, Temyanko L S, Volgunin A A et al 2001 *KSHP OMD* **8** 42-45 (in Russian)
[12] Terentev V F 2000 *Fatigue of metallic materials: Textbook* (Voronezh: Voronezh Technical University Press) 60 p
[13] Filippov A A, Pachurin G V, Kuzmin N A, Matveyev Yu I and Deyev V B 2018 *Proceedings of higher educational institutions. Ferrous metallurgy* **7(61)** 551-556. DOI: 10.17073/0368-0797-2018-7-551-556
[14] Lavrinenko V Yu 2007 *Metalware* **1** 35-37
[15] Ivchenko A V, Ambrazhey M Yu, Machuskaya N D and Kokashinskaya G V 2010 *Metalware* **01(22)** 58-63
[16] Galkin V V 2014 *Press Forging. Metal Forming* **8** 11-14
[17] Galkin V V 2014 *Strengthening Techn. and Coat.* **8** 8-12

IOP Conf. Series: Materials Science and Engineering 1111 (2021) 012042 doi:10.1088/1757-899X/1111/1/012042

Rationale for choosing the sound-absorbing materials for the operator`s cabins of the rail-grinding machine

A A Nazaretov[1], I A Yaitskov[1] and N M Yaitskova[2]

[1] Car and Car Facilities Chair, Rostov State Transport University, 2, Rostovskogo Strelkovogo Polka Narodnogo Opolcheniya sq., Rostov-on-Don, 344038, Russia
[2] Research and Production Center "Labor Protection", Rostov State Transport University, 2, Rostovskogo Strelkovogo Polka Narodnogo Opolcheniya sq., Rostov-on-Don, 344038, Russia

E-mail: aanazaretov@rgups.ru

Abstract. The operation of the rail grinding machines that use noise and vibration-active parts in their structures inevitably leads to the increased levels of vibration and noise that adversely affect their operators. In this field, the specialists from different countries have carried out certain studies. However, the process of forming the vibro acoustic characteristics of the rail grinding machines has not been studied. It should be also noted that theoretical studies of the vibro acoustic dynamics of the rail grinding machines and methods for calculating the vibration and noise spectra are limited, which significantly complicates the substantiation of the engineering solutions to reduce the sound emission intensity of the dominant noise sources. Due to the negative impact of the elevated noise levels on rail grinding machine`s operators, it is necessary for the design stages with appropriate vibro acoustic characteristics that meet requirements of the sanitary standards. The main practically and technologically possible way to align noise emission level is the choice of the sound-absorbing materials and the achievement of the sound insulation, based on the fulfillment of the sanitary noise standards. The article describes the rationale for choosing the sound-absorbing materials for the operator`s cabin of the rail-grinding machine due to the reduction of the air noise component, which is more common than the structural one. A criterion for minimizing acoustic characteristics inside the operator's cabin is determined to select the most rational engineering method of the sound absorption.

1. Introduction

The harmful effects of the vibration and high intensity noise on the labour`s health are well known [1-4]. It should be highlighted that vibration and noise data of various types of the technological machines are presented in works [5-7]. Moreover, the capability of vibro acoustic characteristics to the limiting spectra is not only one of the most important ergonomic indicators, but it is determined as technical condition of the engineering machines of various functional purposes and competitiveness [8-11].

A large number of the scientific works are devoted to the development of the theory and practice of dealing with noise and vibration of the road-building and technological machines [12-15]. However, the problem of reducing vibration and noise levels for different rail grinding machines has not been studied enough, which actually determines the relevance of the materials in this article.

2. Object and methods of research

The functional purpose of the soundproof operator`s cabin and enclosures is to reduce sound pressure levels to sanitary standards. In the overwhelming majority of cases, the operator`s cabins have a shape similar to a thin-walled rectangular parallelepiped exposed to the air and structural noise components. The calculation of the operator`s cabins under the influence of the structural noise component is carried out by the methods of energy balance with a significant vibration effect.

The air noise component is come out of external sources and the calculation of the operator`s cabin elements is done from the position of the required sound insulation value. To fulfill the sanitary and hygienic working conditions of the operators in the cabins, firstly, the air conditioners and fans are installed inside the operator`s cabins according to the parameters of the temperature requirements. In the vast majority of cases, these sources, which are classified as internal, create excess sound pressure levels in the cabins. As it is practically impossible to reduce the noise levels of such sources at the production enterprises, the most reasonable option for fulfilling sanitary standards in this case is a rational choice of sound-insulating materials.

For forming the minimization criterion, it is beyond argument that there is a requirement for maximum reduction of the noise levels at the calculation point in octave bands. However, this condition is necessary, but not sufficient, as the standard curve defines the acceptable noise levels in each of the bands. Therefore, as a necessary and sufficient criterion, it should be recognized one or another deviation degree of the calculated noise levels in octave bands of the spectrum from the normative ones.

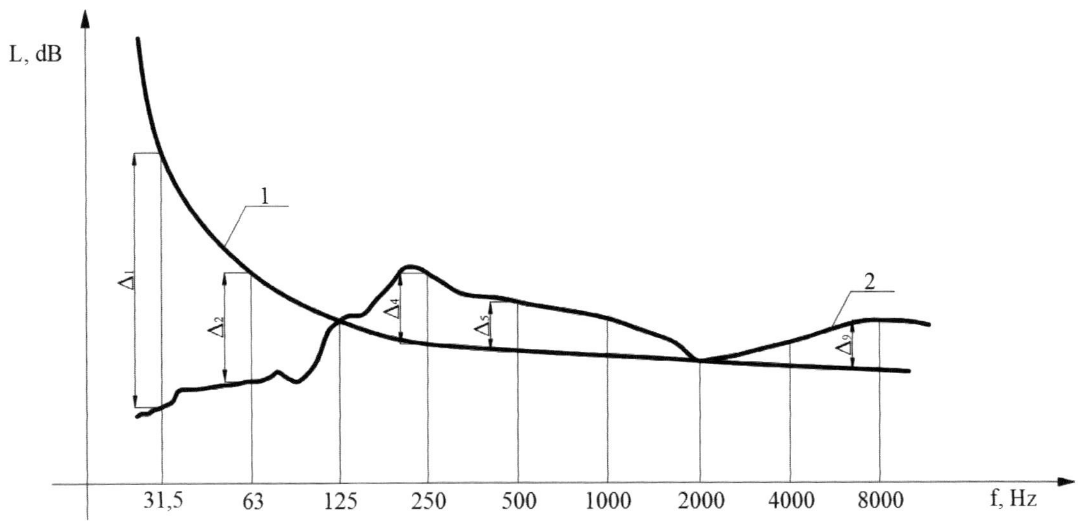

Figure 1. Comparison of the standard and actual noise spectra: 1 – the standard curve; 2 – actual or calculated noise spectrum; Δ_1 … Δ_3 – negative differences $L_{i_p} - L_{i_{stand}}$; Δ_4 … Δ_9 – positive differences $L_{i_p} - L_{i_{stand}}$.

This deviation degree can be considered the sum of the differences between the calculated and standard noise levels for all nine octave bands of the audio range (from 31.5 Hz to 8000 Hz). At the same time, taking into account the different influence degree of sound radiation frequency on the human body (high frequencies are more unfavorable than low frequencies), it is logical to hypothesize that forming the optimization criterion for each octave band, a weight coefficient should be introduced, and the criterion should be chosen as a weighted sum, which value is necessary to minimize:

$$K = min_{U,W}\left(\sum_{i=1}^{9}\left(L_{i_p} - L_{i_{stand}}\right)\theta_i\right); L_{i_p} > L_{stand} \quad (1.1)$$

where $L_{i_{stand}}$ – standard and L_{i_p} – actual or calculated noise levels in *i*-th of nine octave bands of the spectrum; θ_i – the weight coefficient; U – the vector of factors influencing the noise level which are controlled by designer; W – the vector of constructive and technological transitions and physical constants.

In this case, it is obvious to take into account only those spectral components, the levels of which exceed the corresponding levels of the standard curve. The figure 1 shows such a noise spectrum being superimposed on the target curve. Minimization of the criterion is attracted to the actual noise levels to the normative in the $\Delta_i > 0$.

To calculate the weight coefficients θ_i, the following principle is proposed (see Figure 2):

- for middle band of the spectrum, namely for a frequency 500 Hz we take $\theta_5 = 1$;
- a straight line parallel to the ordinate *c-c* axis, we will consider the axis of symmetry, relative to which the normative curve $L_{i_{stand}}$ is mapped to the curve L_{i_p}. The ratio:

$$\theta_i = \frac{L_{i_p}}{L_{i_{stand}}} \tag{1.2}$$

will be considered a weighting factor for the *i*-th band of the spectrum.

The θ_i values are pointed out in Table 1. The weighting coefficients adopted in this way enhance the influence of high-frequency components and weaken the influence of low-frequency ones on the value of the criterion.

Not to dispute the certain conventionality of criterion (1.2), nevertheless, it must be admitted that to a certain extent it objectively assesses the degree of proximity between the calculated and standard values of the noise levels in octave bands.

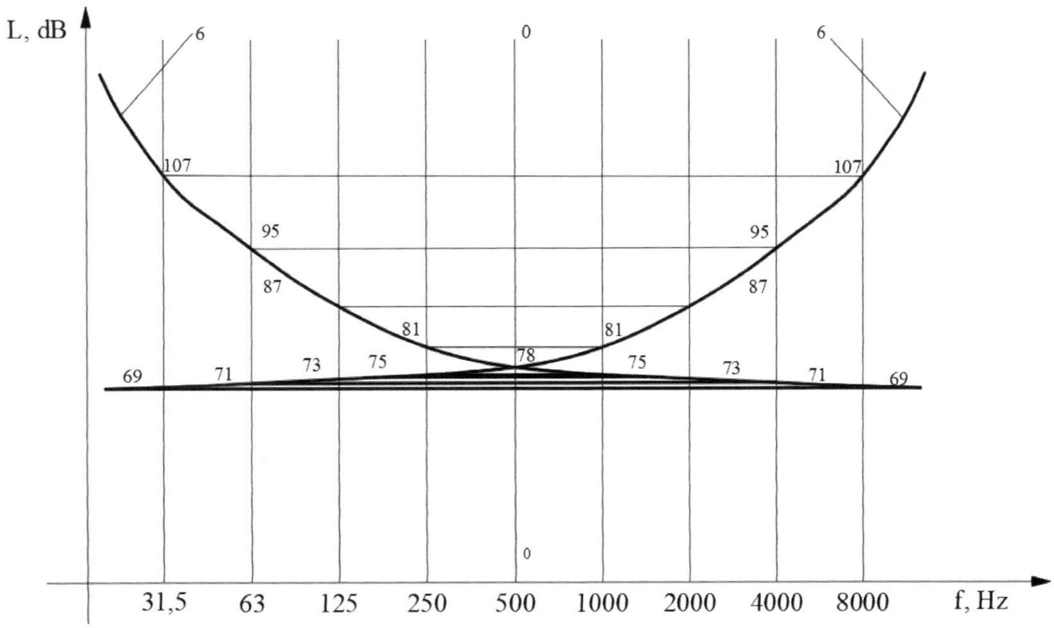

Figure 2. For determination of the weight coefficients: "a" - standard curve (levels $L_{i_{stand}}$); "b" - "anti-standard" curve obtained by symmetric mapping of the curve "a" relative to the c - c axis (levels L_{i_p})

Table 1. The weight coefficients for the terms of the optimization criterion

№ Bands	Geometric average frequency in octave bands	$L_{i_{stand}}$	L_{i_p}	θ_i

3. Analysis of the parameters and controlled variables

According to the research data, the sound pressure levels inside a rectangular volume are determined by the formula:

$$P = \frac{\rho_0 c_0^2 Q_0 \exp(-i\omega t)}{V} \sum_N \frac{\psi_N(P)\psi_N(S) i\omega \Delta_N^{-1}}{(2\pi f)^2-(2\pi f_N)^2}, \tag{1.3}$$

where f_N is the natural frequencies of the internal air volume, which are found by the formula:

$$f_N = \frac{c_0}{2}\left(\left(\frac{n_x}{l_x}\right)^2+\left(\frac{n_y}{l_y}\right)^2+\left(\frac{n_z}{l_z}\right)^2\right)^{0,5}. \tag{1.4}$$

The noise level generated by the source inside the hood is determined by the formula:

$$L = 20lg\frac{P}{P_0}, \tag{1.5}$$

where P_0 is the threshold sound pressure, $2\cdot10^{-5}$ Pa.

In practice, the effective sound pressure $\overline{P}^2$, which is usually measured and formed:

$$\overline{P}^2 = \frac{\rho_0^2 c_0^4 Q_0^2}{2V^2}\sum_N \frac{\omega^2}{(2\omega_N\delta_N)^2+(\omega_N^2-\omega^2)^2}, \tag{1.6}$$

where δ_N is the attenuation factor.

The attenuation factor is determined by the following expression:

$$\delta_N = \frac{c_0}{8V} A_S a_S, \tag{1.7}$$

where $A_S = 2(l_x l_y + l_y l_z + l_z l_x)$ is the area of the inner surface of the hood; a_S – absorption coefficient. The absorption coefficient is:

$$a_S = 1 - \xi^2 \tag{1.8}$$

where ξ is the sound reflection coefficient.

The sound reflection coefficient has the formula:

$$\xi = \frac{i(\alpha+\alpha^{-1}) sha_1' h}{2cha_1' h - i(\alpha-\alpha^{-1}) sha_1' h}, \tag{1.9}$$

where $\alpha = \frac{\rho a}{\rho_0 a_1'}$; $a_1' = \frac{2\pi nf}{c}\left(\sin^2\theta - \frac{c_0^2}{c^2}\right)^{0,5}$; $a = \frac{2\pi nf}{c}\cos\theta$; c is the speed of sound in the cabin material; θ is the angle of incidence of the sound wave on the cabin wall.

Substituting (1.8) into (1.7) and making calculations, we obtain the expression for the absorption coefficient:

$$a_S = \frac{4ch^2 a_1' h + \alpha^2 sh^2 a_1' h \cos^2\theta}{4ch^2 a_1' h + \alpha^2 sh^2 a_1' h}, \qquad (1.10)$$

where h is the cabin wall thickness.

The formula (1.8) is valid for the case of the oblique incidence of a sound wave on the cabin wall. The reflection coefficient of sound in the case of normal incidence on the wall is determined by the expression:

$$\xi = \frac{\alpha - \alpha^{-1}}{\alpha + \alpha^{-1} - 2i\, ctg\, ah}. \qquad (1.11)$$

Then the sound absorption coefficient is obtained by the formula:

$$a_S = \frac{4ctg^2 ah + \alpha^2 \sin^2\varphi}{4ctg^2 ah + \alpha^2}, \qquad (1.12)$$

where $\varphi = acrtg\frac{2\, ctg\, ah}{\alpha}$.

For more exact calculations, the absorption coefficient can be calculated by the formula [192]:

$$a_S = \frac{0,162V}{t_p}, \qquad (1.13)$$

where t_p is the sound reverberation time in the volume, which is easily found experimentally.

The dependence of the performance of a noise source (in this case, an air conditioner or a fan) is set in the following form:

$$Q = S_d \vartheta_{di}, \qquad (1.14)$$

where S_d is a source surface area, m^2; ϑ_{di} is a body vibration velocity, m / s.

Substituting the formulae (1.6), (1.13) into the noise level formula (1.5) and considering the air density $\rho_0 = 1{,}23$ kg/m^3 and the sound speed in air $c_0 = 343$ m/s, we obtain the formula for the noise level created by a point source inside the cabin:

$$L = 20lg\frac{S_d\vartheta_{di}}{V} + 10lg\sum_N \frac{f^2}{\left(\frac{\delta_N f_N}{\pi}\right)^2 + (f_N^2 - f^2)^2} + 200. \qquad (1.15)$$

The formula (1.14) can be used to calculate both the general noise level and the sound pressure levels in octave and three-octave bands. This requires knowing the bandwidth Δfj of the spectrum analyzer that will be used for calculations. Then the sound pressure level in the *j*-th band will be equal to:

$$L = 20lg\frac{S_d\vartheta_{di}}{V} + 10lg\sum_{N_j} \frac{f_j^2}{\left(\frac{\delta_N f_N}{\pi}\right)^2 + (f_N^2 - f_j^2)^2} + 200.$$

where N_j is the number of natural frequencies of the air volume that get into the *j*-th band; f_j is the geometric mean frequency of the corresponding filter band pass.

The noise levels in octave frequency bands depend on a number of parameters which are:

Physical constants: c_0 is the sound speed in air, $c_и$ is the speed of range of the flexural wave in the glazing elements, ρ_0 is the air density, ρ is the density of the glass.

The engine`s vibration levels.

The values of these physical quantities are stored in the database.

Geometric invariable: the dimensions of the enclosure $(l_1 \times l_2 \times l_3)$ and the dimensions of the glazing $(l_4 \times l_5)$, *A* is the internal surface area, *r* is the distance from the noise source to the design point.

Parameters which values are characteristic for each octave band and must be calculated along with solving problem by using parameters of the first two groups: N_1 is the number of natural frequencies of the air volume oscillations; m and n are constant numbers that determine the modulus of oscillation of the glazing element;

Controlled variables, their defined values must provide an extreme criterion and must be found in the design input. These include:

1) The thickness of the glazing elements with definition limits $h_1 \leq h \leq h_2$. This variable is continuous.

2) The absorption coefficients a_s for N_2 facing materials in the form of $N_2 \times 9$ matrix, which gives a_s values for each of the N_2 materials in nine octave bands. The matrix is located in the database. So, this variable is discrete and is set by a set of N_2 values in each octave band.

Consequently, with regard to the found extreme value h, choose such a facing material from N_2 investigated, which would provide the minimum values for the M_2 component over all octave bands:

$$M_2 = 10\lg \sum_N \sum_{m=1}^{\infty} \sum_{n=1}^{\infty} \frac{1}{13,7 f_N A a_s + (f_N - f_{mn})^2}.$$

In connection with the discreteness of the a_s value, the optimum of the M_2 component is found as a result of enumerating 37 specified a_s values in the matrix in each octave band.

To base on the results obtained, we draw up the expressions of the air levels and structured noise for all walls of all components in octave bands and form the objective function (1.1) in at least three similar options for using covering materials.

4. Conclusion

By this means, the reduction of the air noise component, which is more common than the structural one, comes down to two engineering methods:

1. It is rational choice of the sound-absorbing linings on the internal surfaces of the corresponding elements of the cabins.
2. It ensures the required soundproofing of the conforming units of the operator`s cabin.

To select the most rational option for combining the above two methods, the criterion for minimizing the acoustic characteristics inside the operator`s cabin was determined.

References

[1] Yaitskov I A and Chukarin A N 2019 The efficiency of the noise reduction levels of the locomotives and diesel locomotives *Akustika* (Studio D – Akustika s.r.o., České Budějovice) **32** 92-96

[2] Yaitskov I A, Nazaretov A A and Chukarin A N 2020 Analysis of the acoustic characteristics when grinding rails by means of the small mechanization and reducing the noise of the drive systems of the rail grinding machines *Bull. of Rostov State University of Railways* **1(77)** 34-40

[3] Yaitskov I A 2019 Theoretical study of the elements` vibrations of the diesel shunting locomotive cabins *IOP Conf. Ser.: Earth Env. Sci.* **272(2)** 022046. DOI: 10.1088/1755-1315/272/2/022046

[4] Ivanov N I, Kurtsev G M and Elkin Yu I 2005 Noise in the cabins of road-building machines and tractors *Belarusian Railways* **10** 10-15

[5] Dittrich M G 2005 IMAGINE railway noise source model, default source data and measurement protocol, EU-FP6 project IMAGINE report (IMA6TR-050912-TNO01), TNO. p 44

[6] Tholen P 1970 *Schallschutβnahmen beim Einbau von Ver-brennungs motoren in Baumaschinen* (Haus Techn) pp 56-58

[7] Podust S F, Chukarin A N and Boguslavsky I V 2013 *The basics of the vibro-acoustic calculations of domestic electric locomotives* (Rostov-on-Don: Publishing center of DSTU) p 157

[8] Sergienko N P, Kolesnikov I V, Bukharov S N, Pronnikov Y V, Sychev A P and Chukarin A N 2012 *Reduction of noise and vibration of vehicles* (Moscow: Mechanical Engineering) p 297

[9] Pronnikov Y V 2010 Modeling the structural component of noise in the cabs of rolling stock drivers *Bull. RGUPS* **3** 64-68
[10] Ivanov N I 1987 *Fighting noise and vibrations on track and construction machines: 2nd ed* (Moscow: Transport) p 223
[11] Podust S F and Kuklin D A 2013 Modeling of the vibro-acoustic dynamics of a rail on sleepers *Bull. of DSTU* **1-2** 103-111
[12] Kolesnikov I V, Podoust S F, Podoust S S and Chukarin A N 2015 *The ways of noise and vibration reduction when designing, production and operation of the railway rolling stock* (Moscow: VINITI RAS) 216 p
[13] Pronnikov Yu V 2010 On the calculation of the noise component in the cabins of rolling stock drivers created by vibration of the glazing elements *Transport. Safety. Logistics: Mater. of Int. Sci.-Pract. Conf.* (Rostov-on-Don) pp 8-14
[14] Balishanskaya L G, Drozdova L F, Ivanov N I et al. 1992 *Technical acoustics of transport machines: Handbook* (St Petersburg: Polytechnic) p 365
[15] Ivanov N I and Kurzev G 1989 *The Theory and Practice of Construction and Transport Machine Noise Reduction Unikeller Conf.* p 4

IOP Conf. Series: Materials Science and Engineering 1111 (2021) 012043 doi:10.1088/1757-899X/1111/1/012043

Liable Bluetooth tracking technology for enhancement of location-based services

I M Daudov[1], Zh L Gavrilova[2] and V A Kudashkin[3]

[1]Chechen State University, 17a, Dudaev Boulevard, Groznyy, 364015, Russia
[2]Irkutsk National Research Technical University, 83, Lermontov str., Irkutsk, 664074, Russia
[3] Bratsk State University, 40, Makarenko str., Bratsk, 665709, Russia

E-mail: ibr024@mail.ru

Abstract. In recent years, user demand for location-based services has become increasingly urgent. Currently, the main internal positioning technologies include UWB (Ultra-Wide Band), RFID (Radio Frequency Identification), ultrasonic and inertial navigation, optical positioning, Wi-Fi, etc. This article briefly discusses the use of Bluetooth beacons for location, indoor navigation and related algorithms.

1. Introduction

The local positioning and person tracking is an emerging field that has its application in industry, medicine, public services. The great number of positioning are practical and considered for the purpose [1], however wireless communication technologies is the most frequent option [2].

Selection of wireless positioning technology of the transmitting signal in the room includes a network, a mobile terminal and a positioning mode. Wireless positioning in the room is mainly carried out using sensors of a network system. Mobile terminals mainly include smartphones, smart terminals and laptops. The method of posing depends on the current conditions in the room.

Bluetooth technology is widely used in the Internet of Things (IoT), which is the main device when used in low power consumption (LE) mode for beacons, water-free sensors, asset tracking systems, remote controls, performance monitors and alarm systems.

This article will discuss the application of positioning technology using Bluetooth beacons and the implementation of appropriate navigation algorithms.

Beacons with Bluetooth Low Energy (BLE) technology when approaching a client with Apple iPhone (iOS) or Android, display a standard notification on the smartphone or launch a mobile application.

2. Scope of Bluetooth Beacons

Any Bluetooth mobile phone can be used to detect and view BLE devices. One of the most useful beacon services is that they can be used for advertising wherever there are people: in shopping centers, shops, parks and even at events. To use it, there must be a beacon-compatible device [2].

The use of wireless technology to get closer is not news, but with the introduction of low-power Bluetooth features in 2010, beacons are now deployed on a wide scale. This is an important and significant technology for the IoT [3].

Bluetooth transmitted information can be environmental data (temperature, atmospheric-spherical pressure, humidity, etc.), microlocation data (asset tracking, retail, etc.) or orientation data (acceleration, rotation, etc.). These devices are used to transmit data using Bluetooth Low Energy [4] signals.

3. Bluetooth - beacons of different manufacturers

Unlike GPS, beacons can be used to accurately locate the room. There were many applications, including internal navigation, location-based marketing, location-based customer service, customer search, and personalized help. Since the standard Bluetooth LE is used in May, it is well supported on both Android and iOS.

Bluetooth beacons are used to transmit data to neighboring devices. This technology allows devices to perform actions when the device is close to the beacon. Google Eddystone and Apple iBeacon are Bluetooth beacons.

Currently, there are three main market standards for beacons: Apple's iBeacon was introduced in 2013, Network Radius's AltBeacon standard in 2014, and Google's Eddystone appeared in 2015 [2].

Standard iBeacon. Various hardware implementations are iBeacons available on the market in the range of $5 to $30. All these iBeacons have different default settings for iBeacon advertising and transmission frequency. One of such well-known software implementations is available on Android.

Figure 1. IBeacon

The Eddystone standard is capable of supporting four types of packages: Eddystone-URL, Eddystone-UID, Eddystone-EID, Eddystone-TLM. The contents of the URL package are of pe-belt length and apply unique compression schemes to reduce the size of the URL to a limit of 17 bytes. The UID packet has a unique 16-byte beacon ID with a 10-byte namespace. The EID Pa-Ket has a short-lived identifier for May-Ks requiring a higher level of security. The TLM package broadcasts telemetry data about the SA beacon (battery level, time since inclusion, amount of advertising, etc.).

Using the Eddystone Standard, Google developed the Google Nearby service, which allowed Android users to receive beacon notifications without installing any third-party applications. "Google nearby" was installed on Android in 2018.

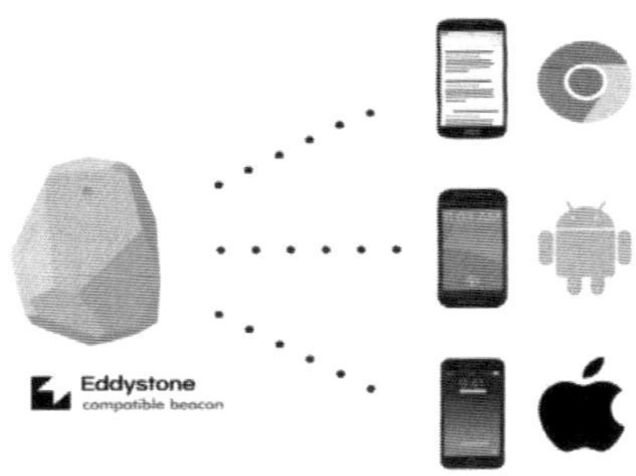

Figure 2. Eddystone

4. Energy efficiency

To implement the system, beacons (or devices that simulate their operation) are needed directly. By Bluetooth Low Energy Beacon we will mean mini-turf battery devices based on BLE, for the transfer of a small amount of static or dynamic information. Such devices are often designed for continuous operation for several years, which is provided by BLE technology, which implies low power consumption, achieves a reduction in data transmission time and plunges the device into sleep mode between packet transmission [7].

Perry Lee, in his book The Architecture of the Internet, argues that the effect of frequent advertising signs affects the life of the lighthouse battery. As a rule, batteries in beacons are lithium-ion cells. CR2032 Cites as an argument the analysis of Aislelabs research (under the international Creative Commons Attribution 4.0 license), in the report "Hitchhiking Guide to Equipment iBeacon: a comprehensive report of Aislelabs. Web March 14, 2015 (https://www.aislelabs.com/reports/beacon-guide/), "the report describes how to analyze the battery life of some ordinary beacons and how to change their re-claw interval (100 ms, 645 ms and 900 ms). They also used different batteries with increased spare power.

The results show that the average life varies between 10 and 15 months, a minimum of 5, a maximum of 36 months depending on the set of micro-schemes, but also, more importantly, due to the advertising interval. Along with the advertising interval, power, without condition, affects the total battery life, but also the number of frames transmitted.

5. Location algorithms

There is a family L_{FS} of collaborative filtering algorithms for analyzing different opinions and views of users and developing personal recommendations - Slope One.

The received signal strength indicator (RSSI) can be translated into between beacons by theoretical or empirical models of radio wave propagation. With openly [6]:

$$L(d) = L_{FS} + 10n\log(\frac{d}{d_0}) \tag{1}$$

where
$d_0 = 1м.$

L_{FS} - free space loss at a distance d_0

n – factor depending on the type of space, the number of obstacles and their material.

The parameter is usually specified by the beacon manufacturers themselves. Parameter n is empirically determined.

To determine coordinates, location algorithms are used - approaches to solving location tasks based on the powers of signals sent by beacons [6].

Algorithms that can be used to determine customer coordinates are:

$$\begin{cases} X_0 = \frac{1}{N}\sum_{j=1}^{N} X_j \\ Y_0 = \frac{1}{N}\sum_{j=1}^{N} Y_j \end{cases}, \tag{2}$$

where N is a number of beacons, X_j, Y_j are coordinates of beacons - Lateration.

The geometric approach is based on calculating the distances between the desired point and at least three more points (Figure 2), with the solution of a system of nonlinear equations. To calculate the coordinates of a custom device, one needs to solve the system of equations:

$$r_j = \sqrt{(X_j - X_0)^2 + (Y_j - Y_0)^2} \quad (3)$$

where r_j - is the distance from the client device to the beacons.

To find the distances, a radio wave propagation model is used, which requires the calibration of parameters that depend on the characteristics of the environment:

$$PL(d) = P_t - P(d) = PL(d_0 0 + n10tg\frac{d}{d_0}, \quad (4)$$

where is the d-distance to the client device, PL (d) is the loss of signal power at distance d, P_t is the transmitter power, P (d) is the signal power at the receiver at the distance d, d_0 is the distance of 1 meter, N is the propagation factor of the signal in the medium.

The first two algorithms are simple to implement and require only knowledge of the location of beacons. However, their disadvantage was low performance, since signal power was not taken into account.

6. Kalman filter. Process and Evaluation

The Kalman filter uses a dynamic model (e.g., a physical law of motion), known control actions, and a plurality of sequential measurements to generate an optimal state estimate. The filter removes measurement noises (random bursts) and produces the result as taking into account the predicted results based on past measurements. The algorithm consists of two repeating phases: prediction and correction. On the first, the prediction of the state at the next moment of time is calculated (taking into account the inaccuracy of their measurement). On the second, the new information from the sensor corrects the predicted value (also taking into account the inaccuracy and noisiness of this information) [7]. The Kalman filter considers the problem of estimating the cost of a discrete time process, which can be characterized by a linear difference equation:

$$x_k = Ax_{k-1} + Bu_{k-1} + W_{k-1} \quad (5)$$

The evaluation also requires the measurement z ϵ Rm as follows:

$$z_x = Hx_k + v_k \quad (6)$$

The random value wk represents the noise of the process, the error between the approach and the real process. Another random variable, vk, represents the noise of the rays, the error between the measurements Hxk and the real value.

Both variables are considered independent of each other, Gaussian and white:

$$p(w) \approx N(O,Q)$$
$$p(v) \approx N(O,R) \quad (7)$$

In this general introduction, the covariance of processes of noise of Q and covariance, and covariance of noise of measurements of R are considered as constant on iterations. However, they can change at each step (each iteration).

Matrix A, of dimension n × n, is a state transition model that relates a state.

In the previous time step k-1 with the state in the current step k. The matrix of B, dimension of n×1, is the operating model which connects the additional operating u entrance ϵ u ϵ R1 with a state. The m × n H matrix is a model of observation that connects co-standing with the measurement zk.

Two definitions are now entered.

The first one:

$-\hat{x}_k^- \in R^n$ is a priori assessment of the state at step k, given the knowledge of the process prior to step k.

The second one is:

$-\hat{x}_k \in R^n$ and relates to estimating a posterior state in step k given by measurement zk.

A priori and posteriori evaluation errors:

$$e_k^- \equiv x_k - \hat{x}_k^-$$

$$e_k \equiv x_k - \hat{x}_k$$

With these errors, the covariance of a priori estimation error and the covariance of posterior evaluation errors can be calculated accordingly as:

$$P_k^- = E\left[e_k^- e_k^{-T}\right]$$

$$P_k = E\left[e_k e_k^T\right]$$

When you output equations for a Kalman filter.

The first step is to find equations that accrue calculating a posterior estimate $\hat{\chi}_k$ as a linear combination of a priori score Z_k and the weighted difference between the actual measure and the prediction $H\hat{x}_k^-$ of the measure calculated using a priori estimate of the state.

$$\hat{X}_k = \hat{X}_k^- + K(z_k - H\hat{x}_k^-) \quad (8)$$

The difference $z_k - H\hat{x}_k^-$ is called innovation and measures the mismatch between the predicted measurement and observed measurement. If the new-maintenance is zero, this means that they are in full agreement, so the predicted measurement is ideal due to a priori assessment of the state, is also perfect.

Matrix K, of dimension n * m, is a Kalman gain that minimizes covariance of aposteriori error R_k. The general expression K in the current time step k;

$$k_k = p_k^- h^t \left(hp_k^- h^t + r\right)^{-1} \quad (9)$$

Considering this expression, some information about the behavior of the Kalman filter can be extracted [8]. The Kalman filter evaluates the process using feedback control. First, the filter evaluates the state of the process at a certain time and then receives feedback in the form of noise measurements. Given this, the equations can be divided into two groups: correction equations and prediction equations. The first of them determine the next time state estimates and covariance errors to obtain a priori estimates for the next time step.

7. Conclusion

The use of iBeacon beacons and Bluetooth Low Energy technology allows easy orientation with almost any mobile device in large buildings and rooms. The development of other uses of these devices was also promising, in particular, to facilitate the orientation of people with disabilities (blind), providing stores with new types of advertising to attract customers and others.

Therefore, there is a need to develop algorithms for processing and filtering incoming data from each beacon. The application of basic space position determination algorithms (triangulation and trilateration) to solve the problem of accurate positioning in a building using beacons is also relevant. Based on this study, it is possible to make a conclusion that today systems that implement positioning functions in the premises are still developing. Manufacturers of such systems had successfully implemented their mobile application platforms. The most popular indoor positioning technology today is the 802.11 radio signal printing method.

References

[1] Bragin D, Kostyuchenko E, Faerman V, Kobzev A, Sidorov I (2021) Comparison of technologies of local patient positioning *International Journal of Advanced Research and Technology* **12(1)** 362-386
[2] Jian Zhang 2016 *Research and implementation of indoor location technology based on iBeacon* (Chengdu: Chengdu university of technology)
[3] Li P 2020 *Architecture of the Internet of Things* (Moscow: DMK Press)
[4] Rui Zhang 2019 *Research on indoor location algorithm improvement based on iBeacon* (Chengdu: university of electronic science and technology)
[5] Bluetooth Low Energy. Retrieved from: http://www.compel.ru/lib/ne/2015/11/3-mayachki-bluetooth-low-energy.
[6] Jian Dong Li and Wang Gen Wan 2020 Research on optimal path planning algorithm based on crowd density *Electronic measurement technology* **43(2)** 38-42
[7] Miniakhmetov R M, Rogov A A and Tsymbler M L 2013 Overview of local positioning algorithms for mobile devices *Bull. of the South Ural State University. Series: Computational Mathematics and Computer Science* 2(2)
[8] 2014 *Navigation in rooms with iBeacon and INS*. Retrieved from: m.habrahabr.ru/post/245325

Influence of plasma arc power on the melting of the envelope layer containing NiCrBSi alloy and the surface coating quality during plasma heating

Van Chie Nguyen[1], A E Balanovskiy[1], S K Kargapol'tsev[2], V E Gozbenko[2,3], Yu I Karlina[1], M G Shtayger[4] and A I Karlina[1]

[1] Irkutsk National Research Technical University, 83, Lermontov Street, Irkutsk, 664074, Russia
[2] Irkutsk State Transport University, 15, Chernyshevsky St., Irkutsk, 664074, Russia
[3] Angarsk State Technical University, 60, Chaykovsky Street, Angarsk, 665835, Russia
[4] MC Mechel Steel, 1, Krasnoarmeysky Street, Moscow, 125167, Russia

E-mail: vgozbenko@irgups.ru

Abstract. The experiments are carried out to obtain coatings by plasma heating of an envelope containing a NiCrBSi alloy with different thicknesses (0.25; 0.5; 1.00 mm) on exposure to different powers (100; 140; 160 A). The resulting coatings have different depths of the hardened layer. With a thin preliminary envelope layer (0.25 and 0.5 mm), defects easily appear under the influence of high power. With a thicker envelope layer (1 mm), the ability of the alloy to diffuse into steel is limited. The low power of the plasma arc of 100 A limits the diffusion and transformation of the molten pool, sufficient for saturation with alloying elements of a thick layer of the envelope (0.5 and 1 mm).

1. Introduction

It is known that under operating conditions carbon steel is faced with factors affecting its quality, such as corrosion, wear, erosion and other mechanical influences. Among the methods for improving the quality of metals, alloying of the surface of metals has contributed to improving the operability and service life of metallic materials. Nickel and chromium based coatings are widely used in various industrial sectors [1]. Many coating methods are popular, for example, chemical, electrochemical deposition and the use of highly concentrated sources. Electrochemical and electrochemical deposition coatings have a more uniform coating or can be easily covered with multiple coats with little impact on the inner layer and are also highly corrosion resistant. However, they have thin thicknesses (about hundredths of a micrometer), even in these methods, water waste is harmful to the environment. Due to stringent requirements for surface preparation before processing, to obtain coatings requires many stages and is time-consuming, with relatively low hardness [2-4].

The combination of chromium and nickel has been repeated many times in a number of studies [5-8]. However, to obtain a protective coating that simultaneously increases the hardness and corrosion resistance, the use of nickel-based alloys (NiCrBSi, NiCrB, NiCr) and the use of a highly concentrated source are considered suitable, since the composition contains not only corrosion-resistant elements, such as Ni, Cr, Si, but also elements that increase the hardening of Ni, Cr, Si, B [4-9]. To overcome some of the disadvantages when using popular highly concentrated sources, such as the cost of equipment in laser overlay weldings [10, 11]; the presence of porosity from a few percent, relatively low adhesion, low productivity with thermal spraying [10-12], electric arc surfacing (TIG) under protective gas is used in many studies [12-14]. In addition, the use of a preliminary envelope layer, as in traditional methods of chemical thermal treatment (CTT) [13-16], recommending a thickness of no more than a millimeter can help increase the efficiency of using molten materials.

We have successfully obtained high-quality cemented coatings by the TIG method with an envelope layer thickness less than a millimetre [16, 17]. In studies [18, 19], the authors used polyvinyl alcohol as a binder forming the applied coating with a thickness of less than 1 millimetre. Plasma treatment with various modification mixtures containing the PR-N80Kh13S2R alloy [19] with thin layers of envelopes showed that a better coating quality was formed than with thick ones. In other works [16, 17, 20] it is shown that the use of silicate glue or water glass to prepare thin layers of the envelope brought better quality coatings. The presence of silicon in the glue increases the self-fluxing ability of the envelope; the silicate glue prevents the alloy from contrasting with air [16, 17, 20]. Many studies using the TIG method consider a wide range of plasma arc power, as for example when working with a current of 60 – 190 A [16, 17].

The purpose of this work is to evaluate the quality of the coating of the NiCrBSi system during plasma heating of mixtures of various envelope thicknesses, including powder alloy PR-N80Kh13S2R and silicate glue with plasma arc current powers (100, 140, 160 A).

2. Materials and methods of study

Plasma heating was carried out after applying the paste on the surface of Steel 3 (St3) and drying the sample. The chemical composition of St3 consists of C (0.14-0.22%), Si (0.05-0.17%), Mn (0.4-0.65%), Ni, Cu, Cr (≤ 0.3%), As (≤ 0.08%), S, P (≤ 0.05%). Powders are used to prepare the paste: alloy PR-H80H13C2R with granules of 40 – 100 microns; glue (silicate); the chemical composition of the PR-H80H13C2R alloy is shown in table 1:

Table 1. The chemical composition of the PR-H80H13C2R alloy

C (%)	Si (%)	Cr (%)	B (%)	Ni (%)	Fe (%)	HRC
0.2-0.4	2.0-2.8	12-14	1.2-1.8	The other	≤ 3	25-35

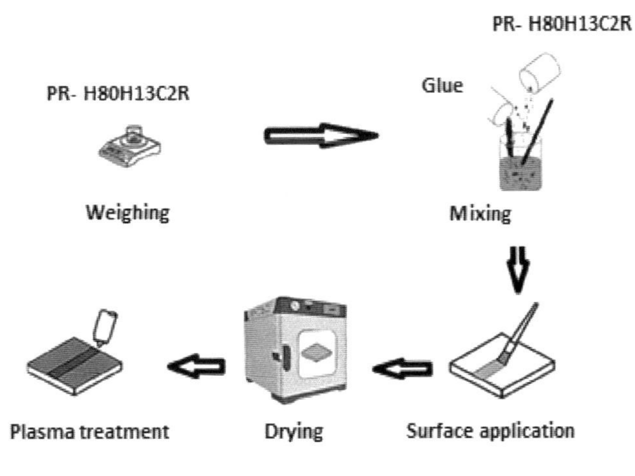

Figure 1. Paste layer manufacturing and plasma treatment scheme.

The order of applying the coating of the experiment is presented in Figure 1. The paste mixture of powder alloy and glue was applied on the surface to obtain a relative thickness of the envelope of 0.25 – 0.50 – 1.00 mm after drying. Plasma modes were: current of 100, 140, 160 A; argon consumption was 10 l/min; the gap between the tungsten electrode and the part is 5 mm; the speed of sample movement is 2.7 mm/s.

The samples after plasma treatment were prepared for metallographic research and microhardness measurement. Before studying the microstructure, samples were etched in a 2% solution of nitric acid in ethyl alcohol. The microstructure was observed with the electro-microscope MET-2. The microhardness measurement was carried out on the hardness testing device of the NMM-G series.

3. Results and discussion

The sample cross-section microstructure types with the resulting coatings are shown in Table 2. We can see that with a thickness of 0.25 mm, all coatings are saturated with alloying agents at different processing capacities. At the same time microfractures occur at 140, 160 A, and there is none of them at 100 A. With a thickness of 0.5 mm, at 100 A the alloy is not completely melted down with the main metal, but forms various sublayers in the hardened surface layer. Using a higher power, for a coating thickness of 0.5 mm, the alloy is melted down with the base metal, but cracks appear again at 160 A. There are no cracks and pores in the coatings of a series of samples with a 1 mm thickness of the envelope layer, but different zones are formed, even chaotic ones. The depth of the surface modified layer at different thicknesses of the smear is presented in Figure 2.

It is obvious that the use of different thicknesses of the envelope and 3 different capacities leads to a strong change in the depth of the resulting coatings. The thicker the layers of pre-applied envelopes, the stronger the mixing down with the main metal at the modes of one power, because of the higher power there is a stronger dilution of the main steel metal with the same thickness of the envelope. High power provides heat transmission sufficient to dissolve alloy powder and support plate in the molten pool and full saturation. Due to this, solid (martensitic) phases are quickly formed, but defects (cracks) also quickly appear during cooling.

The difference in surface layer cooling speeds between plasma treatment with natural cooling and plasma treatment with additional cooling in water may be the cause of this problem. Thicker preliminary layers, for example, 1 mm, significantly prevent the conversion of part of the main steel into a solid solution, leaving enough volume for the NiCrBSi alloy saturation process. It can be seen that in these coatings the boundaries of molten pool surfacing and support plate are uneven, with unevenly distributed sublayers, there appear small areas of saturated phases inside unsaturated areas and, on the contrary, in the microfractures of the saturated sublayer, filled with flows of unsaturated phases.

It is possible to say that coatings No. 1, 3 are of higher quality, corresponding to the modes of: 100 A and 0.25 mm, 140 A and 0.5 mm thick coating (Figure 3). Coating No. 2, corresponding to a 0.5 mm thick envelope and a power of 100 A, has various sublayers: the upper one is lighter, the lower one is darker. The distribution of the microhardness values of the coating No. 2 showed that a large difference between the upper and lower sublayers is due to the unsaturated part of the alloy containing the austenite-forming element Ni, i.e. the power is not enough to distribute heat into depth from the surface of the sample (Figure 2).

Table 2. Plasma coating cross-section types.

Envelope thickness	Power 100 A	 140 A	 160 A
0.25 mm	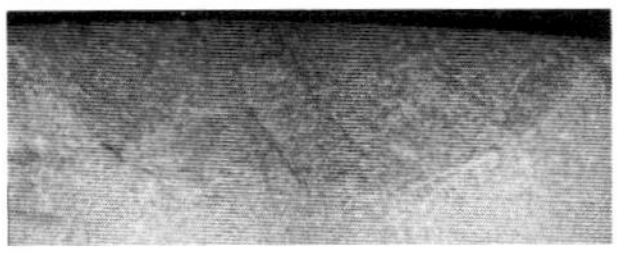 Depth is 1250 μm	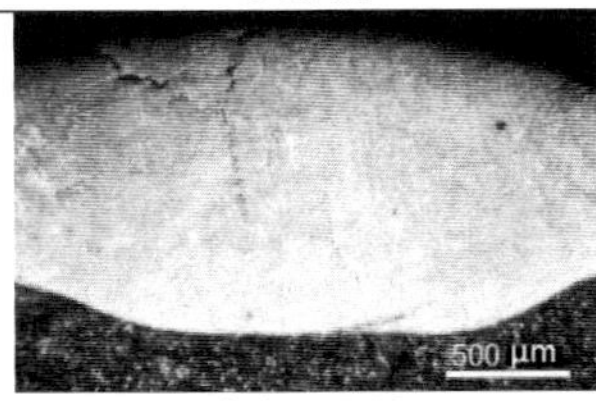 Depth is 1240 μm	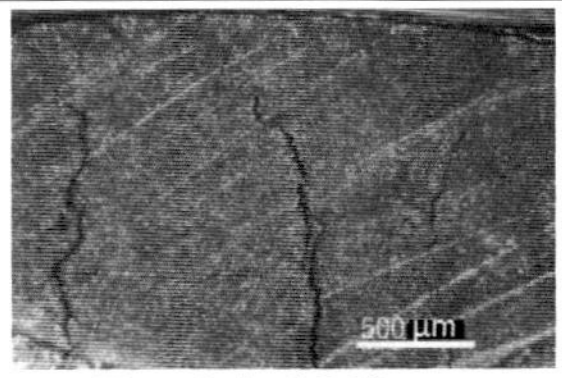 Depth is 1560 μm
0.25 mm	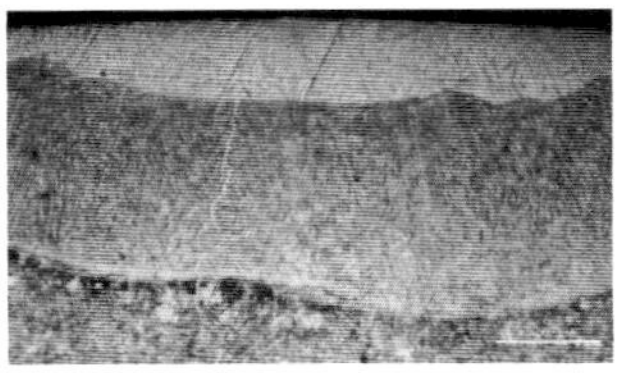 Depth is 1150 μm	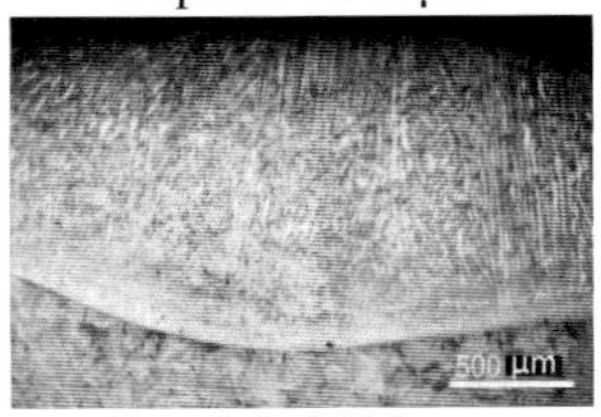 Depth is 1300 μm	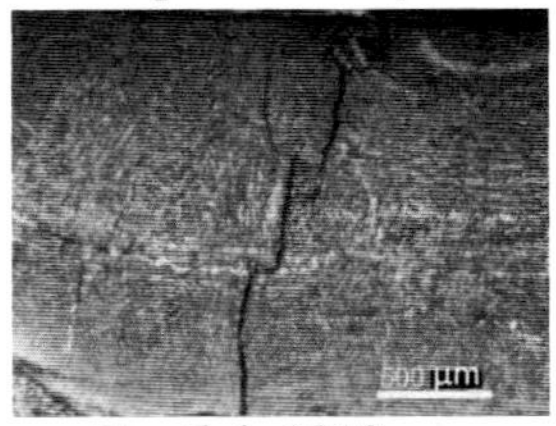 Depth is 1810 μm
1 mm	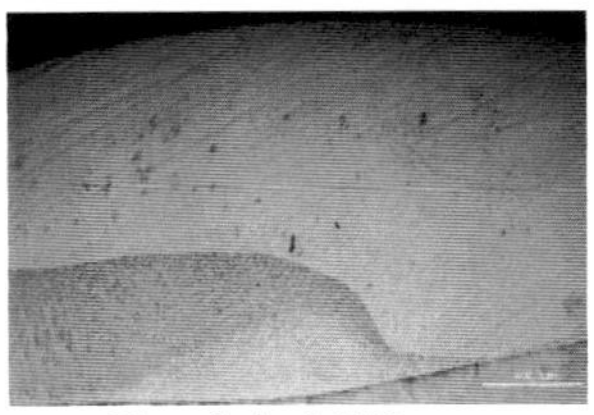 Depth is 1470 μm	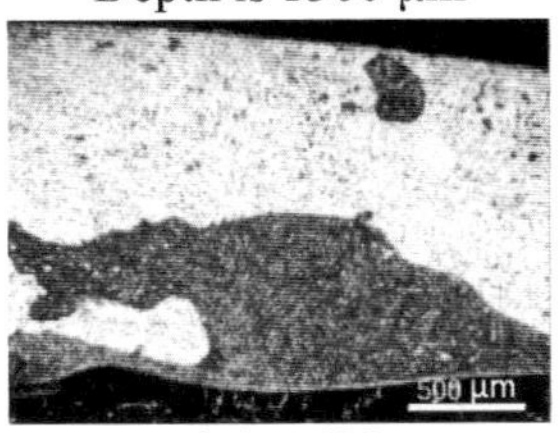 Depth is 1630 μm	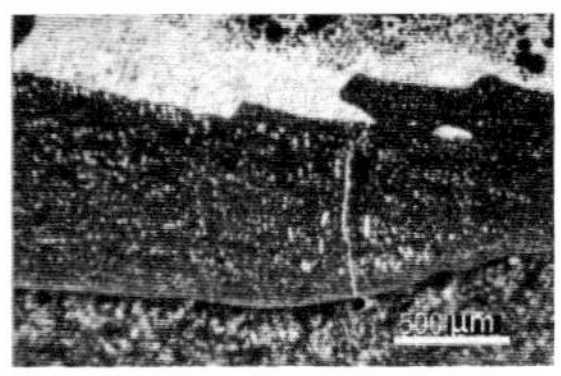 Depth is 2020 μm

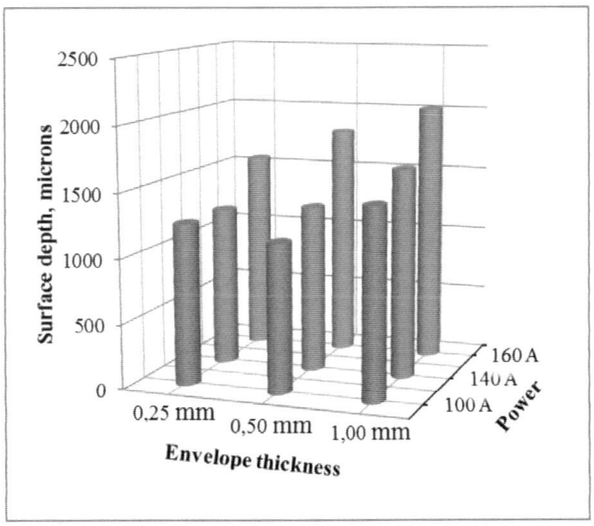

Figure 2. Dependence of the depth of the alloyed layer on the power and thickness of the envelope.

IOP Conf. Series: Materials Science and Engineering 1111 (2021) 012044 doi:10.1088/1757-899X/1111/1/012044

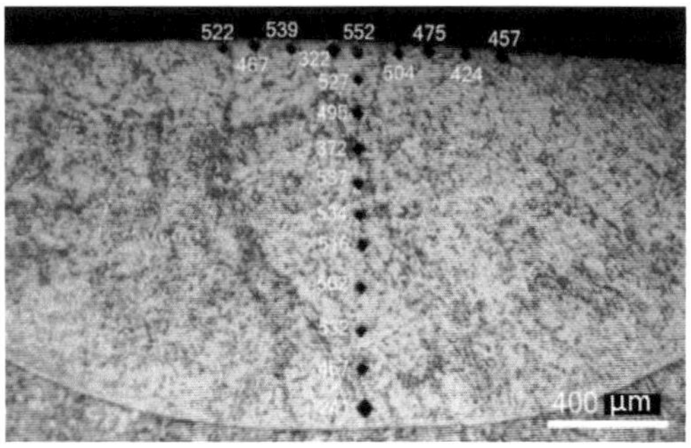

No. 1

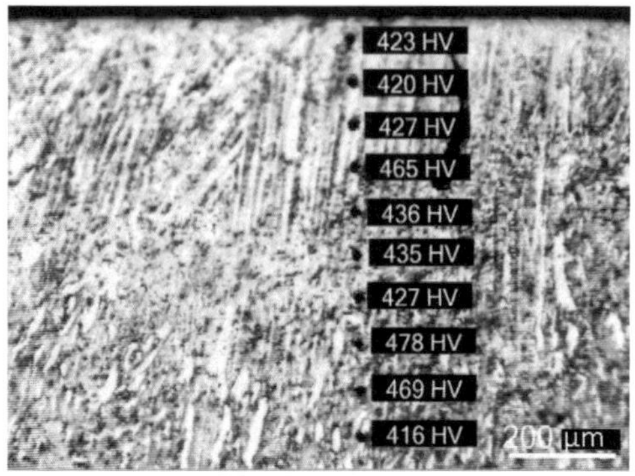

No. 3 - Above

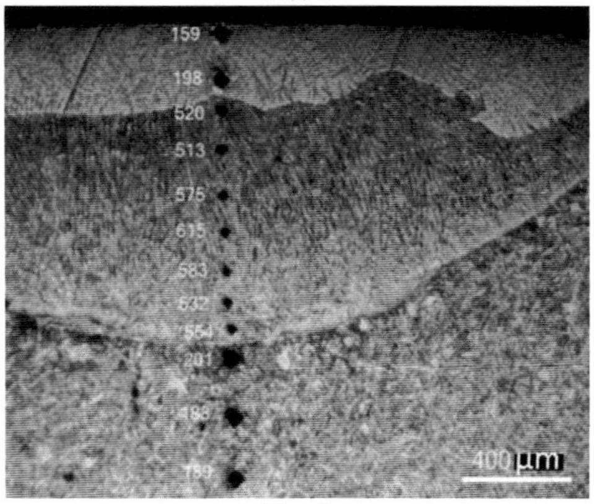

No. 2

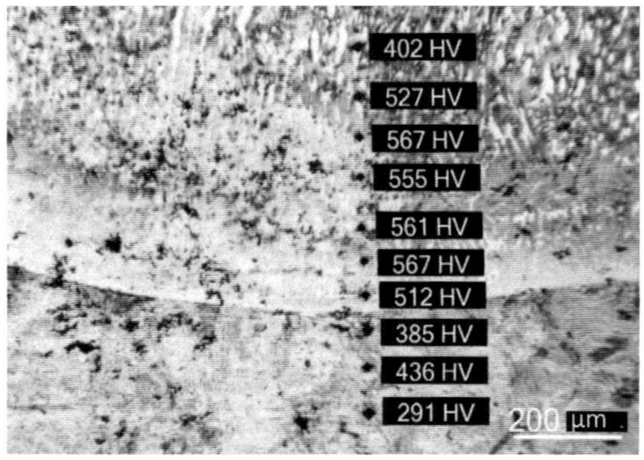

No. 3 - Below

No. 1– 0.25 mm envelope, 100 A; No. 2 – 0.50 mm envelope, 100 A; No. 3 – 0.50 mm envelope, 140 A

Figure 3. Distribution of the microhardness values of the cross-section of samples along the depth from the surface of the coatings.

Figure 4 shows the distribution of microhardness values along the depth from the surface of coatings No. 1, 3. It can be seen from the graph that sample No. 1 has a large difference in hardness between the main coating and the heat affected zone (HAZ), and the microhardness values of the base metal do not differ much from the HAZ zone. In the sample No. 3, the initial values of hardness vary in depth within the range of 400–480 HV, and further in depth, closer to the boundary of the overlay welding bath, it significantly increases within the range of 530–570 HV. In particular, when passing to the HAZ zone, the hardness decreases slowly, even further areas of the base metal are also higher than that of coating No. 1. The difference in the distribution of hardness shows that the parameters of the power of the plasma arc significantly affect the coating quality in terms of the distribution of hardness and phase structures along the layer depth. To clarify the process of coating formation, Figure 5 shows the mechanism.

In the case of only the melting of the NiCrBSi alloy, it is possible to descrine the coating formation mechanism in several steps. In the first step (A), a dried layer of envelope containing a mixture of NiCrBSi alloy and glue (liquid glass) is applied to the surface of the metal. In the second step (B), a high heat flow from the plasma arc partially heats the layer of envelope, and its upper part is melted down as a liquid (partially solid alloy solution is formed during crystallization). In the 3rd step (C), the whole layer of the envelope is melted down in the form of liquid. In the 4th step (E), part of the steel support plate, close to the edges of the envelope, interacts with the liquid. At the same time, the intersecting part of the solid alloy solution starts passing into a steel solution. In the last step (G), parts of the solid alloy solution and steel are mixed together. After cooling, different zones are formed, presented in the type of cross-section (table 2). In general, the coating has 2 sublayers, respectively with the types of zones: the top is the molten alloy; the middle is the mixed saturated sublayer of alloy and steel. The lower part of the surface layer forms a heat affected zone.

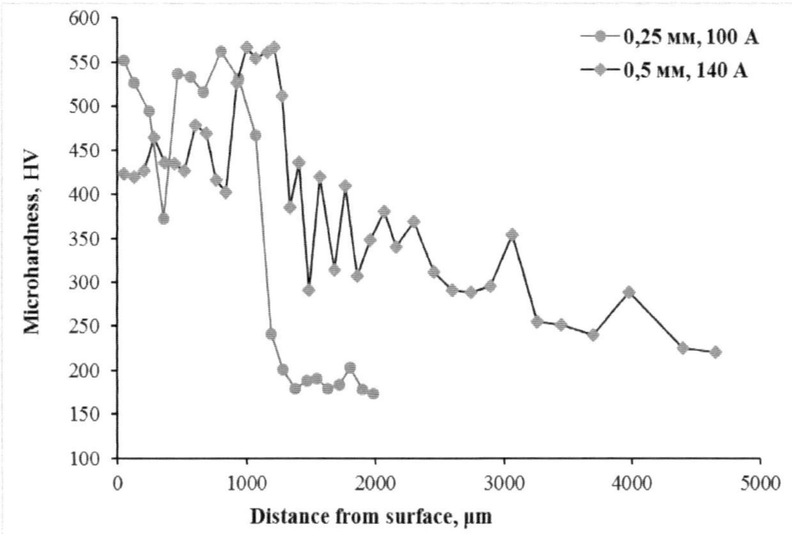

Figure 4. Distribution of the microhardness values along the depth from the surface of the coatings.

4. Conclusion

Coatings of different thickness of the NiCrBSi alloy envelope layer at different plasma arc powers show that:

- With a thin preliminary layer of envelope (0.25 and 0.5 mm) under the influence of a high-power plasma arc, defects occur easily, and with a thicker layer (1 mm) the ability of alloy to be diffused in steel is limited because of the small dissolution of the main metal, which is not enough for the full saturation of the alloy. The small power of the plasma arc by the current intensity of 100 A limits the diffusion and in the small volume of the molten pool, which is not enough for the full melting of the thick layer of envelope (0.5 and 1 mm).
- Microstructure research and micro-solids measurement have shown that coating No.1 corresponding to 0.25 mm thickness, 100 A and No. 3 (0.5 mm, 140 A) are better than the rest of them.

5. Acknowledgments

This paper has been prepared using the equipment kindly provided by the Joint Instrumentation Centre "Baikal Center of Nanotechnologies".

References

[1] Amanov A 2019 Surface engineering-controlled tribological behavior and adhesion strength of Ni-Cr coating sprayed onto carburized AISI 4340 steel substrate *Surface and Coatings Technology* **370** 144-156
[2] Zhina Razaghi, Milad Rezaei and Seyed Hadi Tabaian 2020 Electrochemical noise and impedance study on the corrosion of electroplated Ni-Cr coatings in HBF4 aqueous solution *J. of Electroanalytical Chemistry* **859** 113838
[3] Yousefi E, Irannejad A and Sharafi Sh 2019 Electrodeposition and characterization of nanocrystalline Fe–Ni–Cr alloy coatings synthesized via pulse current method *Transactions of*

Nonferrous Metals Society of China **12-29** 2591-2603

[4] Morks M F and Berndt C C 2010 Corrosion and oxidation properties of NiCr coatings sprayed in presence of gas shroud system *Applied Surface Science* **13-256** 4322-4327

[5] Liang-Yu Chen, Haiyang Wang, Cuihua Zhao and Lai-Chang Zhang 2019 Automatic remelting and enhanced mechanical performance of a plasma sprayed NiCrBSi coating *Surface and Coatings Technology* **369** 31-43

[6] Xu Wang, Dingyong He and Xingye Guo 2019 Hot corrosion behavior of wire-arc sprayed NiCrB coatings *Surface and Coatings Technology* **367** 173-178

[7] Ermolaev A S, Ivanov A M and Vasilenko S A 2013 Laser technologies and processes in the manufacture and repair of gas turbine engine parts *Bull. of PNRPU. Aerospace engineering* **35**

[8] Grigoryants A G et al 2008 *Technological procedures of laser processing* (Moscow) p 650

[9] Hasun A and Morigaki O 1985 *Surfacing and spraying* (Moscow: Mashinostroenie)

[10] Borisov Yu S, Kharlamov Yu A and Sidorenko S L 1987 *Thermal gas coatings from powder materials* (Kiev: Naukova Dumka)

[11] Gurd L M 1985 *Fundamentals of spraying technology* (Moscow)

[12] D Tijo, Manoj Masanta and Alok Kumar Das 2018 In-situ TiC-TiB2 coating on Ti-6Al-4V alloy by tungsten inert gas (TIG) cladding method: Part-I. Microstructure evolution *Surface and Coatings Technology* **34425** 541-552

[13] Bodala Raja Sekhar, Rajesh Kumar Nayak, Sumeet Ranjan Rout and Manoj Masanta 2020 Wear characteristic of TiC coated AISI 1020 mild steel fabricated by TIG cladding method *Materials Today: Proceeding*

[14] Ritesh Prasad, Dipak Tanaji Waghmare, Kamlesh Kumar and Manoj Masanta 2020 Effect of overlapping condition on large area NiTi layer deposited on Ti-6Al-4V alloy by TIG cladding technique *Surface and Coatings Technology* **38515**

[15] Cheng F T, Lo K H and Man H C 2003 NiTi cladding on stainless steel by TIG surfacing process Part I. Cavitation erosion behavior *Surf. COAT: Technol.* **172** 308-315

[16] Balanovskii A E, Grechneva M V, Vu V H and Zhuravlev D A 2017 New plasma carburizing method *IOP Conf. Ser.: Earth Env.* **87** 092003

[17] Balanovskii A E and Vu V H 2017 Plasma surface carburizing with graphite paste *Letters on materials* **7(2)** 175-179

[18] Nguyen Van Trieu 2020 Increasing the hardness of the surface layer of low-carbon steel due to plasma treatment of the modifying coating *Bull. of the Irkutsk State Technical University* **24(1)** 52-63

[19] Miguel J M, Guilemany J M and Vizcaino S 2003 Tribological study of NiCrBSi coating obtained by different processes *Tribology Int.* **36** 181-187

[20] Niranatlumpong P and Koiprasert H 2011 Phase transformation of NiCrBSi – WC and NiBSi - WC arc sprayed coatings *Surface and Coatings Technology* **206** 440-445

IOP Conf. Series: Materials Science and Engineering 1111 (2021) 012045 doi:10.1088/1757-899X/1111/1/012045

Modelling the movement of log trucks with a hydro-mechanical transmission

E A Pitukhin and V I Skrypnik

Petrozavodsk State University, 33, Lenina ave., Petrozavodsk, 185910, Russia

E-mail: Eugene@petrsu.ru

Abstract. The paper proposes a mathematical model of the movement of log trucks with hydro-mechanical transmission, which represents the analytical functional dependence of the speed of the log truck on the covered path. On the basis of real data it is shown that the traction characteristic of hydro-mechanical transmission is best approximated by hyperbolic regression. In this case, the solution of heterogeneous nonlinear differential equation of motion can be obtained by the approximate analytical method through decomposition of a logarithmic component in series. The calculation example showed that the obtained approximate analytical solution deviates from the numerical solution by no more than 0.002%. The obtained model allows one to determine the analytical dependencies for speed and time, which can be used to calculate the modes of movement of log trucks with the hydro-mechanical transmission.

1. Introduction

Logging enterprises in transport operations use log trucks equipped with various types of transmission: mechanical, electrical, hydromechanical, hydrostatic, etc. At the same time, as practice has shown, log trucks are operated not only on public roads and forestry roads of permanent validity, but also on temporary roads, as well as in difficult operating conditions, where the use of log trucks with the hydromechanical transmission is preferable. [1, 2].

There are three hydraulic hybrid drive train architectures, namely: parallel, series and power-split. The parallel architecture assists the conventional drive train by storing and releasing excessive kinetic energy as needed. The series architecture is a hydrostatic transmission with high- and low-pressure hydraulic accumulators placed between two pump/motors as a means to store energy. The power-split architecture splits the input power into two power paths, allowing for both full engine control and efficient power transfer. The power-split architecture is also known as the Hydro-mechanical Transmission (HMT) when hydraulic components are used. Hydro-mechanical transmission are widely used in off-road and agricultural vehicles due to their reliability, smoothness, and high-efficiency power transfer [3–5].

The problems of optimization and control of a hydro-mechanical power split transmission are discussed in [5–9]. In paper [10] summarizes developments, key techniques and trends, analyses problems and difficulties in design and application of hydro-mechanical power split transmission and discusses the methods for solving these problems. At the same time, despite the existence of a lot of work on research of hydro-mechanical transmissions [11–17], the issue of truck traffic modeling with this transmission remains relevant. To resolve this issue, you need information about the dynamic characteristics and modes of movement of log trucks.

2. Materials and methods

The mathematical model of a log truck movement can be represented as a nonlinear heterogeneous differential equation (1) [18]:

$$\frac{G\delta}{g} v \frac{dv}{ds} = T(v), \tag{1}$$

where v – speed of the log truck [m/s];
s – distance covered [m];
G – log truck weight [kg];
δ – rotational mass inertia ratio, [];
$g = 9.81$ – free fall acceleration [m/s^2];
$T(v)$ – traction characteristic of a log truck [kg].

In [18] it was shown that for approximation of the traction characteristic of the hydromechanical transmission hyperbolic dependence of the species is best suited:

$$T(v) = A + B/v, \tag{2}$$

where A, B – hyperbolic regression coefficients.

Taking into account the inhomogeneous part (2), the diffrential equation (1) after integration in quadrature has an analytic general solution of the form:

$$s - \frac{1}{2}\frac{G\delta v(s)^2}{gA} + \frac{G\delta v(s)B}{gA^2} - \frac{G\delta B^2 \ln(Av(s) + B)}{gA^3} + \_C1 = 0, \tag{3}$$

where $\_C1$ – constant integration.

Since the Cauchy problem is being solved, to obtain the private solution of equation (1) it is necessary to set the initial conditions:

$$v(s_0) = v_0 \tag{4}$$

where v_0 – initial speed of the log truck;
$s_0 = 0$ – initial value of the path covered - this assumption is adopted to facilitate obtaining a convenient type of decision (1).

The value of the constant $\_C1$ is obtained from equation (3) after substitution of initial conditions (4) in it:

$$\_C1 = \frac{1}{2}\frac{G\,\delta\,({v_0}^2\,A^2 - 2\,v_0\,B\,A + 2\,B^2\ln(A\,v_0 + B))}{g\,A^3}. \tag{5}$$

Now the required private solution of the log truck movement is represented by the system of equations (3) and (5):

$$\begin{cases} s - \frac{1}{2}\frac{G\delta v(s)^2}{gA} + \frac{G\delta v(s)B}{gA^2} - \frac{G\delta B^2 \ln(Av(s) + B)}{gA^3} + \_C1 = 0 \\ \_C1 = \frac{1}{2}\frac{G\,\delta\,({v_0}^2\,A^2 - 2\,v_0\,B\,A + 2\,B^2\ln(A\,v_0 + B))}{g\,A^3} \end{cases}. \tag{6}$$

Unfortunately, the presence of the speed function under the logarithm makes it impossible to obtain an accurate analytical solution to algebraic equation (6).

Due to the need to obtain speed characteristics of the road train in the form of functional dependencies for further research, as an alternative, it is proposed to find an approximate analytical solution. For this, the logarithmic term in (6) is proposed to be expanded in a Taylor series.

Such an expansion is well known for a function of the form $\ln(x + 1)$ in a vicinity of the point $x = a$

$$\ln(x+1) = \ln(a+1) + \frac{x-a}{a+1} - \frac{(x-a)^2}{2\,(a+1)^2} + \frac{(x-a)^3}{3\,(a+1)^3} + \cdots + \frac{(-1)^n (x-a)^{n+1}}{(n+1)\,(a+1)^{n+1}} \pm \cdots. \tag{7}$$

For our case, under the logarithm there is a function from (3) $Av(s) + B$. The point in the vicinity of which it is necessary to expand the logarithm in a series is determined from the initial conditions (5) $A\,v_0 + B$. As a result we get a substitution:

$$\begin{cases} x + 1 = Av(s) + B \\ a = A\,v_0 + B \end{cases}. \tag{8}$$

We use only the first three terms for the series expansion so that the degree of the approximating polynomial is no higher than the second:

$$\ln(x+1)|_{x=a} = \ln(a+1) + \frac{x-a}{a+1} - \frac{(x-a)^2}{2\,(a+1)^2}. \tag{9}$$

Then, using the replacement of variables (8) in the expansion (9), we obtain the final expression for the logarithmic component of the solution:

$$\ln(Av(s)+B) = \ln(Av_0 + B + 1) + \frac{Av(s) - 1 - Av_0}{Av_0 + B + 1} - \frac{1}{2}\frac{(Av(s) - 1 - Av_0)^2}{(Av_0 + B + 1)^2}. \tag{10}$$

Finally, when expression (10) is added to system (6), we obtain a system of algebraic equations, from which it is possible to express the required speed of the log truck in the form of an analytical functional dependence:

$$\begin{cases} s - \frac{1}{2}\frac{G\delta v(s)^2}{gA} + \frac{G\delta v(s)B}{gA^2} - \frac{G\delta B^2 \ln(Av(s)+B)}{gA^3} + \_C1 = 0 \\ \_C1 = \frac{1}{2}\frac{G\,\delta\,({v_0}^2\,A^2 - 2\,v_0\,B\,A + 2\,B^2\,\ln(A\,v_0 + B))}{g\,A^3} \\ \ln(Av(s)+B) = \ln(Av_0 + B + 1) + \frac{Av(s) - 1 - Av_0}{Av_0 + B + 1} - \frac{1}{2}\frac{(Av(s) - 1 - Av_0)^2}{(Av_0 + B + 1)^2} \end{cases}. \tag{11}$$

Substituting the second and third expressions of system (11) into the first expression, and grouping this expression by powers of $v(s)$, we obtain a quadratic equation for $v(s)$ of the form:

$$k_2 v(s)^2 + k_1 v(s) + s + k_0 = 0, \tag{12}$$

where k_2, k_1, k_0 – coefficients at the corresponding powers of $v(s)$, the values of which are determined by the following formulas:

$$k_2 = -\frac{G\delta}{2gA} + \frac{1}{2}\frac{G\delta B^2}{gA(Av_0 + B + 1)^2}, \tag{13}$$

$$k_1 = -\frac{G\delta B^2\left(\frac{A}{Av_0 + B + 1} - \frac{(-1 - Av_0)A}{(Av_0 + B + 1)^2}\right)}{gA^3} + \frac{G\delta B}{gA^2}, \tag{14}$$

$$k_0 = -\frac{G\delta B^2\left(\ln(Av_0 + B + 1) + \frac{-1 - Av_0}{Av_0 + B + 1} - \frac{1}{2}\frac{(-1 - Av_0)^2}{(Av_0 + B + 1)^2}\right)}{gA^3} + \frac{1}{2}\frac{G\delta({v_0}^2 A^2 - 2v_0 BA + 2B^2\ln(Av_0 + B))}{gA^3}. \tag{15}$$

As a result, one of the two solutions of equation (12), which gives a positive value of the speed of movement, has the form:

$$v(s) = -\frac{1}{2}\frac{k_1 + \sqrt{{k_1}^2 - 4\,k_2\,s - 4\,k_2\,k_0}}{k_2}. \tag{16}$$

Thus, analytical expressions (13)–(16) represent a mathematical model of the speed of a log truck with a hydromechanical transmission. It is obvious that from (16) it is also easy to obtain an expression for estimating the time of movement:

$$t(s) = \frac{s}{v(s)}. \tag{17}$$

3. Results and Discussion

We consider an example of calculating the characteristics of a log truck movement according to the obtained model (13)–(16).

Let us consider the ways of approximation of the real traction characteristic of the log truck. In Figure 1 you can see the initial data of the traction force dependence on the speed of the log truck, as well as their approximation curves.

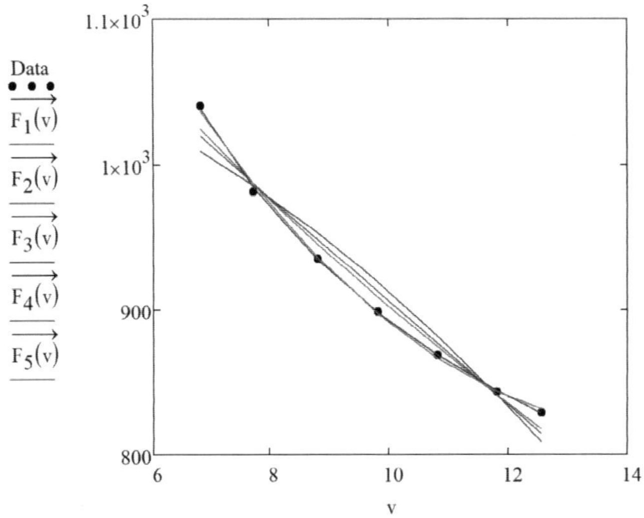

Figure 1. Approximation of traction characteristic of hydromechanical transmission.

For approximation, five types of functions were used, the form of which, together with the results of the approximation quality, is presented in Table 1.

Table 1. Function type and approximation quality indicators

Type of approximation function	Mean Absolute Percentage Error (MAPE), %	Determination coefficient R^2
$F_1(v) = A - B\sqrt{v}$	23.3%	0.981
$F_2(v) = A - Bv^2$	43.6%	0.933
$F_3(v) = A + B/v$	2.0%	0.999
$F_4(v) = A - Bv$	30.5%	0.968
$F_5(v) = A + Bv + Cv^2$	4.7%	0.998

From Table 1 it follows that the best approximation results are obtained using the hyperbolic regression function of the form:

$$F_3(v) = A + B/v, \tag{18}$$

where the corresponding coefficients take the values $A = 580$ and $B = 3119$.

Thus, for model (13)–(16), we set the following calculated parameters:

$$G = 30000\ kg;\ \delta = 1.048;\ g = 9.81\ m/s^2;\ v_0 = 7\ m/s;\ A = 580;\ B = 3119. \quad (19)$$

Taking into account (19), the equation for the speed of the road train (16) takes the form:

$$v(s) = 2.120 + 0.223\sqrt{478.510 + 8.966s}. \quad (20)$$

Figure 2 shows a graph of changes in the speed of the log truck at a distance of 0 to 100 meters. The circles also mark the numerical solution of equation (1). The maximum MAPE between the approximate analytical solution (16) and the numerical solution of equation (1) did not exceed 0.002%. This is a good result for a model that can be used for engineering calculations of log truck driving modes.

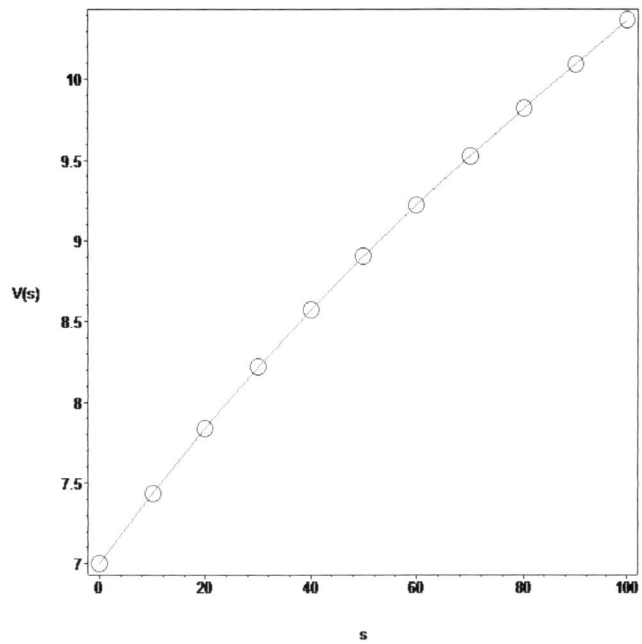

Figure 2. The dependence of the road train speed on the distance covered.

4. Conclusion

As a result of the research, an approximate analytical solution of the differential equation of motion of a log truck with a traction characteristic of a hyperbolic type was obtained. Decomposition in series of logarithmic components of analytical solution allowed one to obtain an approximate analytical solution of differential equations of motion. High accuracy of the obtained solution (mean absolute percentage error of no more than 0.002% from the numerical solution) will allow one to use this solution for scientific research and calculation of motion modes of a log truck depending on changes of model parameters.

References

[1] Girutskiy O I and Tarasik V P 2013 Creation of electronic control systems and diagnosis of hydromechanical transmission of mobile machines: stages, paths and prospects *J. of automobile engineers* **4** 18–23

[2] Nguyen K T 2018 A computer method for calculating the traction-speed characteristics of automotive vehicle equipped with hydromechanical transmission *Universum: Thechnical Science* **3(48)** 19–23

[3] Hayashi T, Saito M, Yoshida Y and Nanri T 2001 Advanced Hydro-Mechanical Transmission with High Durability for Small Utility Vehicles *SAE Technical Paper* 2001-**01**-0876. DOI:

10.4271/2001-010876
[4] Macor A 2011 Optimization of hydro-mechanical power split transmissions *Mechanism and Machine Theory* **46(12)** 1901-1919. doi: 10.1016/j.mechmachtheory.2011.07.007
[5] Ramdan M I and Stelson K A 2016 Optimal design of a power-split hybrid hydraulic bus *J. of Automobile Engineering* **230(12)** 1699-1718. doi: 10.1177/0954407015621817
[6] Li P and Mensing F 2010 Optimization and Control of a Hydro-Mechanical Transmission-based Hybrid Hydraulic Passenger Vehicle *7th Int. Fluid Power Conf.*
[7] Cheong K L, Li P Y and Chase T R 2011 Optimal design of power-split transmissions for hydraulic hybrid passenger vehicles *American control conf.* pp 3295–3300. doi: 10.1109/ACC.2011.5991509
[8] Rossetti A and Macor A 2013 Multi-objective optimization of hydro-mechanical power split transmissions *Mechanism and Machine Theory* **62** 112-128
[9] Ahn S, Choi J, Kim S, Lee J, Choi C and Kim H 2015 Development of an integrated engine-hydro-mechanical transmission control algorithm for a tractor *Advances in Mechanical Engineering*. DOI: https://doi.org/10.1177/1687814015593870
[10] Yu J, Cao Z, Cheng M and Pan R 2019 Hydro-mechanical power split transmissions: Progress evolution and future trends *J. of Automobile Engineering* **233(3)** 727–739. doi: 10.1177/0954407017749734
[11] Guo J and Lai D 1997 The analysis of power split ratio and transmission characteristics of hydro-mechanical transmission *Int. Constr. Chin.* **10** 38–41
[12] Masuzawa K and Ichiryu K 2001 Basic study on input split type hydro-mechanical transmission *Int. Conf. on Fluid Power*
[13] Du J, Yuan S and Wei C 2009 Performance analysis of dual mode hydro-mechanical transmission *Trans. Chin. Soc. Agric. Eng.* **25** 86–90
[14] Xu L, Zhou Z and Peng Q 2012 Drive scheme design and characteristic analysis of multi-range hydro-mechanical CVT *Chin. Mech. Eng.* **23** 125–129
[15] Cronin M 2013 An introduction to hydro-mechanical transmissions *Fluid Power J.*
[16] Jacobson E, Wright J and Kohmäscher T 2011 Hydro-mechanical power split transmissions (HMT) – Superior technology to solve the conflict *Int. exposition for power transmission and technical conf.* pp 125–134
[17] Yong B, Zaimin Z and Shujun Y 2020 Modeling of Power Transition in Full Power Shift of Hydromechanical Transmission *Mathematical Problems in Engineering* **10** 1–14. DOI: https://doi.org/10.1155/2020/5296713
[18] Skrypnik V I, Kuznetsov A V and Pitukhin E A 2019 Derivation of calculated dependencies to determine the speed and time of movement of log trucks based on cars with hydromechanical transmission *Modern High Technologies* **3** 70–76

Use of hydraulic drive of sucker-rod pump 120-6-24 in oil and gas industry as an alternative to import substitution of small-sized pumps

I A Pogrebnaya and S V Mikhaylova

Industrial University of Tyumen, 38, Volgogradskaya str., Tyumen, 625000, Russia

E-mail: ya.irina0607@yandex.ru

Abstract. The purpose of this article is the research aimed at the improvement of the quality of hydraulic equipment, namely, investigation of the use of hydraulic drive of sucker-rod pump 120-6-24 in oil and gas industry as an alternative to import substitution of small-sized pumps. Nowadays, at the fields of Western Siberia, it is necessary to replace outmoded and worn-out equipment with modern equipment of proven efficiency. To a greater extent, the imported equipment is used in oil industry. For Russian industry import substitution is one of the key tasks. This article considers the task of wells operation with sucker rod pumping units with a surface hydraulic drive with an analysis of its performance. As an alternative to import substitution for small-sized pumps, it is recommended to use innovative equipment - a highly efficient hydraulic drive of a sucker rod pumping unit, which provides a general increase in the efficiency of production processes during operation, increasing oil production during well operation in optimal mode. The use of this drive leads to a decrease in the costs of oil and gas production enterprises by reducing the level of oil deficit and prevents equipment failure in oil fields. The method of oil production with a hydraulic drive hydraulic drive of sucker-rod pump 120-6-24 considered in this article is widely applied in oil and gas industry and became a valid alternative to small-sized pumps.

1. Introduction

Oil production using sucker rod pumping units appeared a long time ago, namely a century ago, with the invention of machine tools in the form of pumping chairs. Until recently, the sucker rod pumping units were considered a symbol of oil industry. Nowadays sucker rod pumping units are used less and less. Respectively, global changes in the dynamically developing world require the use of more modern equipment with effective performance. Only the last 20 years a drive have been applied in these pumps, functioning on the basis of a hydraulic drive. However, during the operation of the hydraulic drive, there are ambiguities about the climatic resistance, cost recovery of equipment, repair of hydraulic drive of sucker-rod pump, etc. The hydraulic drive provides pumping of fluid from oil wells according to the usual movement in the plunger of a sucker rod pump (SRP) [1].

The purpose of this article is to consider the prospects of the improvement of the quality of hydraulic equipment by analyzing indicators using import substitution by putting into operation wells of sucker rod pumping units with a surface hydraulic drive.

In this work, we will try to answer the main questions of the alternative to small-size pumps and find out how effective their use is through the example of hydraulic drive of sucker-rod pump.

The issues of the use of hydraulic drive of sucker-rod pump 120-6-24 in oil and gas industry as an alternative to import substitution for small-size pumps was considered by the chief mechanic R.S. Garifullin, Leading Engineers of PJSC TATNEFT L.M. Akhmetzyanov., I. N. Garipov [2].

2. Materials and methods

The Government of the Russian Federation set oil producing companies the task to increase oil production by 2025, for example, the management of PJSC TATNEFT has the task to increase the production of hydrocarbons to 30 million tons by this period. With the current sanctions on the part of foreign countries, it is reckless to hope for such production. Therefore, the import substitution program is paramount at this stage. In this work, we consider the problem of the use of domestic equipment, which is equal to imported equipment in terms of technological characteristics.

In 2011, innovative equipment was developed on the basis of research and development corporation UralNeftService enterprise, a leader in the development of equipment for the production and processing of oil and gas. It was a highly efficient hydraulic drive of a sucker rod pump (hydraulic drive of sucker-rod pump "GERON"), competing with conventional pumping units. The studied hydraulic drive is equal to foreign analogues in terms of reliability and operation. During the development of high-tech equipment, the best experience in the extraction of petroleum products on a global scale was used, progressive views and opinions of leading oil producers were studied. The company engaged in testing the hydraulic drive of the sucker rod pump (HDSRP "GERON") was "Rosneft" oil company, providing its fields. PJSC "Orenburgneft" was chosen as an experimental site, which tested the first hydraulic drives at its wells.

The improvement of hydraulic drives continues to this day. The producer - research and development corporation (RDC) UralNeftService LLC, located in Yekaterinburg, is developing a new generation of hydraulic drives in accordance with the needs and expextations of the market.

At this stage, the developers of alternative equipment for oil workers are engaged in the development, production and implementation of new samples of a number of different models of sucker rod pumping units that meet the needs of a new generation of customers. The priority feature of the hydraulic drive is quite simple assembly and disassembly, much lower power consumption, good automation and intellectualization of oil production process in comparison with a pumping unit. The use of a hydraulic drive allows quick assess to the parameters of well operation, remotel control of technological process in real time with minimal participation of maintenance personnel and equipment, and also increase of the average daily volume of oil production.

According to Alexey Molotkov, General Director of RDC UralNeftService the use of the new equipment turned out to be much easier and cheaper in operation compared to traditional pumping units, which was confirmed by the tests of hydraulic drive of the sucker rod pump. For the assembly of the lightweight and compact structure of "Geron" it is not necessary to use a special foundation, since the support is mounted on road plates or at the wellhead. This can significantly reduce the overall cost of construction and ensure convenience and safety of maintenance.

The hydraulic drive provides pumping of fluid from oil wells according to the usual movement in the plunger of a sucker rod pump (SRP) [3, 4]. Throughout the world, the practice of the use of a sucker rod pump is widespread due to optimal parameters for pumping out well fluid. This increases the filling factor of a pump, especially with high viscosity oil and high gas content in the recovered liquid. Due to the high efficiency of the hydraulic drive, the dynamics of energy consumption is reduced in comparison with electric centrifugal pumps with a flow rate of up to 50 m^3 / day.

In addition, the introduction of hydraulic drives instead of conventional pumping units allows:

- reducing the time of assembly, disassembly of the drive;
- reducing the metal consumption of the mechanism in comparison with traditional drives of pumping units;
- reducing the cost of transporting the drive to oil production sites;
- simplifying the base for the assembly of hydraulic drives.

Nowadays, there are a large number of varieties of the design of hydraulic drives, both Russian and foreign.

The disadvantages of foreign analogues:
- high cost of drives;
- long term delivery of equipment and spare parts;
- high cost of components and parts;
- lack of a sufficient level of knowledge of service personnel.

The disadvantages of domestic hydraulic drives:
- lack of equipment with the required technical characteristics for effective oil production;
- insufficient reliability of equipment during operation in cold weather conditions due to the failure of hydraulic systems.

The analysis of the advantages and disadvantages of the characteristics of this pump, which have not proved themselves to be effective enough during operation, resulted in the necessity to use a sufficiently powerful hydraulic drive [3]. In order to increase the operation of wells, it was decided to use hydraulic drive of sucker-rod pump 120-6-24.

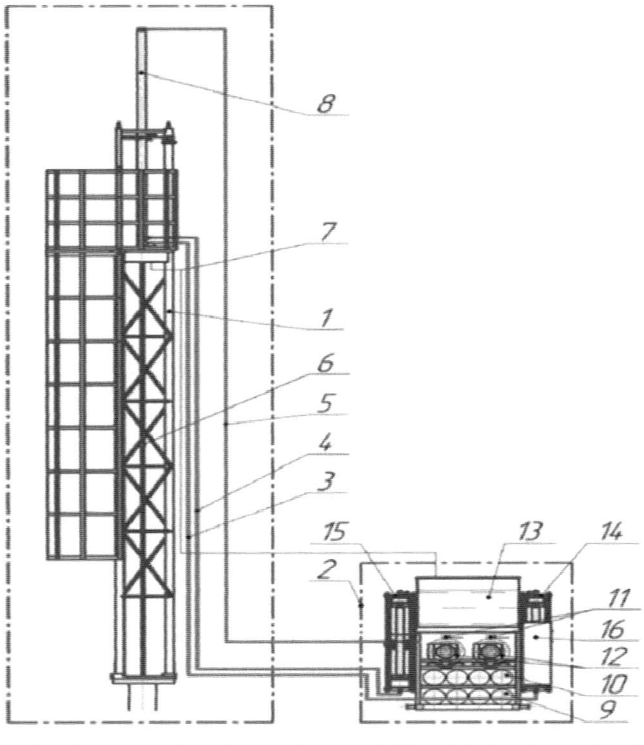

Figure 1. Design of hydraulic drive of sucker-rod pump 120-6-24: 1 - frame, 2 - hydraulic station, 3-4 - main run pipe, 5 – main run-back pipe, 6 - coupling, 7 - drain pipe, 8 - hydraulic power cylinder, 9-10 - nitrogen cylinder, 11 - motor drives, 12 - pumps, 13 - pump reservoir, 14-15 - secondary cylinder GK2-GK3, 16 - operating box.

Operation principle: during operation, power is supplied to the drive from the operating box using electromagnet to lift, the liquid flows through the run pipe into the lower cavity of the secondary hydraulic cylinder rod (2), acting on its lower piston, displaces liquid from the piston cavity (2). Under the influence of the total increase in the gas pressure of the accumulators in the upper piston (2), and created by the hydraulic pump, in the lower piston, the hydraulic power cylinder (HPC1) begins to move upward and, through the wellhead rod and the pump rod column, raises the sucker rod pump piston. The liquid displaced along the line from the cavity of the upper stem (2) returns to the pump.

When approaching the upper position of the rod of the power hydraulic cylinder (PHC1), the flag attached to it sends a signal to the proximity switch. This causes the reverse fluid flow by changing the inclination of the pump washer in the opposite direction. The movement of the liquid in this case is as follows: the pump run pipe is the cavity of the upper rod (2). The rod (2) under the influence of the force on the upper piston begins to move upward, as a result of which the pump piston from the downhole rod moves downward under the action of its own weight and the weight of the pump rod string. The rod of the hydraulic cylinder (2) moves upward, exceeding the resistance of the compressed gas under the action of the total force of the pump in the upper piston and the liquid ejected from the cavity of the rod of the hydraulic cylinder (HC 1). The flag attached to it sends a signal to the lower proximity switch, which leads to a change in the direction of fluid flow through the hydraulic pump. The operation cycle repeats. In order to protect the hydraulic system from overloads, safety valves are provided in the hydraulic system [5].

If the control pressure of the valves in the hydraulic system is exceeded, the oil is drained from the run line into the run-back hole. The alignment of a possible leakage of fluid from the cavity of the hydraulic cylinder rod (HC1) and the lower piston cavity of the auxiliary secondary cylinder (2) occurs through the reverse compensation valve. The stabilization process takes place automatically at the end of the stroke in the upper position through the compensation valve.

Economic feasibility of a sucker rod pump in comparison with an electric-centrifugal pump is reasoned by:

- increased reliability in the presence of mechanical impurities in well fluid
- less sensitivity to salt and asphaltene sediments
- significantly lower energy consumption at flow rates up to 50 m^3 / day.

The advantages over chain drive, pumping units and hydraulic drive with pneumatic balancing are as follows:

-Simple and reliable drive design

-Compact dimensions and ease of transportation.

-Assembly does not require the installation of a special foundation, ordinary road plates or screw piles are enough.

-Less metal construction.

-Reduced time for assembly and disassembly by 4-6 times compared to swinging machines and chain drives.

-Reduced maintenance and maintenance costs, because electronic control system (ECS) continuously monitors the operation.

- Power consumption is reduced:
- up to 50% compared to rocking machines and chain drives;
- up to 30% compared to a hydraulic drive with pneumatic balancing;

-Simplicity and ease of maintenance, due to the placement of the pumping station and ECS in a block box with a removable cover.

-Lack of massive moving and open rotating parts, there is no need for fences. It is safe for others.

-Remote control of the drive operation using telemetry.

-Lack of ropes reduces maintenance and increases safety.

-The unloading of the hydraulic drive is due to the frequency converter, therefore no power reserve is required. It allows significantly energy reduction during operation.

-Easy regulation by the number of swings and the stroke length using ECS allows complete selecting the well flow rate and adjusting the selection depending on the change in flow rate, according to the constant analysis of the dynamometer chart.

-Easy provision of small numbers of swings in low-rate wells. It is possible to install sucker-rod pumping units with a large rod diameter, reducing the number of oscillations, thereby increasing the service life of underground equipment.

-Lack of bending and tensile alternating loads on the power parts of the structure.

The process cyclicity of the pumping by a sucker rod pump contributes to the increase in the permeability of the well bottom zone.

3. Results and Discussion

According to the official reviews of Orenburgneft Company, pilot tests of hydraulic drive sucker-rod pumping unit Geron with an energy recovery unit were carried out at well No. 163 of Pron'kinskoye field, the pilot test results were positive. The main economic effect was obtained by reducing electricity by 27% compared to a similar drive without a recuperation unit. The equipment is recommended for implementation at the facilities of Orenburgneft [6].

In addition, according to the reviews of "Enterprise Kara Altyn", as a result of the experimental operation of the hydraulic drive of sucker-rod pump "Geron"of type HDSRP-80-3.0A with energy recovery, produced by Ural RDC UralNeftService at the Agan oil field in 2015, it was possible to more than double the productivity of the well, due to the competent selection of the speed regime when lifting and lowering the rod of the deep-well pump, as well as the correct setting of the stroke length and the number of double strokes according to the analysis of the deep dynamometer chart [7]. At the same time, electricity consumption, compared to the pumping unit previously operated at this well, significantly decreased (up to 40%). The well was transferred from the periodic mode to the permanent one. Taking into account the positive results of pilot testing, Geron hydraulic drive is recommended for further operation at the fields of company.

The design of "Geron" hydraulic drive proved itself in the most extreme climatic conditions of both Turkmenistan and Colombia, where the temperature rises to $+56^0$. The flow rate of formation fluid with a conventional pumping unit was 4 m^3 / day, and when using the hydraulic drive of sucker-rod pump it was 9 m^3 / day. Under the conditions of the Far North, the studied hydraulic driv proved itself at a temperature of - 55^0. In Nyagan, Khanty-Mansi Autonomous Okrug-Yugra, the flow rate of formation fluid with a conventional pumping chair was 9 m^3 / day, and when using hydraulic drive of sucker-rod pump on this well it was 11 m^3 / day. Good indicators for oil production rates were obtained at Kogalymneftegaz enterprise.

Successful experience of restoring marginal wells, at the Sovetskoye field for well 707 of well platform No. 104 and well 11 of well platform No. 4 of Nizhnevartovskoye field, was obtained using Geron hydraulic drive, as a result of which the flow rate increased several times (the ability to swing the well, thereby select the optimal operating mode).

The positive effect of using hydraulic drives of sucker-rod pump 120-6-24 is expressed in:

- increase in oil production.
- reduction of the level of structural failures.
- reduction of equipment downtime and repair costs.
- optimization of the operating costs of hydraulic systems in comparison with traditional ground drives (pump jack, chain drive).
- annual economic effect is one unit - 97.8 thousand rubles per year roughly estimated.

Table 1. Technical characteristics of hydraulic drives in relation to well parameters (depth, well rate)

Maximum load at the suspension point of the rods, kN	200
Maximum stroke length of the wellhead rod, m	1 ÷ 6
Maximum number of double strokes per minute with 6 m length	4
Maximum number of double strokes per minute with 4 m length	6
Maximum pressure of the working fluid in the hydraulic system, MPa	18
Electric motor type: three-phase, asynchronous 220/380 V, rotation speed 1500 rpm. Maximum power consumption including recuperation, kW / h	2-18
Working fluid tank volume, l	300-800
Working fluid brand	Multigrade low-temperature hydraulic oil of technical regulations 38.101479-00
TBF, not less than, hour	8000
First overhaul period, hour	87500
Full average service life, years	20

In terms of productivity and quality, Geron surpasses the equipment of the leading producers of the USA, Canada and Germany. Table 1 shows the technical characteristics of hydraulic drives.

Hydraulic drive of sucker-rod pump is used not only in Russian oil industry, where the key customers are Rosneft, Lukoil, Tatneft, Orenburgneft, Varyeganneftegaz and Kara Altyn Enterprise.

The companies from Argentina and Iran expressed interest in cooperation with the Urals, a contract was signed with the state concern Turkmennebit (Turkmenneft), cooperation with partners from Azerbaijan, Kazakhstan and other countries continues.

4. Conclusion

Thus, the measures for the introduction of Hydraulic drive of sucker-rod pump, related to the equipment of a new generation, which have been successfully tested both at national and international oil fields provides an overall increase in the efficiency of production processes during operation, increases oil production during well operation in the optimal mode up to 60%, reduces the costs of oil and gas production company by reducing the level of oil deficit to 50% or more, saves up to 30% of electricity due to the frequency converter and prevents equipment failure in oil fields.

All thiese aspects determine the efficiency and prospects of further use of the hydraulic drive, prove the economic feasibility of the use of this type of equipment, leading to the decrease in the cost of producing a barrel of oil [8].

This article discusses a method of oil production using a hydraulic drive of sucker-rod pump 120-6-24, which is widely applied in oil and gas industry and became an alternative to import substitution for small-size pumps.

References

[1] Molchanov A G 2013 *Machines and equipment for oil and gas production* (Moscow: Alliance) 588 p
[2] Garifullin R S, Akhmetzyanov L M and Garipov I N 2018 Improving the efficiency of operating the producing well stock through the use of hydraulic drives *Exposition Oil Gas* **1(61)**
[3] *Submersible pumps for wells: types, characteristics, installation*. Retrieved from: energomir.net/vodoprovod
[4] *Submersible pumps for wells: types, characteristics, installation*. Retrieved from: best-pump.com/progressive-cavity/
[5] *Progressive Cavity Pump*. Retrieved from: bestpump.com/progressive-cavity.
[6] Shadrina A V, Saruev A L, Saruev L A and Kazantsev A A 2008 Methods of researching the technical and economic efficiency of machines for drilling underground wells of small diameters *Izvestiya of Tomsk Polytechnic University* **1**
[7] Pogrebnaya I A and Mikhailova S V 2017 Operation of the associated petroleum gas drying unit on the example of unit No. 4 of the limited liability company "Nizhnevartovsk gas processing complex" *New technologies - for the oil and gas region: mater. of the int. sci. and pract. conf.* (Tyumen: TIU) vol 3 pp 87-90
[8] Charonov V Ya 1996 Economical electric drives for pumping units of marginal wells *Oil industry* **12** 46-48

Improvement of heat transfer of liquid-oil heat exchangers for internal combustion engines

Iu S Prikhodko[1], **D A Akhmetshin**[1], **Ya S Bolkov**[1] **and A N Shishkov**[2]

[1] South Ural State University, 76, Lenina Ave., Chelyabinsk, 454080, Russia
[2] Moscow Polytechnic University, 38, Bolshaya Semenovskaya str., Moscow, 107023, Russia

E-mail: prikhodkoys@susu.ac.ru

Abstract. Removing heat from an internal combustion engine is one of the most promising areas for the development of vehicles and heavy vehicles, especially. The article studies the influence of various parameters of the coolant and the parameters of the heat exchanger on the efficiency of heat transfer when removing heat from an internal combustion engine by an oil-liquid heat exchanger. The influence of "diffuser-confusor effect" on flow turbulization is considered. Literature analysis was carried out during the study. It raised such issues as: the issues of the efficiency of heat transfer from liquid to the wall under various conditions of the flow of the medium, the issues of changing the hydraulic regimes depending on the parameters of the medium, the issues of changing the parameters of the medium from the determining and secondary factors of heat transfer. The data on the influence of the temperature of various oils on their viscosity and density are presented. The influence of the state of the refrigerant on the efficiency of heat transfer is generalized. As a result of the analysis, a number of solutions have been proposed to the safe intensification of heat removal from the internal combustion engine. These solutions can be used for efficient oil cooling in diesel engines, mainly in engine design.

1. Introduction

Internal combustion engines are constantly exposed to thermal effects from the processes occurring in them. Engines are also associated with a variety of equipment that are most efficient in certain temperature ranges. Every engine in a modern engineering unit is equipped with turbochargers that cannot operate beyond a certain temperature range. Diesel fuel becomes thicker when operating temperatures drop significantly, and this fact complicates both starting the engine itself and the engine operation in general. Therefore, the thermal stability of the engine is an important component of its performance. Thermal stability of the engine is ensured by oil cooling systems. Figure 1 shows an oil-liquid heat exchanger used in a number of engines to solve cooling issues.

The oil-liquid heat exchanger is integrated into the lubrication system as an additional element. Its purpose is to provide oil cooling. It is important to do so as engine oil overheating leads to serious problems: the oil viscosity changes, the intensity of its burnout and decomposition increases, and engine performance decreases. Overheated oil cannot provide sufficient quality lubrication of rubbing parts and complicates engine cooling. Moreover, insufficient cooling can result in a variety of damage to the power unit (up to jamming). Oil cooling systems are especially developed on uprated engines, because such engines are subject to significant overheating.

IOP Conf. Series: Materials Science and Engineering 1111 (2021) 012047 doi:10.1088/1757-899X/1111/1/012047

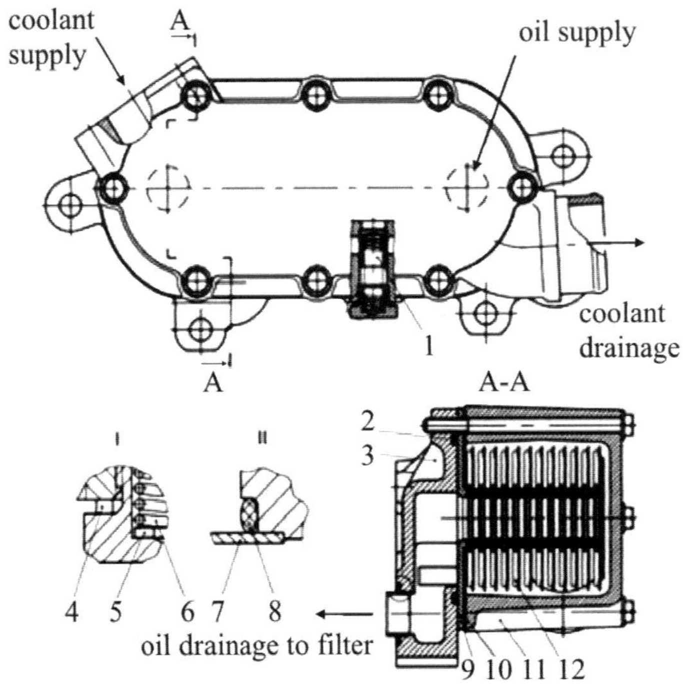

1 – blowoff valve;
2 – o-ring;
3 – corpus;
4 – adjusting washer;
5 – adjusting gasket;
6 – spring;
7 – hub;
8 – o-ring;
9 – flange;
10 – gasket;
11 – corpus;
12 – plate sections.

Figure 1. Oil-liquid heat exchanger (plate type) [1]

There are several approaches to cooling automotive oil. The method of cooling the lubricating oil by means of an oil heat exchanger is discussed in the books [1, 2]. This approach is based on the removal of heat from the heat exchanger due to the flow of coolant circulating in the cooling system of the power unit. As a result of the study, the following indicators were achieved [2]: reduction in overall performance and weight of the lubrication system, ensuring intensive heating of the oil after starting the engine, reducing wear of parts, protection from hypothermia during winter operation. Also, the advantages should be noted [1]: installation of a heat exchanger in any convenient place and needless of long pipelines and many connections. The oil temperature in such a heat exchanger does not drop below the coolant temperature, as a result of which the engine operates in the best temperature mode. The disadvantages of the solution include a malfunction of the lubrication system in case of leakage, a complex design and the need for maintenance. In the conclusion, it is confirmed that the results obtained will increase the resource of the engine and oil, due to operation in a more stable temperature regime. This cooling system is widely used on both diesel engines and gasoline [2]. At present, this approach to oil cooling is widely used on domestic KAMAZ and YaMZ engines [1].

An alternative to a liquid-oil heat exchanger is the use of an air cooler, also called an intercooler. The use of an intercooler for cooling the internal combustion engine is considered in [3]. The disadvantages of using an intercooler include its significant dimensions and mechanical fragility. The intercooler housing is made of aluminum and is easily damaged by the hit of small stones or an easy road accident, and its repair is expensive.

One of the modifications of the liquid-oil heat exchanger for cooling high-power diesel units was presented in the patent [4]. The liquid-to-oil heat exchanger has been modified to eliminate the poor flushing efficiency of the coolant surfaces of the oil heat exchanger inside the radiator coolant manifold, which depends mainly on the performance of the water and oil pumps of the internal combustion engine. Liquid-and-oil heat exchanger includes tube bundle arranged inside cylindrical shell which connects reservoirs provided with branch pipes and partitions mounted perpendicularly relative to tubes; these partitions divide inner space of heat exchanger into cavities and provide for flow of heat- transfer agent from one chamber to another.

After the modification, the heat exchanger started to work as follows. When the rotor rotates from an electric motor in a liquid-oil heat exchanger for internal combustion engines of vehicles, a forced pumping effect of pumping both coolants in countercurrent directions is created. This effect is achieved due to the fact that the liquid-oil heat exchanger for internal combustion engines of vehicles is equipped with pump blades on the inner and outer cylinders of the end parts on the coolant side of the combustion engine. As a result, there was an increase in the level of stability of the oil temperature and a high level of thermal equivalents of coolants at the most severe operating conditions of the internal combustion engine, which do not depend on the crankshaft speed.

2. Scientific novelty

There are two significant advantages of liquid-oil heat exchanger use. Firstly, the oil temperature in the liquid-oil heat exchanger does not fall below the coolant temperature [5]. This means that fewer stresses arise in the engine parts that come into contact with the oil at the same time; so the engine operates in a better temperature condition. Secondly, the heat exchanger can be installed in any convenient place on the engine. This eliminates the need for long pipelines and multiple connections. Disadvantages of a liquid-oil heat exchanger include complex design and the need for maintenance and repair [6].

These problems can be solved by reducing the weight and size of the heat exchanger. To reduce the size of the heat exchanger while maintaining the amount of heat transferred through it, it is necessary to increase the heat exchanger performance [7]. The most promising way to achieve that is to intensify heat transfer [8]. This is an important part of the development of internal combustion engines.

This article discusses the methodology for calculating the optimal flow rate of heat transfer intensification and the most effective design solution when choosing a heat exchanger. This can be achieved by improving the heat transfer coefficient, using a more efficient flow rate of the coolant and improved geometric characteristics of the channel.

3. Results

The key indicator of the efficiency of the heat transfer process is the heat transfer coefficient k, which can be calculated by the formula (1). This is a proportionality constant between the heat flux (the amount of heat passing through a unit area) and the temperature difference (the one between a solid surface and the surrounding fluid), which characterizes the intensity of heat transfer on the surface of a solid body [9].

$$k = \frac{1}{\frac{1}{\alpha_1} + \sum \frac{\delta_i}{\lambda_i} + \frac{1}{\alpha_2}} \tag{1}$$

where α_1 is a heat convection coefficient from the side of fluid to be cooled [W/(m^2·K)]; α_2 is a heat convection coefficient from the side of fluid to be heated [W/(m^2·K)]; $\sum \frac{\delta_i}{\lambda_i}$ is a the total value of the thermal resistance of the wall and contamination [m^2·K/W].

We consider a heated wall or surface being in contact with a fluid flow. When the fluid flows within alternating channels of different cross-sections, a periodic separation of the boundary layer from the surface of the corrugations occurs. This phenomenon is called «diffuser-confusor effect» shown in Figure 2 [10].

The laminar boundary layer grows on the impermeable channel wall within the convergence area due to the viscosity forces. The thickness of the laminar boundary layer is determined by the type of fluid flow and the curvature of the streamlined surface (i.e. corrugation geometric characteristics). When the fluid flows through the diffuser, its velocity decreases, while the pressure increases. The pressure gradient increasing against the flow prevents the longitudinal movement of the fluid in the boundary layer. This backpressure, combined with viscous friction on the wall, slows down the boundary layer. The layer is detached from the surface and begins to flow in the opposite direction at some distance from the protrusion. The backflow appears in the shape of vortex. The combination of these two complex flows results in increased fluid flow turbulence of the fluid flow in the plate heat exchanger flow channel. Turbulent flow is the most efficient flow type for heat exchangers operation.

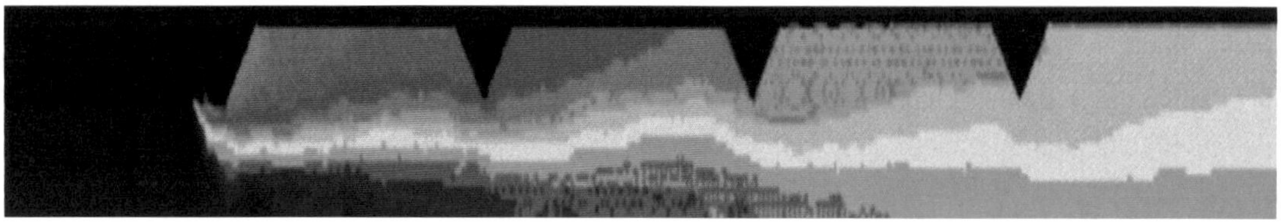

Figure 2. The channel concentration field

It is known that a developed turbulent flow for a stationary isothermal fluid flow in a pipe with a circular profile appears when Reynolds numbers reach Re = 1000. In the flow channels of plate heat exchangers this effect is observed at Reynolds numbers reaching Re = 50-100 [11]. This indicates that the turbulent flow in the plate heat exchanger will develop at lower speeds (as compared to the shell-and-tube heat exchanger under the same conditions). The heat transfer coefficient in the flow path of the plate heat exchanger exceeds the heat transfer coefficient in the tubes of the shell-and-tube heat exchanger about 10 times [12]. This means that a more turbulized flow type will be observed in a plate heat exchanger at the same values of the coolant flow rate. As a consequence, this leads to a higher intensity of the heat transfer process between the channel wall and the moving coolant in a plate-type heat exchanger.

The Reynolds number characterizes the ratio of inertial forces to viscous friction forces in viscous liquids and gases. It is also a similarity criterion for the viscous fluid flow [13]. The Reynolds number can be determined by formula:

$$Re = \frac{p \cdot v \cdot D_h}{\eta} = \frac{Q \cdot D_h}{\nu \cdot A} \tag{2}$$

where p is a fluid density [kg/m^3]; v is a fluid velocity [m/s]; D_h is a hydraulic diameter [m]; Q is a volume flow rate [m^3/s]; ν is a kinematic viscosity coefficient [m^2/s]; η is a dynamic viscosity coefficient [Pa·s].

It is necessary to refer to the concept of the hydraulic diameter for further transformations. It characterizes the measure of the fluid flow passage efficiency by the channel. The hydraulic diameter can be determined by the formula:

$$D_h = \frac{4 \cdot A}{P} \tag{3}$$

where A is a cross-sectional area of fluid flow [m^2]; P is a wetted perimeter of the flow cross section [m].

For rectangular channels, the hydraulic diameter can be determined by the formula:

$$D_h = \frac{4 \cdot a \cdot b}{2 \cdot (a + b)} \tag{4}$$

where a is a channel fill level [m]; b is a channel width [m].

The Reynolds number for a plate heat exchanger can be expressed using the geometric characteristics of the channel and the flow rate of the heat transfer fluid according to formula:

$$Re = \frac{Q \cdot \frac{4 \cdot a \cdot b}{2 \cdot (a + b)}}{\nu \cdot a \cdot b} = \frac{4 \cdot Q}{2 \cdot (a + b) \cdot \nu} = \frac{4 \cdot Q}{P \cdot \nu} \tag{5}$$

where P is a channel perimeter [m].

The resulting dependence relates the Reynolds number to the volumetric flow rate of the liquid (which flows through the cross section of the fluid flow per unit time), as well as to the channel perimeter and kinematic viscosity.

Kinematic and dynamic viscosities are related to each other according to the formula:

$$\nu = \frac{\eta}{\rho} \tag{6}$$

where ρ is a fluid density [kg/m^3].

The dependences of the dynamic and kinematic viscosity of water on temperature and pressure are presented in Figure 3 [14].

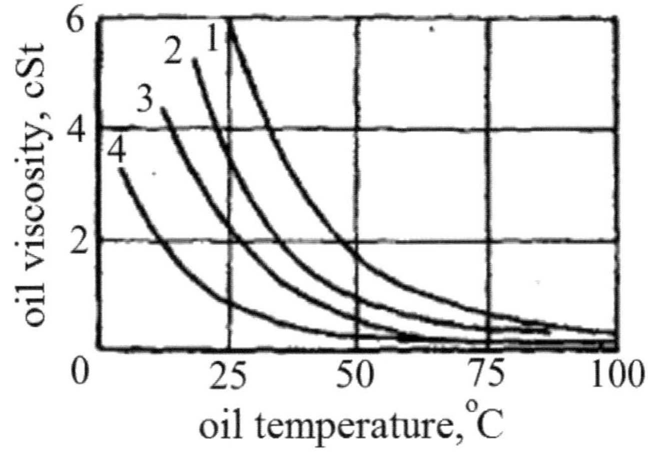

Figure 3. The dependence of engine oil viscosity on temperature: 1 - Aviation oil MK-22; 2 - Autotractor oil "AK-Yu"; 3 - Automobile oil "AC-5"; 4 - Motor oil "AKZp-6"

For a more accurate determination of the density, the American scientist, Doctor of Natural Sciences James Dewey Watson introduced the empirical dimensionless value of ω' [15]. According to Watson, the dependence of ω' on the given parameters is an universal function that is valid for most liquids. The diagram is shown in Figure 4 [15]. The diagram makes it possible to calculate the value of the liquid density for any temperature and pressure. In this case, the critical constants of the fluid and one value of density must be known at arbitrary values of temperature and pressure. The following expression is valid for the data obtained from the Watson diagram:

$$\frac{\rho_1}{\rho_2} = \frac{\omega'_1}{\omega'_2} \tag{7}$$

where indices 1 and 2 characterize the state of the liquid for various parameters of the fluid.

Formula (5) demonstrates ways to more efficiently use a plate heat exchanger in a lubrication system. As it follows from it, the turbulization of the flow increases with an increase in the volume flow rate, but decreases with an increase in the kinematic viscosity of the liquid or the wetted perimeter. Figure 3 shows that the viscosity of oils drops significantly with increasing temperature. Thus, significant cooling of the coolant complicates turbulization of the flow. Therefore, it is necessary to avoid overcooling the oil when removing heat from the internal combustion engine.

Maximization of turbulence occurs at a smaller wetted perimeter. Therefore, to increase the efficiency of the heat exchanger and the engine, preference should be given to the device whose channels have a smaller wetted perimeter when choosing between two devices with an equal heating surface area. A heat exchanger whose plates have a smaller width at the same channel height.

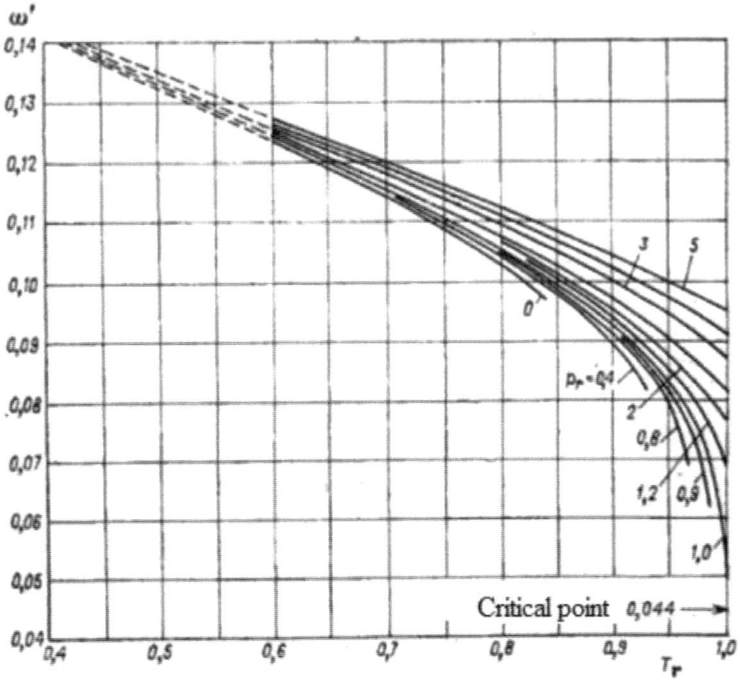

Figure 4. Watson diagram for calculating the density of a liquid [W]

4. Conclusion

The article discusses methods for flow turbulence intensification, which can improve the conditions for heat transfer and indirectly affect the heat transfer coefficient.

The following solutions are presented:

1. reduction of the wetted perimeter of the heat exchanger channel;
2. avoidance of oil overcooling when organizing heat removal from the engine.

These solutions can be used for efficient oil cooling on diesel engines. The use of a plate heat exchanger will simplify the design, maintenance and repair of the heat exchanger. Also, liquid-oil heat exchangers have a long service life, and with regular maintenance and timely flushing and repair, this part will not cause problems, providing effective oil cooling at all engine operating modes. Currently, they are widely used on domestic KAMAZ and YAMZ engines, and liquid metal products have also found application in many modern trucks of foreign production and even in compact passenger car engines.

References

[1] Rosenblit G B 1977 *Heat transfer in diesel engines* (Moscow: Mechanical Engineering) p 216
[2] Kostin A K, Larionov V V and Mikhailov L I 1979 *Thermal intensity of internal combustion engines* (Leningrad: Mechanical engineering) p 224
[3] Shestakov V V, Dik M I and Mingalimov R R 2019 Intercooler with additional air cleaning system for D-260 engine *Mater. of the 64th student sci. and pract. conf. of the faculty of engineering of Samara state agrarian university* pp 32-34
[4] Averkiev L A, RU Patent 2170897C1 (20 July 2001)
[5] Baranovskiy N V, Kovalenko L M and Yastrebenetskiy A R 1973 *Plate and spiral heat exchangers* (Moscow: Mechanical Engineering) p 288
[6] Suxin Q, Jianlin Yu and Gang Y 2017 A review of regenerative heat exchange methods for various cooling technologies *Renewable and Sustainable Energy Reviews* 535–550
[7] Vega M, Almendros-Ibanez J A and Ruiz G 2006 Performance of a LiBr-water absorption

chiller operating with plate heat exchangers *Energy Conversion and Management* **47(18-19)** 3393-3407
[8] Hesselgreaves J E, Law R and Reay D 2016 *Compact Heat Exchangers Selection, Design and Operation* (Butterworth-Heinemann) p 502
[9] Jeonggyun H, Jinhyun K and Honghyun C 2016 Theoretical analysis of thermal performance in a plate type liquid heat exchanger using various nanofluids based on LiBr solution *Applied Thermal Engineering* **108** 1020-1032
[10] Pandey Ch Sh, Prajapati O Sh and Dadhich M 2018 *Analysis of Corrugated Plate Heat Exchanger Using CFD & Taguchi Method* (Lambert Academic Publishing) p 120
[11] Sohel Murshed S M and Lopes M M 2017 *Heat Exchangers Advanced Features and Applications* (IntechOpen) p 171
[12] Kazi S N 2015 *Heat Transfer Studies and Applications* (IntechOpen) p 396
[13] Vidin Yu V, Ivanov V V and Kazakov R V 2018 *Engineering methods for calculating heat transfer problems* (Moscow: Infra-M) p 480
[14] Belkin A P and Stepanov O A 2020 *Diagnostics of heat power equipment* (St Petersburg: Lan) p 240
[15] Moskvichev Yu A, Pavlov O S and Grigorevich A K 2020 *Theoretical foundations of chemical technology. Tutorial* p 272

Modeling the sunflower seeds separation process in air-sieve grain-cleaning machines

I E Priporov, V S Kurasov, E E Samurganov and A B Shepelev

Kuban State Agrarian University named after I. T. Trubilin, 13, Kalinina street, Krasnodar, 350044, Russia

E-mail: priporov.i@edu.kubsau.ru

Abstract. In modern seed-cleaning machines, pneumatic systems have been widely used in recent years. To minimize costs in pneumatic systems of seed-cleaning machines, while increasing performance and improving the functional efficiency, it is necessary to improve their constructive and technological parameters. Improving pneumatic system parameters requires the development of a methodology for modeling the processes of airflow motion for their qualitative characteristics. To improve the efficiency of the vertical pneumatic channel of air-sieve grain-cleaning machines, it has been proposed to modernize it, which will improve the quality of separation of sunflower seed heap in it. The device and technological process of operation of a modernized air-sieve grain-cleaning machine of the MVU-1500 type in sorting a heap of sunflower seeds are presented. The movement of the components of a sunflower seed heap in a vertical pneumatic channel of an air-sieve grain-cleaning machine of the MVU-1500 type is examined and the proportionality coefficients of the airflow for the components of a sunflower seed heap are determined. The obtained values of the theoretical proportionality coefficients of the airflow for the components of a sunflower seed heap vary widely depending on the composition (organic impurity – $0{,}015 \leq k \leq 0{,}061$; seeds of the leading crop – $0{,}120 \leq k \leq 0{,}162$; broken seeds – $0{,}096 \leq k \leq 0{,}115$ and hulled seeds – $0{,}134 \leq k \leq 0{,}153$). Mathematical modeling of the technological process of the improved air-sieve grain-cleaning machine of the MVU-1500 type makes it possible to increase the efficiency of the vertical pneumatic channel by 15-20% without deteriorating the quality of the material itself, as a result of which the specific energy consumption of the process will decrease.

1. Introduction

One of the difficult and important tasks of post-harvest processing is the cleaning of seeds from impurities and sorting. In this regard, it is advisable to create new methods that are more efficient and to improve existing grain and seed cleaning machines [1].

In modern grain and seed cleaning machines, pneumatic systems have been widely used in recent years. To minimize costs in pneumatic systems of grain and seed cleaning machines, and to increase the performance and improve operation quality, it is necessary to improve their design and technological parameters. Improving the parameters of pneumatic systems requires the development of a methodology for modeling air-flow processes for their quality characteristics [2, 3].

Pneumatic systems are used as part of complex grain and seed cleaning machines at all stages of seed cleaning, as well as separate machines – the pneumatic separators [4].

The universal grain-cleaner is equipped with a pneumatic suction system, as it emits less dust into the environment and is better suited for use in closed areas. The pneumatic system of this machine has two pneumatic separation channels to ensure high separation efficiency of the grain heap during secondary cleaning [5-9].

To increase the efficiency of the vertical pneumatic channel of the air-sieve grain cleaning machine, its modernization was proposed, which will improve the quality of separation of the heap of sunflower seeds.

2. Materials

According to the Patent of the Russian Federation (RU 2681493), of MVU-1500 type (Figure 1 a), an air-sieve grain-cleaning machine contains an upper sieve boot 1, a lower sieve boot 2, a common settling chamber connected to a fan (not shown), a pneumatic channel 3 for preliminary aspiration. It also has a pneumatic channel 4 for final aspiration, a feed roller 5, a valve 6, an auger 7 for removing heavy impurities, an auger 8 for removing light impurities, a flap 9 for fine adjustment of the pneumatic pre-aspiration channel. There is also a flap 10 for fine adjustment of the pneumatic channel for final aspiration, a flap 11 for air supply coarse adjustment, an exhaust pipe 12, chutes 13. At the end of the lower sieve boot, under the opening of the vertical pneumatic channel for the final aspiration, there are splitters 14 (Figure 1 b), the inner surface of which has a coating with a low friction coefficient, for example, of fluoroplastic, in an amount of at least three, located at an angle 40-45° in the horizontal plane. Installed at least three splitters so that the width of the pneumatic channel for final aspiration was approximately 1500 mm and, based on design considerations and metal consumption, we obtain the length of the splitter of 500 mm. Splitters 14 (Figure 1 c) are made in the form of combs, the teeth 15 of which have the shape of a truncated cone, the upper base of which is 1/3 of the lower one. Based on design considerations, with a splitter length of 500 mm, we obtain four teeth, each of which is 125 mm. Moreover, the angle in the upper base of the cone is 20-30°, and the lower base of the cone has holes for its attachment to the lower sieve boot 2.

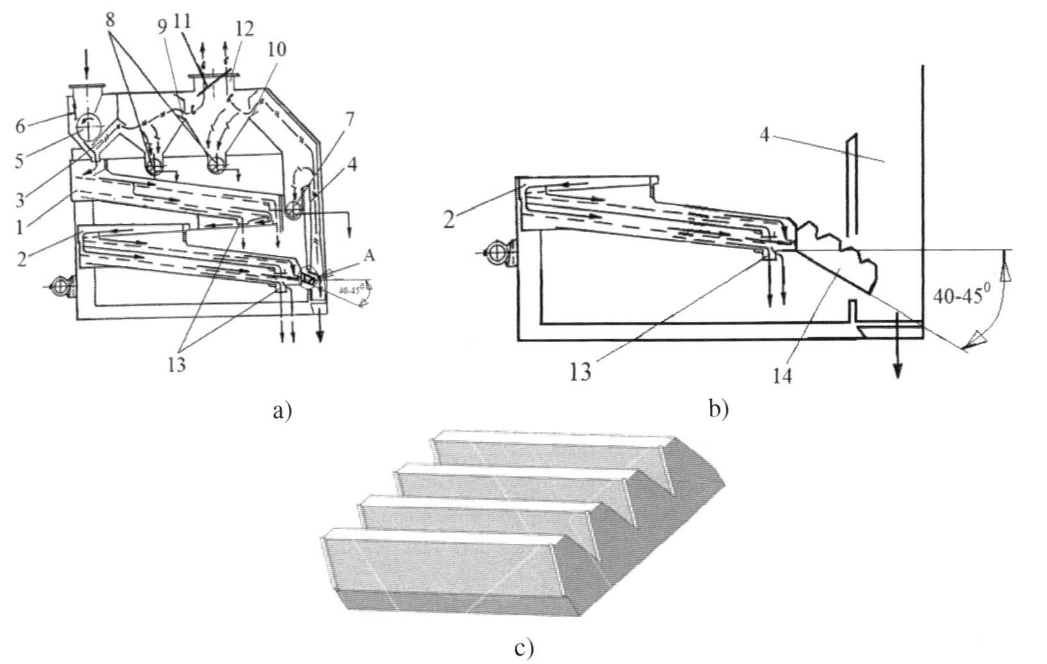

Figure 1. The air-sieve gran-cleaning machine of MVU-1500 type according to the Patent (RU 2681493): a - the machine general view; b - the splitter location; c - the splitter general view

Splitter teeth 14 on the sides have oppositely directed welts 16, the height of which is 1/4 of the width of the pneumatic channel 4 for the final aspiration, and the upper base of the splitter cone 14 has a width equal to 5-6 times the width of the seed. The width of the upper base of the splitter cone is 5-6 sizes of the seed width, because with an average seed width of 7 mm, we get the size of this base – 42 mm, that is, about 5 to 6 sunflower seeds can be placed in the upper base of the cone.

3. Movement of the components of the heap of sunflower seeds in the vertical pneumatic channel of the air-sieve grain-cleaning machine of MVU-1500 type.

When the components of the heap of sunflower seeds descend from the lower sieve boot into the vertical pneumatic channel (Figure 1 b), the following forces act:

F_A – the Archimedean force, kN;

F_{ar} – the air-flow resistance force, kN;

F_{sg}, F_{rg} – the seeds and the refuse gravity force, respectively, kN.

Let us write the Newton second law in the form for:

– the organic impurities (fragments of calathiums and stems):

$$m_{oi}\,\bar{a} = \bar{F}_A + \bar{F}_{\text{ar}} + \bar{F}_{rg}, \tag{1}$$

$$m_{oi}\frac{d\vartheta}{dt} = F_A + F_{\text{ar}} - F_{rg}. \tag{2}$$

If

$$\frac{d\vartheta}{dt} > 0, \tag{3}$$

then the expression (2) will be

$$F_A + F_{\text{ar}} \geq F_{rg}; \tag{4}$$

– of sunflower seeds:

$$m_{ss}\bar{a} = \bar{F}_A + \bar{F}_{\text{ar}} + \bar{F}_{sg}, \tag{5}$$

$$m_{ss}\frac{d\vartheta}{dt} = F_A + F_{\text{ar}} - mg. \tag{6}$$

If

$$\frac{d\vartheta}{dt} \leq 0, \tag{7}$$

then the expression (6) will be

$$F_A + F_{\text{ar}} \leq m_{ss}g. \tag{8}$$

The resistance force (R) at low air-flow rates (Figure 2) in the pneumatic channel for the final aspiration of the air-sieve grain-cleaning machine of MVU-1500 type is determined by the expression:

$$R = k \cdot \vartheta, \tag{9}$$

where k – the proportionality coefficient of the components of heap of sunflower seeds; ϑ – the air flow rate, m/s.

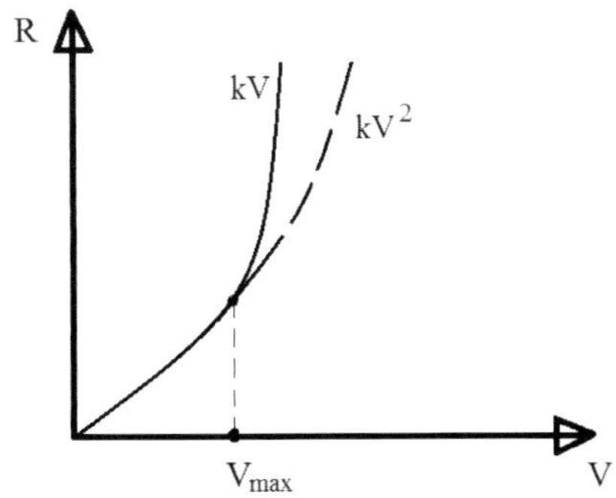

Figure 2. The dependence of the resistance force on the air flow rate

Sunflower seeds placed in a vertical air stream are subjected to the following forces of resistance (R) and gravity (G).

From the expression (9), the proportionality coefficient at low air flow rates will be:

$$k = \frac{mg}{\vartheta_{max}}. \quad (10)$$

The resistance force (R) at high air flow rates (Figure 2.2) in the pneumatic channel for the final aspiration of the air-sieve grain-cleaning machine of MVU-1500 type is determined by the expression:

$$R = k \cdot \vartheta^2. \quad (11)$$

From the expression (11) the proportionality coefficient at high air flow rates will be:

$$k = \frac{mg}{\vartheta_{max}^2}. \quad (12)$$

The force of the air flow in the pneumatic channel for the final aspiration (F_{ar}) and the expression (9) at low rates will be:

$$F_{\text{ar}} \approx R = k \cdot \vartheta. \quad (13)$$

For the organic impurities (fragments of calathiums and stems) of sunflowers seeds the expression (4), according to the expression (13), will be:

$$F_A + k \cdot \vartheta \geq F_{rg}, \quad (14)$$

The Archimedean force for the organic impurities is determined by the expression:

$$F_A = \rho_{\text{ar}} \cdot g \cdot V, \quad (15)$$

Substituting the expression (15) into (14), we perform mathematical transformations and obtain the proportionality coefficient of the air flow in the vertical pneumatic channel at low rates for organic impurities:

$$k = \frac{m \cdot g - \rho_{\text{ar}} \cdot g \cdot V}{\vartheta}, \tag{16}$$

where m – the mass of organic impurity (fragments of calathiums and stems), g. ($m = m_{av}$); ϑ – the organic impurity hovering velocity, m/s; ρ_a – the air density, kg/m^3 ($\rho_a = 0{,}093$ kg/m^3); V – the volume of organic impurity, entering a vertical pneumatic channel, m^3.

The volume of organic impurity or sunflower seeds entering a vertical pneumatic channel will be determined by the expression:

$$V = \frac{4}{3}\pi R_e^3, \tag{17}$$

where R_e – the effective radius of organic impurity, sunflower seeds and other components, m.

Let us determine the proportionality coefficient of the air flow for sunflower seeds according to expressions (8) and substitute the expressions (9), (13) and (15) into it, so it will be:

$$k \cdot \vartheta_s + \rho_{ar} \cdot g \cdot V_s \leq m_{ss} g, \tag{18}$$

from the expression (18) we find the coefficient k at low air flow rates in the vertical pneumatic channel:

$$k \leq \frac{m_{ss} \cdot g - \rho_{ar} \cdot g \cdot V_s}{\vartheta_s}. \tag{19}$$

Substituting the numerical values into the expression (19), we find the proportionality coefficients of the air flow for the components of the heap of sunflower seeds, the calculation results of which are given in Table 1.

Table 1. The proportionality coefficients of the air flow for the components of the heap of sunflower seeds of "Lakomka" variety

Name of component in the heap of seeds	Coefficient value	Reference
Organic impurity	$0.015 \leq k \leq 0.061$	GOST 12037-81 "Seeds of farm crops. Methods for determination of purity and seed lot impurity"
Main crop seeds	$0.120 \leq k \leq 0.162$	
Broken seeds	$0.096 \leq k \leq 0.115$	
Hulled seeds	$0.134 \leq k \leq 0.153$	

4. Conclusion

The obtained values of the theoretical proportionality coefficients of the air flow for the components of the heap of sunflower seeds, as a result of mathematical modeling of the technological process of the operation of the improved air-sieve grain-cleaning machine, allow increasing the efficiency of the vertical pneumatic channel without deteriorating the material quality itself.

References

[1] Pavlyuchenko K V 2016 Experimental study of a pneumatic grain separator with an inclined air flow *Electronic sci. and methodological J. of Omsk State Agrarian University* **3(6)** Retrieved from: http://e-journal.omgau.ru/index.php/2016-god/5 / 29-statya-2016-2 / 375-00125.

[2] Mudarisov S G, Badretdinov I D and Sharafutdinov A V 2009 Modeling the air flow in the pneumatic systems of agricultural machines *Sci. support of sustainable functioning and*

development of the agro-industrial complex: Mater. of All-Russian. Sci. Pract. Conf. with Int. participation pp 116-119

[3] Dorokhov A, Khamuev V G and Lepeshkin K 2018 Modernization of grinding machines of grain-cleaning machines *MATEC Web of Conf.* 05009

[4] Burkov A I, Glushkov A L and Lazykin V A 2020 Calculation of particle trajectories in a pneumatic separation channel by various methods *Agrarian Sci. of Euro-North-East* **21(1)** 62-70

[5] Chebotarev V P 2017 Substantiation and calculation of parameters of a universal grain-cleaning machine *Bull. of the Belarusian State Agricultural Academy* **3** 135-139

[6] Saitov V E, Farafonov V G, Gataullin R G and Saitov A V 2018 Research of a diametrical fan with suction channel *IOP Conf. Ser.: Mater. Sci. Eng.* 012009

[7] Badretdinov I, Mudarisov S, Tuktarov M, Dick E and Arslanbekova S 2019 Mathematical modeling of the grain material separation in the pneumatic system of the grain-cleaning machine *J. of Appl. Engineering Sci.* **17(4)** 529-534

[8] Priporov I E, Kurasov V S, Chebotarev M I, Gumbarov A D and Kuznetsov E V 2020 Capacity of air-and-screen grain cleaner as component of production line of sunflower meal *Int. J. of Emerging Trends in Engineering Res.* **8(1)** 157-163

[9] Nesterenko A V, Leshchenko S M, Vasylkovskyi O M and Petrenko D I 2017 Analytical assessment of the pneumatic separation quality in the process of grain multilayer feeding *INMATEH - Agricultural Engineering* **53(3)** 65-70

Optimization of the sunflower seeds extrusion process in obtaining sunflower cake

I E Priporov and A B Shepelev

Kuban State Agrarian University, 13, Kalinina street, Krasnodar, 350044, Russia

E-mail: priporov.i@edu.kubsau.ru

Abstract. One of the promising directions is the use of the design of the extruder working bodies, where the implemented method of compaction, movement with the timely withdrawal of sunflower cake in a bulk form, is applied. Based on the results of experimental studies using the planning method, a mathematical model of the extrusion process was obtained, which describes the specific energy consumption of the extruder in obtaining sunflower cake. The centerpoints of the experiment are located in the study area, which makes it possible to establish the optimal parameters of the factors for their various combinations. The analysis of two-dimensional sections in interaction with each other makes it possible to determine the optimal values of all the factors under study: the material temperature at the exit - t = 133.67 °C; the extrusion auger winding pitch - h = 7.58 mm; the matrix die hole area - S = 323,89 mm^2. With these values of the factors, the minimum specific energy consumption of the improved extruder E = 69.06 kW-h/t compared to serial ones is ensured due to the implementation of the intake and the extrusion augers connected in sequence in the form of a catenoid. This includes the joints of the sections in the form of a parabola according to the normal distribution law, the tops of the conical head in the form of a parabola, which shows that the material is not damaged when leaving the conical head, and affects its mass and density.

1. Introduction

Extrusion is one of the most promising methods for preparing food for feeding. In the process of extrusion [1], the process material is simultaneously influenced by pressure and temperature, as a result of which the structure of the feed fiber changes, starch is dextrinized to glucose, the feed is sterilized, the inhibitors of the digestive tract are inactivated, and the palatability is improved [2, 3].

The current trend in the development of extrusion technologies is to ensure the maximum reduction in the extrusion process energy consumption with improving the quality of the extrudate [4].

In the production of extruded feed products, the quality indicators of the product, as well as the energy consumption of the extrusion process, depend both on the properties of the process material and on the design and operating parameters of the extruder [5-8]. One of the promising directions is the use of the design of the extruder working bodies, where the implemented method of compaction, movement with the timely withdrawal of sunflower cake in a bulk form, is applied.

2. Materials

Experimental studies were conducted using an improved extruder according to the Patent of the Russian Federation (RU 2693072) (Figure 1).

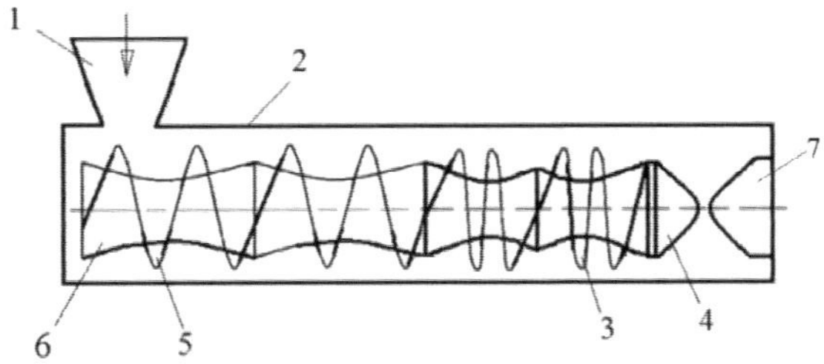

Figure 1. The protein feed extruder:1 – charging hopper; 2 – body; 3 – extrusion auger; 4 – auger head; 5 – intake auger; 6 – extrusion auger section; 7 – clamping sleeve

3. Results

Based on the results of experimental studies, using the planning method, a mathematical model of the extrusion process was obtained, which describes the specific energy consumption of the extruder in obtaining sunflower cake:

$$y_s = 70{,}641 - 1{,}2x_1 - 2{,}192x_2 - 4{,}8x_3 - 1{,}75x_1x_2 - 0{,}25x_1x_3 - 0{,}75x_2x_3 + 4{,}375x_1^2 + \\ +13{,}376x_2^2 + 4{,}375x_3^2. \quad (1)$$

To study the factors' impact on the optimization parameter, Table 1 was built.

Table 1. Factors impacting on the quality of sunflower cake production by an improved extruder, their symbols and levels of variation

Factors	Symbols		Variation step	Variation level		
	coded	natural		–1	0	+1
Temperature of the material at the exit from the extrusion auger of the die [9], ^{0}C	x_1	t	25	110	135	160
Extrusion auger winding pitch, mm	x_2	h	2	6	8	10
Matrix die hole area [10], mm^2.	x_3	S	100	200	300	400

The analysis of the mathematical model (1) was conducted by the method of finding the extremum in space, that is, by the method of processing response surfaces in two-dimensional space [11].

After calculating the regression coefficients, the hypothesis of the second order model adequacy was tested. For this, the corresponding sums of squares, variances and F-test were calculated. For the mathematical model (1), the tabular value of the F-criterion is $F_{0{,}05} = 4{,}74$, and the calculated value will be:

$$F_{calc} = \frac{10{,}8805}{5{,}582} = 1{,}95.$$

Since the tabular values of the Fisher criterion are greater than the calculated value, then model (1) is adequate as a second-order polynomial.

The coordinates of the maximum point are obtained by differentiating the model (1) and solving the systems of the equations [11].

The maximum point coordinates (coded values) of the surfaces will be:

$$x_1 = 0{,}1751;\ x_2 = 0{,}1092;\ x_3 = 0{,}5629.$$

Substituting the obtained values of the results of the coded values into the equation (1), we determine the minimum specific energy consumption E = 69,07 kW-h/t.

As a result of the calculation, the following coefficients were obtained for finding the specific energy consumption of the extruder:

$$B_1 = 4{,}375;\ B_2 = 13{,}329;\ B_3 = 4{,}422.$$

The obtained coefficients for the family of regression equations in canonical form for the specific energy consumption of the extruder will have the following form:

– the sunflower cake temperature at the exit from the extrusion auger of the die and the extrusion auger winding pitch (Figure 2 a):

$$y - 69{,}065 = 4{,}375X_1^2 + 13{,}329X_2^2, \quad (2)$$

or

$$\frac{(X_1)^2}{\frac{y-69{,}065}{4{,}375}} + \frac{(X_2)^2}{\frac{y-69{,}065}{13{,}329}} = 1;$$

– the sunflower cake temperature at the exit from the extrusion auger of the die and the matrix die hole area (Figure 2 b):

$$y - 69{,}065 = 4{,}375X_1^2 + 4{,}422X_3^2, \quad (3)$$

or

$$\frac{(X_1)^2}{\frac{y-69{,}065}{4{,}375}} + \frac{(X_3)^2}{\frac{y-69{,}065}{4{,}422}} = 1;$$

– the extrusion auger winding pitch and the matrix die hole area (Figure 2 c):

$$y - 69{,}065 = 13{,}329X_2^2 + 4{,}422X_3^2, \quad (4)$$

or

$$\frac{(X_2)^2}{\frac{y-69{,}065}{13{,}329}} + \frac{(X_3)^2}{\frac{y-69{,}065}{4{,}422}} = 1.$$

Substituting different values of the responses in (2) – (4), we obtain the equations of the corresponding contour curves of the response surfaces: for (2) this is the family of ellipses, and for (3) and (4) – the curved surfaces of the minimax type. The graphical calculation results are shown in Figure 2.

The analysis of the given two-dimensional sections $X_1 - X_2$ (Figure 2 a) shows that with a decrease in the factors X_2 (the extrusion auger winding pitch) and X_1 (the material temperature at the exit), the specific energy consumption indicator decreases and has a minimum value of E = 70 kW-h/t.

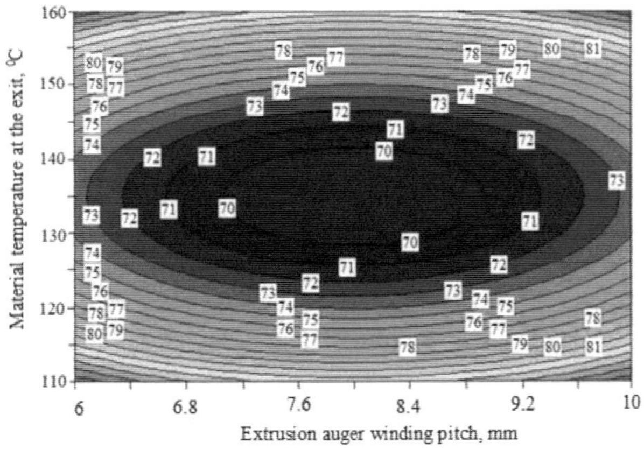

a

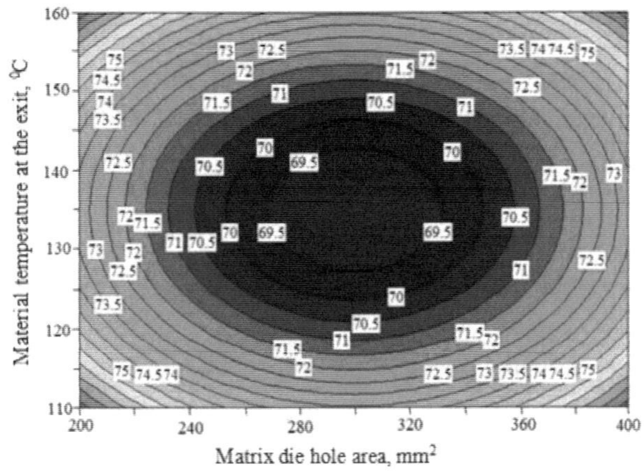

b

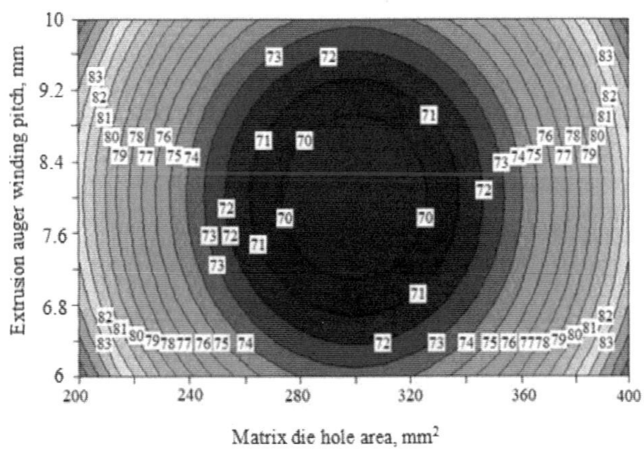

c

Figure 2. Two-dimensional sections to study the X_1, X_2 and X_3 factors impact on the specific energy consumption of the extruder: *a* – the material temperature (x_1) at the exit and the extrusion auger winding pitch (x_2); b – the material temperature (x_1) at the exit and the matrix die hole area (x_3); *c* – the extrusion auger winding pitch (x_2) and the matrix die hole area (x_3)

The analysis of the two-dimensional sections $X_1 - X_3$ (Figure 2 b) and $X_2 - X_3$ (Figure 2 c) shows that the factor X_1 (the material temperature at the exit) and X_3 (the matrix die hole area) fixed at the zero level determine the optimal values:

– for the X_1 factor (the material temperature at the exit) t = 133,67 °C;

– for the X_2 factor (the extrusion auger winding pitch) h = 7,58mm;

– for the X_3 factor (the matrix die hole area) S = 323,89 mm^2 at the minimum specific energy consumption of the extruder E = 69,06 kW-h/t.

The X_2 factor (the extrusion auger winding pitch) has the same impact on the optimization parameter as the X_3 factor (the matrix die hole area), since the contour curves are circular, which shows the uniform impact of each factor on the optimization parameter.

4. Conclusion

The centerpoints of the experiment are located in the study area, which makes it possible to establish the optimal parameters of the factors for their various combinations.

The analysis of two-dimensional sections in interaction with each other makes it possible to determine the optimal values of all the factors under study:

X_1 – the material temperature at the exit t = 133.67 °C;

X_2 – the extrusion auger winding pitch h = 7.58 mm;

X_3 – the matrix die hole area S = 323.89 mm2.

With these values of the factors, the minimum specific energy consumption of the improved extruder E = 69,06 kW-h/t compared to serial ones is ensured due to the implementation of the intake and the extrusion augers connected in sequence in the form of a catenoid, and the joints of the sections in the form of a parabola according to the normal distribution law, as well as the tops of the conical head in the form of a parabola, which affects that the material is not damaged when leaving the conical head, and affects its mass and density.

References

[1] Korotkov V G, Popov V P, Solovykh S Yu, Voloshin E V and Antimonov S V 2018 Optimization of the processing of grain raw materials on a screw press extruder *Bull. of Orenburg State Agrarian University* **4(72)** 186-189

[2] Timofeeva D V, Korotkov V G, Popov V P et al 2013 Optimization of the process of transformation of the aggregate state of grain raw materials during extrusion processing *Bread products* **8** 46-48

[3] Kabdusheva A, Kurmanov A, Khasenov Y, Sapa V and Amantayev M 2020 The effect of parameters on the performance efficacy of a single screw feed extruder using a design experiments and response surface methodology *Bulgarian J. of Agricultural Science* **26(2)** 492-497

[4] Popov V P, Martynova D V, Antimonov S V et al 2017 Theoretical substantiation of the energy- and resource-saving design of a screw press extruder for the production of high-quality feed products *Bull. Orenburg State Agrarian University* **6(68)** 107-109

[5] Timofeeva D V, Kishkilev S V, Popov V P et al 2015 Study of the transformation of the structural and mechanical properties and chemical composition of protein-starch-fiber-containing raw materials in the channel of a single-screw press extruder *University complex as a regional center of education, science and culture: Mat. of All-Russian sci. and methodical conf.* pp 1007-1014

[6] Li H, Xiong Z and Li X 2019 Optimization of the extrusion process for the development of extruded snacks using peanut, buckwheat, and rice blend *J. of Food Processing and Preservation* e14264

[7] Fouad D M, Moataz A, Salem H G and El-Garaihy W H 2019 Multi-objective optimization of multi-channel spiral twist extrusion process using a response surface approach and finite element analysis *Mater. Sci. and Technology* **2018** 1470-1477

[8] Kryuchkov D I and Zalazinsky A G 2017 A hybrid modeling system designed to support decision making in the optimization of extrusion of inhomogeneous materials *AIP Conf. Proc. "Mechanics, Resource and Diagnostics of Materials and Structures, MRDMS 2017: Proc. of the 11th Int. Conf. on Mechanics, Resource and Diagnostics of Materials and Structures"* 040028

[9] Pakhomov V I, Braginets S V, Alferov A S, Gaidash M V and Stepanova Yu V 2016 Studies of the process of extruding a mixture of grain concentrates with crushed green mass of leguminous herbs *Bull. of Don State Technical University* **2(85)** 154-159

[10] Novikov V V, Konovalov V V, Inozemtseva L V and Belyaev D V 2010 Experimental substantiation of rational parameters of the modernized extruder KMZ-2U *The Volga Region Farmland* **4(17)** 48-51

[11] Melnikov S V, Aleshkin V R and Roshchin P M 1980 *Planning an experiment in the study of agricultural processes* 168 p

Developing optoelectronic devices based on AFV properties to control the physical and chemical parameters of oil products

N Rakhimov, V Sh Mukhametshin and K T Tyncherov

Ufa State Petroleum Technological University, Branch of the University in the City of Oktyabrsky, 54a, Devonskaya St., Oktyabrsky, Republic of Bashkortostan, 452607, Russian Federation

E-mail: nemrah@mail.ru, info@of.ugntu.ru

Abstract. This article deals with the development of optical devices based on anomalous photovoltage films. We propose technology for the manufacturing of the optical radiation receiver based on anomalous photovoltage films and optoelectronic sensors to control the physical and chemical parameters of substances and materials. The photoreceiver is a semiconductor device that registers optical radiation and transforms the input optical signal into the output electrical signal of the photodetector. The goal of this work is to facilitate automation and improve labor productivity, as well as miniaturize and improve the quality of oil product analysis and develop a new type of telemetric sensors. The use of generator photoreceivers based on multilayer composition (MLC) as sensors facilitates the fast and accurate oil product measuring and monitoring without contact even for their significant quantities.

1. Introduction

The efficiency improvement for the development of oil reserves that are difficult to recover largely depends on the parameter control for the processes in oil beds and the changes in properties of produced fluids [1–8]. It is also crucial to increase the quality of oil product analysis using telemetric sensors.

2. Materials and Methods

The development of metering and monitoring using optics in the oil and gas industry is fraught with the emergence of various optoelectronic devices. These optoelectronic devices have various functions in nondestructive inspection systems. The generic photoreceivers (photodiodes, photoconductors, phototransistors, etc) in the optocouplers are used to control just one parameter of the object. If a generator MLC photoreceiver is used, the metering circuit becomes versatile [9, 10].

Anomalous photovoltage (APV) films are basically special converters that transform the light input into the anomalously high photovoltage [11]. This fact and the study of external impacts on the properties of films became the basis for the development of a range of optoelectronic devices.

The APV films were examined using a powdered cadmium telluride in vacuum with the substrate temperature of 200–250 °C. The film thickness was between 0.7 and 1.5 um.

The reverse side of the APV film was exposed to an artificial lighting source with the intensity of $\sim 8 \cdot 10^{-2}$ W/cm^2 and the power of 300 V/cm^2. During the experiment, we established that the reflection of light leads to the anomalous photovoltage effect. We assumed that if the reverse side of the glass or quartz substrate was treated with Ag reflecting coating, the value of the APV effect shall change. The application of the silver reflecting coating on the reverse side of the glass substrate was done under the pressure of 10^{-4} mm Hg and the temperature of 250–600°C. These conditions were created to reduce the possible oxidation of the silver reflecting coating.

It is known [1] that the matrix APV receiver transforms the distributed radiation flux (optical image) into an electrical potential relief, creating on a glass-metal plate with discrete anisotropic conductivity, consisting of a glass base and thin covar wires piercing it. On one side of such a glass-metal plate, a thin layer of platinum is applied by cathodic sputtering, after which special holes are created with the help of a covar etchant.

A photosensitive semiconductor film is applied to the surface of the plate prepared in this way by thermal evaporation in a vacuum. This film generates APV under illumination. The deposition is carried out at an angle to the normal, as a result of which a discrete photovoltaic element in the form of an APV film is formed on the lateral surface of each hole, contacting one end with a covar fiber (output of an elementary photocell) and the other end contacting with a common platinum electrode on the surface of the plate. The system of such microphotocells forms a photodetector matrix, which is an autonomous sensor of multielement optical information.

With this in mind, the film production process was carried out as follows:

1. On one side of the glass substrate positioned at 45° to the molecular beam direction, a 0.2 um thick layer of cadmium telluride (CdTe) was applied under the temperature of 420-600 °C and the pressure of 10^{-5} mm Hg.
2. On the reverse side of the glass substrate, a 1 um thick reflecting layer of silver was applied using the thermal evaporation method under the substrate temperature of 250–300°C and the pressure of 10^{-4} mm Hg.

3. Results

Example 1. The photosensitive layer was applied by the thermal evaporation of crystalline cadmium telluride under the temperature of 400 °C in vacuum and the pressure of 10^{-4} mm Hg on the glass substrate. The substrate was positioned at 45° to the direction of the molecular beam. The thickness of the film was 0.2 um. Then the reverse side of the glass substrate was coated with a reflecting layer of silver using the thermal evaporation method under the substrate temperature of 250°C and the pressure of 10^{-4} mm Hg. The film thickness varied within 1 um. The value of anomalous photovoltage under the luminance of 10^4 lx, in this case, was about 210–215 V.

The operating principle of any optron is based on the transformation of the electric signal into optical radiation [9, 10]. Thus, the photon acts as the information medium and facilitates a rather high optoisolation of the optron input and output.

This led us to formulate the main problem, i.e. the development of an APV film-based optron. This device must have the same properties as the emitter-photodiode, emitter-photocell, and other similar optrons.

We must provide that the developed APV-film optron can amplify even very low signals. During the experiments conducted, we found out that optrons with direct optical coupling based on the APV effect facilitated a steady signal amplification in the optical radiation receiver (ORR). Such optrons are optoelectronic voltage transformers with electromagnetic coupling.

In general, depending on the functional capabilities of the ORRs, they are divided into amplitude (integral) ones, which respond only to the average illumination value of their photosensitive surface, and coordinate-sensitive (informational) ones, which respond to the spatial distribution of the optical signal. Coordinate-sensitive ORRs, in turn, are subdivided into selective transformational, position-sensitive and two-dimensional coordinate-sensitive ones.

Selective transformational ORRs are used for image analysis and scanning, which converts a spatially distributed optical signal into a temporal distribution of an electrical video signal. Scanning can be based on a wide variety of physical phenomena using, as a rule, linear-progressive scanning of the image.

The authors investigated the possibility of applying the technology developed to control the physical and chemical parameters of various substances and materials using the APV effect. To this end, high-quality APV films were produced based on CdTe admixed with In, Cu, Al, and Se. The operating principle of the APV film is shown in the figure.

The authors also studied the operating principles of the optical radiation receiver. It was synthesized from thin semiconductor materials and shaped like a wedge. Its spectral range turned out to be pretty wide: from UV to IR.

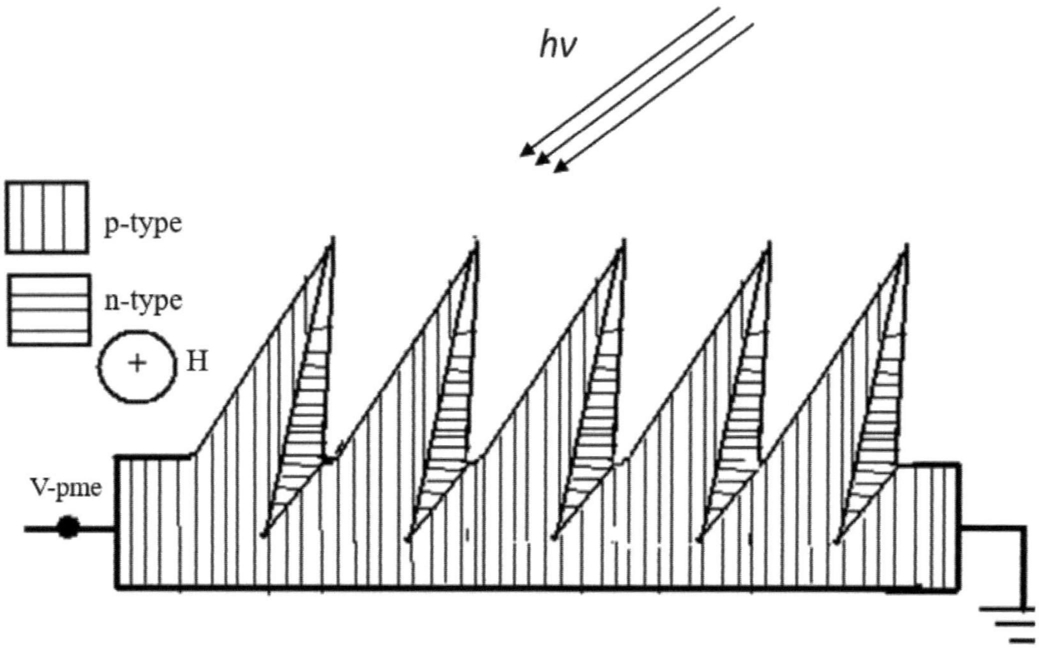

Figure 1. The operating principle of the anomalous photovoltage film

The number of anomalous photovoltage films in the photoreceiver was calculated. Varying the layer production technologies, we could change the structure of transmission electrons to obtain, for instance, the heterojunction, for which the offset of continuity band was equal to the difference between the electron affinities of two materials.

The produced optical radiation devices based on the APV effect have very high photoconductivity. The spectral range of photosensitivity indeed covered the range from the UV to IR regions [11, 12].

We also studied the sensitivity of the devices to the magnetic fields. In multiple layer elements, anomalously high photomagnetic effects were observed.

As we established, the optoelectronic devices with MLC elements can be used to meter and research the topology of magnetic fields and objects. Generic photoreceivers need a power supply in their circuits to operate, and MLC photoreceivers with the APV effect do not require a power supply. Such photoreceivers become voltage sources when exposed to light.

4. Conclusion

The optical radiation elements in optoelectronic devices become autonomous photoreceivers with light excitation. Using such devices in nondestructive inspection and oil product analysis systems, it is possible to measure mechanical geometrical parameters of various objects and to identify the composition and structure of materials.

The multilayer photoreceiver reacts to light polarization. This property can be used to identify defects, evaluate surface roughness, and measure the thickness of complex structure objects. Besides, the exposure to light creates internal fields in MLCs, which gives them new photoelectronic properties.

Using MLC as the basis, it is possible to develop a matrix photoreceiver without any external power supplies working as an image converter (for various objects and items) into the electric potential pattern. In this case, the MLC-based optoelectronic device will operate as a control device for the product's appearance.

As it turned out, one of the most promising optical methods for non-destructive testing is an optoelectronic method based on an emitter and an APV receiver, which eliminates an external power source for an ORR, reduces weight and dimensions, and provides complete electrical isolation between the "light-emitting diode - ORR" circuits.

The optoelectronic primary converters using APV receivers as autonomous ORR open extensive opportunities in the development of controlling and metering devices for physical and chemical parameters of compounds and materials, which is completely confirmed by the existing extensive experience of using APV receivers.

References

[1] Miller G A 2007 Some Perspectives on Various Methods of Oil Shale Extraction, Piceance Basin, Colorado *27th Oil Shale Symp. Colorado School of Mines Golden (Colorado, 15-19 October 2007)* p 14

[2] Akhmetov R T, Mukhametshin V V, Andreev A V and Sultanov Sh Kh 2017 Some Testing Results of Productive Strata Wettability Index Forecasting Technique *SOCAR Proceedings* **4** 83–87. DOI: 10.5510/OGP20170400334

[3] Mukhametshin V V 2017 Eliminating uncertainties in solving bottom hole zone stimulation tasks *Bulletin of the Tomsk Polytechnic University. Geo Assets Engineering* **328(7)** 40–50

[4] Akhmetov R T and Mukhametshin V V 2018 Estimation of displacement coefficient with due account for hydrophobization of reservoir using geophysical data of wells *IOP Conf. Ser.: Earth Env.* **194(6)** 062001. DOI: 10.1088/1755-1315/194/6/062001

[5] Kimos A 2011 Engineering Natural Gas from U.S. Shale Deposits. Retrieved from: http://buildipedia.com/aec-pros/engineering-news/extracting-natural-gas-from-shale

[6] Batalov D A, Mukhametshin V V, Dubinskiy G S and Andreev V E 2018 Laboratory grounding of waterproofing sealant based on acrylic polymers *IOP Conf. Ser.: Earth Env.* **194(4)** 042003. DOI: 10.1088/1755-1315/194/4/042003

[7] Rogachev M K, Mukhametshin V V and Kuleshova L S 2019 Improving the efficiency of using resource base of liquid hydrocarbons in Jurassic deposits of Western Siberia *J. of Mining Institute* **240** 711-715. DOI: 10.31897/PMI.2019.6.711

[8] Akhmetov R T, Kuleshova L S and Mukhametshin V V 2019 Application of the Brooks-Corey model in the conditions of lower cretaceous deposits in terrigenous reservoirs of Western Siberia *IOP Conf. Ser.: Mater. Sci. Eng.* **560(1)** 012004. DOI: 10.1088/1757-899X/560/1/012004

[9] Grundmann M 2020 *Nano-optoelectronics – concepts, physics, and devices* (Berlin)

[10] Sakai K 2005 *Terahertz optoelectronics* (Berlin)

[11] Rakhimov N R and Seryoznov A N 2005 *Coordinate-sensing optical receiver* Patent RF № 2246779 bul no 5

[12] Rakhimov N R, Umarova G A and Zakirova K D 2019 Studying optoelectronic detectors for fatigue parameters control on machine-building metal surfaces *J. of Phys.: Conf. Ser.* **1333(6)**

IOP Conf. Series: Materials Science and Engineering 1111 (2021) 012051 doi:10.1088/1757-899X/1111/1/012051

Development of digital models and information systems for machine-building enterprises

K Yu Zhigalov[1], Kh A Daudov[2] and A V Gorban[3]

[1]V.A. Trapeznikov Institute of Control Sciences of Russian Academy of Sciences
[2]Chechen State University, 17a, Dudaev Boulevard, Groznyy, 364015, Russia
[3]Irkutsk National Research Technical University, 83, Lermontov str., Irkutsk, 664074, Russian Federation

E-mail: kshakalov@mail.ru

Abstract. The paper considers the use of artificial intelligence systems based on neural systems for the automatic construction of a digital model of a machine-building enterprise. For this purpose it is proposed to split the task into two parts: automated design of a digital model of a terrain and three-dimensional terrain objects. Besides, for the automated construction of three-dimensional objects, they must be previously classified. Accordingly, we solve three problems simultaneously. As a result of the described techniques, the Geoinformation System can be used to automate the enterprise plan in real time.

1. Introduction

Digital terrain models are now widely used everywhere. They can be used for a wide range of tasks, for example, to build routes of equipment and unmanned aerial vehicles. However, at present, both the relief and the terrain as a whole are changing quite quickly. New buildings and engineering structures appear [1]. This is especially true in cities and suburbs. In this regard, the tasks of automated construction of three-dimensional maps, which are essentially the DSM, are quite relevant.

The construction of the DSM can be divided into two large stages:

- Construction or update of the terrain;
- Construction of terrain models and their placement on the constructed terrain.

2. Construction of digital terrain models

The following input data can be used to perform this task:

- vector maps of the area;
- raster maps of the area;
- aerial laser scanning data;
- ALS data of a given area;
- ground laser scanning data;
- total station data;
- GPS/GLONASS receiver data.

It shall be noted that more than one data source is frequently used. Let us consider the main methods of constructing the DTM currently used:

Construction of a basic DTM according to vector maps:
Under office-compiled conditions, scanned raster images of maps are first geopositioned. For this purpose points are placed on the edges of a map and their spatial coordinates taken on margin design of each sheet are entered. After that, the margins are cut off and the sheet is transformed according to the coordinates. Then the diagonals are vectorized (this is a fairly simple operation, which consists in circling the contours with lines and then assigning them heights). The GIS builds a digital relief model by interpolation over the vectorized contours. If the maps are too old, and there has been any activity in the area recently, there is a need to specify the surface. To do this, surveyors go to the field and make additional measurements using laser scanners or total stations. In parallel, the obtained data is linked to characteristic relief points using GPS/GLONASS receivers [2, 3]. When surveyors return from the field, all this data is transferred to GIS, geo-positioned and tied to the existing DTM. The following procedure is performed to recalculate the terrain model at the points of its deviation. This method is suitable for the construction and design of road sections with a small length (up to 20-30 km).

Construction of the basic DTM according to aerial laser location data:
To do this, a flight route is being built along the entire length of the future route. Then an aircraft with a laser scanner on board flies through it. The laser scanner makes shooting of a terrain with a laser rangefinder and an aerial camera. It should be noted that the aerial laser scanning data (including aerial photography) are already geo-positioned and can be directly loaded into GIS. Since the point clouds contain a number of errors before constructing the DTM, they are filtered for gross errors. Then, according to them, by interpolation, the DTM is built and covered with an aerial photograph for clarity. A point cloud is no longer needed and is excluded from future use in GIS [4].

This method is quite expensive, accordingly it is used mainly in the design and construction of road sections of several tens of kilometers in length.

Construction of the base DTM according to the ALS:
To do this, a flight route is being built along the entire length of the future road. Then an airplane with a camera on board flies along it and performs aerial photography. Under office-compiled conditions, all photos are stitched together by means of matching characteristic points on each picture. This is followed by orthophototranformation of pictures. After that, the sewn sheets must be geopositioned. Geo-positioning is done due to the same characteristic points with the following options:

- field definitions of geospatial coordinates of these points using GPS/GLONASS receivers;
- obtaining coordinates from vector/raster maps.

In both cases, points from transformed photos are assigned the value of reference points from maps or GPS receivers, after which all images in the geospatial coordinate system are recalculated. Further, in the stereo mode, operators place heights on the images [5]. The DTM construction is then started.

This method is somewhat cheaper than the air laser scanning, but the work takes a longer time. It makes sense to use this method when designing and constructing large area facilities.

Construction DTM according to projects
The vector project is loaded into GIS as an additional layer. Geo-positioning of the project into a spatial coordinate system is usually required. This procedure is performed using datum points and mapping them to the basic DTM. Further, over the entire length of the route, excesses or heights are placed relative to the DTM (taken from the plan references). Based on the obtained data, a digital model of the layer is being built. This operation is either repeated for the remaining layers, or copies of this layer are made with the required design excess [6].

3. Construction of terrain objects

As mentioned earlier, this task in its automated solution consists of 2 stages:

To learn how to find objects (classify them) using an artificial intelligence system;

To get the most appropriate objects from the repository and bring them as close as possible in geometry to existing objects.

As part of the study, a complete algorithm for identifying local objects was developed, including introductory data of various types and specification of context information through machine reading methods. The algorithm provides for an automated neurointelligence system with reinforcements.

The classifier method based on a random forest completed with the CNN light network is used here as the basis of the system for determining objects according to snapshots. Some algorithms and methods included in this general algorithm require further study in the following research areas:

• Training reinforcement subsystem;
• Data refinement subsystem from various sources;
• Topographical survey data loading subsystem;
• Validation system of terrain objects determination.

For correct training of the neural AI system, it is necessary to create and refine a classifier.

Taking into account several options for possible data acquisition for the recognition system, data for analysis are collected and classified from different positions:

• Aerial photography (meaning images from both an orbit and aircrafts);
• Land survey.

For each type of data, a multi-stage classifier was created, according to which, in the future, AI training bases were formed.

1. For processing high-altitude survey data, the classifier contains the following units:

• Vegetation (shrubs and trees);
• Constructions (buildings, structures, etc.);
• Infrastructure facilities (roads, railways, squares, parking lots, transformer platforms, airfields, bridges, power lines);
• Land surfaces (fields, beaches);
• Water (rivers, lakes);
• Factories.

2. To process ground survey data, the classifier is modified to include the following units:

• Vegetation (shrubs, trees);
• Buildings (1, 2, n- storey buildings);
• Factories;
• Infrastructure (bridges, roads, tunnels);
• Other infrastructure (fences, pillars, electric poles).

The difference in classifiers is explained by the impracticality of using some classes for some types of information (for example, it is impossible to recognize road signs on altitude survey) and vice versa. Besides, on land survey it is possible to determine the height of buildings (in storeys) and according to studies, on altitude survey it is only possible to separate high-rise buildings from low-rise ones.

All objects that fall into classes must be clearly visible in snapshots. Since the task of recognizing cartographic objects is not entirely trivial (the comparison was made by a set of images in the libraries TensorFlow, IMAGINET (http://www.image-net.org) and others, it turned out that the ratio of images of people, objects, clothes in relation to houses, plants, etc., is approximately 200 to 1). In some classes, there were no training images in these libraries. This comparison suggests that there are currently no full training libraries (like TensorFlow). The primary assembly of the library was carried out in the manual mode, at least 100 photographs, based on different directions of the training systems:

• Aerial photography (footage from satellites, aircraft, UAVs, quadrocopters). The following applications were used to get snapshots:

- Google Earth Pro 7.3.3.7786. Sampling of images from 2014 to 2020;
- SASPlanet 200606 (https://sasplanet.ru);

The images were uploaded to various territories with maps for more convenient testing of recognition quality by comparing objects outlined by neurointelligence (tested by the system) and already mapped in the manual mode;

• Land survey (data collected from machines and on foot). For the test, pictures were taken from the Yandex Panorama using a parser in the network. The data selection condition is similar to space imagery. In this case the data was checked manually.

For the automated collection of the library, we used a specially designed parser. All classes were created as a tree file directory (in the root of the file tree of the main classes with subclasses inserted into them). A parser found and processed open photos of the library on the Internet (including using Google, Yandex) according to a written (contextual) description and placed the corresponding pictures in folders as *.jpeg files.

Searching and downloading (parsing) of pictures from the Internet for subsequent training was carried out using automatic algorithms and manually (50/50). For these purposes, we used the Yandex.Pictures tool (https://yandex.ru/images/), where in the search results next to most images you can see the link “similar”, which opens a page with images similar to the one available when clicking. Accordingly, we used both currently existing image search directions – the sample picture was searched manually for meta-information, and similar images using image search technology for their content and their automatic download.

In all libraries, the images must follow the following principle: in each class, at least 1400 images that meet the following requirements: image file type: *.JPEG, resolution up to 1920x1200, file size up to 2 MB.

Next, you must manually mark all snapshots and conduct a training procedure for the artificial intelligence system. There is a lot of information about this in the Internet, hence we will not focus our attention on it here.

Upon completion of the training, we will get a system that can identify objects at the input information (in our case, snapshots) and classify them. If there is a finished library of such three-dimensional objects, the AI is able to select terrain objects from this library of the model (by recognition) and place them on the previously built DTM.

4. Conclusion

Pictures of the Moscow region were used as tests, the percentage of guessing objects and their correct placement ranged around 89-94%, which for the current stage of research is quite sufficient. In the future, it is planned to train AI to create textures of real objects according to incoming shots, geometrically select similar objects in the library and place prepared textures directly on them. Thus, the maximum visual similarity of the built digital models of the terrain to real objects will be achieved.

References

[1] Klyuev R V, Bosikov I I, Egorova E V, Gavrina O A 2020 *Sustainable Development of Mountain Territories* **V.12 3(45)** 418-427

[2] Klyuev R, Bosikov I, Gavrina O, Madaeva M, Sokolov A 2021 *Advances in Intelligent Systems and Computing* **1258** 262-271

[3] Klyuev R, Turluev R, Khadzhiev A 2020 *International Ural Conference on Electrical Power Engineering (UralCon)* INSPEC Accession Number: **20034001** 324-329

[4] Bovkun A, Korodyuk I 2019 Analysis of the development of small innovative enterprises in the construction industry. *IOP Conf. Ser.: Mater. Sci. Eng.* **667**

[5] Skorobogatova Y A, Bovkun A S, Ivanov M Yu, Shilova O S 2020 The role and place of small innovative enterprises in the construction industry in the modern economy of single industry towns. *IOP Conf. Ser.: Mater. Sci. Eng.* **880**

[6] Repinskiy O D, Konyukhov V Yu, Bovkun A S, Schupletsov A F 2021 Improving the competitiveness of Russian industry in the production of measuring and analytical equipment. *Journal of Physics: Conference Series* **1728**

IOP Conf. Series: Materials Science and Engineering 1111 (2021) 012052 doi:10.1088/1757-899X/1111/1/012052

Modern equipment for core sampling while drilling wells in the fields of Tomsk region

N N Saveleva and S N Shed

Tyumen Industrial University, 38, Volodarsky St., Tyumen, 625000, Russia

E-mail: nnsavelieva@yandex.ru

Abstract. The drillers face the task to increase the volume of work on exploratory wells to the Lower Cretaceous and Jurassic basement of the fracture intervals in Tomsk region. Paleozoic basement Pz and Kv are the objects with a complex geological structure. The main purpose of each coring sampling from all types of wells is to obtain direct information about the rocks that compose the objects of interest in the fields, which is necessary for the solution of a set of problems of prospecting, exploration and production of hydrocarbons. Research tasks are to identify the features of the development of fracture intervals, the conditions of rock deposition, the type of rocks, the fracture orientation, permeability, porosity, saturation of the pore space with fluid and to forecast promising areas in the Paleozoic basement. The paper describes the work on core sampling in the Paleozoic deposits of Tomsk region. The core sampling equipment and the results obtained at the Severo-Kalinovskoye field are presented. According to the results of the analysis of the cores, the conclusions are drawn about the further stages of drilling for the assessment of the deposits of the Paleozoic basement in Tomsk region.

1. Introduction

To start operating new fields, it is necessary to carry out additional exploration of hydrocarbon reserves. Nowadays, the Paleozoic basement is unexplored in Tomsk region. The Paleozoic deposits of Western Siberia can contain significant hydrocarbon resources. However, the question of how to find and extract them efficiently is still open.

Oil exploration work in Western Siberia, including on deposits of the pre-Jurassic complex, began in the 1950s. In 1954, the first deposit was discovered in Tomsk region. Numerous oil and gas shows from the Paleozoic were obtained during drilling on the structures of the South Minusinsk intermountain area. The second stage of oil and gas exploration in the Paleozoic deposits dates back to the 70s-80s, XX century. Thus, in 1974, a powerful oil spring was obtained at the Maloichskoye field in Novosibirsk region. Later, due to purposeful drilling into pre-Jurassic formations, the deposits in Paleozoic sediments were also discovered in Tomsk and Tyumen regions.

In the pre-Jurassic complex, the total initial geological reserves of oil, which are currently state-owned, comprise to 400 million tons in Western Siberia, including 174 million tons belonging to Tomsk region [1]. However, these numbers may increase significantly, because the main problem of Paleozoic reserves is that they have not yet been truly estimated.

The oil of the pre-Jurassic complex is considered hard to recover. At the same time, it is also hard to find: the technologies have a low degree of maturity.

In Tomsk region the most extensive experience in the development of pre-Jurassic oil in Russia was accumulated, most of the Paleozoic deposits were discovered and all possible types of basement deposits were represented. In 2014, the region received the status of a test site for the introduction of the latest technologies for prospecting, exploration and development of unconventional sources of hydrocarbon raw materials, which include the Paleozoic [2].

The expansion of prospecting and exploration objects in the basement of oil and gas basins of Western Siberia and other oil and gas provinces of the Russian Federation corresponds to the transformation stage of the paradigm of oil and gas geology, characterized by the active development of the theory of the deep origin of hydrocarbons, which allows substantiating the Paleozoic deposits as promising in terms of oil and gas deposits exploration.

2. Materials and methods

The higher the information content obtained in the process of well site construction (logging, coring, testing, etc.), the higher the reliability of the assessment of oil and gas fields. The most effective way to obtain this information is core sampling [3]. This is due to the fact that core is the main direct source and carrier of information about rocks, providing the opportunity for visual and direct study of their properties. Therefore, the improvement of the quality of coring is an important task of exploratory drilling in order to improve the quality of the used core sampling equipment.

At the moment, the drillers have a task to increase the volume of work on drilling exploratory wells to the Lower Cretaceous and Jurassic basements of the fracture intervals in Tomsk region. The Paleozoic basement Pz and Kv are the objects with a complex geological structure. However, there are many questions and unsolved problems, especially during the work with coring, what is an urgent scientific and technical problem. This is explained by insufficient knowledge of this fracture in Tomsk region, limited actual material obtained during core sampling from wells, incomplete study of the properties of reservoir fluids, the lack of prediction of the reservoir properties of the geological section of the Paleozoic basement Pz and Kv, low efficiency in the development and development of deposits. The search for modern methods of isolated core sampling has shown that the use of this core sampling equipment can increase the percentage of removal and increase the metering of coring.

3. Results and discussion

The main purpose of each core sampling from all types of wells is to obtain direct information about the rocks that compose the objects of interest in the fields, which is necessary for the solution of a set of problems of prospecting, exploration and production of hydrocarbons. The set of problems is extensive and may include the following tasks such as: the identification of the features of the development of fracture intervals, the conditions of deposition of rocks, the type of rocks, the fracture orientation, permeability, porosity, saturation of the pore space with fluid and forecast of promising areas in the Paleozoic basement.

Drilling companies use Boart Longyear, LF-90, LF-230, LM-75, and other drilling units drill rigs for core drilling. For deep and directional core drilling projects, FORACO BF830 multipurpose drilling rigs are also used. The annual volume of core drilling significantly exceeds 100,000 m. The range of depths of work varies from several tens of meters to 2,500 m.

For core drilling, high-quality drill pipes and tools with a removable coring (double core barrel) of the best world producers are used. The use of high-quality drilling tools, regular maintenance, scheduled preventive and overhaul of drilling rigs, high qualifications and motivation of personnel allow the drilling speed to not fall below 2,000 linear meters in the project portfolio as a whole for 1 machine per month. The core recovery is traditionally not lower than 95%. For sections usually characterized by poor core recovery (less than 90%), triple core drill is used, as well as other specialized drilling technologies. In order to study the spatial strike of rocks and their fracturing, an oriented core sample is taken at the request of a customer.

The drilling rig operators of leading drilling companies independently perform inclinometric surveys in wells using magnetic and optical inclinometers of the Reflex EZ-Shot, EZ-trak and Reflex MAXIBOR II types. For additional geophysical surveys and hydrogeological surveys of wells, time-tested specialized subcontractors are usually invited. Geological support of core drilling is also carried out with the involvement of qualified geological contractors.

The equipment for core sampling while drilling wells in Tomsk region on the Paleozoic basement Pz and Kv. Core sampling is a complex technological operation that requires a lot of time and money, the correct selection of the type of core sampling tool and the type of drilling bits (Figure 1).

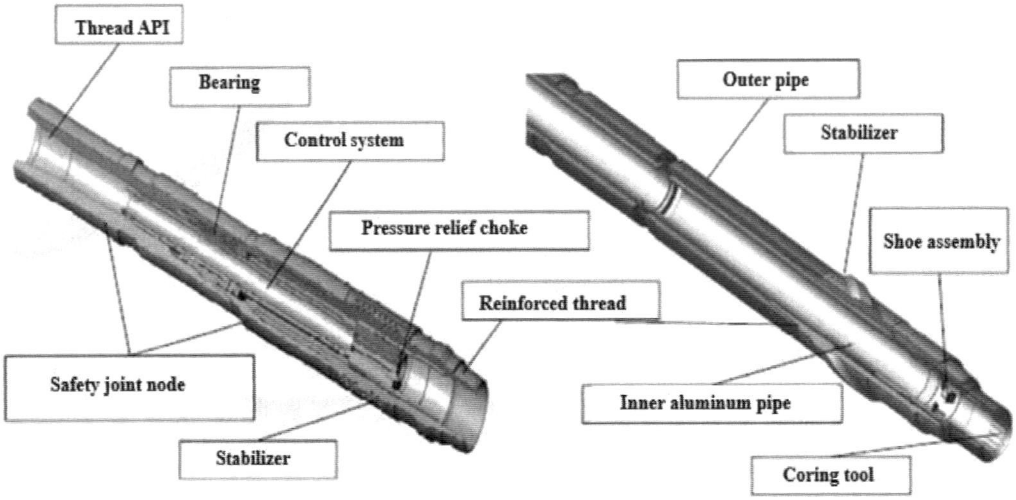

Figure 1. Systems:
a) control internal assembly system; b) core positioning system

In order to solve the set tasks, a core sampler UK-127/80 was used, developed by VNIIBT-BI with a length of 6.0 m and 8.0 m using an isolated technology with the use of disposable fiberglass and steel core receivers and drilling bits produced by Russian factories both UDOL and BURSERVICE.

Basic information and technical data of the coring device UK-127/80 [4]

1. The coring device UK-127/80 (hereinafter referred to as the device) of one-two-section design is intended for core sampling in rocks of all categories when drilling wells with a rotary method or bottomhole motors using cutting drill bits reinforced with diamond-hard alloy plates, synthetic or natural diamonds. The coring device includes a sealed filled disposable coring pipe for filling it with cuttings along the annular space.
2. The device is produced in a single-section version UK-127 / 80.000, which is the base for the assembly of devices consisting of two or more sections. Technical characteristics are shown in Table 1.

Table 1. Core technical characteristics

1. Diameter of the lower and upper stabilizers, mm	$137^{-0,2}$
2. Nominal body diameter, mm	$130^{\pm 3}$
3. Nominal core diameter, mm	80
4. Maximum length of the sampled core, m	6,0+ 8,0 or (12,0+18,0)
5. Temperature range up to	200^0C

Table 2. System of core sampling and standard size of drilling bits for coring.

Type of system of core sampling	Diameter of applied drilling bits, mm		
	142,9	152,4	155,6
UK-127/80	+	+	+

The reliability of the results is confirmed by actual core sampling from well No. 609 of the West Luginetskoye field and wells No. 37 and No. 91 of the Severo-Kalinovoye field. The results of the work of UK-127/80 device in the fields are presented in Table 3.

Table 3. Results of the work of UK-127/80 device

№ of well	Area, deposit	Layer designation	Coring interval, m	Drilling progress, m	Core recovery	
					m	%
37	Severo-Kalinovoye	Pz and KV	3080-3180	46,5	43	92,5
91	Severo-Kalinovoye	Pz and KV	3051-3081,9	29,4	20,95	71,3
609	West Luginetskoye	Pz and KV	2748-3060	173,8	127	73
		Total:		249,7	190,95	77

The high level of the reliability of the assessment of oil and gas fields depends on the level of the information content of the core. Therefore, the improvement of the quality of the core is an important task for exploration drilling, as well as for sidetracking. In addition to the informative one, the problem of the improvement of the quality of the core also has an economic aspect, which consists in the fact that the low quality of the core requires drilling additional wells with its sampling.

That is why, when core sampling with a small diameter, UDOL developed U-109/67 system of core sampling, the development of which in the field of engineering and technology of coring, due to their high efficiency, gained fame and recognition. In the context of their scientific and technical level, the applied systems of core sampling and the isolated core sampling technology are superior to national and international analogues and can significantly increase the information content of the core material, due to the increase in core recovery up to 90-100%. It is necessary to note that the operational characteristics of the existing core barrels of UKS-U-109/67 series of various sizes and the ease of maintenance and high reliability do not require special adaptation of the drilling crew when using them. Moreover, during the work, the technique and technology of coring is constantly being improved.

Equipment for core sampling during sidetracking

There are many different systems and types of coring tools for core sampling during sidetracking (Table 4).

Table 4. Core sampling devices used for sidetracking coring

Type of technical means	Casing diameter, mm			
	139,7	146	168	178
UKSU-109/67	+	+	-	-
	Diameter of drilling bits, mm			
	120,6/67	122/67 123/67 126/67	142,9/80 152,4/80	154/80 156/80

The use of an internal core pipe located in the steel body of the core device has two functions: it improves the quality of the core material, the material of which is physically supported during the extraction of the core and the internal core pipe serves as a protection system for the core raised to the surface. The material for the production of internal core pipes can be PVC, ABS polymer plastic, aluminum, steel, fiberglass and other composite materials. In practice, the most commonly used are steel and fiberglass pipes. Internal core pipes are inserted inside a traditional coring machine and held in place by a coring tool assembly and frictional force.

4. Conclusion

In practice, disposable internal core-receiving fiberglass receivers are used, produced by kompozitmash, Perm. The use of a disposable internal core pipe allows increasing the safety of the core by performing all operations with the core after extraction from the well without removing the core from the pipe. Internal coring pipes usually have a sectional design with a long section reaching 6-12 m. During core sampling, a two-section device is used and should be used. Fiberglass pipes can be used for temperatures between 80-100°C. Fiberglass pipes can be used up to temperatures of 120°C and 180° C using a special high temperature resin. However, in the process of coring, especially during sampling in the weathering crust and the Paleozoic, core plugging often occurs, both in the assembly and inside the core-receiving soil dump. There are a lot of reasons leading to the destruction of the core, locking during drilling and lifting of the tool.

In order to avoid the rise of the system of core sampling, after the incoming core is locked, Kompozitmash plant in Perm has developed an anti-locking telescopic system with one lock during the selection process for this type of device. This system makes it possible, after a lock has occurred, to fix core receiver and ensure that they can be triggered at the required moment and will allow the continuation of core sampling when a lock occurs in the core receiver. In this case, the diameter of the sampled core corresponds to the diameter, as indicated in the certificate of the drilling bit.

The paper describes the work on core sampling in the Paleozoic deposits of Tomsk region. The results are presented at the Severo-Kalinovskoye field. Nowadays the core is being studied in the laboratory. According to the results of the core analysis, the analysis will be carried out and a strategy for further core sampling in the Paleozoic deposits of Tomsk region will be developed.

References

[1] Zubova E A 2019 On increasing the innovative potential of oil and gas enterprises at the present stage *Science and business: ways of development* **7(97)** 105-107

[2] Goncharov A V 2000 *Features of the geological structure of the Paleozoic oil and gas promising deposits of the Nyurolskaya depression, Western Siberia:* candidate dissertation (Moscow)

[3] Punanova S A and Shuster V L 2018 A new look at the prospects for oil and gas content of deep-seated pre-Jurassic deposits of Western Siberia *Georesursy* **20(2)** 67-80. DOI: 10.18599 / grs.2018.2. p.67-80.

[4] Saveleva N N 2020 *Oil and gas equipment. Well equipment: a tutorial* (Tyumen: TIU) 102 p

[5] Saveliev Ya V and Savelieva N N 2019 Automation of industrial processes and every life *IOP Conf. Ser.: Mater. Sci. Eng.* 012068

Comparative study of state of charge estimation algorithms for Lithium-Ion Battery

N Shamarova[1], P Komarnicki[2] and C Wenge[2]

[1] Irkutsk National Research Technical University, 83, Lermontov Street, Irkutsk, 664074, Russia
[2] Magdeburg-Stendal University of Applied Sciences, 2, Breitscheidstr. Magdeburg, 39114 Germany

E-mail: shamarovanataliia@gmail.com, k15@istu.edu

Abstract. Batteries are known to have found full applications in industries ranging from small electronic devices, mobile phones to complex systems such as electric vehicles and energy storage systems. The last one is characterized by the presence of large batteries with more than hundreds or thousands of batteries, so the process of estimation of their parameters becomes even more critical, demanding and complex. The evaluation of the state of batteries should be understood as finding the exact value of the following values: state of charge, capacity and internal resistance. The last two parameters are used to assess the health of batteries, to find available power and amount of energy respectively. Estimation the state of charge remains one of the most difficult tasks in the study of battery conditions. It is also one of the most important parameters that affect the correct operation of the battery, namely reliability, safety, performance. The main task of this paper is to present methods for reliable testing and evaluation of lithium batteries under different scenarios and conditions

1. Introduction

The active development of generation based on renewable energy sources requires the use of electric energy storage systems. This is due to the stochastic nature of generation from such sources [1-5]. Lithium-Ion Batteries are the most common for power systems with a large share of generation from renewable energy sources. Optimization of energy storage is required. A large number of methods and models for assessing battery state of charge are currently available in the literature. In general, they can be divided into two large groups: electrochemical methods and empirical methods [6, 7]. Electrochemical methods are highly accurate, but they can only be applied in laboratory conditions because they are highly computationally complex.

The most widely used methods are based on empirical measurement, an engineering approach that allows online measurement because they do not require large computational resources, while their accuracy can hardly be proven in practical use. Unlike other parameters as the current voltage, which can be measured directly with measuring meters, the state of charge cannot be measured directly, this is due to the fact that this parameter is sensitive to thermodynamic and kinetic processes occurring inside the battery. The state of charge must be determined in a thermodynamic equilibrium state, but in practice the available useful energy is described under polarization conditions and will be a function of numerous factors regulated by kinetic regimes with electrochemical, mechanical and thermal deviations. Moreover, these factors are often interrelated, making measurement even more difficult, making empirical methods of correlation-based assessment ambiguous and difficult to understand [8]. This paper has considered two widely used empirical approaches to assessing battery charge status, such as the ampere clock method and the open circuit voltage method. The residual capacity was also determined and the Coulomb efficiency factor was calculated.

2. Methods

Charge status (SOC)

The evaluation of the state of charge using the empirical method is mainly carried out using the Coulomb account method, also called the ampere-hour counter. Its implementation requires knowledge of the battery charge at the first moment t_0. The following formula describes the state of charge:

$$SOC(t) = SOC(t_0) - \frac{\eta_i}{Q}\int_{t_0}^{t} i(t)dt \tag{1}$$

where *SOC (t)* - indicates the state of charge of the battery at the moment *t*, η_i - Coulomb efficiency factor (for charge η_i <1 and for discharge η_i =1) *Q* - initial cell capacity, *i(t)* - current flowing through the battery at moment t.

This is the common and least computationally demanding method of measuring SOC. The pendant counting method is part of many methods for assessing the state of charge of lithium-ion batteries. Although its simplicity in implementation, this method has a number of significant disadvantages [9]:

- the need to know the battery charge status at the start of time t0 to correctly calculate the integral value.
- overestimating or underestimating the true value of the charge state may occur due to the denominator value of the nominal capacity.

However, the method can be reasonably accurate if there are good current measurements and sufficient recalibration points.

Open circuit voltage method (OCV)

The assessment of the state of charge can be made using a predetermined relationship between the open circuit voltage and the battery charge state. However, this algorithm requires a lengthy experiment in which the energy storage device must be decommissioned over an extended period of time to determine the above relationship, making this approach impossible to apply in real time. This type of battery charge measurement is more applicable in the laboratory. In the case of lithium-iron phosphate (LiFePO4), final balance can be achieved after the first 60 minutes after relaxation.

It is worth considering that, for LiFePO4 batteries, the OCV curve has a long constant part and also has a hysterical effect, which leads to significant deviations in the assessment of charge status [8, 10].

However, the SOC-OCV curve is ideal for the same batch of lithium-ion batteries, allowing experimental curves for online measurements to be used [11]. In addition, the OCV hysteresis can usually be neglected at medium and high temperatures [12]. Direct assessment of SOC based on OCV has very low computational complexity and relatively high accuracy.

Capacity

Determining the actual capacity of a battery pack is one of the most important testing methods and is usually the first step in determining its condition. The nominal capacity of a battery is exponentially dependent on the load current, as described by Pakert's empirical formula (2), which was presented by B. by Pakert in 1897 [13]. The inability to accurately account for this effect leads to significant errors in the determination of nominal capacity, especially at high discharge currents of the accumulator. The value of full capacity is determined by laboratory tests for a specific cell design.

$$C = I^n T \quad (2)$$

where: C is the theoretical capacity of the battery expressed in Ah, T is the discharge duration (time) in h, and n is the Pukert number constant for a given battery. In practice, the value is between 1.01 and 1.4 and for an ideal battery n = 1 [13].

3. Parameter measurement

Capacitive test

The battery cell used for testing is a standard 100 Ah LiFePO4, with a maximum charge voltage of 3.6 V and a minimum discharge voltage of 2.8 V.

The battery is discharged at various direct currents of 1C, 2/3C and 1/3C to determine the actual capacity of the battery. The results of the experiment are shown in Figure 1. The discharge rate of the battery has a significant effect on the voltage curve - as the battery increases, the voltage curve shifts down noticeably. When the discharge current is high (100 A), the discharge curve will deform and the battery capacity will noticeably decrease.

Table 1 shows the results of the test, from which it can be seen that the highest actual battery capacity was extracted at the lowest current.

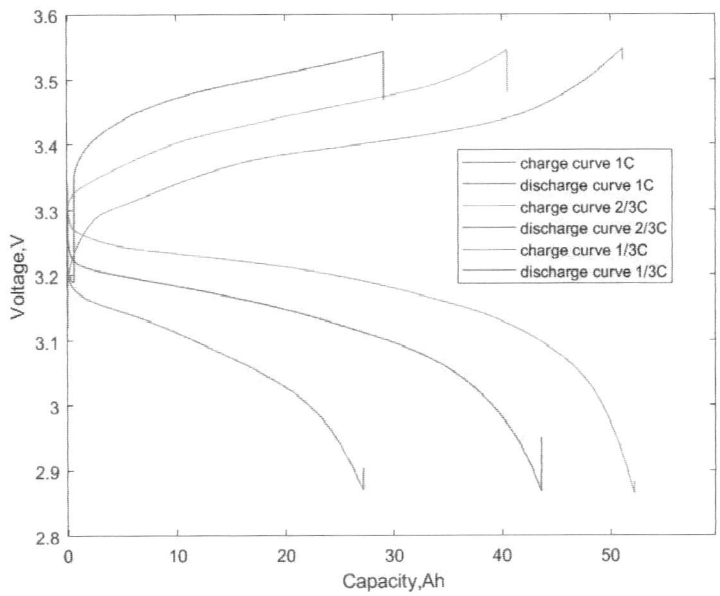

Figure 1. Results statistic capacity test

Table 1. Results of static capacitive test

Current (A)	Capacity-Charge (Ah)	Capacity-Discharge (Ah)	Average
1C	29,11	27,15	28,13
2/3C	40,48	43,56	42,02
1/3C	51,14	52,18	51,66 (max)

A 10 year old battery was used for the test, but the number of charge cycles was small, so the reason for the loss of capacity is more related to the cell self-discharge. The degree of capacity loss was about 50% according to the test results. The calculated actual capacity was used to assess the state of the charge using the ampere-hour method.

OCV Test

The OCV-SOC correction curve is often used to correct the current integral errors during the ampere-hour charge evaluation. This curve is used when the battery is at rest (not charging or discharging) for a sufficiently long time (30-60 minutes) and it is assumed that the battery voltage at the terminals is approximately equal to the OCV value [10]. A cell charge and discharge test with an impulse direct current of 100A was used to remove the OCV. The duration of the pulse phase is 25 s while the relaxation phase is 30 s. Figures 2-3 shows the results of voltage measurements during a complete experimental discharge and charge under the action of a pulse current. The open circuit voltage corresponds to the points marked with a marker in the diagrams below. The OCV-SOC curve is obtained by averaging the curve of charge and discharge as shown in Figure 4. The interpolation of the curve can be obtained by using a polynomial of sixth degree.

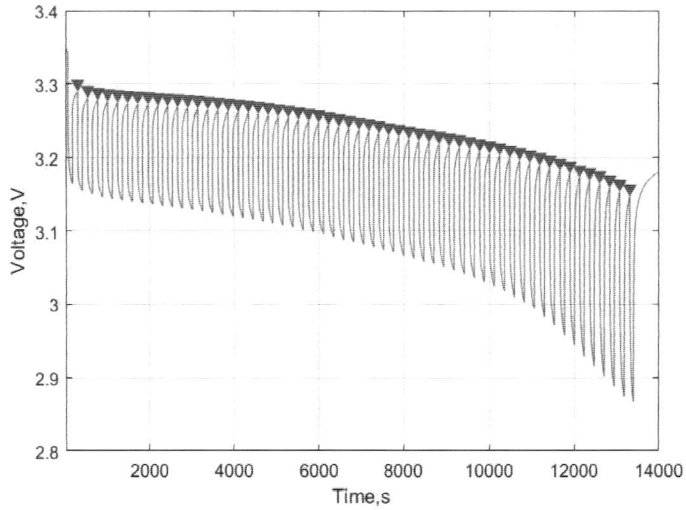

Figure 2. OCV-Discharge test

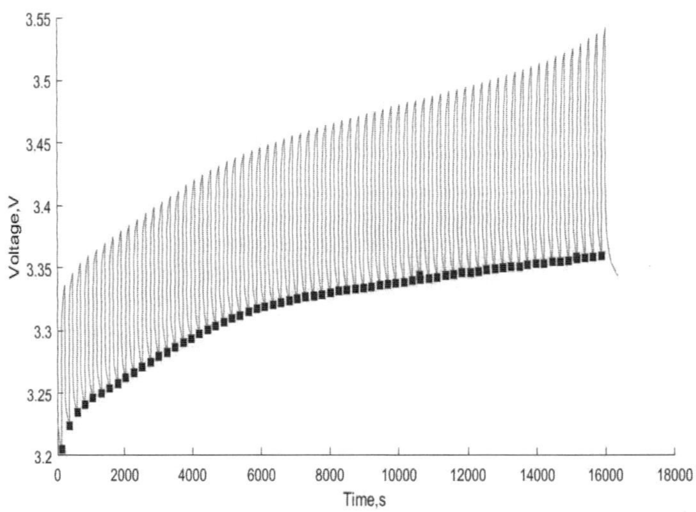

Figure 3. OCV-Charge test

As it can be seen from the graph, the distance between the discharge and the OCV curve is large enough to reflect the hysterical effect that could lead to a charge error in the future. In fact, the hysteresis is related to the relaxation duration of the stress, with the hysteresis level decreasing as the relaxation period between pulses increases due to diffusion of lithium ions inside the cell.

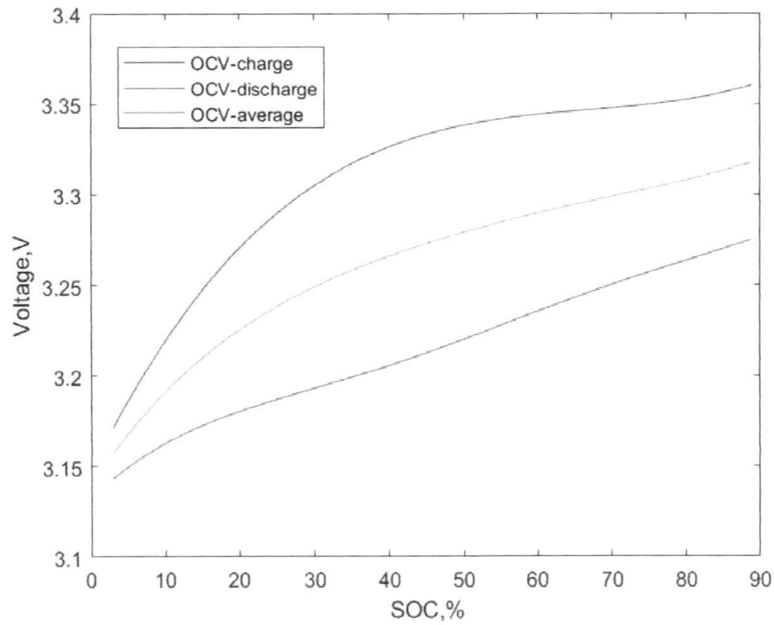

Figure 4. OCV-SOC relationship

4. Conclusion

The analysis of the state of the battery, by determining the residual capacity and coulomb efficiency, confirms the previously mentioned battery age, however, the battery can still be used with alternative energy sources or as an uninterruptible power supply. The accuracy of the method for the estimation of the amperes of a watch depends on a correctly set residual capacity. As the lifetime of the battery increases, it decreases significantly; late calibration may lead to an error in the determination of the charge and, as a result, irreversible damage to the battery. The charge assessment can also be assessed using the OCV-SOC ratio curve, but the hysterical effect must be taken into account.

Acknowledgments

The research was carried out within the state assignment of Ministry of Science and Higher Education of the Russian Federation (project code: 0667-2020-0039)

References

[1] Del Carpio Huayllas T E, Ramos D S and Vasquez-Arnez R L 2011 Microgrid systems: Current status and challenges *IEEE/PES Transmission and Distribution Conf. and Exposition: Latin America, T and D-LA* **2010** 5762853

[2] Suslov K V and Voropai N I 2011 The microgrid concept and challenges in small isolated regions of Russia *CIGRE 2011 Bologna Symp. - The Electric Power System of the Future: Integrating Supergrids and Microgrids*

[3] Hatziargyriou N, Yadusky W, Taylor J, Bak-Jensen B, Barlow M, Joos G, Rotheram A and Baitch A 2018 The impact of battery energy storage systems on distribution networks *CIGRE-Int. Council on Large Energy Systems*

[4] Suslov K, Piskunova V, Gerasimov D, Ukolova E, Akhmetshin A, Lombardi P and Komarnicki P 2019 Development of the methodological basis of the simulation modelling of the multi-energy systems *E3S Web of Conf.* **124** 01049. DOI: 10.1051/e3sconf/201912401049
[5] Suslov K, Solonina N and Gerasimov D 2018 Assessment of an impact of power supply participants on power quality *Int. Conf. on Harmonics and Quality of Power, ICHQP.* DOI: 10.1109/ICHQP.2018.8378836
[6] Jiang S 2011 A parameter identification method for a battery equivalent circuit model *Technical report, SAE Technical Paper*
[7] He H, Xiong R and Fan J 2011 Evaluation of Lithium-Ion Battery Equivalent Circuit Models for State of Charge Estimation by an Experimental Approach *Energies* **4(4)** 582–598. DOI: 10.3390/en4040582,2011
[8] Li Z, Huang J, Liaw B Y and Zhang J 2017 On state-of-charge determination for lithium-ion batteries *J. of Power Sources* **348** 281–301
[9] Khandorin M M 2014 Estimation of the residual capacity of a lithium-ion battery in real time *Electrochemical power engineering* **14** 65–693
[10] Baccouche I, Jemmali S, Manai B, Omar N and Amara N 2017 Improved OCV Model of a Li-Ion NMC Battery for Online SOC Estimation Using the Extended Kalman Filter *Energies* **10(6)** 764
[11] Zhang C, Jiang J, Zhang L, Liu S, Wang L and Loh P A 2016 Generalized SOC-OCV Model for Lithium-Ion Batteries and the SOC Estimation for LNMCO Battery *Energies* **9(11)** 900
[12] Zheng Y, Ouyang M, Han X, Lu L and Li J 2018 Investigating the error sources of the online state of charge estimation methods for lithium-ion batteries in electric vehicles *J. of Power Sources* **377** 161–188
[13] Komarnicki P, Haubrock J and Styczczynski J 2018 *Elektromobilität und Sektorenkopplung – Infrastruktur und Systemkomponenten* (Springer Verlag)

The development of microwave installations for separating the fluff from skins of rabbits in the agricultural enterprises of low power

E Shamin[1], G Novikova[2], O Mihkailova[3], M Prosviryakova[4], N Kirillov[5], V Storchevoj[6] and S Zaitsev[7]

[1] Institute "Food technologies and design", branch of Nizhny Novgorod state engineering and economic university, 13, Gornaya str., Nizhny Novgorod, 603062, Russia
[2] Laboratory of energetic and technological equipment, Nizhny Novgorod state engineering and economic university, 22a, Oktyabrskaya str., Knyaginino, 606340, Russia
[3] Chair "Infocommunicational technologies and net systems", Nizhny Novgorod state engineering and economic university, 22a, Oktyabrskaya str., Knyaginino 606340, Russia
[4] Chair "Electrification and automation", Nizhny Novgorod state engineering and economic university, 22a, Oktyabrskaya str., Knyaginino, 606340, Russia
[5] Chair "Morphology, midwifery and therapy", Chuvash state agricultural academy, 29, Karla Marksa str., Cheboksary, 428003, Russia
[6] Chair "Electric drive and electrical technologies", Russian State Agrarian University – Moscow Agricultural Academy named after K. A. Timiryazev, 7, Listvenichnaya Alleia str., Moscow, 127550, Russia
[7] Chair "Mechanization, electrification and automation of agricultural production", Chuvash state agricultural academy, 29, Karla Marksa str., Cheboksary, 428003, Russia

E-mail: ds17823@yandex.ru

Abstract. The article is devoted to the development of technology and technical means designed to weaken the retention force of hair in the dermis of rabbit skins due to selective dielectric heating of hair bulbs under the influence of an ultra- high frequency electromagnetic field (UHFEMF). Microwave technologies and more than 20 installations with UHFEMF sources and non-traditional resonators have been developed that provide continuous collection of fluff from rabbit skins. Resonators have been tested that ensure the separation of downy raw materials from the skin of rabbit skins in a continuous mode when implementing a set of reasonable design and technological parameters. Novelty of technical solutions of developed technologies and microwave installations with non-traditional resonators, such as: cylindrical (2655770, 2651594, 2680543, 2682890); semi-cylindrical (2018101622, 2703918); toroidal (2674605, 2699709); coaxial (2655748, 2651593); biconic (2715341, 2018117136) proved by patents of the Russian Federation. A technical specification has been developed for creating a radio-hermetic microwave installation with low-power magnetrons and a biconic resonator, which ensures the separation of down raw materials from rabbit skins in a continuous mode at low operating costs.

1. Introduction

Analysis of research results on the implementation of methods for processing animal and animal skins allows us to identify a priority microwave technology for collecting fluff from rabbit skins.

The scientific problem is to reduce operating costs for collecting fluff from rabbit skins while maintaining quality at low-capacity agricultural enterprises.

The aim of the work is to separate the decontaminated down raw materials from rabbit skins by developing technology and ultra-high -frequency continuous installations with low operating costs and maintaining electromagnetic safety.

The existing technologies of primary processing of animal and rabbit skins are reflected in the works of many authors [2, 4-6, 11]. Many scientists are engaged in the implementation of the basics of the theory of electromagnetic fields and waves in the development of innovative technologies and technical means [1, 7-9].

2. Materials and research methods

Analyzing various approaches to collecting hair from animal skins and studying the specific features of the impact of the ultra-high frequency electromagnetic field (UHFEMF) on raw materials, namely selective heating of raw materials components, depending on their electrophysical parameters, we developed a scientific innovation idea. At the same time, taking into account the accumulated experience of the research team of the university, the results reflected in the works of foreign and domestic scientists, we have developed more than 20 installations and technologies that implement the main criteria for designing microwave installations. The effective design of the working chamber of microwave installations was chosen by analyzing the parameters of the electrodynamic system with non-traditional resonators: spherical, cylindrical, coaxial, toroidal, ellipsoid, biconic, etc.

We calculated and visualized the distribution of the electromagnetic field and the power flux density in the developed non-traditional resonators using the CST Microwave Studio program, as well as evaluated their own q-factor. Theoretically, we calculated the intrinsic q-factor of the developed unconventional resonators as the ratio of double capacity to their surface area, taking into account the skin layer.

For experimental research, laboratory samples of low-power magnetron installations (OM75S 31, M24FA-410A, etc.) with air cooling and corresponding design versions of resonators were made.

Three-factor experiments of type 23 on the theory of active planning were performed with the help of laboratory installations and effective modes of operation of continuous microwave installations with non-traditional resonators were identified. Qualitative indicators were evaluated by studying the physical and chemical parameters of skin and its bacterial contamination.

3. Results and discussion

The method of implementation of innovative technology and technical means intended for processing rabbit products is shown in Figure 1.

The problem of the number of rabbits based on the introduction of scientific and technological progress remains unresolved. In small-scale agricultural enterprises, the waste-free technology of rabbit production is not sufficiently developed today. Having carried out monitoring of technological and technical problems, and ways to solve them, we found that it is not possible to achieve high-quality dressing of rabbit skins with minimal costs under the conditions of rabbit farms. We propose to collect fluff from the skins of rabbits that have increased down productivity, for example, from rabbits of the «White giant» breed, and not to dispose of the skins. The use of down not only reduces the cost of meat products, but also provides additional sources of raw materials for the felt industry, while it can be used in the production of granular feed from combined raw materials of plant and animal origin. Therefore, the development of innovative technology and technical means using ultra-high frequency electromagnetic field (UHFEMF) energy to separate downy raw materials from rabbit skins at reduced operating costs is actual.

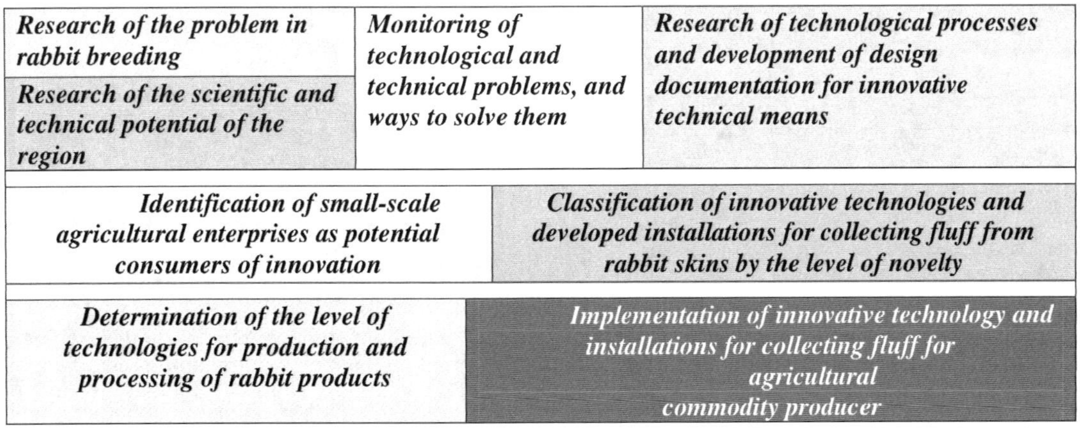

Figure 1. The method of implementation of innovative technology and technical means intended for processing rabbit products in the farms

Taking into account scientific and technical potential in the region and in Russia by promising industries «Rabbits» (the demand for rabbit meat in Russia 300 thousand tons per year), studies of technological processor data fluff from the skins of rabbits and developed design documentation for microwave installation with different working chambers (resonators). The developed installations with non-traditional resonators implement the following criteria:

1. The unit provides continuous operation while maintaining electromagnetic safety for service personnel.
2. The intensity of the electric field in the volume resonator is greater than the critical one, at which the viability of the bacterial microflora of the vegetative form stops in the raw material.
3. The working chamber, which includes resonators, provides repeated exposure to UHFEMF with a process life of less than 0.5.
4. The performance of the microwave installation is regulated by using several low-power air-cooled magnetrons and changing the power of each generator.
5. The configuration of the volumetric resonator ensures maximum intrinsic q-factor and uniform distribution of the electric field in the raw material, taking into account the depth of penetration of waves.
6. The dose of UHFEMF exposure to raw materials is adjusted depending on the size of rabbit skins. The duration of the impact is regulated by the speed of movement of transporting mechanisms, and the specific power – the volume of loading and power of generators.
7. The installation contains nodes that ensure the absorption of brine or sourdough mezdra skin and combing, collecting fluff raw materials.
8. With repeated exposure to UHFEMF, the skin with a mezdra fabric impregnated with brine or sourdough is endogenously heated no more than 45–50 °C.
9. The intensity of heat generation is controlled by regulating the power of microwave generators depending on the volume of loading.
10. Humid air is removed while maintaining the radio-tightness of the installation.
11. The beyond the waveguides limit the emission to the required limits during transfer and unloading skins.

A model of the mechanism of the technological process of separating down from rabbit skins is shown in Figure 2. An innovative idea lies in the weakening of the retention force of hair in the hair follicles due to selective endogenous heating of the dermis of the skin, the mezdra side of which is impregnated with a special sponge or when the mezdra, smeared with vegetable oil, is covered with moistened rock salt. Homogenized fermented mixture (sourdough) in UHFEMF under the influence of endogenous heat reduces the strength of hair retention in the bulb, their separation from the dermis occurs in a few minutes, depending on the state of the original skin. Features of mustard powder are based on the hydrolyzing effect of acids formed from starch, a significant content of fat in it, and the action of enzymes in the skin.

COLLECTING SETS OF RAW WITH THE SKINS OF RABBITS
Soaking the mezdra side of the skins with a certain concentration of brine or sourdough (a homogenized fermented mixture of rye flour, water, yeast, mustard powder and salt)
Selective endogenous heating of multi component raw materials (hair, skin, mezdra tissue, hair bulbs, brine or sourdough) in UHFEMF
Weakening of the hair retention force in the bulbs due to selective endogenous heating, disinfection of down raw materials in high-tension EP and collection using combing nodes and a pneumatic pump. Quality control

Figure 2. A model of the mechanism of the technological process of separating down from rabbit skins

Histomorphological assessment of the degree of dewatering of rabbit skins, the mezdra side of which is covered with vegetable oil and moistened rock salt, showed that when exposed to UHFEMF, the hair bulb is destroyed. The epidermis layer is completely loosened, resulting in a weakening of the retention force of the hair in the dermis, and with a slight pressure on the suction of the pneumatic pump is separated quite easily. At the same time, the strength properties of down raw materials are preserved.

Two technologies have been developed to implement this microwave technology.

1. Collection of decontaminated downy raw materials in the process of rotation of boxes with skins, the mezdra side of which is impregnated with brine or sourdough in resonators containing pegs. The principle of operation of the unit is based on ensuring the weakening of the retention force of down in the hair follicles in the process of selective exposure to UHFEMF, combing the fluff fibers while preserving their marketable appearance, collecting them and transporting them pneumatically. For example, this technology is implemented by a microwave installation (Figure 3) with a semi-cylindrical resonator and mobile tubes for collecting fluff from rabbit skins (patent No. 2699194) [1, 10].

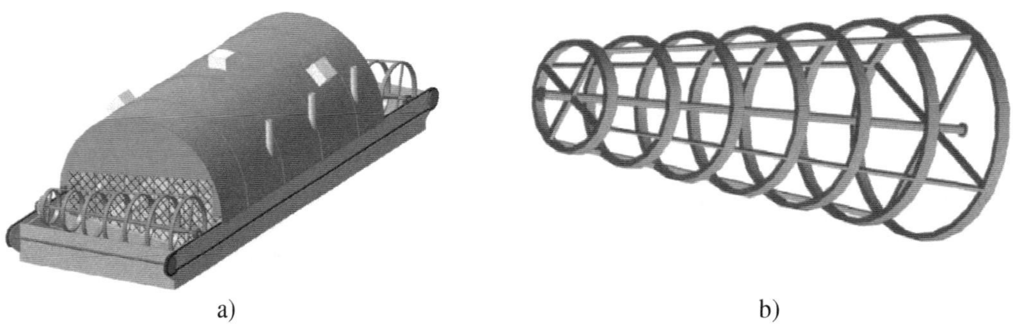

a) b)

Figure 3. Microwave installation with a semi-cylindrical resonator and mobile rules for collecting fluff from rabbit skins: a) installation; b) dielectric rule

The resonator chamber is formed by a horizontally positioned semi-cylinder and a rectangular tray. The tray is mounted for movement and rotation around its axis dielectric pravilki (with the possibility of movement and rotation around its axis), made in the form of a frame assembled from rods, forming the shape of a truncated cone, which are attached to the rims of appropriate diameter, and the center pravilki laid-axis (Figure 3 b). The ends of the rule axes are located in the retainer rollers located above the conveyor chains installed in the skids. The skids are mounted on top of each other in parallel, along the side surfaces of the pallet.

2. Collection of decontaminated downy raw materials in the process of continuous exposure to UHFEMF on expanded skins in the form of plates with moistened mezdra brine or sourdough. Most of the developed microwave installations implement this technology as the most effective. For example, in a conveyor microwave installation, the fluff is separated from the unfolded rabbit skins (Figure 4).

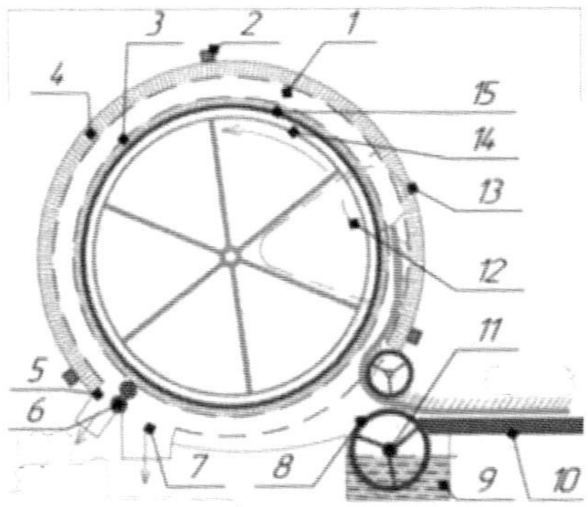

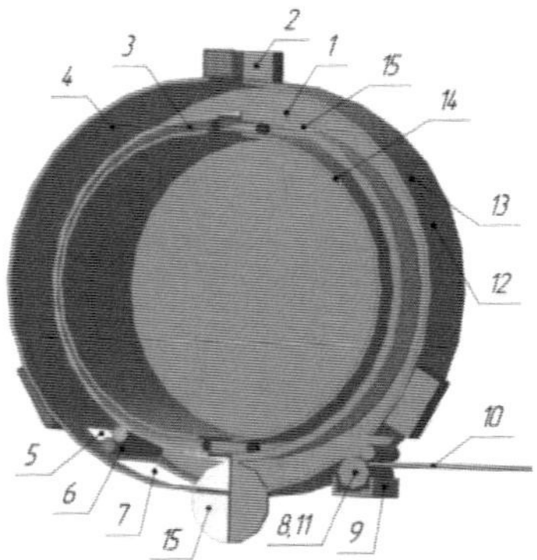

Figure 4. Conveyor microwave installation: 1 – coaxial resonator; 2 – magnetrons; 3 – perforated dielectric drum; 4 – dielectric pegs; 5 – slot for a pneumatic pump; 6 – dielectric rollers; 7 – slot for unloading the skin; 8 – dielectric rollers; 9 – sponge bath; 10 – table for laying skins; 11 – electric drive of the drum, 8 rollers and 6 rollers; 12 – hatch with an observation window; 13 – external cylinder; 14 – inner cylinder, 15 – toothed crown

The conveyor microwave installation contains a horizontally located coaxial resonator 1, represented as a system of two coaxially arranged outer 13 and inner 14 cylinders with common bases, with radii differing by more than a quarter of the wavelength, and a length equal to a multiple of half the wavelength. The outer cylinder has three slots along the side surface and 4 dielectric flexible pegs on the inner surface. In the annular space between the cylinders, a dielectric perforated drum 3 with dielectric pegs is coaxially installed. The size of the annular gap between the dielectric pegs corresponds to the thickness of the skins. Magnetrons 2 are installed on the side surface of the outer cylinder with a 120-degree shift along the perimeter. On one end side of the resonator there is an electric drive 11 of the drum 3, transporting rollers 8 and dielectric rollers 6, and on the other – a hatch 12 with an observation window. Along the perimeter of the drum 3 there is a toothed crown 15. The slot 5, which is located in front of the rollers 6, is connected through the air outlet to the pneumatic pump. A storage tank is attached to the other slot 7, and between these slots along their radial rollers is installed so that they cover the gap between the dielectric pins. Under the transporting rollers 8, located in front of the third slot, a bath 9 is installed so that the lower leading roller is immersed in the sponge, and a table 10 is installed at the level of the gap between them. The average perimeter of the coaxial resonator 1 should be a multiple of half the wavelength.

Removal of hair from rabbit skins immediately after slaughter is possible by soaking with brine 5% concentration and exposure to UHFEMF dose equal to 0.2 kWh/kg. After more than 2 hours after slaughter, it is more effective to spray a sourdough from a fermented mixture of rye flour, water, yeast, mustard powder or rock salt on the mezdra side of the skin, when moving through the resonator.

The manufactured and tested microwave installation with an unconventional resonator for separating the hair from the skin of rabbits is shown in Figure 5, and the technical characteristics of the installation are shown in Table 1.

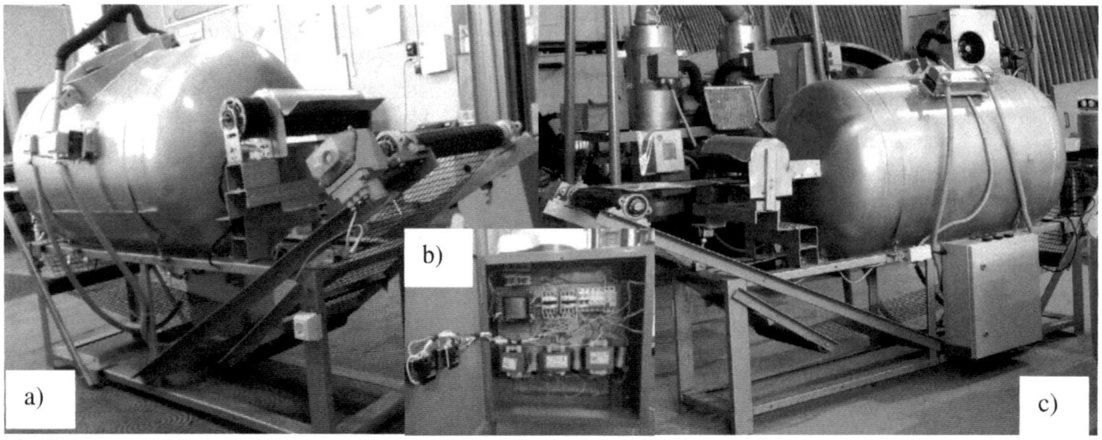

Figure 5. Microwave installation for separating down from rabbit skins: a, b) installation in different laboratories; c) control cabinet

Table 1. Technical characteristics of microwave installation

Power consumption of four generators, kW	4,8
Plant capacity, kg / h; PCs/h	15–20; 34–45
Power of the conveyor motor, kW	0,25
The speed of movement of the conveyor, m/s	0,015–0,2
Pneumatic pump power, kW	0,25
The power of the fan for cooling the magnetron, kW	0,25
Power consumption of the microwave installation, kW	5,55
Specific energy costs, kWh / kg	0,3
Overall dimensions, m	1,2×1,0×2,15

4. Conclusion

Studies of the pattern of distribution of the EP tension in non-traditional resonators show that in a cylindrical resonator, the EP is evenly distributed over the entire working area. The tension reaches 4 kV/cm, and the q-factor is within 7000, in a toroidal one – 7 kV/cm. The q-factor is 9000, in a biconic one – up to 6 kV/cm, and the q-factor is 7000.

A truncated biconic resonator with a volume of 350 l with a q-factor of 9000 with a magnetron power of 3200 W will provide an electric field strength of 1.2–1.5 kV/cm in a volume of up to 55 l in the area of the cone bases and reduce the total microbial number by half.

The symmetrical biconic resonator, which has a sufficiently high intrinsic q-factor, is able to implement the main criteria for designing microwave installations, namely: continuity of the technological process due to open vertices, ensuring electromagnetic safety and high EP intensity sufficient for disinfection of downy raw materials; to increase the efficiency of the installation to 0.7. With a capacity of 35-45 PCs/h, the power consumption of 5.55 kW, energy costs for the technological process of separating down from rabbit skins and collecting is 0.3 kWh/kg.

Acceleration of the weakening of the retention force of hair in the shit of skins can be carried out by exposure to microwave UHFEMF, if the mezdra side is impregnated with a homogenized fermented mixture of rye flour, water, yeast, mustard powder, or rock salt on top of vegetable oil.

References

[1] Tushar G, Huacheng Z, Ashim K D and Kama H 2015 Microwave drying of spheres: Coupled electromagnetics-multiphase transport modeling with experimentation. Part II: Model validation

and simulation results *Food and Bioproducts Processing* **96** 326. DOI: 10.1016/j.fbp.2015.08.001

[2] Tianyi S, Zhijun Z, Jingxue H, Shiwei Z, Xiaowei W and Wenqing Z 2020 Sensitivity analysis of intermittent microwave convective drying based on multiphase porous media models *Int. J. of Thermal Sci.* **153** 106344. DOI: 10.1016/j.ijthermalsci.2020.106344

[3] Shamin E A, Novikova G V, Mikhailova O V and Belova M V, RF Patent No. 2699194, Microwave installation for the separation of hair from rabbit skins when in brine (3 September 2019)

[4] Campañone L A and Zaritzky N E 2005 Mathematical analysis of microwave heating process *J. of Food Eng.* **69(3)** 359. DOI: 10.1016/j.jfoodeng.2004.08.027

[5] Guido S J, Martin D V, Andrzej I S and Georgios D S 2012 On the effect of resonant microwave fields on temperature distribution in time and space *Int. J. of Heat and Mass Transfer* **55(13–14)** 3800. DOI: 10.1016/j.ijheatmasstransfer.2012.02.065

[6] Resurreccion F P Jr, Luan D, Tang J, Liu F, Tang Z, Pedrow P D and Cavalieri R 2015 Effect of changes in microwave frequency on heating patterns of foods in a microwave assisted thermal sterilization system *J. of Food Eng.* **150** 99. DOI: 10.1016/j.jfoodeng.2014.10.002

[7] Budd C J and Hill A D 2011 A comparison of models and methods for simulating the microwave heating of moist foodstuffs *Int. J. of Heat and Mass Transfer* **54(4)** 807. DOI: 10.1016/j.ijheatmasstransfer.2010.10.022

[8] Jinghua Y, Junqing L, Yuan X, Yang Y, Huacheng Z and Kama H 2019 An approach for simulating the microwave heating process with a slow-rotating sample and a fast-rotating mode stirrer. *Int. J. of Heat and Mass Transfer* **140** 440. DOI: 10.1016/j.ijheatmasstransfer.2019.06.017

[9] Wenjia Z, Donglei L, Juming T, Shyam S S, Barbara R, Huimin L and Fang L 2015 Dielectric properties and other physical properties of low-acyl gellan gel as relevant to microwave assisted pasteurization process *J. of Food Eng.* **149** 195. DOI: 10.1016/j.jfoodeng.2014.10.014

[10] Shamin E A, Belova M V, Mikhailova O V, Novikova G V, Zaitsev P V and Sharonova T V 2019 Developing multigenerator plants for separating rabbit wool from conies *J. of Environmental Treatment Techniques* **882**

[11] Mohammad R H, Do Y B and Prashanta D 2010 Analysis of microwave heating for cylindrical shaped objects *Int. J. of Heat and Mass Transfer* **53(23–24)** 5129. DOI: 10.1016/j.ijheatmasstransfer.2010.07.051

Positioning automation when assembling panel structures

M V Kovalevich and E S Shemonaeva

Moscow Aviation Institute (National Research University), 4, Volokolamskoe highway, Moscow, 125993, Russia

E-mail: kovalevichmv@mai.ru

Abstract. The article discusses the process of assembling low-stiffness panel structures typical of the aerospace industry. Due to the low stiffness of the incoming elements, the assembly of such structures differs from the assembly of structures typical of general mechanical engineering. To switch to a robotic assembly of such structures, it is necessary to comply with special requirements both to the technology and to the design of the elements included in the assembly. The paper investigates the process of positioning automation during the assembly of panel structures using industrial robots. The article describes a method of verification calculation for accuracy, which allows establishing whether the proposed positioning method meets the specified requirements for the aerodynamic contour of the product. Requirements for structural elements have been formulated, which will allow positioning during assembly with the required accuracy. The results obtained in the course of the study can be used in the design of new structures, the assembly of which will be carried out using automation tools. In addition, the developed requirements must be taken into account when switching to new assembly technologies for existing structures. The results of the study can help to improve the efficiency of assembly of flexible structures used in the aerospace industry. It was also found that it is necessary to conduct further research in terms of the influence of deflections of the structure from its own weight and during the supply of the tool that performs the setting of joints.

1. Introduction

Today, industrial robots are widely used in assembly operations in mechanical engineering. Many mechanical engineering products are classified as structures of high rigidity, which have pronounced assembly bases on the elements of mating parts during assembly.

However, there are also such designs that are characterized by insufficient independent rigidity of the incoming elements. For example, these include panel structures that are common in the aerospace industry. Such structures acquire the required rigidity after the establishment of power relationships - installation of fasteners. Thin-walled, lightweight panel structures, as a rule, are not self-sufficient in terms of basing. Therefore, they are collected in special assembly devices, which are, first of all, a source of additional bases [1, 2].

Assembly devices are characterized by a high cost due to high metal consumption, extremely high requirements for manufacturing accuracy with large dimensions, as well as their narrow specialization. Modern aircraft technology is aimed at simplifying the design and reducing the number of special assembly tooling.

The transition to robotic assembly of aircraft structures [3-8] imposes special requirements both on technology and on the design of the elements included in the assembly.

IOP Conf. Series: Materials Science and Engineering 1111 (2021) 012055 doi:10.1088/1757-899X/1111/1/012055

2. Formulation of the problem

Currently, the measurement-assisted assembly and its various variations are actively developing in the world [4-8]. The essence of the method lies in the positioning of the assembled elements using automation and under the control of measuring systems and, then, in keeping them in a given position during the process of making connections, as shown in Figures 1 and 2. Automatically controlled columns are used as positioners. A more flexible, but also more complex, scheme seems to be where industrial robots are used as positioners.

The positioning process is a complex problem associated with the problems of ensuring the accuracy of a given aerodynamic contour and the rigidity of structural elements.

Features of a measurement-assisted assembly using robotics are as follows.

1. It is necessary to pre-assemble the panels in a simplified assembly device - a pallet. The same pallet is used to position the panel in subsequent assembly steps.

2. The positioning accuracy of an industrial manipulator will not allow obtaining the accuracy required in the design of aviation technology. In this regard, the position of the robot is corrected using a laser tracker. In this case, the coordinates must be taken directly from the panel.

Figure 1. Collaborative panel positioning

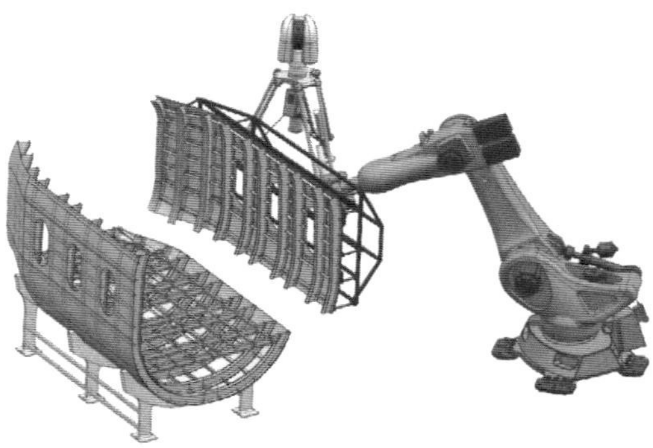

Figure 2. Automatic panel positioning

3. Calculation methods

Reasoning from theoretical premises, the main consequence of nanopowder introduction into the melt should be refinement of the macro- and microstructure, as the powder particles must serve as nuclei of new grains. Figures 1 and 2 show photos of the microstructure of cast samples from the bronze of the lead-tin bronze grade, both modified by SDP of aluminium oxide and without modifier addition. The phase composition of the bronze under study represented in photos of the microstructure is a solid solution of tin in copper, lead inclusions and eutectoid inclusions based on electron compound. The ability to meet high requirements is determined by a verification calculation for accuracy.

The task of calculating the accuracy at the current stage is to determine the expected deviation from the theoretical contour at an arbitrary point of the side panel of aircraft.

The error of the functional size ωL is determined [1]:

$$\omega_L = \Delta_L \pm K_1(1 + K_2 + K_3)\frac{\delta_L}{2} \quad (1)$$

$$\Delta_L = \sum_{i=1}^{m-1} \Delta_i \quad (2)$$

$$\Delta_i = BO_i - HO_i \quad (3)$$

$$\frac{\delta_L}{2} = \sqrt{\sum_{i=1}^{m-1} \left(\frac{\delta_i}{2}\right)^2} \quad (4)$$

$$\frac{\delta_i}{2} = \frac{BO_i + HO_i}{2} \quad (5)$$

where K_1 is the risk coefficient; K_2 - coefficient that takes into account the error caused by deformation when making connections; K_3 - a coefficient that takes into account the additional error that occurs when fixing elements; Δ_L - middle of the tolerance range; Δ_i - middle of the tolerance zone at each stage of the transfer of dimensions; BO_i and HO_i - upper and lower deviation at each stage of size transfer; $\frac{\delta_L}{2}$ - standard deviation; $\frac{\delta_i}{2}$ - standard deviation at each stage of size transfer.

The calculation results can be formatted graphically in the form of an image of the tolerance field shown in Figure 3.

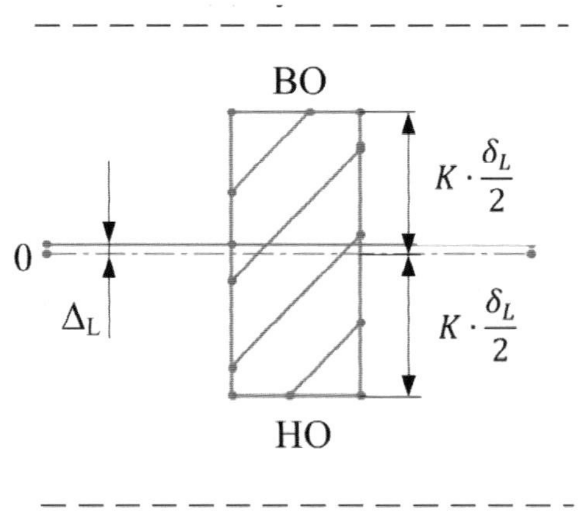

Figure 3. Tolerance field with the location of deviations

For the task at hand, we take the following values of the coefficients: $K_1 = 2.37$, $K_2 = 0.3$, $K_3 = 0.3$.

When assembling by this method, there are some imprecisions associated with the assembly process, as well as with the transfer of dimensions from the electronic model to the product. The sources of these imprecisions with their values are shown in Table 1.

As a result of calculating the final imprecision, the expected deviation from the theoretical contour is ±1.32 mm.

Table 1. Imprecisions arising from the assembly by the coordinate-positional method

Priority	Dimension transfer stage	Dimension transfer imprecision (*BO*, *HO*), mm
1	Positioning by industrial robot	±0.1
2	Measurements with a laser tracker	±0.05
3	Deformation of the structure under its own weight	±0.3

4. Development of basic design requirements

Implementation of technology for existing structures entails reworking structures or designing new structures. In this regard, it is necessary to define and formulate requirements for constructions, the fulfillment of which will allow the effective application of a measurement-assisted assembly.

The requirements can be divided into two groups: for the basing process and for the joining process.

4.1. Requirements for the construction basing process

The requirements group determines the conditions that must be provided for in the design for the implementation of basing during assembly.

4.1.1. Weight of the structure. Restrictions on the weight of the product structure are imposed by the carrying capacity of an industrial robotic arm. The weight of the transition frame between the structure and the industrial robot must be taken into account.

We consider an aircraft in which the masses of the left and right side panels, taking into account the frame elements, are approximately 85 kg and 65 kg, respectively. The weight of the top panel with frame elements is 100 kg. Frame weight is approximately 80 kg. The weight of the attachments is 100 kg. The assessment of the lifting capacity of industrial manipulators was carried out on the basis of an analysis of the characteristics of the KUKA KR 1000 Titan robot. The maximum load capacity of this industrial robot is 1300 kg for palletizing operations. When holding the load in a certain position, the carrying capacity of the robot is estimated at 1000 kg. Thus, this value sets the weight limits.

4.1.2. Weight of the structure. The presence of platforms for the laser reflector. Determination and correction of the position of the panel in space is carried out using measurements performed by a laser tracker. To eliminate the influence on the accuracy of the panel positioning of errors associated with the adapter frame, the reflectors of the laser tracker must be installed directly on the panel. For this, the panel design must provide for appropriate places for their installation.

4.1.3. The rigidity of the structure. In general, the requirement is similar to the structural rigidity requirements for pre-assembly in an assembly fixture. Low stiffness can primarily affect the accuracy of mating contours in the joint zones. A diagram of the influence of deflection on positioning accuracy is shown in Figure 4.

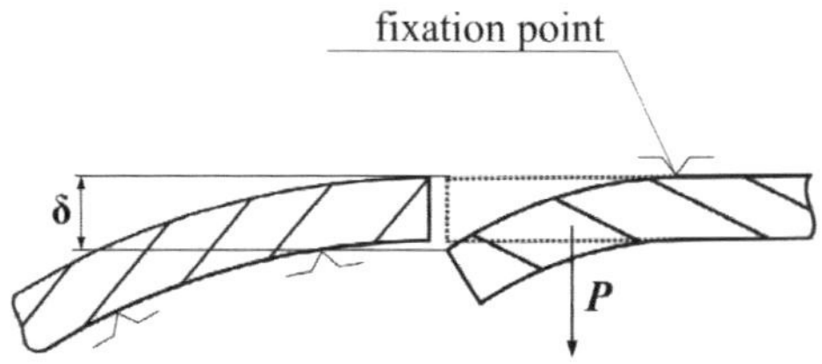

Figure 4. Scheme of deflection influence on positioning accuracy: δ - deformation; *P* - weight of the panel area

4.2. Requirements for process jointed with industrial manipulators
The requirements group that must be provided for in the design to make joint during assembly.

4.2.1. The use of special fasteners in the structure of the unit. To implement the assembly method, special one-sided fasteners are required. Fasteners must be more durable than existing core rivets.

4.2.2. Approaches to the places of connections. The location of the seams should not interfere with the supply and movement of the multifunctional riveting head used in automatic riveting. A general view of the currently most common KUKA multifunctional head is shown in Figure 5.

4.2.3. The rigidity of the unit joints. The connection is made at a local power load from the equipment on the structure of the compartment. The rigidity of the structure at the place where the connection is made must be sufficient so that deformation from loads does not interfere with the quality of the connection. The initial data for the calculation are the load from the supplied tool and the permissible deformation. A diagram of the occurrence of a deflection in the area of the connection is shown in Figure 6.

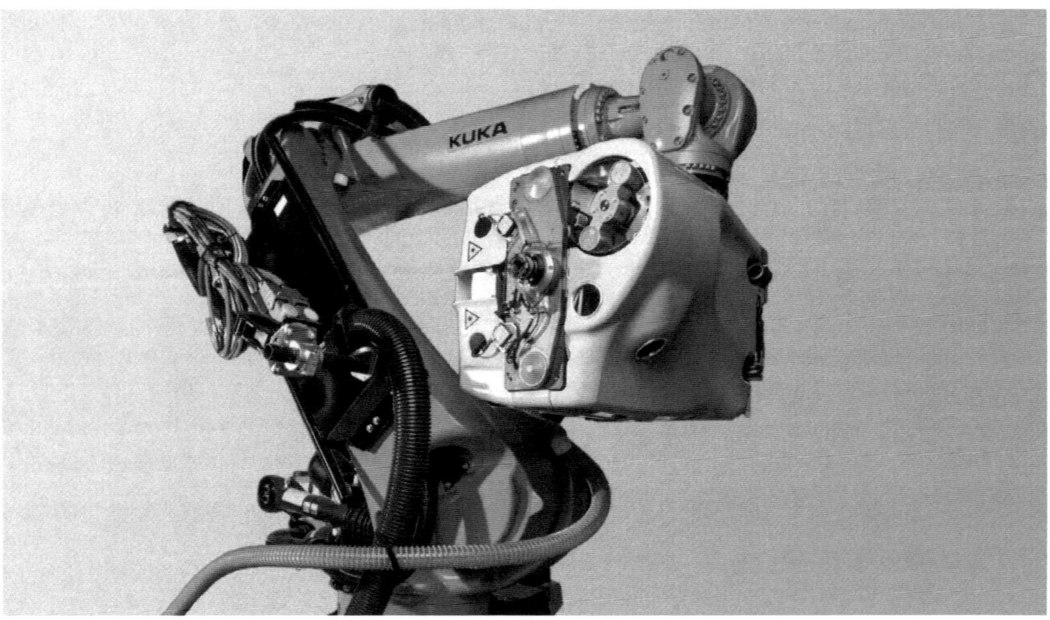

Figure 5. General view of the KUKA multifunctional head

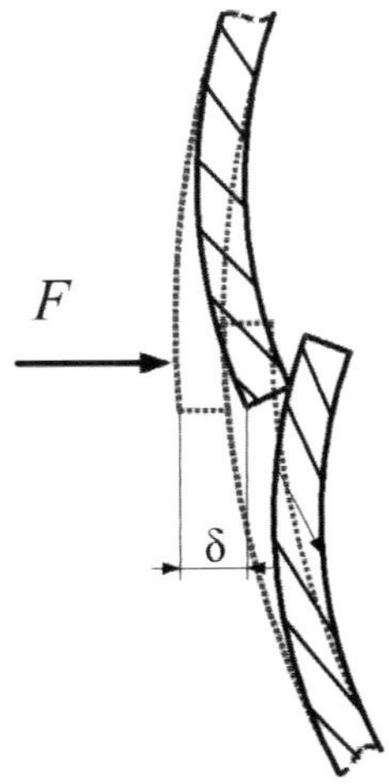

Figure 6. Scheme of the occurrence of deflection at the place of the connection: δ - deformation; *F* - tool impact force

5. Conclusion

As a result of the studies carried out, it was found that in order to achieve the required accuracy with the automated assembly of structures of low rigidity, it is necessary to comply with a number of requirements.

- The mass of the structure, including assembly equipment, should not exceed the lifting capacity of an industrial robot.
- The design of the product to be assembled must provide for areas for the installation of reflectors for laser measuring devices.
- The rigidity of the structure must be sufficient and exclude the deflection of the structure under its own weight.
- Using specialized one-sided fasteners.
- Open approaches for an industrial robot to the joints.
- The rigidity of the joints of the structure must be sufficient and exclude the deflection of the structure from the supply of the tool.

References

[1] Pekarsh A I, Tarasov U M and Krivov G A 2006 *Modern technologies for aircraft assembly* (Moscow: Agraf-press) p 304
[2] Guseva R I 2013 *Features of the airframe assembly technology* (Komsomol'sk-na-Amure: KnASTU) p 133
[3] Muelaner J E, Martin O C and Maropoulos P G 2013 Achieving Low Cost and High Quality Aero Structure Assembly through Integrated Digital Metrology Systems *Procedia CIRP* **7** 688-693
[4] Maropoulos P G, Muelaner J E, Summers M D and Martin O C 2014 A new paradigm in large scale assembly - research priorities in measurement assisted assembly *Int. J. Adv. Manuf. Technol.* **70** 621–633
[5] Kayani A and Gray I 2009 *Shim for Arrangement Against a Structural Component and a Method of Making a Shim* Airbus UK Ltd. patent application number: US20080211185

20080916 patent number: US2009100791 (A1)

[6] Kayani A and Jamshidi J 2007 Measurement assisted assembly for large volume aircraft wing structures *4th Int. Conf. on Digital Enterprise Technology* pp 426-434

[7] Muelaner J E, Kayani A, Martin O and Maropoulos P G 2011 Measurement assisted assembly and the roadmap to part-to-part assembly *7th Int. Conf. on Digital Enterprise Technology* pp 11-19

[8] Summers M 2005 Robot Capability Test and Development of Industrial Robot Positioning System for the Aerospace Industry *SAE Technical Paper* pp 1-13

IOP Conf. Series: Materials Science and Engineering 1111 (2021) 012056 doi:10.1088/1757-899X/1111/1/012056

Correlation analysis to determine the relationship between the geometric dimensions and the electromagnetic moment of a switched reluctance motor

A V Shevkunova, A S Shapshal, T Z Talakhadze and S A Shapshal

Rostov State Transport University, 2, Rostovskogo Strelkovogo Polka, Narodnogo Opolcheniya sq., Rostov-on-Don, 344038, Russia

E-mail: llh@rgups.ru

Abstract. An electric drive is an indispensable part of any mechanism. Without it, the the functioning of any industry is impossible, especially in the railway. It allows you to automate technological processes and increase their efficiency. For efficient railway transport, it is necessary to increase the level of technical development of traction rolling stock, which is possible through the use of promising types of electric machines, one of which is a switched reluctance motor. To create a highly efficient and technological switched reluctance electric drive, it is necessary to have information about the relationship between the geometric dimensions of its magnetic system and the average value of the electromagnetic moment. The purpose of this article is to determine the strength and direction of the relationship between the average value of the electromagnetic moment and the changing geometric values of the tooth-groove zone of the object under study. The numerical value of the electromagnetic moment was obtained using the stochastic method in the MATLAB application package, which directly interacted with the FEMM version 4.2 program, which is based on the finite element method. The research carried out in the article and the results obtained allow us to conclude that there is or is not an appropriate connection between the considered elements of the magnetic system design and the value of the electromagnetic moment.

1. Introduction

Currently, regulated automated electric drives are used in most areas of energy and industry, including the railway industry. The regulated electric drive is the most widespread at this time and one of the most effective energy-saving and resource-saving technologies known at this time.

The rapid development of power electronics has given humanity a number of important discoveries, without which it is currently impossible to imagine the modern production and operation of regulated electric drives. Powerful controlled power switches – transistors that can switch currents up to 600 A at a voltage of up to 1200 V with a frequency of 30 kHz or more, as well as processor tools that provide the creation of «artificial intelligence» – have become available.

Among all the relatively new electric drives, where the frequency converter and control system are integral parts, we can single out one – switched reliance drive (SRD).

The switched reluctance machine (SRM) currently forms the basis of a modern SRD of industrial production, determines the technical and economic possibilities of wide application of electric traction in railway [1] and automobile transport [2, 3], provides appropriate power functions and operating modes of electric drives of machine equipment, complex robotic systems [4, 5] and other technological equipment of automated production. Epy aunctional scheme of the switched reliance drive traction rolling stock is shown in Figure 1.

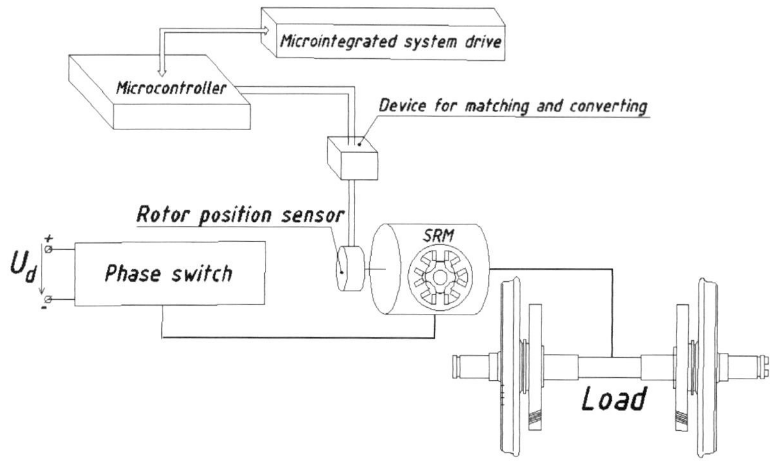

Figure 1. Functional scheme of the switched reliance drive.

Output parameters such as electromagnetic torque and shaft speed are usually used to match the mechanical characteristics of the SRM and the production facility. These parameters are determined by the configuration of the active part of the SRM, which in turn consists of the yoke and stator teeth, the yoke and rotor teeth, and the stator winding. To create a SRD that is competitive on the world market and has high technical and economic indicators, it is necessary to have reliable, scientifically based information about the relationship and degree of influence of the geometric variables of the magnetic system and the average value of the electromagnetic moment.

Quite often, in the course of research, it is necessary to determine or exclude the existence of a relationship between different values of the process. Variables can be linked by a functional (hard-deterministic) or stochastic (probabilistic, statistical) relationship. As you know, in a functional relationship, each value of one independent attribute corresponds to a strictly defined value of another dependent attribute. In the case of a probabilistic relationship between the observed variables, the value of the factor attribute corresponds to the set of possible values of the effective attribute, i.e. when the value of one value corresponds to a certain distribution of another value. A special case of stochastic dependence is the correlation dependence, which allows you to determine the strength and direction of the relationship between variables. This is the nature of the dependence and have the geometric dimensions of the fragments of the magnetic system and the average value of the electromagnetic moment.

The purpose of this article is to determine the tightness, the direction of communication, and the trend of changes in the average value of the electromagnetic moment under the influence of varying values of the geometric dimensions of the elements of the active part of the SRM. Also, the results obtained make it possible to further determine the degree of influence of various design parameters of the tooth-groove zone of the active part on the average value of the electromagnetic moment.

2. Results and Discussion

The term «correlation» (from lat. correlatio-relation, relation, dependence) was first applied by J. Cuvier in 1806. The first mathematical description of the method was made by O. Brave in 1846. Currently, correlation analysis is a fairly widespread, flexible and effective tool for processing statistical data. This method has found its application in various fields of science and technology [6–8].

In this article, a three-phase SRM with a 12/8 magnetic system configuration (stator teeth – 12, rotor teeth – 8), designed for an electric traction drive of an electric train, with a rated power of 300 kW and a torque equal to 1.99 kN·m, was used as the object under study.

Variations in the numerical values of various design variables of the magnetic system SRM were used as a factor parameter, and the average value of the electromagnetic moment was used as a result parameter. The geometric dimensions, the values of which were considered separately, are shown in figure 2.

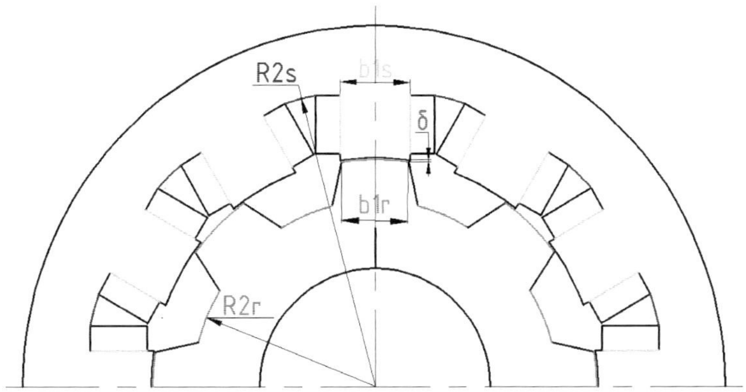

Figure 2. Fragment of the magnetic system under study.

The study was carried out with the parameters that have the greatest influence on the formation of the numerical value of the moment. This information was taken from earlier dissertation research with the NTI-1200 traction motor, which also has a 12/8 configuration. Thus, the relationship between the average value of the electromagnetic moment and the geometric values of the following parameters of the magnetic system was considered:

- Slope of the side surface of the stator tooth, deg. (*b2s*) (figure 1 does not show this element, since *b2s* = 0).
- Radius of the stator in the grooves, mm (*R2s*).
- Stator tooth width, mm (*b1s*).
- Radius of the rotor along the grooves, mm (*R2r*).
- Air gap, mm (δ).

Motor dimensions such as the external diameter of the stator, the length of the magnetic core in the axial direction, and the shaft diameter are often specified in the technical specification, since it is not uncommon to integrate a SRM magnetic system into the housing of an asynchronous motor or DC motor. However, if these dimensions are not defined in the terms of reference, it is also possible to include them in the parameters under study as variable variables. The values of the values shown in table 1 were used as restrictions for calculating the numerical values of the above geometric parameters.

Table 1. Limitations of the considered geometric parameters.

Geometrical parameter	Original size, mm	Limitations, mm
Slope of the side surface of the stator tooth, deg. (*b2s*)	0	0...10
The radius of the stator in the grooves, (*R2s*)	301	296…306
Stator tooth width, (*b1s*)	73	68…75
Radius of the rotor along the grooves, (*R2r*)	188	183…193
Air gap, (δ)	2	1…3

The data sets required to build this correlation were obtained using an optimization algorithm based on the Monte-Carlo method implemented in a software product designed to automate and optimize the design of the active part of the SRM used in previous studies [9, 10].

The Monte-Carlo method was chosen because it differs from other methods in its simplicity and generality, as well as its guarantee of finding the global extremum. The disadvantage of this method is slow convergence, but this disadvantage can be partially eliminated by modifying the method [11, 12]. The distribution of the random variable was assumed at a uniform probability density. The optimization algorithm and processing of the results of field calculations were carried out using the Matlab application software package, which, in turn, directly interacted with the FEMM 4.2 program for modeling and calculating magnetic fields.

The sample size after calculations for each case consisted on average of 200 values. As a result, scattering diagrams and regression functions were constructed, which are shown in the corresponding figures. The constructed scatter plots allow you to clearly see the nature of the relationship between variables. Each figure shows linear regression equations and their approximation confidence values R^2 (coefficient of determination). Some of the figures also present alternative equations that are polynomial in nature, due to the fact that the coefficient of determination of the linear regression function took a lower value for the distributions under consideration. All variants of trend lines available in Excel (exponential, logarithmic, power) were considered, but the values of R^2 when using them differed from the linear function by no more than 2 %, so their further consideration and application is impractical.

On all the charts under study, the orange dashed line shows a linear regression line, and the blue dotted line shows a polynomial trend line. Also, the coordinate with the highest value of the moment variable is highlighted in red on each scatterplot.

The first dependence under consideration was a function with variables such as the angle of inclination of the side surface of the stator tooth and the average value of the electromagnetic moment (Figure 3). At the same time, the value of the air gap remained unchanged.

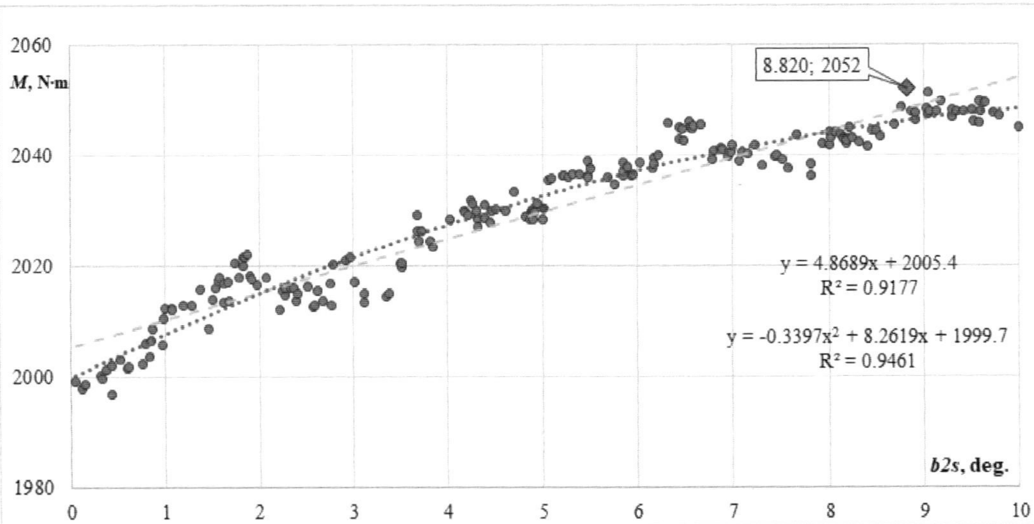

Figure 3. Graph of the correlation between the electromagnetic moment (*M*) and the angle of inclination of the side surface of the stator tooth (*b2s*).

The graphic relationship of the structural element of the magnetic system - the slope of the side surface of the stator tooth and the electromagnetic moment, is shown in figure 3. The maximum value of the moment is obtained at the point highlighted in red with the value $b2s$ = 8.82. According to the initial design of the object under study, the slope of the side surface of the stator tooth was 0 degrees. After the optimization calculation, the value of this structural element was 10 degrees.

The magnitude of the accuracy of the approximation for this nonlinear dependence is maximum for the equation of a curve line polynomial regression. The correlation coefficient for this data set takes a value equal to r = 0.96 and indicates that this relationship can be characterized as «direct» and close (strong).

The next geometric parameter of the stator, depending on the value of the moment, was considered *R2s* (stator radius along the grooves, mm) (Figure 4). In this case, the maximum average value of the moment is equal to the size of the stator radius along the grooves, which practically borders on the accepted numerical restrictions. For this case, the optimal point has moved down from the original size. In the initial version, *R2s* = 301 mm, and after the optimization calculation – 296.43 mm.

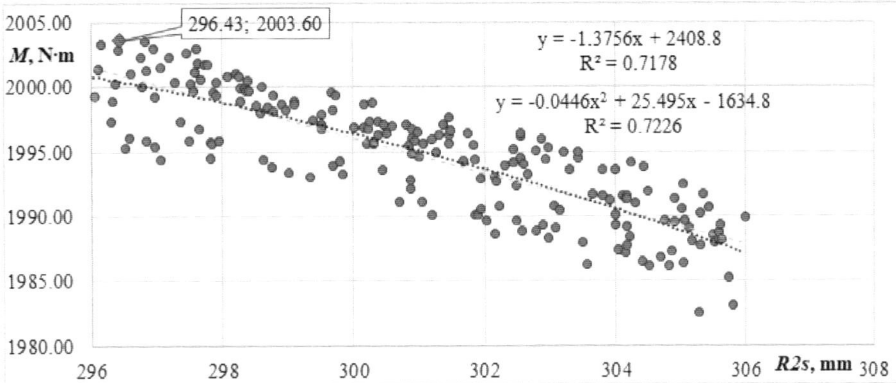

Figure 4. Graph of the correlation between the electromagnetic moment (*M*) and the stator radius along the grooves (*R2s*).

The approximation confidence value for this dependence takes a rather low value. It shows only 33 % of the variations in M values due to the variability of *R2s* sizes. The correlation coefficient for this data array takes a value equal to r = -0.84. The specified value of the coefficient according to the scale of Chedoke indicates a fairly strong negative linear relationship. According to previous studies with SRM 12/8, the *R2s* parameter played a significant role in the formation of the moment value, so to establish a more accurate relationship, it is recommended to calculate with a large number of random throws.

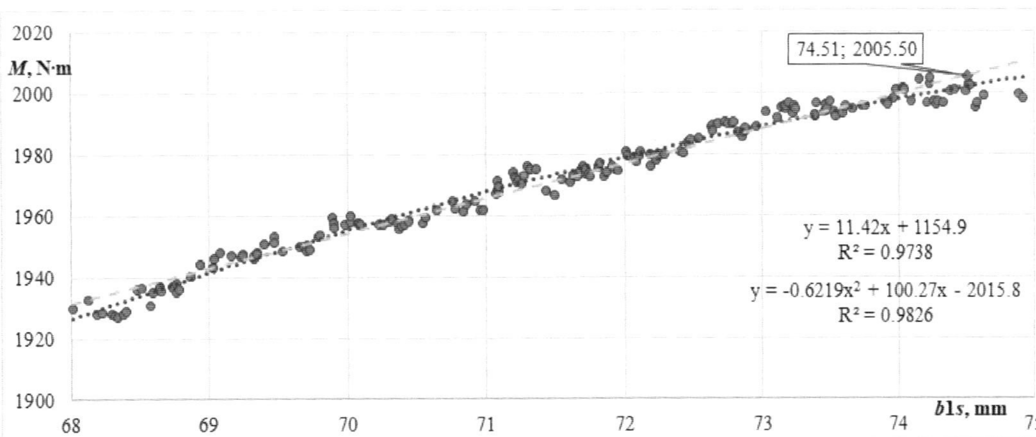

Figure 5. Graph of the correlation between the electromagnetic moment (*M*) and the radius of the stator tooth width (*b1s*).

The third data set studied was the numerical values of the variables *M* from the values of the stator tooth width (Figure 5).

The presented correlation has the values of the coefficients of determination for linear and polynomial functions that are almost identical (97.3 and 98.2 %), differing only from each other by 1 %. Since in this case the value of R^2 for the linear regression function is quite high, it is highly likely that this dependence can be considered as linear, and the correlation coefficient, which is used only for linear relationships, indicates a «direct», positive and very strong, almost functional relationship (r = 0.96). As can be seen from this picture of the distribution of variables, the maximum value of the moment is reached when the stator tooth width is 74.51 mm. This value exceeds the size of the original design (73 mm) by 2 %.

Further, the geometric parameter of the rotor was considered (Figure 6) and the statistical dependence under study was a set of variables *R2r* (radius of the rotor along the grooves, mm) and *M*.

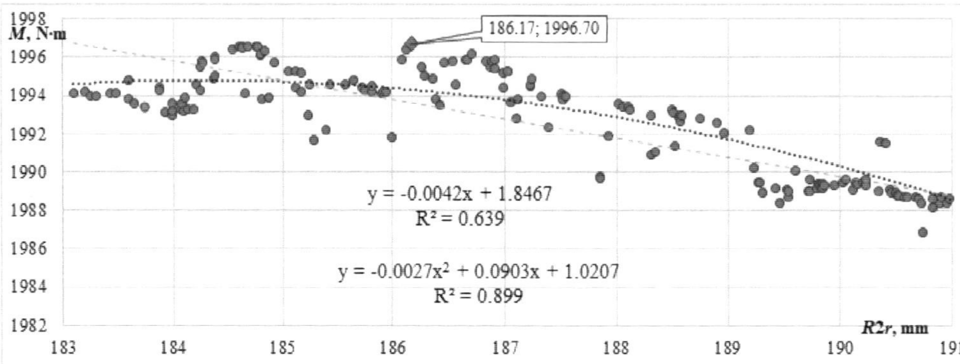

Figure 6. Graph of correlation between the electromagnetic moment (*M*) and the rotor radius along the grooves (*R2r*).

The quality of the regression model for a graphically represented distribution of the observed parameters is almost 30 % better when using a polynomial trend line. The value of the coefficient of determination R^2 = 0.89 can be interpreted as follows: 90 % of the variance of the resulting factor (*M*) is explained by this regression equation, and the remaining part, which is 10 %, is explained by other factors of variance. For such a value R^2, the proportion of the explained variation *R2r* is high enough and the constructed regression model with a high degree of probability has good approximating properties. When considering this relationship as linear, we can say that the direction of the correlation is negative (r = -0.86), and the tightness of the interaction of variables can be defined as strong, but rather close to moderate. The highest value of the electromagnetic moment is observed when the radius of the rotor along the grooves is equal to 186 mm (less than the original 2 mm), which indicates such a structural change as a narrowing of the rotor yoke.

The last correlation field considered was a distribution consisting of variables of the electromagnetic moment (*M*) and the air gap (δ) (Figure 7).

The maximum value of R^2 is 0.9994 if the regression line is linear. A value of 99.9 % indicates that there is a functional relationship, i.e. those 99.9 % of changes in the first variable (*M*) are caused by variations in the second variable (δ). The linear regression equation describes the relationship of these variables with sufficient accuracy. The quality of the model with this value R^2 is good and can be used in the future to predict the values of the performance indicator. The value of the correlation coefficient (r = -0.996) indicates a linear functional relationship in which all the values of the studied variables are located almost on the same straight line. The relationship between *M* and δ can be defined as negative and practically functional (very strong). The maximum torque is achieved when the air gap is 1.01 mm, which is less than the initial 0.99 mm (from δ = 2 mm). The decrease in the value of δ is natural, since the value of the electromagnetic moment is inversely proportional to the value of the air gap. However, creating a SRM with a very small air gap will lead to technological difficulties and problems in further operation (for example, an increase in the degree of unevenness of the air gap under different poles).

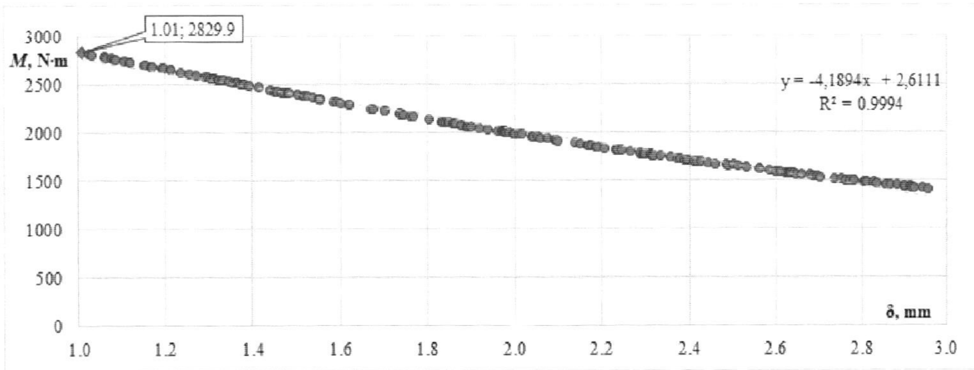

Figure 7. Graph of the correlation between the electromagnetic moment (*M*) and the air gap (δ).

Figure 8 shows the magnetic system fragments to illustrate the changes that the geometric fragments under consideration have undergone. The original geometry is shown in black, and the resulting geometry is shown in blue.

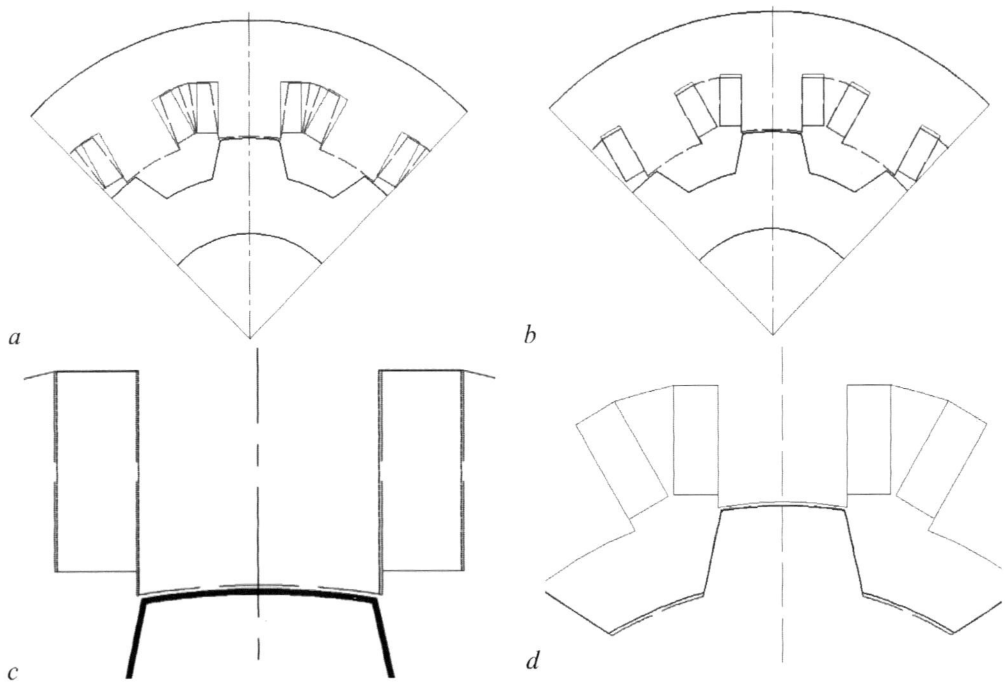

Figure 8. Comparison of the initial geometry of the magnetic system and the one obtained after optimization calculation: *a* – angle of inclination of the side surface of the stator tooth (*b2s*); *b* – stator radius along the grooves (*R2s*); *c* – radius of the stator tooth width (*b1s*); *d* – rotor radius along the grooves (*R2r*).

3. Conclusion

To create a SRM with high energy parameters, in particular a high value of the electromagnetic moment, it is necessary to carry out a comprehensive optimization of the geometric parameters of the stator and rotor. In turn, the FEMM program allows you to take into account the features of calculating electric machines of this type. This program guarantees the consistency of conflicting requirements when the extremum of the target function is reached, for example, the maximum of the average value of the electromagnetic moment.

The use of correlation analysis allowed us to determine the strength and direction of the relationship between the studied structural elements of the active part SRM 12/8 and the average value of the electromagnetic moment. Based on the results of the analysis, it is possible to judge the existence of a corresponding correlation of these parameters.

The results of calculations of the average value of the electromagnetic moment when varying a separate structural size allowed us to see the degree of influence of each parameter of the magnetic system on the formation of the moment value.

References

[1] Daldaban F and Ustkoyuncu N 2010 *Energy Conversion and Management* **51(3)** 465–469
[2] Xiang C, Liu F, Liu H, Han L and Zhang X 2016 *J. of Sound and Vibration* **371** 277–294
[3] Xudong G, Risha N, Chengyu J, Xudong W and Yongqin Z 2018 *Computers & Electrical Engineering* **70** 914–930
[4] Divandari M, Rezaie B and Ranjbar Noei A 2019 *Computers & Electrical Engineering* **70** 106472
[5] Paramasivam S and Arumugam R 2005 *Energy Conversion and Management* **46(9–10)** 1365–1378
[6] Mirzaee Ghazani M and Khosravi R 2020 *Physica A: Statistical Mechanics and its Applications* **560** 125172
[7] An C, Zhu R, Wang X, Long Y, Lu Y, Chen Y and Zhong H 2020 *Annals of Nuclear Energy* **142** 107283
[8] Zhu Q 2020 *J. of Process Control* **95** 32–44
[9] Petrushin A D and Kashuba A V 2018 *Russian Engineering Research* **38(9)** 705–706
[10] Petrushin A D and Shevkunova A V 2018 *IOP Conf. Ser.: Mater. Sci. Eng.* **313(1)** 012009
[11] Krupenev D, Boyarkin D and Iakubovskii D 2020 *Reliability Engineering & System Safety* **204** 107171
[12] Acebrón J 2019 *Applied Mathematics and Computation* **362** 124545

Methods for determining endurance limits for welded joints of marine stationary platforms

I V Starokon

Russian state university of oil and gas named after I.M. Gubkin, 65, Leninskiy pr., Moscow, 119991, Russia

E-mail: aps@gubkin.ru

Abstract. Today, oil, gas and other useful resources are being actively extracted at offshore fields using fixed offshore platforms. According to Federal Law No. 116 On Industrial Safety of Hazardous Production Facilities, offshore platforms are classified as hazardous production facilities, accidents at which are accompanied by significant damage. To prevent accidents at these facilities, an industrial safety examination is carried out, during which their resource is determined. The key parameter for determining the resource is the value of the endurance limit (or fatigue limit). Various theories are currently being applied to determine the fatigue limit. However, calculations on these theories give different results. The purpose of this article is to study various methods for determining endurance limits using welded joints of fixed offshore platforms as an example and choosing a method that allows you to obtain the most accurate value of this limit.

1. Introduction

It is well known that there are potentially recoverable hydrocarbon reserves on the territory of the oceans (covering an area of seventy-five million square kilometers). Offshore development has a long history and tradition. For example, in the area of the Japanese city of Izumosaki, an artificial island was built on which drilling rigs and other equipment necessary for oil and gas production were located. This construction dates back to 1870. Around 1900, the first deviated wells were drilled off the coast of California. The development of offshore fields was actively conducted after the end of the Second World War. So, for example, since 1947, deposits have been actively developed in the Gulf of Mexico. At that time, the first drilling platforms on a pile foundation were built there. Gradually, with the development of technology, the oil and gas production infrastructure was increasingly moving away from the coast. At the beginning of the year 2000, confirmed offshore gas reserves rose to fifty-three billion cubic meters and accounted for 33% of the world's gas reserves. The potential resources of Russian offshore oil and gas fields can be compared with the potential reserves of the largest offshore fields in the world. If we recalculate these reserves for fuel equivalent, it turns out that Russia possesses more than one hundred billion tons of fuel equivalent. Of this amount, 70% of oil and gas resources are located at depths of less than 100 meters, which greatly facilitates their production.

As it was mentioned before, nowadays, the world is actively producing oil, gas and other useful resources in offshore fields. During the development and operation of oil and gas fields on the shelf, various structures are used. These structures include floating drilling rigs, self-elevating drilling rigs, offshore stationary platforms and many types of other special structures. Of particular interest are fixed offshore platforms (FOPs). Offshore fixed platforms have a very long history of use. One of the first in the history of FOPs was the so-called Guy Manson forts shown in fig. 1, which were built in 1942 at the mouth of the Thames to protect against German aviation and are still in operation for a variety of purposes.

Figure 1. Guy Manson Forts

In the USSR, fixed offshore platforms also have a long history of application. Back in 1949, the development of the Oil Stones field began, which are located on the Caspian Sea at a distance about 40 km east of the Absheron Peninsula. For the extraction of hydrocarbon resources, steel flyovers were built on which all the necessary equipment was located. Nowadays, FOPs are very often used for the extraction of hydrocarbon resources, metals, minerals and other minerals in offshore fields by both domestic and foreign companies. In Russia, fixed offshore platforms are located in various regions. So, for example, in the Sakhalin region, PJSC Gazprom operates the platforms: Molikpak, Piltun-Astokhskaya-B, and Lunskaya-A. FOPs "Berkut" and "Orlan" are operated by Rosneft. In the Caspian Sea, PAO Lukoil operates LSP-2 and the D-6 platform. The FGUP RK Chernomorneftegaz company operates about 30 offshore stationary platforms on the Black Sea shelf. Offshore stationary platforms are used for production at the largest offshore fields around the world. An example would be the Khyberniya platform, located in the Atlantic Ocean near Canada. In the Gulf of Mexico in operation are more than four thousand of these structures. Those. it can be confidently noted that FOPs are used in the practice of world oil and gas production (both by domestic and foreign companies) and have a long history of application.

In accordance with Federal Law No. 116 On Industrial Safety of Hazardous Production Facilities, fixed offshore platforms are classified as hazardous production facilities (HPF). According to this law, HEP is characterized by the fact that for such facilities there is a high risk of an emergency, which can be accompanied by significant economic damage and lead to massive loss of life. For example, during the accident of the Alexandr Kinnell platform, more than 120 people died.

Nowadays there are many platforms (both in Russia and abroad) with a long service life. FOPs are operated in difficult offshore conditions. In these conditions, the platform is exposed to a variety of loads. Examples of such loads can serve as wave, wind, ice, from naval vessels, seismic, from the movements and others as a result of the impact of these loads arise variable stress, which are known from the classical theory of mechanics of destruction lead to the accumulation of fatigue damage in welded elements and connections of the platforms, which can provoke an emergency situation. Of particular importance the problem of the accumulation of fatigue damage comes out for assessing the life of repaired welded joints of an offshore platform.

Currently, the repair of these compounds is based on different technologies which include the installation of reinforcing pads, terminologically or special metal inserts. The resource of welded joints is based on the diagrams of fatigue, but to date the diagrams of fatigue for the repaired welds are missing. For such diagrams the key value is the fatigue limit (also called in the literature the endurance limit). In the normative-technical documentation describes various methods of finding the limits of fatigue, which give completely different results. Therefore, the aim of this paper is to study the accuracy of different methods for the determination of endurance limits with regard to its definition for welded joints of offshore fixed platforms, which will continue to determine the endurance limit for each of the known technology of repair welds.

Nowadays, a regulatory and technical framework has been developed which contains fatigue diagrams for new welded joints of offshore stationary platforms. This documentation includes the foreign regulatory framework [1, 2] and the Russian one [3]. According to the regulatory documents [1–3], the endurance limit of welded joints of offshore platforms is in the range from 20 MPa to 35 MPa, if we operate in terms of the range of stresss and a diagram is constructed accordingly. And from 10 MPa to 18MPa, if the diagram is built for the amplitudes of alternating stresss. Therefore, we will consider its value that will be close to the amplitudes of alternating stresses of 10-35 MPa as a target indicator for methods of calculating endurance limits. We also note that the basic endurance limit of a welded joint without taking into account the residual stresses, roughness and other factors described below is determined depending on the temporary tensile strength and is assumed to be 234 MPa. It should be noted that oil and gas production is carried out under difficult conditions [4, 5], and factors such as anticorrosion coatings [6–9], kinetics of austenite [10] and other factors described in works [10-15].

2. Results

It is generally accepted that endurance (or fatigue) is determined on the basis of laboratory tests. Moreover, the results obtained in the laboratory must be correlated with the actual welded joint, taking into account the scale factor, the surface roughness of the platform welded joint, stress concentration, residual stresses in the welded joint, and other factors. The objective of this article is to choose a comparison method, which in the technical literature is also called scaling or transfer, the laboratory value of the endurance limit on real welded joints of the offshore platform.

Let us analyze how various methods of calculating the endurance limit allow us to obtain a result close in value to 10-35 MPa. As an example, in a we consider the results of experiments performed by the author and presented in the table 1. The objective of the experiment was to determine the fatigue life of tubular welded joints of an offshore stationary platform. As a result of the study of the project, it was found that during the construction of offshore platforms for welding elements made of steel 09G2S (D), electrodes E50A of the UONI-13/55 grade in accordance with SAUS 9467-75 were used. The tensile strength of such a weld should be 490 MPa.

Table 1. Test results of welded joints

Stress amplitude σ_a, [MPa]	**Sample №1**	**Sample №2**	**Sample №3**	**Sample №4**	**Sample №5**	**Sample №6**	**Number of cycles, *N***
115	4332	3975	4672	4348	4301	4359	**4332**
82	10253	9071	10798	10803	10164	10504	**10253**

Various methods are proposed for determining the fatigue limit of a weld metal:

- **Method number 1.** Goltsev Method:

$$\sigma_{-1} = \sqrt{\frac{\sigma_1^2 N_1 - \sigma_2^2 N_2}{N_1 - N_2}}, (1)$$

Where: σ_{-1} - endurance limit, MPa; σ_1 and σ_2 - certain level of test stress values, MPa; N_1 and N_2 – number of cycles before fracture for the corresponding stress level.

- **Method №2.** Muratov's Method:

$$\sigma_{-1}=\frac{\sigma_1\sqrt{N_1}-\sigma_2\sqrt{N_2}}{\sqrt{N_1}-\sqrt{N_2}}, (2)$$

Where the values are the same as in Eq. 1

- **Method №3.** Panteleev's Method:

$$\sigma_{-1}=\sqrt{\frac{\sigma_1^2\sigma_2^2(N_2-N_1)}{\sigma_1^2N_2-\sigma_2^2N_1}}, (3)$$

Where the values are the same as in Eq. 1

We will carry out calculations and calculate the endurance limit values in accordance with these methods and present them in table 1.

Table 2. Endurance limits obtained by the methods of Goltsev, Muratov and Panteleev

Number of method	1	2	3
Value of the endurance limit σ_{-1}	44,3	38,8	27,3

However, the endurance limits defined in Table 1 do not take into account the following important factors that determine the functional relationship between laboratory tests and real samples:

1. The proportionality of the size of the samples and real objects;
2. Asymmetry of stress cycles [16, 17];
3. The value of residual stresses.

We define the amplitudes of the first principal stresses for an alternating cycle by the equation:

$$\sigma_{1a}^{p}=\frac{\sqrt{6}\alpha\sigma_{-1}}{K+[M_a/(1-R_{\sigma_0}^{res})](\eta_1+R_{\sigma_0}^{res}\eta_2)}, (4)$$

Where: σ_{-1} endurance limit, endurance strength, obtained by one of three methods, mentioned above, and $R_{\sigma^0}^{res}$ is obtained by equation:

$$R_{\sigma^0}^{res}=\frac{(\sigma_{min}^0)_{res}}{(\sigma_{max}^0)_{res}}=\frac{\sigma_{1\,min}+\sigma_{2\,min}+\sigma_{3\,min}+\sigma_{1\,res}+\sigma_{2\,res}+\sigma_{3\,res}}{\sigma_{1\,max}+\sigma_{2\,max}+\sigma_{3\,max}+\sigma_{1\,res}+\sigma_{2\,res}+\sigma_{3\,res}}, (5)$$

Where $\sigma_{1\,res}, \sigma_{2\,res}, \sigma_{3res}$ – first, second and third (main residual) stresses; $(\sigma_{min}^0)_{res}, (\sigma_{max}^0)_{res}$– octahedral stresses, defined as the sum of residual and stresses caused by the sum of all the loads; $\sigma_{1\,max}, \sigma_{2\,max}, \sigma_{3\,max}$ maximum principal stresses of a cycle; $\sigma_{1\,min}, \sigma_{2\,min}, \sigma_{3\,min}$ minimum principal stresses of a cycle, and α is obtained by the equation:

$$\alpha=\frac{0{,}249\sigma_v+2{,}5}{0{,}35\sigma_v+70}, (6)$$

Where: σ_v temporary resistance to endurance, MPa. The remaining values are given in [15–20].

The author developed a model of an offshore platform in the Structure CAD software package presented in fig. 2. This model allows to determine with high accuracy the stress acting in welded joints, taking into account the different directions of action of the wave load. Using the example of an offshore platform located in the Black Sea region, the maximum possible wave impact once a hundred years was simulated. As a result, the following data were obtained characterizing the range of stresses in the welded joints of the platform. Data are given in the table 2.

Figure 2. Platform model in the StructureCAD software package

Table 3. Stress amplitude in welded joints of the offshore platform

Load combination	Direct exposure to wave load			Inverse wave load		
	σ_1	σ_2	σ_3	σ_1	σ_2	σ_3
K1	160	12,4	2,1	1	-	-
K2	194	17,8	3,6	-33	7,8	1,2

Table 4. The endurance limits adjusted by the Klykov method obtained by the methods of Goltsev, Muratov and Panteleev

Number of the method	1	2	3
Value of the endurance limit σ_{-1}	21,6	18,9	34,3

We correlate the endurance limit obtained by the three methods described above with real welded joints of the platform and the results are summarized in table 3, having obtained data on the stress state of welded joints and acting according to the method developed by Klykov [18].

3. Conclusion

Calculations by Method №1 and Method №2 gave a result close to the point stated in the standards. The result calculated by method №3 is close to the point stated in the standard. It should be noted that all methods showed good convergence of the results and, according to the author, can be applicable in assessing the endurance limits of welded joints of fixed offshore platforms. The choice of the method of calculating the endurance limit will be especially important when determining the resource of welded joints of offshore platforms, taking into account the accumulated damage [18, 19] and the selected repair technology.

References

[1] DNV-RP-C103. 2008 *Recommended Practice. Fatigue Calculation of Marine Steel Structures* (Norway: DNV)
[2] DNVRP 2A-WSD 2005 *Recommended practice for the planning, design and construction of offshore stationary platforms-calculation of permissible stresses* (New York: American Bureau of shipping)
[3] 2014 *Rules for the Classification, construction and equipment of floating drilling rigs and offshore stationary platforms* (St. Petersburg: Russian Maritime Register of Shipping)
[4] Guseynov Ch 2020 In Arctic - new technical means and technologies for the development of oil and gas fields in long-term freezing deepwater areas *Mater. Sci. and Engineering* **734** 012174
[5] Bezkorovayniy V, Bayazitov V and Bobov D 2018 Management of the Design and Construction of Offshore Oil and Gas Facilities with Bim Base *Mater. Sci. and Engineering* **463(4)** 042056
[6] Efimenko L, Kapustin O, Gusev V, Merkulova A and Nikiforov N 2014 Evaluation of structural and phase composition and corrosion resistance properties of coatings made by laser surfacing *Chemical and Petroleum Engineering* **50** 402-411
[7] Efimenko L 2015 Methodological fundamentals of the evaluation of the weldability of structural materials using oil and gas pipes as an example *Welding Int.* **9(28)** 727-731
[8] Efimenko L, Elagina O, Ramus A, Volkov I, Ladyzhanskii A, Rakhmanov D, Konovalov A and Kurkin A 2015 Repair features of mechanical and corrosion damage in a main gas pipeline under pressure *Chemical and Petroleum Engineering* **50** 749-757
[9] Elagina O, Efimenko L and Volkov I 2015 Study of the strain ageing on corrosion failure and submicrostructure of low-carbon steels for gas-distriution pipelines *Chemical and Petroleum Engineering* **51** 433-437
[10] Efimenko L, Kapustin O, Merkulova A, Subbotin R, Puyko O and Shchkalova A 2016 Features of transformation kinetics of austenite in heat treatment of stamped and welded components made of 10G2FBYU steel *Welding Int.* **30** 472-478
[11] Efimenko L and Ramus A 2016 Effect of the morphology of structure on the resistance of welded joints of high-strength pipe steels to brittle fracture *Metal Sci. and Heat Treatment* **57** 559-563
[12] Efimenko L, Kapustin O, Ramus A and Ramus R 2016 Control of softening processes in the heat-affected zone during welding of high-strength steels *Metal Sci. and Heat Treatment* **58** 435-441
[13] Efimenko L, Kapustin O, Merkulova A, Subbotin R, Puyko O and Shchkalova A 2016 Features of transformation kinetics of austenite in heat treatment of stamped and welded components made of 10G2FBYU steel *Welding Int.* **30** 472-478
[14] Makarov G 2014 Modernization of systems of pipeline transport of gas and oil *Welding Int.* **28(9)** 744-747

[15] Starokon I 2019 Varieties and kinetics of fatigue cracking of fixed offshore platforms for oil and gas production *Mater. Sci. and Engineering* **734** 012055
[16] Antonov A 2014 Investigation of fields of residual stresses in welded structures *Welding Int.* **28** 966–969
[17] Starokon I 2020 Features of fatigue processes occurring in restored welded joints of offshore fixed platforms *Mater. Sci. and Engineering* **734** 012171
[18] Starokon I 2020 Resource of repaired welded compounds of marine oil and gas structures *Mater. Sci. and Engineering* **734** 012193
[19] Starokon I 2019 On the results of studying the influence of vibration-fluctuations processes on the resource of welded joints of marine stationary platforms for oil, gas and other useful minerals on the shelf *J. Akustika* **285** 148-151

Adjustment of mud pump parameters by the frequency conversion method

R I Suleimanov, M Ya Khabibullin, A Yu Davydov and L M Zaripova

Ufa State Petroleum Technological University, Branch of the University in the City of Oktyabrsky, 54a, Devonskaya St., Oktyabrsky, Republic of Bashkortostan, 452607, Russian Federation

E-mail: rustamsul@rambler.ru, info@of.ugntu.ru

Abstract. The article describes the process of adjusting the main parameters of a mud pump by the method of frequency conversion of the current in the electric circuit feeding the pump electric motor. This technical solution creates conditions for more efficient operation of the mud pump, reducing the consumption of electrical energy and increasing the values of operation parameters.

1. Introduction

The development of deep wells depends on a technological regime that determines the efficiency of the well drilling process (the axial force on the borehole bottom, rotation frequency of the tool and the circulation regime influenced by the pump design of the drilling rig) [1–12].

Studies show that it is important to improve the theoretical foundations of scientific ideas about the laws of wear and hydraulic action processes that are created in a piston mud pump during its operation and its operating indicators.

The commonly used pump for drilling deep wells is a NBT-600 pump. This pump is used in 3-6 classes of drilling rigs. The piston pump NBT-600 is a three-piston design of a one-way principle of operation. The nomenclature contains several pairs of "piston-cylinder bushing" with different diametrical dimensions from 120 to 180 mm with a multiplicity of 10 mm with 120 double strokes per minute, the theoretical capacity varies from 20 to 45 l/s, the outlet pressure ranges from 12 to 25 MPa , the stroke length of the pumping rod is 250 mm.

The performance of the NBT-600 pump is calculated by formula:

$$Q = \frac{z \cdot \left(\frac{\pi \cdot D^2}{4} \right) \cdot S \cdot n}{60}, \qquad (1)$$

where z – number of cylinder liners, this pump has three ones; n – number of double strokes of the pumping rod, n = 120 min^{-1}, D – pump piston diameter; S – pumping rod stroke, S = 0.25 m; efficiency of the NBT-600 pump– η=0.81.

The pump outlet pressure is calculated by formula:

$$P = \frac{N_{pr} \cdot \eta}{Q}, \qquad (2)$$

where N_{pr} – hydraulic pump power, Вт; η – pump efficiency; Q – pump capacity, m^3/sec.

Table 1. Main parameters of the NBT-600 pump

Piston diameter, mm	Pump discharge pressure, MPa	Pump capacity, m^3 / s
120	25.0	0.0198
130	21.2	0.0232
140	18.3	0.0269
150	15.9	0.0309
160	13.9	0.0352
170	12.4	0.0397
180	11.1	0.0445

2. Materials and methods

The application of the method for adjusting the main parameters of the mud pump by the method of frequency conversion of the current supplied to the asynchronous electric motor in the electric circuit allows replacing the electric DC drive. The speed control of a DC electric motor is carried out in various ways, but practice shows that the most vulnerable link is the electric motor, since it has low reliability and a relatively high cost. Sparks emanate from the brushes and the collector becomes quickly worn out. The use of this electric motor is prohibited in explosive and dusty areas.

DC electric motors are weaker than induction electric motors in terms of reliability and simplicity of design solutions, since there are moving contact elements in the design of DC electric motors. One of the positive qualities of the design of asynchronous motors is small overall and diametrical dimensions and their weight, low cost with the same values of drive power.

One of the main disadvantages of asynchronous electric motors is the complex process of adjusting the speed. Traditional methods are variation in the applied voltage or introduction of a certain amount of resistances into the electric circuit in the winding of the electric drive.

Historically, the issues of frequency conversion of electric current supplied to the electric drive began to be dealt with in the middle of the first half of the twentieth century, but it was impossible to achieve efficient operation of electric motors using the frequency control. Frequency converters had a high cost, which did not allow their introduction into the frequency control system of an electric drive. Today, with the high development of technological solutions in the field of electronics and electrical engineering, the use of highly efficient computerized processors, the approach to the solution of regulation control using frequency converters has allowed global manufacturers to provide consumers with products at a more optimal price.

The methods used to change the frequency of rotation of structural elements provide for the use of mechanisms of various types: hydraulic and turbo couplings, resistors integrated into the moving or static part of the electric motor, electromechanical converters, mechanical variators, static frequency converters.

These mechanisms are not able to create the required quality of speed control, while installation and maintenance works incur significant costs.

Practice has shown that static frequency converters are suitable for the high-quality control of an asynchronous electric motor.

The method of rotation speed adjustment involves changes in the frequency f_1 of the supply voltage:

$$\omega_0 = \frac{2\pi \cdot f_1}{p}, \quad (3)$$

and at a constant number of pole pairs p, the angular velocity of the stator magnetic field changes.

The similar method provides a smooth change in rotational speed in a large wide range, while the mechanical characteristics show stable rigidity.

The proposed adjustment reduces power losses, since there is no increase in the slip effect of an induction motor.

In order to create operating conditions for an asynchronous motor with higher energy indicators, such as efficiency, overload factor, power factor, it is necessary to pay attention to the input voltage.

The use of a controlled electric drive is accompanied by efficient energy savings and provides new capabilities of systems and objects. The regulation of a parameter of the technological chain saves electrical energy and is used in devices where it is necessary to regulate the speed parameters of movement and change its performance.

The efficiency of frequency adjustment is shown in Figure 1.

The highest efficiency of frequency conversion is obtained at technological facilities where fluid is pumped and transported (e.g., at facilities where a piston pump is operated). The most commonly used method of changing pump performance is "throttling". This method uses control valves and gate valves, but it does not fulfill modern tasks.

The use of a gate valve when changing the liquid flow does not provide any useful work, the gate valve simply restrains the flow. The frequency control, in which the pressure or flow rate changes in the required values, decreases energy consumption and reduces the loss of pumped liquid.

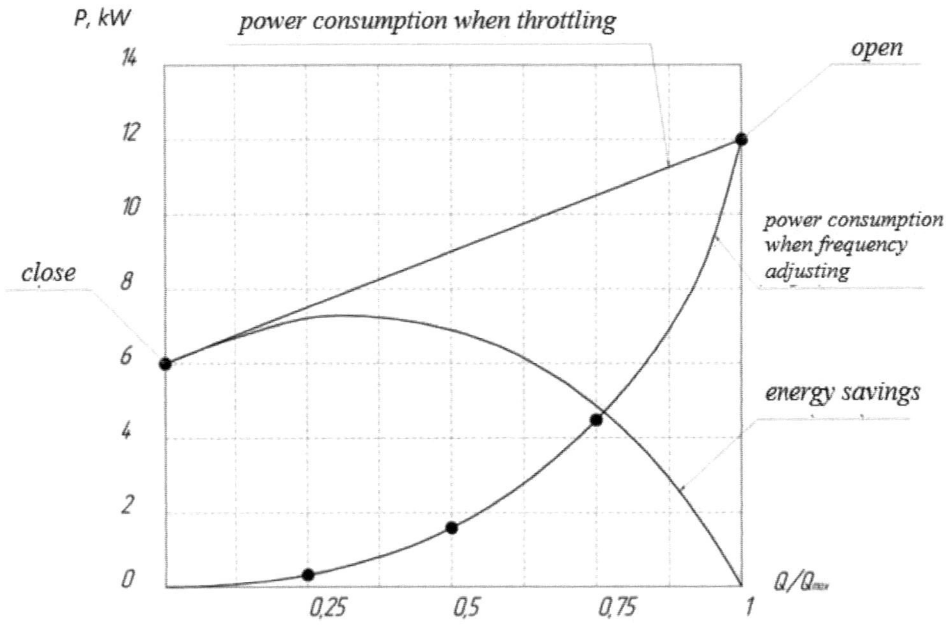

Figure 1. Comparison of frequency regulation and throttling

3. Results

The frequency converter in the NBT-600 piston pump saves electrical energy. This phenomenon can be seen in the cubic dependence of the drive power, which varies with the change in speed of the electric motor (Figure 2).

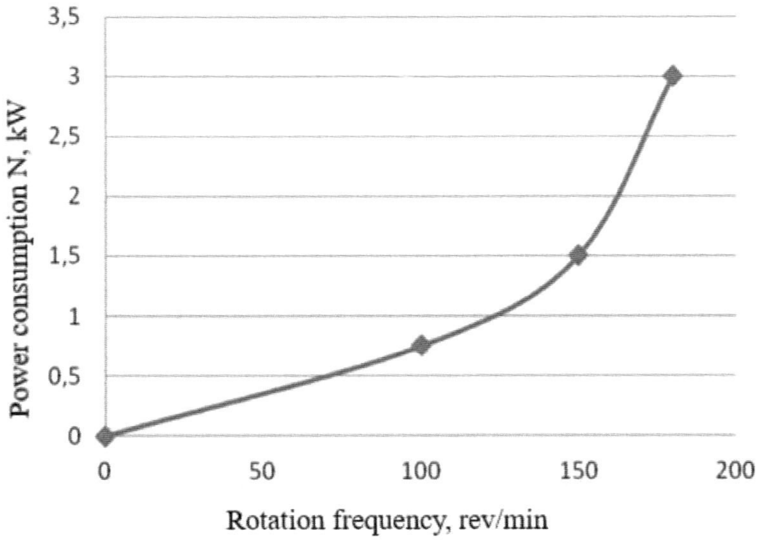

Figure 2. Dependence of the drive power on the speed of the electric motor (H-n) using a frequency converter

When using the "throttling" method, the dependence is shifted to the left along the operating characteristic of the working device, which creates the required flow rate, while the drive power is constant, and the flow rate decreases (Figure 3).

Figures 4 and 5 show the characteristics obtained with the frequency adjustment.

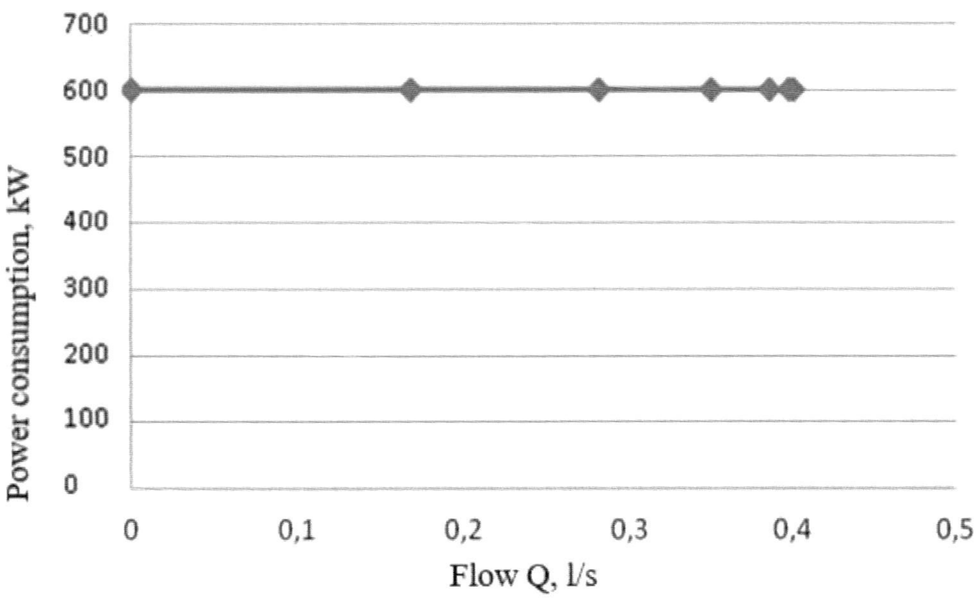

Figure 3. Graphical dependence of the pump flow on the drive power (Q-H)

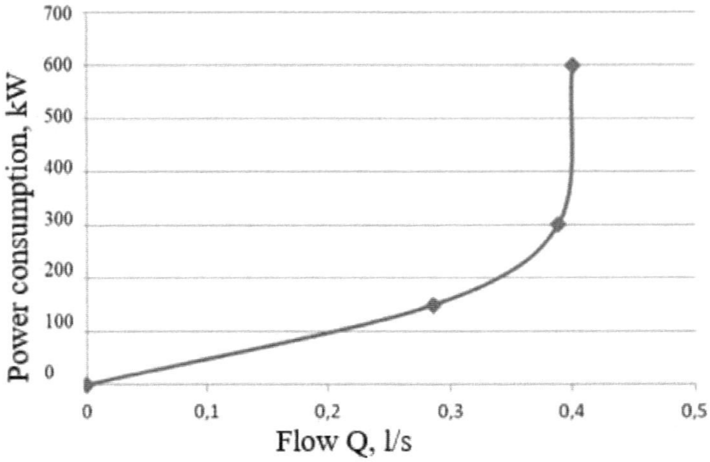

Figure 4. Graphical dependence of the pump flow on the drive power (Q-H) using an inverter

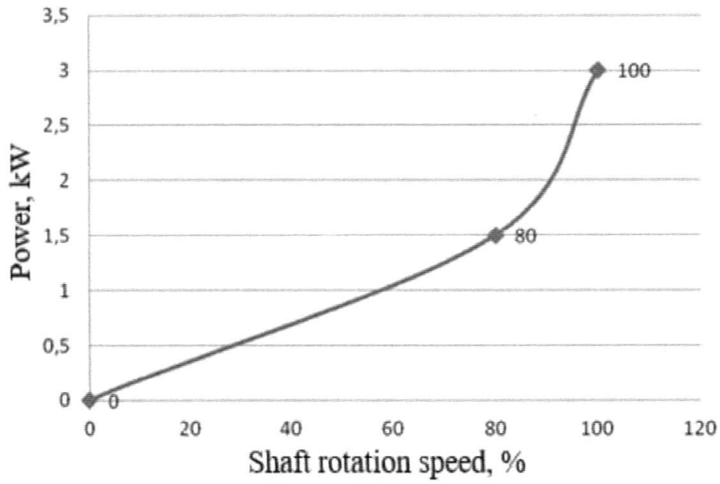

Figure 5. Graphical dependence of the shaft rotation speed on the drive power using an inverter

Changing the speed of movement of the pump elements with the frequency converter, the graphical dependence moves along the operating characteristic to the left side, the required flow rate for the technology is provided, the drive power will require much lower values than with the "throttling" method.

It can be seen that the drive power realized by the frequency converter significantly decreases with a change in feed, when compared with the "throttling" method. The drive power of the piston pump is expressed in a cubic dependence on the speed.

Initially, drive power of the piston pump was 3 kW, but with a 20% decrease in speed by the frequency control, it decreased to 1.5 kW, which is reflected in significant savings in electrical energy.

In the electric drive in an uncontrolled mode, the constancy of drive power is ensured by the use of a sleeve-piston cylinder pair of different diametrical dimensions. This mode is due to the dependence of discharge pressure P, capacity of the piston pump Q and diametral size of the piston D.

Points 7, 6, 5, 4, 3, 2, 1 (Figure 5) show the limiting maximum pressure at which it is possible: 1) to ensure the strength of the pump parts with different diameters of the piston group; 2) ensure the highest possible drive power. The drilling technological process (Figure 5) begins from point 0 at the outlet pressure p <p7 using a piston with a diameter of D7. Further drilling with a piston with a D7 diameter should not be carried out above point 7, since the piston pump drive will experience an excessive load, and the output pressure will be higher than the maximum allowable; therefore, at point 7, the piston with a diameter D7 is replaced with a piston with a diameter D6. The feed of the piston pump will sharply drop to the pressure corresponding to point 8, and reduce the outlet pressure of the piston pump in proportion to the square of the pump feed. During the subsequent drilling with D6 piston, the output pressure will increase to p6 (the maximum permissible value), at which it is necessary to change the piston with a diameter of D6 to a piston with a diameter of D5, etc. 0–7, 8–6, 9–5, 10–4, 11–3, 12–2, 13–1 (Figure 6); as a result, we can approach the equality P × Q = const, but the drive power will be not fully used.

With a variable drive, the preservation of equality P × Q = const under the same conditions is ensured by increasing the output pressure and decreasing the drive speed, but this will decrease the volumetric efficiency of the pump and the flow.

The operation of the piston pump in a variable drive mode is carried out in the following sequence. The well drilling process starts from point 0 (Figure 6). With increasing penetration, the outlet pressure increases and goes to p7 (point 7). With a variable speed drive, the piston is not changed at p7, because the speed of the drive at the nominal shaft torque is reduced by the frequency control. As a result, the feed of the piston pump decreases, while the output pressure is stable and equal to the allowable piston. This action will be carried out up to point 8 '. A further decrease in rotation speed has low efficiency, the pump flow decreases, it is necessary to replace the piston with D7 into the piston with D6. With this mode, the drive frequency reaches the nominal value, and with an increase in penetration, the outlet pressure rises to P6 (point 6), which is the limiting value for the piston with D6. A further decrease in the drive speed will ensure the transition of the mode from point 6 to point 9', after which the piston is replaced to the one with D5, etc. As a result, the output and feed of the piston pump will change in the following order: 0-7-8'-6-9'-5-10'-4-11'-3-12'-2-13'-1.

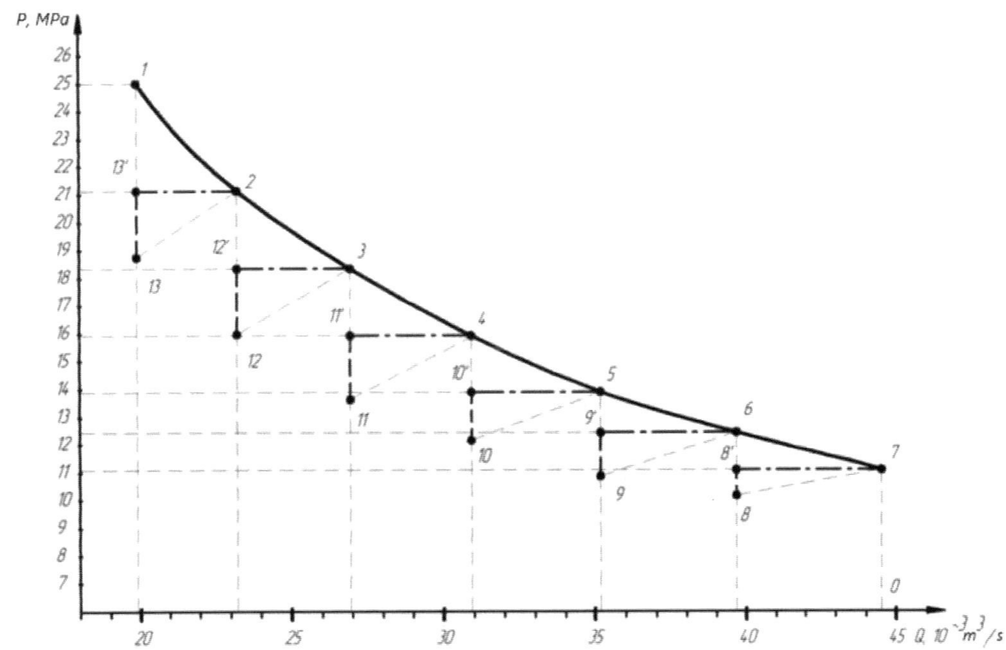

Figure 6. Characteristics of the piston pump NBT-600: — · — · · – variable drive (0-7-8'; 6-9'; 5-10'; 4-11'; 3-12'; 2-13'; 1); – – – – fixed drive (0-7; 8-6; 9-5; 10-4; 11-5; 12-2; 13-1); Q – pump capacity, P - pump outlet pressure

When comparing fixed and variable speed drives, the total downstream pressure loss (Figure 6) are: ΔP = ΔP8-8' + ΔP9-9' +ΔP10-10' +ΔP11-11' + ΔP12-12' +ΔP13-13' = 0,95 + 1,61 + 1,79 + 2,29 + 2,44 + 2,48 = 11,56 MPa.

4. Conclusion

The full use of power of the piston pump with a variable drive ensures that the nominal pump flow is maintained throughout the entire cycle of well construction.

References

[1] Singh V P, Pandey S K, Agrawal S and Singh N 2018 Impact of Communication Delay on Load Frequency Control of Interconnected Power System *5th IEEE Uttar Pradesh Section Int. Conf. on Electrical, Electronics and Computer Engineering, UPCON 2018* 8596987

[2] Popescu D, Dinu R C and Bratu C 2017 Fuzzy control of cooling water pumps related to a power plant *11th Int. Conf. on Electromechanical and Power Systems, SIELMEN 2017* pp 541-546

[3] Gehrke O and Richert T 2017 Use cases for integrated electrical and thermal energy systems operation and control with a view on simulation tools *52nd Int. Universities Power Engineering Conf., UPEC 2017* pp 1-6

[4] Tyncherov K T, Ikhsanova F A and Selivanova M V 2018 Ranking objects by the main component method and highlighting the most informative object parameters *Radioengineering* **9** 185-192

[5] Suleimanov R I, Zainagalina L Z, Khabibullin M Ya, Zaripova L M and Kovalev N O 2018 Studying heat-affected zone deformations of electric arc welding *IOP Conf. Ser.: Mater. Sci. Eng.* **327(3)** 032053. DOI: 10.1088/1757-899X/327/3/032053.

[6] Almukhametova E M, Shamsutdinova G F, Sadvakasov A A, Tyncherov K T, Petrova L V and Stepanova R R 2018 Modeling development of Fyodorovsky deposit *IOP Conf. Ser.: Mater. Sci. Eng.* **327(4)** 042100. DOI: 10.1088/1757-899X/327/4/042100.

[7] Tyncherov K T, Mukhametshin V Sh, Paderin M G, Selivanova M V, Shokurov I V and Almukhametova E M 2018 Thermoacoustic inductor for heavy oil extraction *IOP Conf. Ser.: Mater. Sci. Eng.* **327(4)** 042111. DOI:10.1088/1757-899X/327/4/042111

[8] Khuzina L B, Mukhametshin V Sh, Tyncherov K T and Shaikhutdinova A F 2018 On the choice of the oscillators' installation site *Int. J. of Civil Engineering and Technology* **9(9)** 1952-1959

[9] Khabibullin M Ya and Suleimanov R I 2018 Selection of optimal design of a universal device for nonstationary pulse pumping of liquid in a reservoir pressure maintenance system *Chemical and Petroleum Engineering* **54(3-4)** 225-232

[10] Valeev A S, Kotenev Yu A, Mukhametshin V Sh and Sultanov Sh Kh 2019 Substantiation of the recovery of residual oil from low-productive and heterogeneous formations in Western Siberia by improving the waterflood system using gas and water-gas impacts *IOP Conf. Ser.: Mater. Sci. Eng.* **560(1)** 012204. DOI: 10.1088/1757-899X/560/1/012204

[11] Kuleshova L S, Kadyrov R R, Mukhametshin V V and Akhmetov R T 2019 Auxiliary equipment for downhole fittings of injection wells and water supply lines used to improve their performance in winter *IOP Conf. Ser.: Mater. Sci. Eng.* **560(1)** 012071. DOI: 10.1088/1757-899X/560/1/012071

[12] Kuleshova L S, Kadyrov R R, Mukhametshin V V and Safiullina A R 2019 Design changes of injection and supply wellhead fittings operating in winter conditions *IOP Conf. Ser.: Mater. Sci. Eng.* **560(1)** 012072. DOI: 10.1088/1757-899X/560/1/012072

IOP Conf. Series: Materials Science and Engineering 1111 (2021) 012059 doi:10.1088/1757-899X/1111/1/012059

Selection of optimal operating modes for a block cluster pumping station

R I Suleimanov, M Ya Khabibullin, L Z Zainagalina and L M Zaripova

Ufa State Petroleum Technological University, Branch of the University in the City of Oktyabrsky, 54a, Devonskaya St., Oktyabrsky, Republic of Bashkortostan, 452607, Russian Federation

E-mail: rustamsul@rambler.ru, info@of.ugntu.ru

Abstract. The article discusses a method for optimizing the model of a modular cluster pumping station using a variable frequency drive. Thanks to this technology, the energy consumption of the pumping units of the installation are sharply reduced. The equipment service life increases as a result of the application of this technology.

1. Introduction

At present, waterflooding is a well-established method of developing and enhancing oil recovery, applicable practically under all geological, physical and technical and technological conditions.

The use of waterflooding as the primary method is determined by the highest final oil recovery coefficient of the water-driven regime in comparison with others [1–6].

The main functions of the RPM system are as follows [7, 8]:

- preparation and provision of the required volume of water injection into the reservoir;
- control of the volume of injected water;
- reliable and economical operation throughout the entire life of the field.

Increasing the energy efficiency of the reservoir pressure maintenance system at the current high cost of electricity and the tendency to increase it against the background of a high percentage of water cut in good products means a decrease in current economic operating costs and an increase in the economic efficiency of the oil production process [9].

2. Materials and methods

Pumping units BKNS are the main energy-producing and at the same time energy-consuming elements of the pressure maintenance system. The type and number of installed pumping units determine the head and productivity of the BKNS [10, 11].

In the overwhelming majority, pumping units BKNS are represented by a group of multistage sectional centrifugal pumps of the type CNS 180, CNS 240, CNS 300, CNS 500 in an amount of 4 to 6 units.

The use of centrifugal sectional pumps is due to their significant advantages over other pumps [12, 13], the main of which are:

- relatively low cost, since the construction uses steel, cast iron and polymeric materials;
- ease of maintenance and operation;
- a wide range of performance regulation while maintaining a sufficiently high-efficiency value;

- stable operation of pumps in sequential and parallel operation on a single pressure water conduit.

The share of pumping units in the total balance of electricity consumption of the reservoir pressure maintenance system is 80–90%, and the cost of water injection in the cost of oil production is 26–35%.

The number of switching on-off per day of pumping units in the pumping station with tanks reaches 40-50 and in some cases 100. This number of switching-one for high-power units is unacceptable; therefore, in installations with units with a capacity of more than 150-250 kW, regulation of the operating mode of individual pumps is used, included in the BKNS.

An increase in the supply of the water supply network leads to an increase in the frequency of switching on the units and a decrease in the duration of pauses. The reason is that when the supply of the water supply network is increased, the volume of liquid in the tank is quickly triggered. In this case, the liquid level reaches the lower position more quickly, and as a result, additional pumping units are switched on more often [14].

The deliberate change in the flow and head of the pumps following the changing operating conditions of the system is called regulation.

Losses in pumps are associated with irrational operating modes of the system, which are characterized by a discrepancy between the parameters of the pump group and the parameters of the hydraulic network and the changing conditions of the oil field development technology [15].

Such modes of operation cause an increase in the share of non-productive losses of electricity in the technological system and increased specific consumption of energy parameters for water injection.

The reasons for the occurrence of electricity losses in the pumping units of the BKNS technological system of pressure maintenance can be conditionally divided into two groups [16]:

1) the first group – power losses determined by the type of pumping units in operation, individual characteristics of the hydraulic network and field equipment, adopted by the applied field development scheme.

Reducing these losses of the first group is possible:

- carrying out computational and technical developments to improve the flow path of the pumps, which will ensure high energy performance of the machine;
- improvement of materials and the inner surface of pipelines;
- improving the quality of calculations for the selection of technological equipment.

2) the second group – electricity losses determined by irrational operating modes of the pumping units of the BKNS, are considered as additional, which can be reduced by improving the existing approaches to the design of new and modernization of the operating RPM technological systems.

The reasons for the high share of electricity losses in the pumping units of the BKNS of the technological system are the low level of controllability of the electric drive of the pumps and the low adaptability of the BKNS to the changing conditions of the oil field development technology. The amount of power consumed by the pumping unit depends on its individual energy characteristics, which decrease during operation due to operational wear. The pump performance is also significantly influenced by the frequency, conditions and quality of repairs, which are not always carried out satisfactorily. Subsequently, this leads to a difference in hydraulic and energy characteristics of pumping units operating in parallel, which causes excessive energy consumption by pumping units with a lower capacity.

Pressure maintenance systems mainly operate on the principle of a centralized system, in which BKNS are united by one mode of water injection through high-pressure pipelines and water supply to injection well pads. The maximum head is determined from the condition of pumping the required volume of water into the wells with the highest injectivity.

The pressure in injection wells is controlled in the following ways:

1) fitting (flow limiter),

2) throttling with gate valves.

3) regulation of the pump flow, in which part of the liquid flow from the outlet of the BKNS is diverted to its inlet through the bypass, which is a special pipeline with a regulator valve.

4) step regulation of the pump flow, which is realized by inputting or outputting one or more pumps from the system.

The considered methods of flow control provide only technological parameters. They do not take into account energy parameters, which reduces the possibility of increasing the energy efficiency of the technological system for maintaining reservoir pressure.

Table 1. Comparative analysis of methods for regulating the performance of pumping units BKNS

Regulation method	Energy loss level	Compliance with the injection plan	Efficiency	Depreciation of equipment	Notes
Throttling on the pressure side of the pump unit	high	high	low	high	Wear of pumping equipment and valves. An inevitable increase in non-productive power losses
Installation of chokes at the injection wellhead	average	low	low	high	
Bypass application	high	low	low	high	
Step control: on / off pumping units	high/ average	average	high	high	Wear of both pumping and electrical equipment. Limited number of communications in the absence of soft start systems.
Changing the number of pump impellers (pumps)	low	average	low	low	Reasonable in cases of pumping units operation for a long time in a constant flow mode
Frequency-controlled electric drive	low	high	high	low	Wide range of performance control of pumps, pumping units

3. Results

The task of optimizing the control of the operating modes of the BKNS of the reservoir pressure maintenance system is associated with the need to take into account the many continuous technological, hydraulic and operating parameters of the elements of the entire system and can be implemented in two ways:

1) joint control of operating modes of pumps BKNS and hydraulic network,

2) increasing the level of controllability of the BKNS, i.e. more adaptive and operational behaviour of the system.

The optimization process consists of convergence of the hydraulic characteristics of the network and pumps of the BKNS while minimizing non-productive power losses. Each state of the hydraulic network and a set of planned tasks for injection into injection wells has its optimal mode of operation of the BKNS. That is, the position of the operating point on the flow-pressure characteristic of the hydraulic network, while the specific power consumption is minimal.

Optimization of regulation of operating modes of pumps BKNS is implemented in the following directions:

a) simultaneous control of the operating modes of the hydraulic system and pumps,

b) increasing the flexibility of regulating the parameters of the cluster pumping station.

The methods for converging the characteristics of the BKNS and the operating point of the hydraulic network are as follows:

a) change in the number of impellers in one or more pumps;

b) changing the characteristics of the hydraulic network;

c) the use of a frequency-controlled electric drive of pumping units BKNS.

Frequency control provides the most significant operational regulating effect on pump performance, eliminating power losses. The figure shows the flow-pressure characteristic of the CNS 180-1900 and the characteristics of the hydraulic network, the optimal operating point on the graph is point A1, the flow rate is 180 m^3/h, the head is 1900 m.

In practice, when the pressure maintenance system is operating, a situation arises with a decrease in the supply of fluid from the hydraulic network. In this case, the BKNS operator regulates the fluid supply to the BKNS intake by closing the shut-off valve. As a result, the characteristics of the hydraulic network will change and point A1 will turn into point A2. However, the system will immediately experience energy losses (A2-A3) with an increase of 8.2% (1.55 MPa) since this action limited the capabilities of the hydraulic network. When using frequency control, for example, when the speed is reduced by 5%, point A1 will move to point A3 with a constant characteristic of the hydraulic network, which will not result in energy losses.

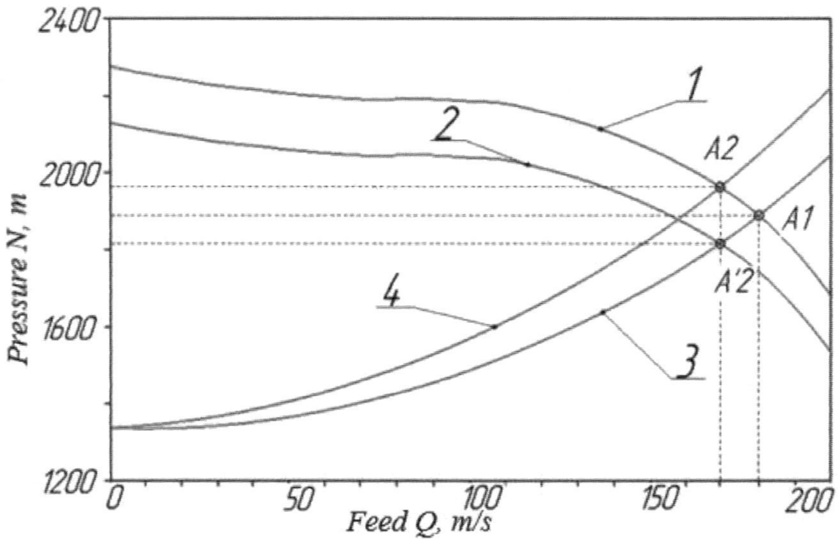

Figure 1. Flow-pressure characteristic of the CNS 180-1900: line 1 – the pressure characteristic of the pump at the rated speed, n = 3000 rpm; line 2 – pressure characteristic of the pump with a 5% reduced rotation frequency, n = 2850 rpm; line 3 – characteristic of the pipeline at the full opening of the gate; line 4 – characteristic of the pipeline with a 10% decrease in the degree of opening of the gate

4. Conclusion

The use of a frequency-controlled electric drive provides an increase in the efficiency of energy consumption of the technological system for maintaining reservoir pressure and the operability of the BKNS, which favourably affects the resource of the equipment.

References

[1] Rogachev M K and Mukhametshin V V 2018 Control and regulation of the hydrochloric acid treatment of the bottomhole zone based on field-geological data *J. of Mining Institute* **231** 275-280. DOI: 10.25515/PMI.2018.3.275.

[2] Rogachev M K, Mukhametshin V V and Kuleshova L S 2019 Improving the efficiency of using resource base of liquid hydrocarbons in Jurassic deposits of Western Siberia *J. of Mining Institute* **240** 711-715. DOI: 10.31897/PMI.2019.6.711

[3] Kuleshova L S, Kadyrov R R, Mukhametshin V V and Safiullina A R 2019 Design changes of injection and supply wellhead fittings operating in winter conditions *IOP Conf. Ser.: Mater. Sci.*

Eng. **560(1)** 012072. DOI: 10.1088/1757-899X/560/1/012072

[4] Kuleshova L S, Kadyrov R R, Mukhametshin V V and Akhmetov R T 2019 Auxiliary equipment for downhole fittings of injection wells and water supply lines used to improve their performance in winter *IOP Conf. Ser.: Mater. Sci. Eng.* **560(1)** 012071. DOI: 10.1088/1757-899X/560/1/012071

[5] Mukhametshin V V and Andreev V E 2017 Search and argumentation of decisions aimed at increasing the efficiency of bottom-hole zone stimulation in oil accumulations with challenged reserves *SPE Russian Petroleum Technology Conf. (16-18 October 2017, Moscow)* pp 1–23. DOI: 10.2118/187785-MS

[6] Akhmetov R T, Mukhametshin V V and Kuleshova L S 2019 Simulation of the absolute permeability based on the capillary pressure curves using the dumbbell model *J. of Phys.: Conf. Ser.* **1333(3)** 032001. DOI: 10.1088/1742-6596/1333/3/032001

[7] Martinez V and Ascendo F 2018 A new practical water injection system in offshore fields *Proc. of the Annual Offshore Technology Conf.* vol 3 pp 1761-1769

[8] Silva G, Correia B, Cunha A, Santos B and Lima A 2017 Water injection for oil recovery by using reservoir simulation via CFD *Int. J. of Multiphysics* **11(1)** 83-96

[9] Singh V P, Pandey S K, Agrawal S and Singh N 2018 Impact of Communication Delay on Load Frequency Control of Interconnected Power System *5th IEEE Uttar Pradesh Section Int. Conf. on Electrical, Electronics and Computer Engineering, UPCON 2018* 8596987

[10] Khabibullin M Ya and Suleimanov R I 2018 Selection of optimal design of a universal device for nonstationary pulse pumping of liquid in a reservoir pressure maintenance system *Chemical and Petroleum Engineering* **54(3-4)** 225-232

[11] Almukhametova E M, Shamsutdinova G F, Sadvakasov A A, Tyncherov K T, Petrova L V and Stepanova R R 2018 Modeling development of Fyodorovsky deposit *IOP Conf. Ser.: Mater. Sci. Eng.* **327(4)** 042100. DOI: 10.1088/1757-899X/327/4/042100.

[12] Suleimanov R I, Zainagalina L Z, Khabibullin M Ya, Zaripova L M and Kovalev N O 2018 Studying heat-affected zone deformations of electric arc welding *IOP Conf. Ser.: Mater. Sci. Eng.* **327(3)** 032053. DOI: 10.1088/1757-899X/327/3/032053.

[13] Tyncherov K T, Mukhametshin V Sh, Paderin M G, Selivanova M V, Shokurov I V and Almukhametova E M 2018 Thermoacoustic inductor for heavy oil extraction *IOP Conf. Ser.: Mater. Sci. Eng.* **327(4)** 042111. DOI:10.1088/1757-899X/327/4/042111

[14] Shaidullina R M, Amirov A F, Muhametshin V Sh and Tyncherov K T 2017 Designing Economic Socialization System in the Educational Process of Technological University *European J. of Contemporary Education* **6(1)** 149–158. DOI: 10.13187/ejced.2017.1.149

[15] Khuzina L B, Mukhametshin V Sh, Tyncherov K T and Shaikhutdinova A F 2018 On the choice of the oscillators' installation site *Int. J. of Civil Engineering and Technology* **9(9)** 1952-1959

[16] Gabdrakhmanova K F, Izmailova G R, Larin P A, Vasilyeva E R, Madjidov M A and Marupov S R 2018 Nomogram method as means for resource potential efficiency predicative aid of petrothermal energy *J. of Phys.: Conf. Ser.* **1015(3)** 032036.

Mathematical modeling of the sunflower cake grinding process

A B Shepelev and I E Priporov

Kuban State Agrarian University, 13, Kalinina street, Krasnodar, 350044, Russia

E-mail: shepelevab@mail.ru

Abstract. The design of an improved sunflower cake grinder is proposed, which consists in the modernization of knives in knife blocks in the form of an Archimedes Arbelo curvilinear triangle with saw-edged teeth along the perimeter of the entire surface of the cutting edge. This will allow obtaining cake in a crumbled form, depending on the type of the farm animal by constructing a mathematical model that considers physical and mechanical properties of the material, which describes the dependence of its design parameters on the factors under study that affect the technological process of its production and thereby improve the grinding quality. The article presents the mathematical model that allows determining the optimal values of the design parameters of the improved grinder. The improved sunflower cake grinder allows reducing its specific energy consumption compared to series-produced ones without deteriorating the quality of the material and to increase its performance. Based on the experimental studies conducted, the improved sunflower cake grinder allows it to be used in technological lines for the production of mixed feed with its adding in a crumbled form. The obtained design parameters of the improved grinder will be used in the further modernization of series-produced grinders for concentrated feed produced by Russian enterprises for various types of farm animals, which will allow replacing imported concentrated feed with Russian ones and thereby not depend on them due to the current situation in the world.

1. Introduction

The dynamic development of modern animal husbandry as a branch of the national economy is impossible without the use of concentrated feed. Various groups of farm animals need concentrated feed, which has significant differences both in the composition of the initial components and in granulometric characteristics [1]. In organizing the full feeding of animals, the rational use of concentrated feed is very important [2]. A significant share in the prime cost of concentrated feed is the expenses for grinding. In this regard, the research is aimed at reducing the energy intensity in their production [1, 3].

2. Materials

Optimization of their design parameters is a perspective direction for reducing the energy intensity when grinding sunflower cake in a bulk form and increasing the grinder performance [1, 4-6].

To reduce the energy intensity of the process and increase the performance of sunflower cake grinders in a bulk form, the design of the grinder (Figure 1) according to the Patent of the Russian Federation (RU 2648392) has been proposed, which has a charging hopper 2, a damper 3, a body 1, a frame 9, an electric motor 10, a rotor 8 with discs 4 and 5, knife packages 6 with knives 7 in the form of flat geometric shapes, formed by circles.

The grinder has an even number of knife packages with knives installed parallel in a staggered arrangement, the surface of the cutting edges of which is made in the form of an Archimedes Arbelo curvilinear triangle with saw-edged teeth along the perimeter of the entire surface of the cutting edge. The height of these teeth is not less than the thickness of the process material. Thus, the teeth inclination angle is no more than 30°, and the teeth of the knives of the subsequent knife packages are directed in the opposite direction from the knives of the previous knife packages and at the same angle, and the knives are installed with the possibility of adjusting the distance between them depending on the size of the process material. Knives are selected for various feed, both in pelleted and crumbled forms.

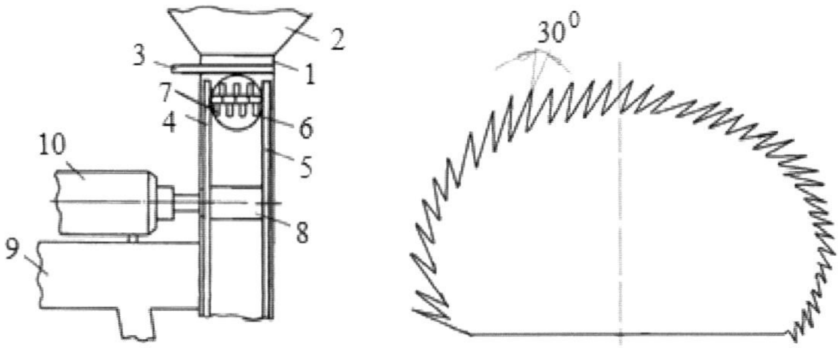

Figure 1. The sunflower cake grinder: a – scheme, b – knife-cutting face

The technological process of the grinder, according to the Patent (RU 2648392), is conducted as follows.

When grinding the material, for example, the obtained sunflower cake, it is poured into the charging hopper 2 and the electric motor 10 is turned on, after which the damper 3 is opened. The crushed product contacts knives 7 in the form of an Archimedes Arbelo curvilinear triangle with saw-edged teeth along the perimeter of the entire surface of the cutting edge, which will improve the quality of grinding, and the entire surface of the knife cutting edge participates in the technological process and sliding cutting takes place, while the energy requirement decreases.

3. Results

Considering the above, a three-factor experiment was conducted, which describes the dependence of the design parameters on the factors under study:

$$(E, W) = f(x_1, x_2, x_3),$$

where E – grinding process energy intensity, kW-h/t; W – grinder performance in obtaining crumbled sunflower cake, t/h; x_1 – crumbled sunflower cake thickness, mm; x_2 – tooth inclination angle on the knife packages, deg.; x_3 – number of knife packages, pcs.

Based on the results of the three-factor experiment, the regression equations were determined in coded form:

– grinder performance in obtaining bulk sunflower cake

$$y_s = 1{,}426 + 0{,}1308x_1 - 0{,}11x_2 - 0{,}023x_3 - 0{,}38x_1x_2 + 0{,}064x_1x_3 + 0{,}09x_2x_3 + 0{,}119x_1^2 + \\ +0{,}018x_2^2 - 0{,}23x_3^2; \quad (1)$$

– grinder energy intensity in obtaining bulk sunflower cake

$$y_s = 2{,}096 + 0{,}0696x_1 - 0{,}03x_2 + 0{,}196x_3 - 0{,}188x_1x_2 - 0{,}336x_1x_3 + 0{,}24x_2x_3 - 0{,}056x_1^2 + +0{,}095x_2^2 - 0{,}16x_3^2, \quad (2)$$

which were validated on the basis of Fisher test [7] and were:

– grinder performance

$$F_{calc} = \frac{0{,}425}{0{,}103} = 4{,}13,$$

– grinder energy intensity

$$F_{calc} = \frac{0{,}7037}{0{,}172} = 4{,}09.$$

The tabular value of the Fisher criterion at 5% significance level was $F_{0{,}05} = 4{,}78$. The mathematical models (1) and (2) validation have been confirmed.

As a result of processing the full-factor experiment, the regression equations in the canonical form (2-7) of dependency were obtained:

– grinder performance on the material thickness and tooth inclination angle (Figure 2):

$$y - 1{,}4 = 0{,}119X_1^2 + 0{,}0259X_2^2, \quad (2)$$

or

$$\frac{(X_1)^2}{\frac{y-1{,}4}{0{,}119}} + \frac{(X_2)^2}{\frac{y-1{,}4}{0{,}0259}} = 1;$$

– grinder performance on the material thickness and the number of knife packages (Figure 3):

$$y - 1{,}4 = 0{,}119X_1^2 - 0{,}2379X_3^2, \quad (3)$$

or

$$\frac{(X_1)^2}{\frac{y-1{,}4}{0{,}119}} + \frac{(X_3)^2}{\frac{y-1{,}4}{-0{,}2379}} = 1;$$

– grinder performance on the tooth inclination angle and the number of knife packages (Figure 4):

$$y - 1{,}4 = 0{,}0259X_2^2 - 0{,}2379X_3^2, \quad (4)$$

or

$$\frac{(X_2)^2}{\frac{y-1{,}4}{0{,}0259}} + \frac{(X_3)^2}{\frac{y-1{,}4}{-0{,}2379}} = 1;$$

–grinder energy intensity in obtaining bulk sunflower cake on the material thickness and tooth inclination angle (Figure 5):

$$y - 2{,}14 = -0{,}056X_1^2 + 0{,}1426X_2^2, \quad (5)$$

or

$$\frac{(X_1)^2}{\frac{y-2{,}14}{-0{,}056}} + \frac{(X_2)^2}{\frac{y-2{,}14}{0{,}1426}} = 1;$$

- grinder energy intensity in obtaining bulk sunflower cake on the material thickness and the number of knife packages (Figure 6):

$$y - 2{,}14 = -0{,}056X_1^2 - 0{,}2076X_3^2, \quad (6)$$

or

$$\frac{(X_1)^2}{\frac{y-2{,}14}{-0{,}056}} + \frac{(X_3)^2}{\frac{y-2{,}14}{-0{,}2076}} = 1;$$

- grinder energy intensity in obtaining bulk sunflower cake on the tooth inclination angle and the number of knife packages (Figure 7):

$$y - 2{,}14 = 0{,}1426X_2^2 - 0{,}2076X_3^2, \quad (7)$$

or

$$\frac{(X_2)^2}{\frac{y-2{,}14}{0{,}1426}} + \frac{(X_3)^2}{\frac{y-2{,}14}{-0{,}2076}} = 1.$$

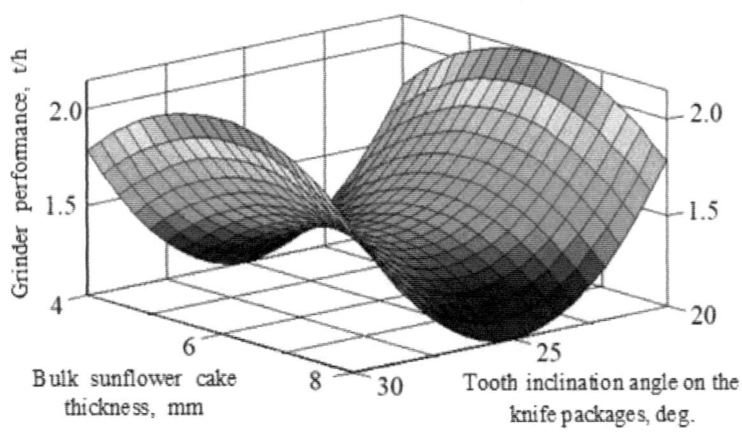

Figure 2. The dependence of the grinder performance on the bulk sunflower cake thickness and the tooth inclination angle on the knife packages

IOP Conf. Series: Materials Science and Engineering 1111 (2021) 012060 doi:10.1088/1757-899X/1111/1/012060

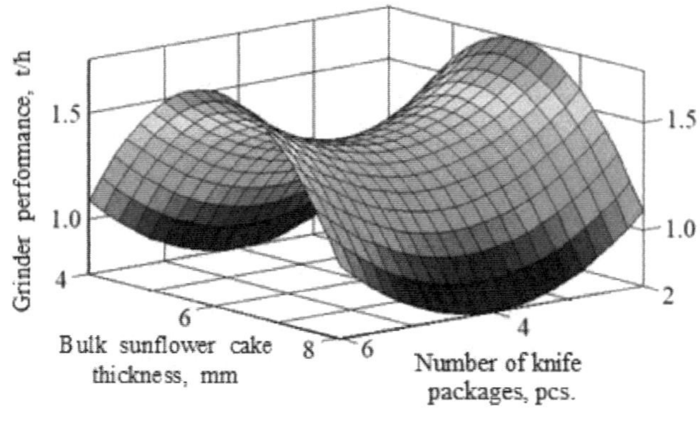

Figure 3. The dependence of the grinder performance on the bulk sunflower cake thickness and the number of knife packages

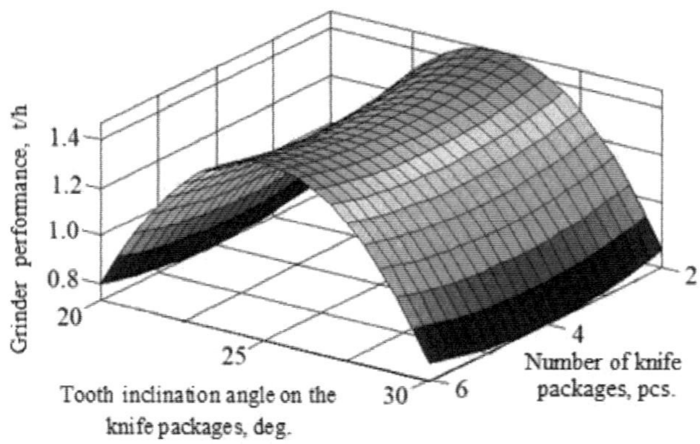

Figure 4. The dependence of the grinder performance on the tooth inclination angle on the knife packages and the number of knife packages

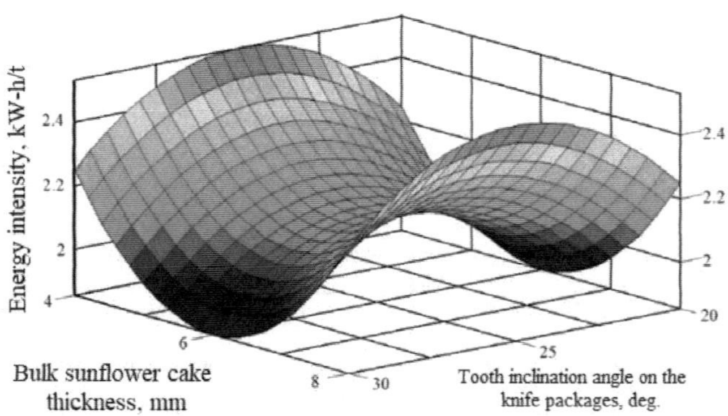

Figure 5. The dependence of the grinder energy intensity on the bulk sunflower cake thickness and the tooth inclination angle on the knife packages

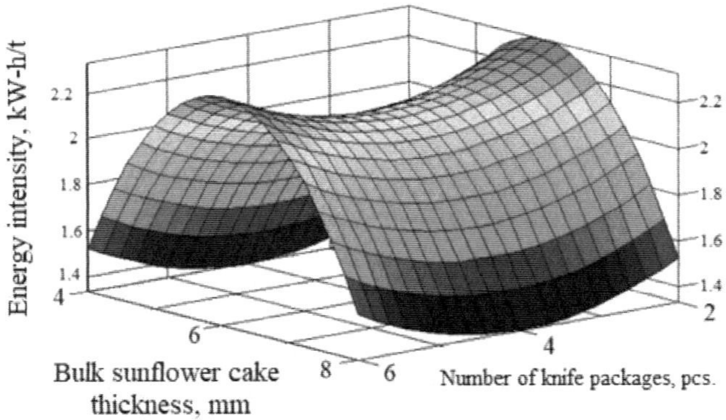

Figure 6. The dependence of the grinder energy intensity on the bulk sunflower cake thickness and the number of knife packages

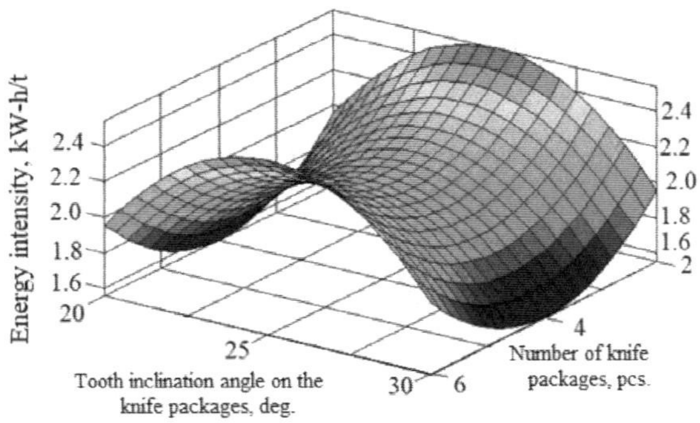

Figure 7. The dependence of the grinder energy intensity on the tooth inclination angle on the knife packages and the number of knife packages

The regression coefficients of the canonical form have the same signs, hence the response surface curves refer to the type of minimum for the equations (2) and (6) [6].

The system of contour curves equal to the performance and energy intensity of the grinder has the form of ellipses (Figure 2, Figure 6). In this case, the large axes of the ellipses are located according to the factor X_1 (the material thickness), which is explained by the larger, than factor X_2 (the tooth inclination angle of knives on the knife packages), influence on the performance and energy intensity of the grinder.

Figures 2 and 6 show the optimal values of the factors are observed with the following limits of their variation: the material thickness – 6 mm and the tooth inclination angle of knives on the knife packages – 25 degrees with a minimum of the grinder energy intensity – 2.1 kW-h/ t and its maximum performance of 1.4 t/h.

The regression coefficients of the canonical form have different signs, hence the response surface curves refer to the type of the minimax for the equations (3) - (5) and (7) [5].

The system of contour curves, equal to the performance and energy intensity of the grinder, has the form of a paraboloid.

Figures 4 and 7 show the optimal values of the factors are observed with variations: the tooth inclination angle is 25 degrees and the number of knife packages is 4 pcs. with a minimum of the grinder energy intensity of 2.15 kW-h/t and its maximum performance of 1.4 t/h.

4. Conclusion

The scientifically grounded and developed sunflower cake grinder, for which the Patent of the Russian Federation was received, has the knives in the form of an Archimedes Arbelo curvilinear triangle with saw-edged teeth affecting the sunflower cake and obtaining it in a crumbled form. Knives in the knife packages are selected for various feeds, depending on the type of farm animal, to obtain it in granular and crumbled forms.

The use of mathematical modeling of the sunflower cake grinding process to a bulk form makes it possible to determine the optimal values of the design parameters of the improved grinder: the tooth inclination angle on the knife packages is 23.48...25.76 degrees, the number of knife packages is 4...6 and the physical and mechanical properties of the material: cake thickness – 5.42...6,02 mm.

The improved sunflower cake grinder allows reducing its energy intensity to 2.14 kW-h/t and increasing its performance to 1.4 t /h without loss of quality.

Based on the experimental studies conducted, the improved sunflower cake grinder allows applying it in the modernization of technological lines for the production of compound feeds in its bulk condition.

References

[1] Savinykh P A, Palitsyn A V and Ivanov I I 2017 Study of a rotary-centrifugal feed grain grinder with various working bodies *Dairy Bull.* **2(26)** 119-129

[2] Khapov Yu S 2017 Development of a feed chopper-mixer *Modern research and developments* **4(12)** 518-521

[3] Sabiev U K and Pushkarev A S 2016 Comparative analysis of devices for grinding grain materials *Bull. of Omsk State Agrarian University* **1(21)** 221-226

[4] Apazhev A K, Shekikhachev Y A, Hazhmetov L M, Fiaphev A G, Shekikhacheva L Z, Hapov Y S, Hazhmetova Z L and Gabachiyev D T 2019 Scientific justification of power efficiency of technological process of crushing of forages *J. Phys.: Conf. Ser.* 55002

[5] Savinyh P, Shirobokov V, Fedorov O and Ivanov S 2018 Influence of rotary grain crusher parameters on quality of finished product *Engineering for Rural Development* **17** 131-136

[6] Ushakov Y, Shakhov V, Asmankin E and Naumov D 2019 Theoretical study results of product flow management process in hammer-type shredder working chamber *Engineering for Rural Development* 185-191

[7] Melnikov S V, Aleshkin V R and Roshchin P M 1980 *Planning an experiment in the study of agricultural processes* 168 p

Features of information visualization in coal quarrying

K Yu Zhigalov[1], R R Turluyev[2] and T Ya Druzhinina[3]

[1]V.A. Trapeznikov Institute of Control Sciences of Russian Academy of Sciences
[2]Chechen State University, 17a, Dudaev Boulevard, Groznyy, 364015, Russia
[3]Irkutsk National Research Technical University, 83, Lermontov str., Irkutsk, 664074, Russian Federation

E-mail: kshakalov@mail.ru

Abstract. Today, such areas as coal mining, construction of large facilities, providing enterprises with coal (crushed stone, sand, etc.) reserves, etc., face a problem of providing all employees of a coal mining company with equally understandable information. This is bound to different qualifications of the majority of employees. However, as practice shows, the meaningful use of geographic information systems requires certain skills. Based on the aforesaid, it is possible to conclude that in order to be equally able to use the same information system by different employees, its maximum visualization and simplification of the graphical user interface are required. The results of the studies made it possible to test the presentation pattern by grouping the main functions of the software.

1. Introduction

Recently, the coal mining industry has been faded into insignificance among developers and integrators of software complexes (geographic information systems). However, the industry is strongly underestimated by manufacturers and integrators, which affects the development of relevant software solutions.

The collection and presentation of data at coal mining facilities is now carried out according to established technology. Information is collected in the following ways:

- Tacheometric measurements;
- Stereoscopic aerial photography;
- Surface laser scanning;
- Aerial laser scanning.

Collection, storage, processing and subsequent presentation of data in coal companies is mainly carried out using outdated GIS systems. Most of their functions are displayed either on a 2D or on a pseudo-three-dimensional map (a map represents a flat-surface substrate). Three-dimensionality in such systems is achieved by placing separate simplified 3D models of various objects on a substrate [1, 2]. As mentioned earlier, in terms of organization and functionality the work in these systems is extremely difficult for non-experts.

All this makes it very difficult for senior and middle managers (not directly related to coal mining) to access most of the information stored in GIS. Accordingly, there is a need for modern multi-module geographic information systems that can be scaled both by the structure as a whole and by the interface in particular based on various kinds of tasks.

2. Graphical interface

As part of the ongoing study, a program with an experimental graphical user interface was developed, in which scalability is achieved by combining functions into modules activated separately depending on their purpose [3]. The connection of these modules to the core of the system is arranged only if they are accessed from the corresponding program menu, which significantly saves the system resources and allows them to be distributed to visualize the environment and work with the database. Such an implementation has one disadvantage associated with long waiting period when first accessing the necessary module after starting the program.

3. Example of implementation

Let us consider the implementation of this GIS concept using the example of an open-pit coal mining.

In a specific case, there is a variety of different company divisions engaged in the work of quarries (logistics, surveyors, mine survey staff, management, etc.). With varying degrees of confidence all of them are quite familiar with standard plans and maps, not to mention GIS systems [4, 5].

For employees having insufficient understanding of plans, maps and GIS, the software complex looks as simple and intuitive as possible (Figure 1).

Figure 1. Simplified interface of the test software complex.

In this case, the employees of the company can:

• Examine the 3D model of the entire object;

• Examine map/aerial photography from the "top";

• Perform various measurements;

• Examine measurements that occurred at the site over time, both discretely and in dynamics;

• Make comparisons between planned and actual production (by comparing reference models with actual ones at specific point in time);

• Examine attributive information by object;

• Mark/highlight/move objects.

Based on the list it is clear that the functionality described above is more than enough to ensure the maximum informational content for managers at the top and middle levels of the organization. In this case, the user receives a complete visually clear picture of the object in all possible angles and with all the necessary data (Figure 2).

Figure 2. Simplified interface with additional information output.

This mode of operation provides the user with the following features:

1. It is easy, quick and intuitive to understand the operation of the software complex for a non-expert user;

2. Significantly facilitate the client of the program, thus making it possible to use Internet technologies and browsers to launch the application from any location that has access to the Internet without the need to download and install the software. This aspect indicates the possibility of cross platform application;

3. Restrict functionality for secure remote presentations and reports [6].

The following function modules are added to the professional program layout:

• creating and loading 3D models. This module supports all the main formats of 3D model programs (3Ds MAX, AutoCad, etc.);

• editing attributive data. Editing is carried out by means of adding certain data on the object by clicking the right mouse button;

• editing data. Both vector and raster data are edited. Besides, it can be used to vectorize scanned or photographed materials;

• working with the DB. It represents a form and query designer that allows not only populating the database, but also creating new tables for each individual object in particular and the project as a whole;

• performing calculations. Allows calculations of various kinds, such as orthophotriangulation, interpolation, data filtering, etc.;

• modeling of motion of equipment and objects. This module is optional and is used to design automated control systems of facilities on site (excavators, cranes, and other systems);

• automated monitoring and control. Currently, this module may not be used in mines, however, the development of efficient production is not possible without automated control systems. Such systems make it possible to significantly increase production and obtain more accurate real-time data on the state of the object;

• etc.

All of the above modules are connected to the core of the system only after their first call, provided that there is a client with data libraries installed in it. These modules allow using the functionality of geographic information systems and simple graphic editors in the application to coal sections of both open and closed types [5].

4. Conclusion

In conclusion, it should be noted that the approach to the construction of GIS systems described above can be implemented at other types of facilities (construction, oil and gas production, etc.). The introduction of next-generation GIS technologies will increase the efficiency of using databases and make systems available to a greater number of users, such as middle and senior managers. This will significantly increase their awareness and speed of making important decisions. It is worth noting the fact of obtaining data in real time, which is also beneficial for the object management system as a whole.

References

[1] Klyuev R V, Bosikov I I, Egorova E V, Gavrina O A 2020 *Sustainable Development of Mountain Territories* **V.12 3(45)** 418-427
[2] Klyuev R, Fomenko O, Gavrina O, Turluev R, Marzoev S 2021 *Advances in Intelligent Systems and Computing* **1258** 272-281
[3] Klyuev R V, Bosikov I I, Tavasiev V Kh, Gobeev M A 2020 *IOP Conference Series: Materials Science and Engineering* **918** 012124
[4] Bovkun A, Korodyuk I 2019 Analysis of the development of small innovative enterprises in the construction industry. *IOP Conf. Ser.: Mater. Sci. Eng.* **667**
[5] Skorobogatova Y.A., Bovkun A.S., Ivanov M. Yu., Shilova O.S. 2020 The role and place of small innovative enterprises in the construction industry in the modern economy of single industry towns. *IOP Conf. Ser.: Mater. Sci. Eng.* **880**
[6] Repinskiy O D, Konyukhov V Yu, Bovkun A S, Schupletsov A F 2021 Improving the competitiveness of Russian industry in the production of measuring and analytical equipment. *Journal of Physics: Conference Series* **1728**

Models and classifications of secure resource management methods in distributed information communication networks

R R Turluyev[1], M Yu Ivanov[2] and G M Beregova[3]

[1] Chechen State University, 17a, Dudaev Boulevard, Grozny, 364015, Russia
[2] Bratsk State University, 40, Makarenko str., Bratsk, 665709, Russia
[3] Irkutsk National Research Technical University, 83, Lermontova str., Irkutsk, 664074, Russia

E-mail: ibr024@mail.ru

Abstract. This paper analyzes the models and classification of secure resource management methods in distributed information communication networks in order to identify the most optimal model.

1. Introduction

Despite many problems that arise in organizing the interaction of objects in distributed information systems, the main one remains the task of protecting a share when organizing its division among many competing processes (resource management task) [9-11, 24-26]. This task is divided into two key subtasks.

The first subtask relates to a conflict that occurs when processes share a share. In this case, if the algorithm is chosen incorrectly, there is a threat of self-destruction of such a resource. For example, multiple accesses of a database may result in incorrect results, and in the worst case, complete database destruction. When accessing a common communication channel, a situation is possible when the LAN stops transmitting data, spending all the time to resolve the conflict [4].

Thus, the system can be destroyed without any malicious intent for natural reasons hidden in it itself. The second sub-task is related to malicious intent on the part of persons wishing to access information hidden from them, or wishing to destroy the information (telecommunication) system [12–14]. In this case, we have to talk about the attack, and, if it succeeds, about the intervention (forced introduction) of certain alien objects that disrupt the normal operation of the system.

When solving these problems, the choice of the control model is of fundamental importance, since many key parameters depend on it, primarily the safety and effectiveness of control algorithms [20]. Thus, the classification of control models according to certain characteristics is of great importance. For example, there is traditionally a classification of models on the principle of the presence or absence of a control center [8, 11, 21-23].

This classification is conceptual and can only help to develop the most general principles for the construction of the system. Therefore, it is interesting to find classifications and models that provide the most informed solutions when choosing a management method.

2. Analysis of models

ISO/OSI Model

The **ISO/OSI** model (seven-tier open systems architecture) is developed on the basis of the concept of interaction of user processes (application processes) through the transport network using the messaging method (Fig. 1) [15].

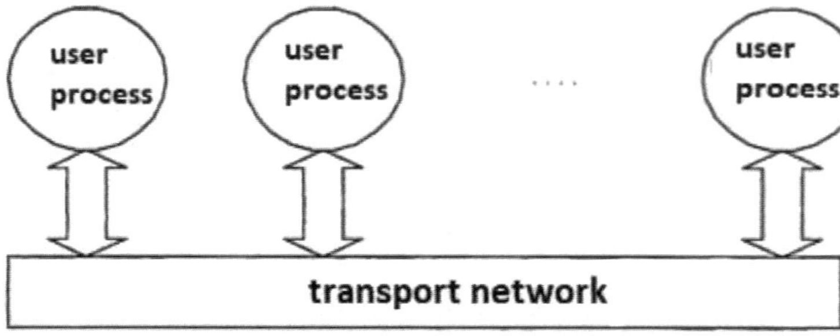

Figure 1. Transport Network Messaging Method

At the same time, the task of the transport network is to deliver messages from one point to another without any interpretation, as well as masking the message delivery mechanism (i.e., smoothing out various implementations of network segments, or ensuring transparency). Thus, according to the concept, the network appears to be homogeneous to user processes [1, 2].

The ISO/OSI model (Figure 2) is a detail of the concept considered. In addition to detail, the model arranges processes vertically (interface interactions) and horizontally (protocol interactions). Let us consider the main features of the ISO/OSI model (hereinafter simply models). As can be seen from the figure, at least two systems are required for its consideration. In the figure, these two systems are designated as System A and System B.

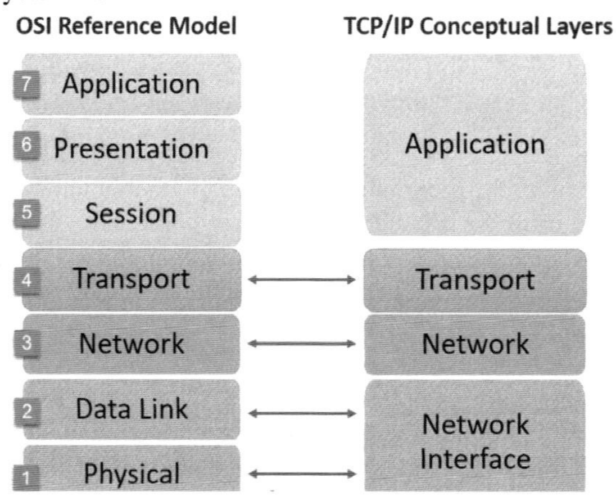

Figure 2. OSI model levels

The lower four layers (transport-physical) according to the concept form a transport network transparent to high-level processes (application-session). Protocols for the transport network are currently standardized (for example, the IP or X.25 protocol family).

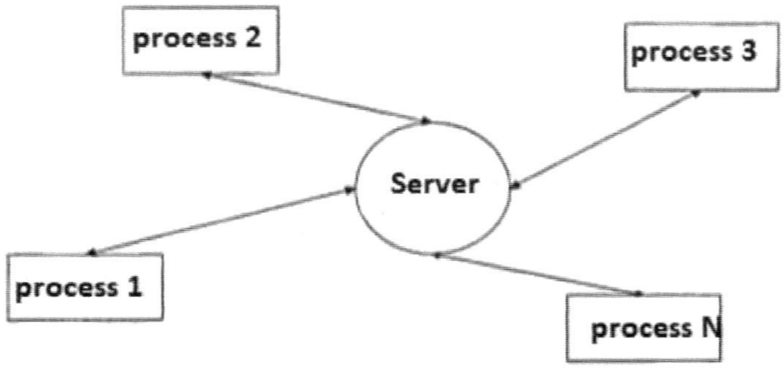

Figure 3. Mechanisms for interaction of protocol processes

The mechanisms of process interaction in accordance with these protocols are well described in the literature [2, 4, 5] and in the documentation on standards (RFC and CCTT), so we will not touch them. We note only that the transport layer protocol is called end-to-end. Processes of this level provide a network-wide interface (transport layer interface) within the systems.

Monitoring (throughout the session) of the correct use of resources includes (if necessary) maintaining the interaction statistics of the participating processes of the session.

The main difficulty in organizing a session is solving the problem of building a logical star of processes (Fig. 3), i.e. building an algorithm for mutual exclusion during multiple access (MD) of processes to a shared resource (server) in time sharing mode [3, 18, 19]. This task is difficult even for centrally managed, concentrated systems; in distributed systems, it is complicated by at least an order of magnitude. The goal of the task is to eliminate the conflict that arises during CBM, and, as a result, to ensure the safe management of common resources (CR).

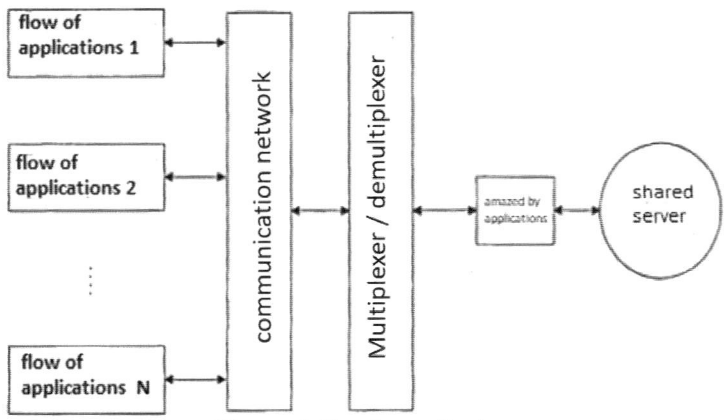

Figure 4. Hoor's monitoring

Moreover, the work provides a method that allows you to organize the storage of history in the memory area of client processes. In this method, after executing each elementary command, it is proposed to send the client only the data (result), but also the current state of the server process (context). At the same time, the client sends the server not only a command with the necessary data, but also a saved context. Thus, the server resumes service from the point of execution of the last client command. The disadvantage of the method is the impossibility of using it in distributed telecommunications systems

(PTCS) with a low data rate.

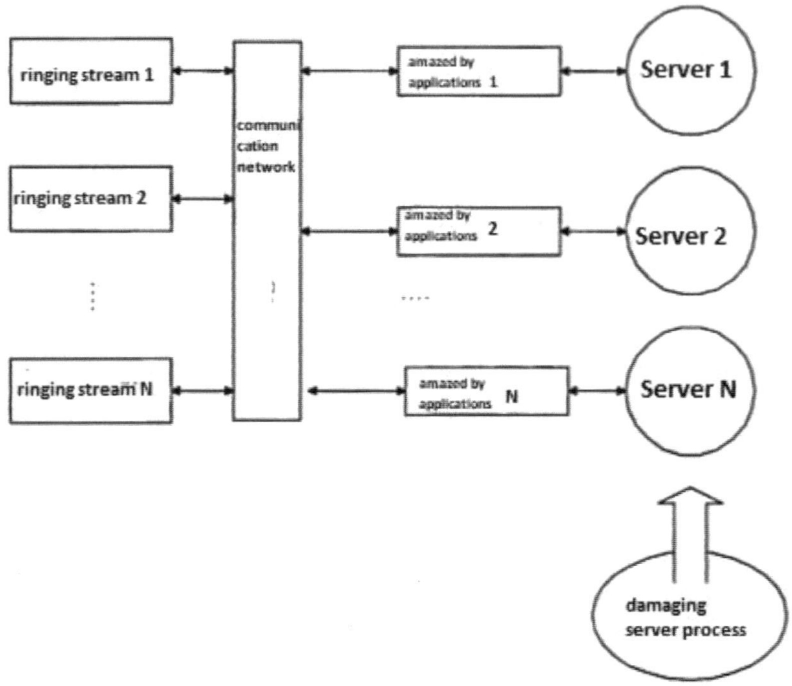

Figure 5. Parallelization of the server

The following process interoperability is based on server parallelization (Figure 5). At the same time, a certain generating process is constantly present on the server machine. Each time a new client is connected, this process starts a copy of the server so that each client has its own query queue.

Thus, the storage of the history and the preservation of the context is assigned to the operating system. The scheme is very popular due to the simplicity of its implementation, and is used in a wide class of client-server systems (for example, in the WWW - worldwide web). However, due to the absence of a common query queue in the schema, it becomes necessary to solve access to a share - for example, a database, and transaction synchronization problems come to the fore. A typical situation for such systems is a denial of service, when all the time is spent on resolving the conflict [4–6], or incorrect maintenance (selling two tickets for the same place is the Siren system).

The last, highest level of the ISO/OSI model - the application - allows us to solve any problems in case they cannot be solved using standard OS tools. For example, in UNIX and Microsoft Windows, the CBM problem is solved using the resource blocking flag, which actually leads us to random service algorithms with all the resulting negative consequences (congestion of systems, denial of service, and so on). Therefore, it is necessary to place the appropriate algorithms (operation with FIFO, HPF, RR queues and so on) at the application level in safe RIS and PTCS.

3. Analysis of classifications

Basic models. We consider the classification of systems based on the principle of interaction of command flows with data flows (Flynn classification [20]). This classification involves four models - SISD (Single Instructions, Single Data), SIMPLE (Single Instructions, Multiple Data), MISD (Multiple Instructions, Single Data) and MIMD (Multiple Instructions) Initially, models were developed to analyze the principles of instruction synchronization in multiprocessor systems, and they are currently used in this area.

Let us enter some definitions.

A flow (according to Flynn) will be called a process if it requires some (possibly remote) resource for its execution.

A resource means any object of the system (data, procedures, other processes), if it can be provided by the system at the disposal of one or more processes. Let us say that the task of managing a resource exists if there is competition between several processes for a certain resource. We consider the applicability of Flynn models within the definitions entered [7].

The SISD model (Figure 6) is conflict-free due to the presence of only one process (thread, processor) and cannot reflect the essence of the task of managing a resource due to the lack of competition in it. However, SISD can serve as an excellent model of the Hoora monitor - some privileged operating system (OS) process that is responsible for correctly separating the resource guarded by the monitor.

In fact, all the tasks that occur in the CBM are somehow reduced to ordering requests (i.e., placing parallel processes in some sequentially serviced queue) in order to facilitate process synchronization. Strictly speaking, the main task in solving the problem of mutual exclusion in a parallel environment is just to reduce all fairly complex interactions to the SISD model.

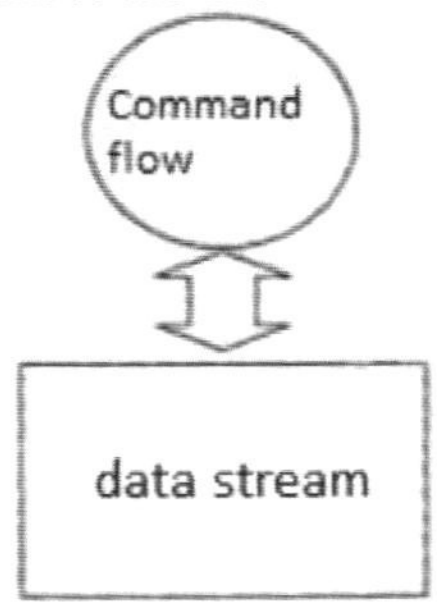

Figure 6. SISD Model

The SIMD model (Figure 6) assumes many processes running the same algorithm. Moreover, it is implicitly understood that at each point in time, all processes execute the same command, each with its own data stream. It will be appreciated that such scheme cannot be implemented on a single processor machine and requires a multiprocessor.

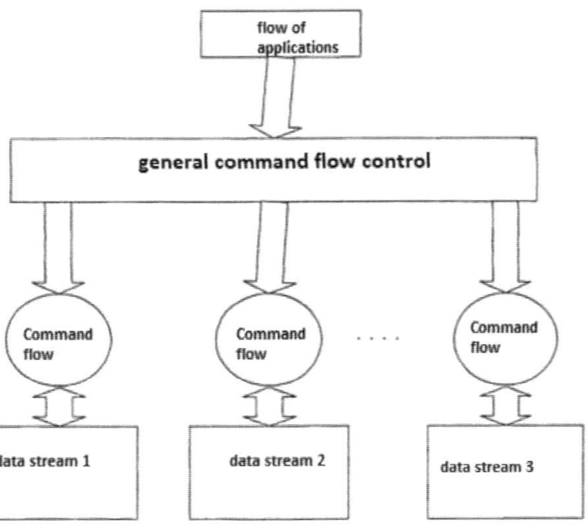

Figure 7. SIMD Model

Within such model (which is true for other classical multiprocessors in which each processor performs only one process), the terms "process" and "processor" are essentially identical.

The model cannot be used to solve a resource management problem for two reasons. Firstly, unlike SISD, there is no share in it. Secondly, in RTCS, for topological reasons, it is impossible to synchronize the actions of several processes so that the commands in these processes are executed synchronously.

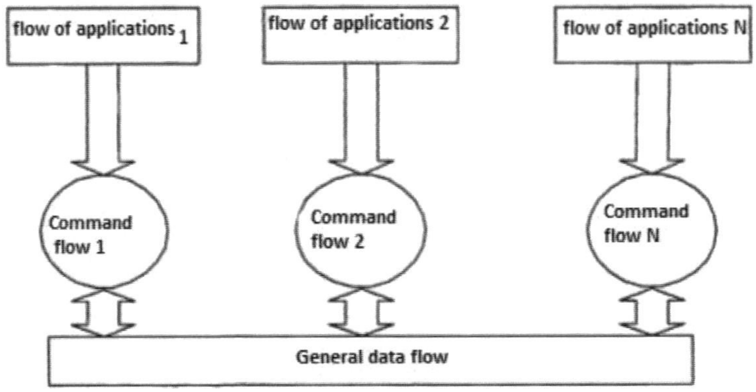

Figure 8. MISD Model

MISD model (Figure 8). Flynn defined this model as an "empty class" because at the time of classification creation there were no systems suitable for MISD. Currently, a system having explicit MISD features is known. This is a class of pipeline processors where many different processors perform operations on common data. In this case, the actions of the processors must be interconnected. Moving from pipeline processors to multiple processes in RTCS and from common data to a shared resource, we see that MISD is, in fact, ideal for solving our problem. In fact, MISD has both features that determine the task of managing a resource - the presence of many processes and their competition for a shared resource [16, 17].

4. Conclusion

Today's technologies provide a range of tools to control access to distributed resources. There are two groups of technologies used to provide security in distributed systems. The first group provides partial authentication, communication protection and independence of the access control mechanism from the applied communication technology. This group includes Kerberos, SESAME, and GAA API systems. This allows you to use any communication protocol, but application developers have to make significant efforts to integrate security technologies with the main communication mechanisms.

Here are the main results obtained during the work.

1. Models and classifications of distributed telecommunication systems construction were analyzed. Based on it, a resource monitoring classification (RMK) is proposed, which allows you to identify the main relationships between the processes and resources of the system at the early stages of design.

2. Analysis of distributed resource sharing algorithms in RTCS is performed. It has been found that the weak point of these algorithms is the procedure for delivering broadcast messages in a distributed environment.

3. Based on the studies conducted, the need was identified for the creation of reliable high-speed algorithms and software and hardware for managing the safe division of distributed resources.

4. For cluster RITCS with DNU and VOC, control algorithms are proposed using the communication channel not only as a means of transmitting information, but also as a common resource necessary for synchronizing distributed processes. This approach allows you to reduce overhead on management, making them independent of the number of subscribers of the distributed system.

5. An extended Shannon model is proposed, taking into account the specifics of data transfer and processing in RITCS.

References

[1] Butrimenko A V 1981 *Development and operation of computer networks* (Moscow: Finance and Statistics) 256 p
[2] Vine Yu and Paluoya R 2013 Specialized local data network of a distributed control system. In book: *Local computer networks (Theses of reports of the All-Union Conference)* (Riga: IEVT) pp 274-278
[3] Dedkov A F and Schers A L 1984 Language tools for organizing distributed processes in local networks. In book: *Local computer networks (Theses of reports of the All-Union Conference)* (Riga, EEVT) pp 146-148
[4] Prangishvili I V, Podlazov B C and Stetsyura G G 2012 *Local microprocessor computing networks* (Moscow: Science)
[5] Prangishvili I V and Stetsyura G G 1980 *Microprocessor systems* (Moscow: Science) 236 p
[6] Zimmermann H 1980 OSI Reference Model - the ISO Model of Architecture for Open Systems Interconnection *IEEE Trans. Commun.* **28(4)** 425-432
[7] Illarionov Yu A and Monakhov M Yu 2004 *Safe management of resources in distributed information and telecommunication systems,* Monograph (Vladimir: VlSU) 212 p
[8] Illarionov Yu A 2015 *Introduction to the theory of information security. Comprehensive protection of informatization facilities. Book 8. Tutorial* (Vladimir: VlSU) 88 p
[9] William A 2011 *System programming in Windows 2000* (St. Petersburg: Peter) 624 p
[10] Dijkstra E 1972 Interaction of sequential processes. In the book: *Programming languages* (Moscow: World) pp 9-86
[11] Cyclitzis D and Bernstein F 1977 *Operating systems* (Moscow: World) 336 p
[12] Levin M 2001 *Methods of hacker attacks* (Moscow: Educational book plus) 222 p
[13] Lukatsky A 2001 *Detection of attacks* (St. Petersburg: BHV-Petersburg) 624 p
[14] Lukatsky A V 2000 *Year 1999: a look at the information security market* (BYTE, JVb6-7)
[15] Material from the National Library named after N. E. Bauman Network model OSI 2015
[16] Zimmermann H 2013 OSI Reference Model - the ISO Model of Architecture for Open Systems Interconnection *IEEE Trans. Commun.* **28(4)** 425-432
[17] Davis D et al 1984 *Computing networks and network protocols* (Moscow: World) 536 p
[18] Erkhov E 1998 Scenarios of attacks on banking systems in the Internet *Analytical Banking J.* **7**
[19] Hoor K 2013 Record processing. In book: *Programming languages* (Moscow: World) pp 278-342
[20] Karpikov A V, Aliev R I and Babyr N V 2020 An analysis of the effectiveness of hydraulic fracturing at YS1 of the Northern field *IOP Conf. Ser.: Mater. Sci. Eng.* **952** 012036
[21] Klyuev R, Bosikov I, Gavrina O, Madaeva M and Sokolov A 2021 Improving the energy efficiency of technological equipment at mining enterprises *Advances in Intelligent Systems and Computing* **1258** 262-271. DOI: 10.1007/978-3-030-57450-5_24
[22] Klyuev R, Fomenko O, Gavrina O, Turluev R and Marzoev S 2021 Energy indicators of drilling machines and excavators in mountain territories *Advances in Intelligent Systems and Computing* **1258** 272-281. DOI: 10.1007/978-3-030-57450-5_25
[23] Sidorenko A A, Dmitriyev P N and Ivanov V V 2020 A study of gas drainage methods efficiency in Kotinskaya mine in Russia *ARPN J. of Engineering and Applied Sciences* **15(4)** 530-535
[24] Mazakov E B, Matrokhina K V and Trofimets V Y 2021 *Traffic management at the enterprises of the mineral industry. Advances in raw material industries for sustainable development goals* (London: CRC Press) pp 528
[25] Kortiev L I, Klyuev R V, Totchiev F G, Vaneev S D and Babunska N I 2020 Temporary and operational transport roads in construction of transfer linear structures and their scientific and

technical support *IOP Conf. Ser.: Mater. Sci. Eng.* **952** 012023. DOI: 10.1088/1757-899X/952/1/012023

[26] Makhovikov A and Kryltcov S 2016 Development of secure and reliable communications system for corporate sector over mobile telecom networks *Asian J. of Information Technology* **15(24)** 5120-5125

Increasing the efficiency of the wireline in the drilling rig lifting complex

R I Suleimanov, M Ya Khabibullin, L M Zaripova and A Yu Davydov

Ufa State Petroleum Technological University, Branch of the University in the City of Oktyabrsky, 54a, Devonskaya St., Oktyabrsky, Republic of Bashkortostan, 452607, Russian Federation

E-mail: rustamsul@rambler.ru, info@of.ugntu.ru

Abstract. The article deals with optimizing the work of the wireline system of the drilling rig. It reveals the characteristic failures and breakdowns of the elements of the wireline system. Based on the analysis the authors proposed to optimize the wireline system of drilling rig BU 300 EUK with 5x6 string-up by forcibly changing the involved the crown block and the wireline block sheaves to more optimal 4x5 string-up.

1. Introduction

Drilling of exploration wells involves repeated downhole operations. This specificity results from a periodic study of exposed horizons based on extractable rocks such as cores. In this rhythm the lifting complex of the drilling rig works for endurance. This action especially affects the operability of the wireline system and in particular the technical condition of the wirelines [1-5].

In turn, the wireline system itself provides a sharp reduction in the specific load on the lifting complex. The wireline operates in the optimal mode of operation rather than in the extreme mode, which provides some significant advantages: it reduces the braking and traction loads acting on such a line winding element as the draw-works drum. This fact significantly reduces the size and weight of the elements of the drilling rig lifting complex by reducing the diameter of the lines and diameters of traction sheaves and the draw-works drum winch, reducing the area of cross-sections of the lifting shaft, transmission shafts, and brake elements.

The nature of kinematic transformations in the wireline system creates a number of drawbacks: 1) the wireline undergoes a large number of bends during the movement of the wireline block during the downhole operations, 2) the wireline has constant contact with the sheaves of the wireline system, 3) the wireline gets wound and swelled from the draw-works drum at high speed, 4) there is a significant increase in the length of the traction element and the speed of the wireline winding on the draw-works drum. The wireline wraps on the draw-works drum in several layers due to the considerably longer length of the wireline, depending on the multiplicity of the wireline system tackle, and the small technological dimensions of the draw-works drum.

The imperfection of the wireline system, with all its advantages, definitely affects the durability of the wirelines, which are the most vulnerable elements in this case. The main reason for the wireline failure is metal fatigue and mechanical wear due to sheaves work and multilayer winding on the draw-works drum.

The analysis of the operation of wirelines at various drilling enterprises shows that they do not pay attention to the issue of careful operation of the lines and do not account for the indicators of wireline runtime. This usually results in the following consequences: 1) an increase in the consumption of lines, 2) incomplete development of the technical resource of lines, 3) an increase in the number of downtimes, and accidents associated with the wirelines operation.

The practice of domestic and foreign wireline operations at large volumes of well construction shows that safe, competent, and efficient operation of wireline can provide a significant economic effect, which will grow with increasing volumes of directional drilling and depth of well construction. A quantitative indicator of the efficiency of a wireline system is the durability and work performed by a wireline on raising or lowering drilling tools, drilling and casing strings. The method of determining the runtime of the wireline system and its elements, practiced in drilling companies, does not account for the parameters of the lifting complex and is usually used only to determine a comparative characteristic in such operating conditions. The use of new knowledge-intensive structures of wirelines, new elements of the lifting complex, as well as modern systems of control and monitoring raises the question of reviewing the methods of objective assessment of the wireline runtime. This makes the issues of assessing the wireline runtime relevant [6-12].

2. Materials and methods

The wireline system of the drilling rig performs tripping operations, which take a significant part of the time in the general cycle of well construction.

During exploration drilling, the number of tripping operation cycles is many times higher as compared to deep production drilling, as there is periodic excavation of the rock (core) sample to the bottom surface. It means that tripping operation, as a technological operation, is the most important target for the search of methods to optimize the well drilling process.

There are two ways to optimize wirelines system of the drilling rig:

1) changing the design of the lifting complex of drilling rigs: preparation and implementation of new constructions of the lifting complex,

2) elaborating on the methods of parameters calculation: search for the regime and design parameters of the optimal value for the wireline systems of drilling rigs.

Optimizing the parameters of the downhole lifting complex has its own peculiarities: 1) a huge range of measures and actions determining the technological process of well installation; 2) operation of the equipment included in the lifting complex, in conjunction with each other; 3) a significant number of designs and a variety of influencing factors.

These features determine the number of variables to be determined. Optimization of the downhole lifting complex of drilling rig goes in two directions:

1) parameter optimization;
2) structure optimization.

The variation in the parameters of the downhole lifting equipment of the drilling rig is limited by the listed requirements:

1) maintenance of technical equipment performance at the level of the value specified in the requirements of the technological process;
2) possibilities of technical realization for object of execution required by the customer;
3) application of safe methods of plant maintenance and operation.

Such conditions result from imposing restrictions on the following parameters:

1) maximum lifting speed;
2) maximum speed of tool descent;
3) maximum speed of line winding on draw-works drum;
4) maximum allowable hook force for the selected installation;
5) maximum drive power;
6) maximum weight of drilling equipment;
7) complex universal capability.

The technology for optimizing the technological process of the downhole lifting complex provides further improvement of the units, increases their efficiency.

Application of the practical results obtained from optimizing the parameters of the downhole lifting complex will ensure the improvement of technical and economic indicators of the drilling exploration well.

A steel line is a non-repairable product whose durability and lifetime depends on the following parameters: 1) wireline quality; 2) structural parameters of lifting force elements, load and features of well drilling, and correct exploitation.

The main criterion for the rejects of wirelines during operation is the number of broken wires at the laying step determined by the visual inspection of the most worn sections of the wireline.

During the tripping operations in the wireline system, the line has to pass through tackle sheaves many times, which eventually causes fatigue damage and wear of outer wires.

When the line passes through the sheave, the wires suffer double bend and wear out through the contact of the wireline with the sheaves. A significant part of the wireline wraps on the draw-works drum in several layers. The wireline undergoes multiple loads, whose number is equal to the number of drill string stands, and the nature of the load is unsteady.

The number of wireline loads, the number of bends on the blocks, the multi-layer winding damage factor, and other wear factors depend on many parameters of the lifting system and can be optimized.

Such parameters include the multiplicity of the tackle and the diameter of the sheaves of the wireline system, the derrick height and the stand length, the diameter, and length of the draw-works drum.

The maximum number of double bends in a wireline section depends on the following parameters: derrick height; tackle multiplicity; length of a standard drill pipe stand. The section of the double bend undergoes wear on the sheaves, i.e. its durability determines the durability of the entire wireline in string-up.

Possible ways to increase the durability of the wirelines are as follows:

1) applying wireline bypasses or slipping and cutting in the wireline system, so wirelines supplied to drilling rigs are longer than required for the rigging of the wireline system, the wireline system has a fixed wireline section fastening device for performing the bypasses;

2) straightening and pre-drawing of wirelines during manufacture, which significantly reduces the laying stress and residual torque in the line, pre-drawing of wirelines increases their durability;

3) increasing manufacturing accuracy and development of new designs of wirelines;

4) developing and introducing of new wireline lubricants;

5) introducing devices that ensure the lubrication of wirelines during operation;

6) specifying the requirements for traction sheaves and screw threading on the draw-works drums;

7) using coated traction sheaves;

8) improving the design of the draw-works drums and screw threading to improve the conditions of multi-layer wireline winding;

9) improving the kinematic schemes of the wireline system, introduction of the system with two running ends, etc;

10) reducing vibrations in the running section of wireline with the help of wireline stabilizers and dampers;

11) ensuring the necessary tension of the running section when winding the line on the draw-work drum;

12) reducing dynamic loads by applying control systems to the drive and brake of draw-works.

The sequential interaction of the wireline with the sheaves that it bends in the wireline system results in a discrete, i.e. multiple of an integer, distribution of the bends along the length in the string-up, and an increase in the number of bends from one section to the next occurs gradually, increasing by one. The analysis of the obtained expressions determining the number of double bending of the wirelines on the sheaves shows that there are ratios of the derrick height, the height of the lift and the multiplicity of the string-up, at which the total number of bends of the wireline can be minimized for all runs. Compliance with these ratios will increase wireline durability.

The choice of the optimal ratio of derrick height, string-up multiplicity, and elevator lifting height can increase the durability of wirelines by 10-15 %.

The wireline durability at work on sheaves depends on the law of the fatigue curve. When determining the durability or the number of wireline bends on the blocks, it is necessary to bring the wireline load to a constant equivalent load level.

3. Results

When lifting long drill strings, the total number of tripping operations will decrease with increasing stand length. The number of lifting system loads will also decrease since the number of stands with a longer length will reduces in the string. Therefore, the assessment of the wireline durability according to the above criterion requires considering the total number of wireline bends on the sheaves.

When lifting the string, the running (movable) end of the wireline carries the highest tension, the fixed end of the wireline carries the lowest tension.

The tension of the running end of wireline is determined by the formula

$$P_{RN} = \left(\frac{Q_h + P_{eq}}{\beta^{m-1}}\right) \cdot \beta^m \cdot (\beta - 1), \tag{1}$$

where Q_h is the maximum load on the hook, N;

$$Q_h = q \cdot L_w, \tag{2}$$

where q is the weight of the 1st running meter of the drilling tool, q=300 N/m (according to GOST 16293-89 "Complete drilling rigs for production and deep exploration drilling. The main parameters"); P_{eq} is the weight of constantly lifting equipment (hook of drilling, wireline block), P_{eq}=5 t = 5000 kg = 50 kN; β is coefficient of sheave resistance, β= 1.03; m is the multiplicity of tackle (the number of working strings of the wireline system), $m = 2в‹ ... Z_{p\check{y}p} = 2в‹ ... 5 = 10$(for equipment 5x6).

$$Q_h = 300 \cdot 2000 = 600000 \text{ H} = 600 \text{ kN}.$$

The tension of the fixed end of the wireline is determined by the following formula

$$P_{wb} = \left(Q_h + P_{eq}\right) \cdot \frac{\beta - 1}{\beta \cdot (\beta^{m-1})}. \tag{3}$$

$P_{FE} = 1{,}03^{10} \cdot (1{,}03-1) \cdot (600000+50000) / (1{,}03^{10}-1) = 78467$ N.

Determining the tension of working strings:

$$S_1 = P_{RN} \cdot \frac{1}{\beta} = 78467 \cdot \frac{1}{1{,}03} = 76181 \;\; H; \tag{4}$$

$$S_n = S_{n-1} \cdot \frac{1}{\beta}. \tag{5}$$

The tensions of the other strings are similarly calculated: S_2 = 73962 N, S_3 = 71807 N, S_4 = 69716 N, S_5 = 67685 N, S_6 = 65714 N, S_7 = 63800 N, S_8 = 61942 N, S_9 = 60137 N, S_{10} = 58386 N.

$P_{FE} = (600000+50000) \cdot (1{,}03-1) / 1{,}03 \cdot (1{,}03^{10}-1) = 56682$ N.

Determining the efficiency of the wireline system

$$\eta_{ws} = \frac{\beta^{m}-1}{m \cdot \beta^{m} \cdot (\beta - 1)} = \frac{1{,}03^{10}-1}{10 \cdot 1{,}03^{10} \cdot (1{,}03-1)} = 0{,}853. \tag{6}$$

Number of stands in the string:

$$q_l = \frac{L_W}{ICTPIP_{\mu}C\ddagger}, \tag{7}$$

where L_w is the length of the well; L_{st} is the length of the drill string stand.

$$q_{l=24} = 2000/24 = 83; \; q_{l=27} = 2000/27 = 74.$$

The total number of bends of the most loaded wireline section of all stands:

$$n_l = 2в‹ ... q_l в‹ ... n_{max}, \tag{8}$$

where 2 considers the doubling of the number of line bends due to the double track of the wireline block.

$$n_{l=24??} = 2 \cdot 83 \cdot 5.6 = 929; \; n_l = 2 \cdot 74 \cdot 6.3 = 932.4 .$$

The number of bends made by the wireline for both options differs by no more than 3%.

Changes in stand height by 0.5 ... 1 m lead to changes in the optimal stand length. Determining the optimal length of the stand should consider the accurate movement of the wireline block during the tripping operation (Figures 1-4).

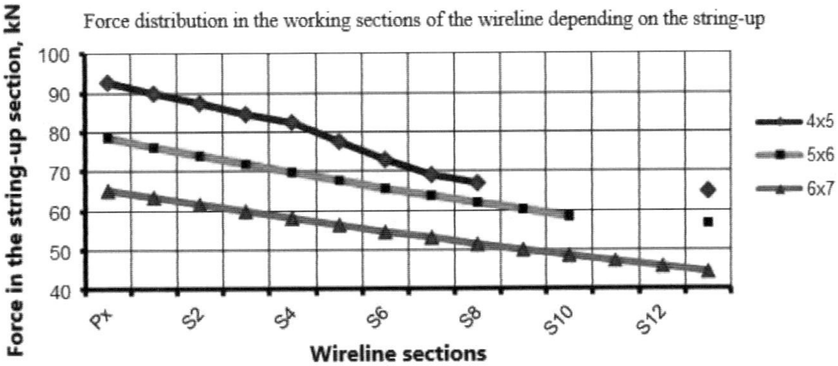

Figure 1. Force distribution in the working sections of the wireline depending on the string-up

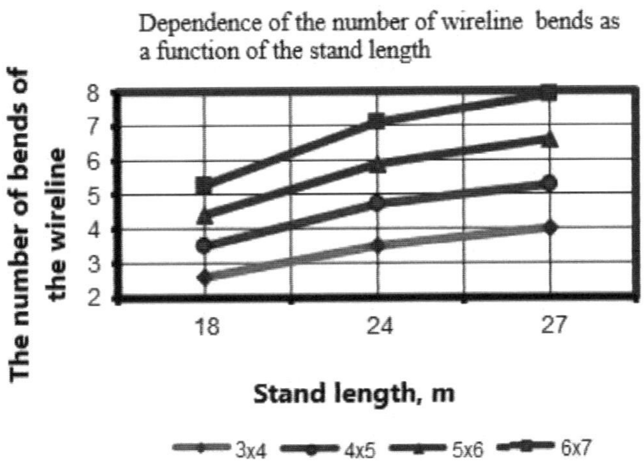

Figure 2. Dependence of the number of wireline bends as a function of the stand length

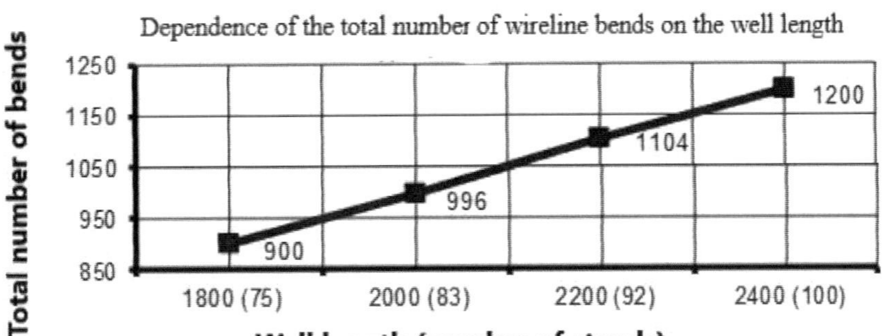

Figure 3. Dependence of the total number of wireline bends on the well length

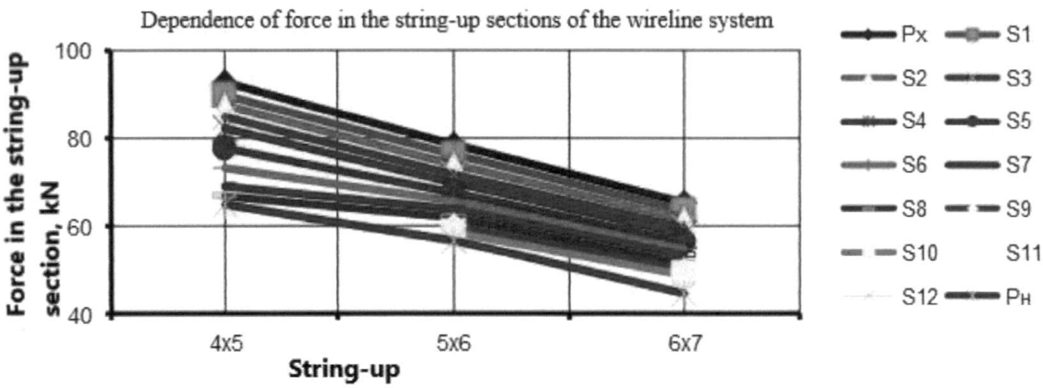

Figure 4. Dependence of force in the string-up sections of the wireline system

4. Conclusion

The choice of the optimal ratio of derrick height, multiplicity of tackle, and elevator lifting height increases the durability of the wirelines by 15 ... 20 %. Determining the optimum bypass length when working on wireline and using the spare length of the line from the drum, as well as timely bypass based on a full assessment of wireline runtime increases the technical life of the lines and their durability. Estimation of the runtime of wirelines with the account of the lifting complex operation parameters, the spreading of wear of the wireline over its entire length in the wireline string-up with regard to its multiplicity of tackle and the real load during all cycles of tripping operations, which contributes to the efficiency of the wireline operation, identification and elimination of factors that have a significant impact on the wear index to optimize the operation of the wear with the use of the bypass method.

References

[1] Hua Jian, Zhou Sizhu and Hu Lei 2015 Structural dynamics modification for derrick of deep well drilling rig based on experimental modal test and frequency sensitivity analysis *Advances in mechanical engineering* **7(9)** 1687814015605746

[2] Luo Ju, Li Liang-Gang and Yi Wei 2014 Working Performance Analysis and Optimization Design of Rotary Drilling Rig under on Hard Formation Conditions *Int. (China) Geological Engineering Drilling Technology Conf.* **73** 23-28

[3] Shaidullina R M, Amirov A F, Muhametshin V Sh and Tyncherov K T 2017 Designing Economic Socialization System in the Educational Process of Technological University *European J. of Contemporary Education* **6(1)** 149–158. DOI: 10.13187/ejced.2017.1.149

[4] Rogachev M K and Mukhametshin V V 2018 Control and regulation of the hydrochloric acid treatment of the bottomhole zone based on field-geological data *J. of Mining Institute* **231** 275-280. DOI: 10.25515/PMI.2018.3.275.

[5] Zeigman Yu V, Mukhametshin V Sh, Khafizov A R and Kharina S B 2016 Prospects of Application of Multi-Functional Well Killing Fluids in Carbonate Reservoirs *SOCAR Proceedings* **3** 33–39. DOI: 10.5510/OGP20160300286

[6] Sun Youhong, Shi Yuanling and Wang Qingyan 2018 Study on speed characteristics of hydraulic top drive under fluctuating load *J. of petroleum science and engineering* **167** 277-286

[7] Almukhametova E M, Shamsutdinova G F, Sadvakasov A A, Tyncherov K T, Petrova L V and Stepanova R R 2018 Modeling development of Fyodorovsky deposit *IOP Conf. Ser.: Mater. Sci. Eng.* **327(4)** 042100. DOI: 10.1088/1757-899X/327/4/042100.

[8] Rogachev M K, Mukhametshin V V and Kuleshova L S 2019 Improving the efficiency of using resource base of liquid hydrocarbons in Jurassic deposits of Western Siberia *J. of Mining Institute* **240** 711-715. DOI: 10.31897/PMI.2019.6.711

[9] Tyncherov K T, Mukhametshin V Sh, Paderin M G, Selivanova M V, Shokurov I V and Almukhametova E M 2018 Thermoacoustic inductor for heavy oil extraction *IOP Conf. Ser.: Mater. Sci. Eng.* **327(4)** 042111. DOI:10.1088/1757-899X/327/4/042111

[10] Yakupov R F, Mukhametshin V Sh and Tyncherov K T 2018 Filtration model of oil coning in a bottom water-drive reservoir *Periodico Tche Quimica* **15(30)** 725-733

[11] Zainagalina L Z, Suleimanov R I, Gabdrakhimov M S and Khabibullin M Ya 2018 Determining oscillating system dynamic parameters of a near-bit junk pulper *Advances in Engineering Research* **157** 642-645

[12] Khabibullin M Ya and Suleimanov R I 2019 Automatic packer reliability prediction under pulsed transient flooding of hydrocarbon reservoirs *IOP Conf. Ser.: Mater. Sci. Eng.* **560(1)** 012024. DOI: 10.1088/1757-899X/560/1/012024.

IOP Conf. Series: Materials Science and Engineering 1111 (2021) 012064 doi:10.1088/1757-899X/1111/1/012064

Application of the least square method for calculating the optimal well profile

F K Usmanova and L Z Usmanova

Ufa State Petroleum Technological University, Branch of the University in the City of Oktyabrsky, 54a, Devonskaya St., Oktyabrsky, Republic of Bashkortostan, 452607, Russian Federation

E-mail: usmanova_15@mail.ru

Abstract. The article offers a new approach to calculating a wellbore profile of a directional well. The calculation of the well trajectory based on inclinometric studies using correlation analysis is considered. The coordinates of measurement points are calculated automatically by building a comparison table in Microsoft Excel. Based on the results of calculations, the functional relation the most fully characterizing the well path is selected. We propose the least square method for calculating the optimal well profile. According to revealed functional relation, the length of the well is determined by integration based on the calculated values of displacements. The estimation of wellbore deviation from the well profile design represented as a parabolic dependence at a certain drilling interval was performed.

1. Introduction

In the course of well design, an important task is to determine financially advantageous conditions for the drilling process, in particular, to build an optimal well profile. The exact trajectory of drilling tool penetration in the rock depends on its control and correction, on-line calculation of the spatial coordinates of the well. The wellbore deviation from planned position may lead to reduction in flow rate, oil recovery index, need of drilling additional wells to extract the remaining oil [1 - 5]. There may be a possible leakage of the casing, an increased probability of poor cementing of wells, complicated oil extraction, an extension of the wellbore, the difficult elimination of accidents, etc. [6]. In many cases, controling the wellbore position is important not only for achieving optimal coordinates, but also for completing the well with the preferred trajectory of entry into the productive formation [7].

At present, there are quite a large number of mathematical calculation methods, that assist in modeling the position of the wellbore in space. These methods include tangential, balanced tangential, the method of average angels, the method of radius of curvatures, the method of minimal curvature, and their combinations [8]. According to [9] the world practice recognises the method of minimal curvature as the most accurate method, since using it the trajectory of the well is the closest to the objective form. The well sections represented in this method as a series of spherical arcs and straight lines are quite simple and do not carry a large number of restrictions associated with data output from the definition area [9]. All existing methods are based on simple geometric interpretations that can be mathematically processed. Calculations are widely used in three-dimensional visualization of the drilling process, so it can be optimized while maintaining safety. However, the literature sources are not systematized, are incomplete, and require additions (mathematical justification). These features make programming controversial [9]. The development of equipment and technology, especially computing one, assists in searching for more accurate and universal methods for calculating and predicting the optimal well profile. It is obvious that a reliable calculating algorithm is more complex, analyzing, and providing full consideration of the well profile as a function. Correlation analysis is successfully used in various fields, especially for obtaining predictive data.

The article aims to verify the possibility of applying correlation analysis for the design and prediction of the optimal well profile. To achieve this aim, the coordinates and parameters of the well profile are calculated employing the least square method. It is necessary to solve a number of tasks: to determine displacement and length of the wellbore; to determine the accuracy of implementation of the well profile design; to evaluate the deviation of the wellbore from the well profile design represented as a functional relation at a certain drilling interval.

2. Research methodology

Technological processes having a wide variety, are characterized by a set of parameters that reflect certain properties, modify over time and in space. Changes in these parameters are often interdependent and mutually dependent. Sometimes the relation between parameters is very close, and in some cases this relation is not detected at all. The closer the relation is, the more accurate decisions are made and the easier it is to manage the processes. In mathematical statistics, they distinguish two types of dependence functional (strictly deterministic) and statistical. In our study, we consider functional dependence which is always expressed by formulas. It is complete and accurate since the list of factors and mechanisms of their influence on the variable in the form of an equation is usually known. We used the least square method which is the basis of correlation analysis [10 - 12] to determine this type of equation.

Inclinometry technology provides measurements of azimuth a, the zenith angle b, and length of the wellbore from the wellhead to each measurement point at each point of the wellbore. The calculation of the drilling trajectory means to find coordinates of measurement points in the Cartesian coordinate system in relation with the wellhead and the drilling point of the side hole. Otherwise, calculate the vertical depth Z of the measurement point, its horizontal displacements X and Y in the North-South and in the East-West directions [13].

Projections *dx, dy, dz* of the elementary interval *dl* of the wellbore on the coordinate axes have the form:

$$dx = dl \cdot sina \cdot sinb; \; dy = dl \cdot cosa \cdot sinb; \; dz = dl \cdot cosb. \tag{1}$$

By integrating these elementary increments along the length of the interval, you can determine the increments of the coordinates Δx, Δy, Δz of the wellbore for the interval Δl between the measurement points:

$$\Delta x = \int_0^{\Delta l} sina\ sinb\ dl; \; \Delta y = \int_0^{\Delta l} cosa\ sinb\ dl; \; \Delta z = \int_0^{\Delta l} cosb\ dl. \tag{2}$$

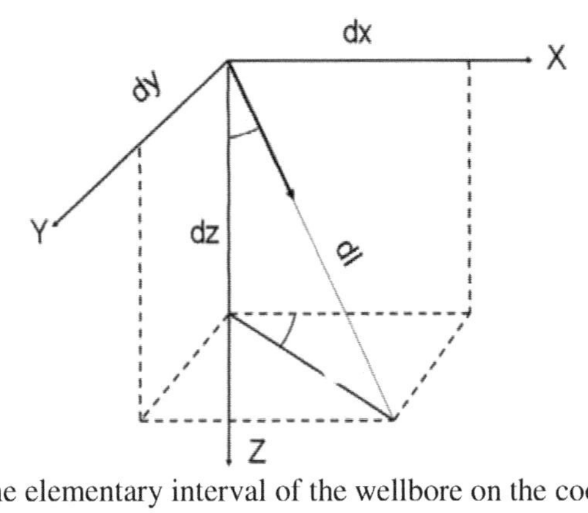

Figure 1. Projections of the elementary interval of the wellbore on the coordinate axes

In the simplest methods, the results of measuring the zenith angle and azimuth at two adjacent points are used to calculate the increments of the wellbore coordinates. In more accurate methods, the results of measuring at three or more points are used. This assists to reduce errors in constructing an actual wellbore profile using the same measuring tools [13].

Let us consider using the minimum curvature method for calculating the coordinates and trajectory parameters of the actual directional well profile. The zenith angle and azimuth were measured using an inclinometer with a magnetic azimuth detector [13]. The initial data for calculating the well profile are described in table 1.

Table 1. Initial data for calculating the well profile

Coordinates of the well profile			Zenith angle, deg.	Azimuth, deg.
X, m	*Y*, m	*Z*, m		
246.0	60.0	1473.0	14.50	102
240.1	84.8	1569.7	15.00	105
233.2	110.4	1666.1	15.75	105
225.3	137.0	1762.2	16.50	108
216.4	164.4	1858.0	17.00	108
206.6	192.8	1953.3	18.00	110
195.9	222.2	2048.3	18.50	110
184.9	252.4	2143.0	19.00	110

3. Results

Based on the obtained coordinates, horizontal displacements A for each profile point from the vertical line of the wellhead are calculated according to the formula:

$$A = \sqrt{x^2 + y^2}. \quad (3)$$

The coordinates of the profile points in the horizontal plane are given in table 2.

Table 2. Coordinates of the profile points in the horizontal plane

Coordinates of the well profile			Displacement
X, m	*Y*, m	*Z*, m	A, m
246.0	60.0	1473.0	253.2
240.1	84.8	1569.7	254.6
233.2	110.4	1666.1	258.0
225.3	137.0	1762.2	263.7
216.4	164.4	1858.0	271.7
206.6	192.8	1953.3	282.6
195.9	222.2	2048.3	292.6
184.9	252.4	2143.0	321.9

Well profiles are curves that are similar with a part of a parabola, so we can assume that the correlation $Z = z(A)$ has the form:

$$z(A) = a \cdot A^2 + b \cdot A + c. \quad (4)$$

In this case, the parameters *a, b,* c are easiest to determine from the so-called system of normal equations that meet the requirements of the least squares method (LSM). It is possible to write this requirement as $\sum(z - z_i)^2 \rightarrow min$ and determine at what values of parameters *a, b,* c the sum of squares of deviation z_i from $z(A)$ will be minimum. After finding private derivatives of the specified sum for *a*, *b* and *c* separately and equating them to zero, it is easy to write down a system of equations, whose solution will give parameters of the required function. The system of equations for determining the second order function is as follows:

$$\begin{cases} nc + b\sum A + a\sum A^2 = \sum z \\ c\sum A + b\sum dA^2 + a\sum A^3 = \sum Az \\ c\sum dA^2 + b\sum dA^3 + a\sum dA^4 = \sum dA^2 z\,. \end{cases} \quad (5)$$

Microsoft Excel can help to search for unknown *a, b, c* and plot the resulting functional dependence. In the solution process, apart from the parabolic dependency graph, there were also found graphs of other dependencies that approximate the data in Table 2. Figures 2-4 shows the graphs.

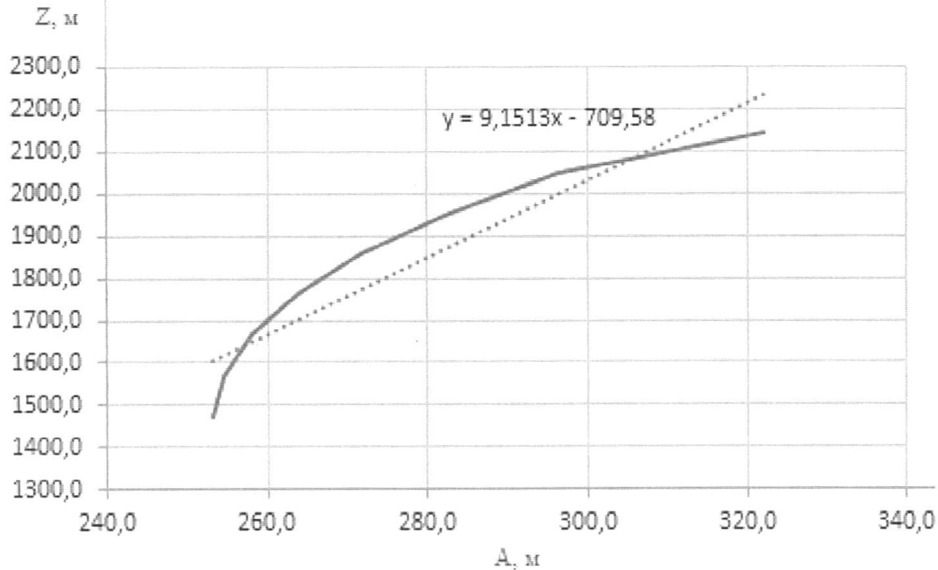

Figure 2. Graph of linear dependence

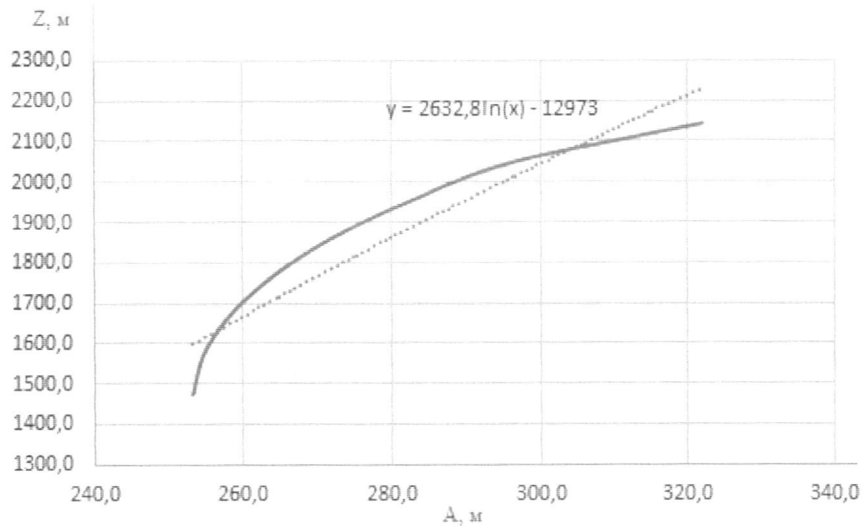

Figure 3. Graph of the logarithmic function

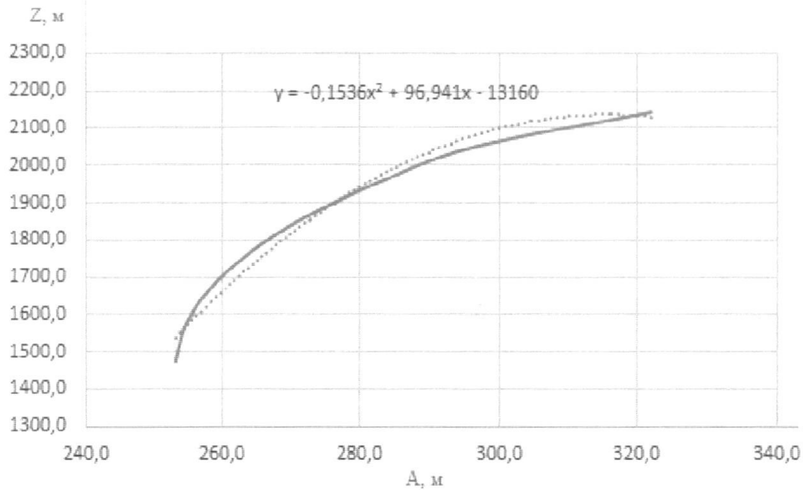

Figure 4. Graph of the parabolic function

The comparative analysis of the presented charts has shown that the data are most clearly approximated by the parabolic function (Fig. 4). Since the vertical axis is directed downwards when designing the well profile (Fig. 1), the searched function $Z = Z(A)$ is in the area of negative values. Therefore, this function is a functional dependence:

$$Z(A) = 0,1536\,A^2 - 96,941A + 13160. \quad (6)$$

The mathematical course gives a formula for calculating the length of a curve arc from the point M(a); c) to the point N(b); d).

$$L = \int_a^b \sqrt{1 + (y(x))^2}dx \quad (7)$$

Using (6), (7) and integration methods, we will find the wellbore length with displacement in the interval from 253 to 322 m:

$$L = \int_{253}^{322} \sqrt{1 + (0,3072\text{Pђ} - 96,941)^2}dA \approx 748\ (m).$$

4. Conclusion

The estimation of wellbore deviation from the well profile design represented as a parabolic dependence at a certain drilling interval from 1500 to 2200 m, was not more than 30 m. This conclusion requires detailed consideration. When conducting further research of more coordinates, it is possible to select the most accurate function with an error tending to zero. It is necessary to consider technical capabilities of drilling tools, drill string bottom layout, geological section of the well, etc.

The research has shown that correlation analysis allows using different mathematical methods in modeling the optimal well profile. It is applicable when other methods do not provide more complete information about the well profile and provide timely data for wellbore corrections.

References

[1] Rogachev M K, and Mukhametshin V V 2018 Control and regulation of the hydrochloric acid treatment of the bottomhole zone based on field-geological data *Journal of Mining Institute* **231** 275-280 DOI: 10.25515/PMI.2018.3.275

[2] Rogachev M K, Mukhametshin V V and Kuleshova L S 2019 Improving the efficiency of using resource base of liquid hydrocarbons in Jurassic deposits of Western Siberia *J. of Mining Institute* **240** 711-715. DOI: 10.31897/PMI.2019.6.711

[3] Yakupov R F, Mukhametshin V Sh, Khakimzyanov I N and Trofimov V E 2019 Optimization of Reserve Production From Water Oil Zones of D3ps Horizon of Shkapovsky Oil Field by Means of Horizontal Wells *Georesourses* **21(3)** 55–61. DOI: 10.18599/grs.2019.3.55-61

[4] Stenkin A V, Kotenev Yu A, Mukhametshin V Sh and Sultanov Sh Kh 2019 Use of Low-Mineralized Water for Displacing Oil From Clay Productive Field Formations *IOP Conf. Ser.: Mater. Sci. Eng.* **560(1)** 012202. DOI: 10.1088/1757-899X/560/1/012202
[5] Valeev A S, Kotenev Yu A, Mukhametshin V Sh and Sultanov Sh Kh 2019 Substantiation of the recovery of residual oil from low productive and heterogeneous formations in Western Siberia by improving the waterflood system using gas and water-gas impacts *IOP Conf. Ser.: Mater. Sci. Eng.* **560(1)** 012204. DOI: 10.1088/1757-899X/560/1/012204
[6] Levinson L M, Akbulatov T O, Levinson M L and Khasanov R A 2013 *Construction and navigation of complex wells. Textbook* (Ufa) 157 p
[7] Liu X and Shi Z 2001 Improved Method Makes a Soft Landing of Well Path *Oil & Gas J.* **48**
[8] Dorovskikh I V, Zhivaeva V V and Vorobiev S V 2010 *Construction of the design and actual profiles of the well: guidelines* (Samara: Saratov State Technical University) 50 p
[9] Sawaryn S J and Thorogood J L 2005 A Compendium of Directional Calculations Based on the Minimum Curvature Method *SPE Drilling and Completion*
[10] Gmurman V E 1999 *Probability theory and mathematical statistics* (Moscow: Higher school)
[11] Shaidullina R M, Amirov A F, Muhametshin V S and Tyncherov K T 2017 Designing Economic Socialization System in the Educational Process of Technological University *European J. of Contemporary Education* **6(1)** 149–158. DOI: 10.13187/ejced.2017.1.149
[12] Gabdrakhmanova K F, Usmanova F K and Khramov G A 2016 Probabilistic and Statistical Approach as a Means of Predicting the Efficiency of Hydraulic Fracturing *Eastern European J. of advanced technologies* **3(8(81))** 30-36
[13] Bliznyukov V Yu, Kim A S, Povalikhin and Kane S A 2014 *Calculation and correction of the well trajectory during drilling: method. instructions* (Ukhta: USTU) 36 p

IOP Conf. Series: Materials Science and Engineering 1111 (2021) 012065 doi:10.1088/1757-899X/1111/1/012065

Transmission type parallel to ship with hybrid power plant

R Yu Dobretsov, S B Dobretsova and A I Vasiliev

Peter the Great Saint Petersburg Polytechnic University, 29, Polytechnicheskaya str., Saint Petersburg, 195251, Russia

E-mail: vasiliev.ai@edu.spbstu.ru

Abstract. The kinematic scheme of a two-stream transmission is considered, which makes it possible to implement the principle of using two engines (including those using different physical principles and having different energy characteristics) and to implement the idea of design a hybrid power plant, including one adapted for operation on a driverless ship. The result of the research is the structural diagram of the transmission, the kinematic diagram of the summing gear in its composition and proposals for the implementation of the main units of the summing gear. The gearbox allows for smooth connection or disconnection of selected motors to the payload. The considered power plant combines an internal combustion engine and a traction motor. The main benefit is increased reliability through redundant motors. Improvements in efficiency are expected by ensuring consistent engine performance. The composition, the kinematic diagram, the principle of the transmission are proposed. The equations describing the kinematic and force relations in the summing gear are considered. The used method of kinematic and force analysis has been tested in the machine-building industry in relation to planetary gears as part of highly loaded transmissions of wheeled and tracked vehicles. References are given to technical solutions that make it possible to implement the proposed transmission based on the technological capabilities mastered by Russian manufacturers. The proposed scheme seems to be promising for use on ships for various purposes, including military boats, in the power plant of which a gas turbine engine is used.

1. Introduction

The use of a pair of energy sources (heat engines) with the same operating parameters today can be considered as a reliably tested approach in shipbuilding. As an example, we can cite the principle of using diesel engines combined into a “diesel-gear unit” [1].

The development of this principle implies that the prime movers can use different physical principles. For example, the engine used in the normal mode of motion can operate according to the diesel thermal cycle, and in the forced modes the ship also uses the additional energy of a less economical gas turbine power plant [2]. For example, an economical "diesel" can be used on the march, and the summation of the energy of a diesel engine and a gas turbine - in an extreme mode.

The principle of using “electromechanical transmission” on ships is also classic, when the energy of a heat engine is converted completely into electrical energy and is realized with the help of a propulsion device [1–3]. This approach is currently justifying itself from the standpoint of energy efficiency, since the "electric transmission" of the ship is quite successfully competing with the mechanical transmission.

However, the principle of parallel use of a heat and electric motor opens up more possibilities than sequential conversion of thermal energy into electrical energy:

An important advantage of a mechanical transmission is the required presence of a mechanical connection between the thermal (or electric) engine and the propulsor. Refusal of such a connection (for example, principles [4, 5]) determines the potential vulnerability of the control object to external influences.

A main weakness of driverless aquatic and terrestrial vehicles is the lack of prompt repair when possible in penetration of failures. At a critical moment there are no qualified engineers, able to solve the problem. For the power plant, this situation means the loss of mobility and inability to complete the task. Therefore, the use of a parallel-type hybrid power plant consisting of an internal combustion engine (ICE) and a traction electric motor can increase the ship's autonomy by providing duplication of the internal combustion engine and a traction electric motor.

As a side positive effect, one can consider the possibility of implementing a forced mode, when the internal combustion engine and the traction electric motor operate in parallel at maximum power. The question of the efficiency of such a power plant is debatable: much depends on the quality of control algorithms and the actual situation. An obvious disadvantage of the concept is the complication and increase in the cost of the power plant, an increase in its mass. This disadvantage is compensated by the expansion of the operational properties of the hybrid power plant.

This article researches a variant of the kinematic transmission scheme that ensures the operation of a parallel-type hybrid power plant.

2. Materials and methods

The object of the research was the structural diagrams of hybrid power plants and kinematic diagrams of summing gearboxes adapted for use on ships. At various stages of the work, theoretical methods (synthesis, abstraction, generalization, deduction, analogy, computer modeling) and empirical methods (description, comparison) were used.

3. Results of theoretical research

Transmission scheme for hybrid propulsion parallel type.

A simplified kinematic diagram of the transmission is shown in Figure 1. The predecessors of such a technical solution are proposals [6, 7] in the field of designing tracked and wheeled vehicles.

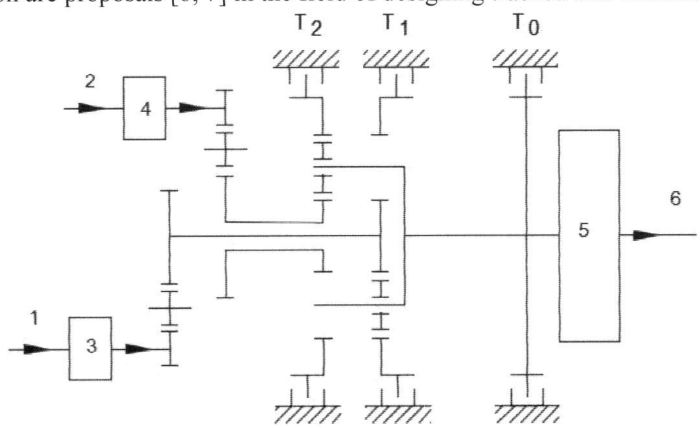

Figure 1. Simplified kinematic transmission diagram: 1, 2 - power supply from engines; 3, 4 - gearboxes of the input transmission; 5 - reducer of the output transmission; 6 - power diversion to the load; T_0, T_1, T_2 – controls.

The power plant consists of two engines (the power supply from them is shown – 1, 2) and the systems serving them.

One engine is a diesel internal combustion engine with an electronic control system (for example, Common Rail), the other is an asynchronous traction electric motor.

The use of a diesel engine with electronic injection control will make it easier to integrate into the on-board network, for example, using CAN protocols.

The use of an on-board network is inevitable on an unmanned vessel, when it is required to provide control of the electronic control system over all important components and assemblies.

It is possible to use any other heat engine (running on gas, petrol, fuel oil, etc.; a diesel engine with individual high-pressure fuel pumps, with pump injectors; a gasoline rotary piston engine; a gas turbine engine, etc.), if its system control can work as part of the on-board network.

The supply voltage of the traction electric motor is advantageously chosen high, according to the experience of designing military tracked vehicles («Puma», Germany) - about 750 V. With this supply voltage, the traction electric motor will be quite compact.

The motors are connected through 3, 4 gearboxes.

For internal combustion engines, it is advisable to provide at least a reverse in the gearbox. Other modes of operation of the gearbox can be borrowed from the practice of designing ships: A hybrid power plant can be built on the basis of an already developed structure based on an internal combustion engine.

For a traction electric motor, it is rational to use a gearbox with two gears to reduce its weight and size characteristics.

The combination of the power streams from the motors takes place in a gearbox consisting of two planetary gear sets.

Planetary gears are characterized by an internal gear ratio or kinematic parameter (gear ratio from the sun gear to the epicyclic gear when the carrier is stopped). The practice of designing power transmissions based on planetary gear sets [8] narrows the range of values of the kinematic parameter for simple planetary mechanisms (Figure 1) to approximately (–1.6) … (–4.5). There are other designs of planetary mechanisms, including those with a positive value of the kinematic parameter, but in Figure 1 the option seems to be optimal from the point of view of ensuring maximum torque reduction.

The gears of the rows are preferably made involute spur teeth (with a sufficiently high efficiency, the axial loads on the bearings of the supports are reduced).

Sun gears are connected to the motors. The common carrier allows one to take power. The control is carried out by turning on the brakes (T_1 and T_2) associated with the epicycles of the corresponding planetary gear sets. If the brake is on (there is no slipping in it), the selected motor is connected to the load. The gearbox controls can be configured as hydraulic disc brakes. For capacities up to about 750 kW in domestic transport engineering, there are proven design solutions and materials [8]. The best solution seems to be the use (with engine power up to 1500 kW) of a steel-metal-ceramic friction pair on a copper basis. Such discs have high durability, there is no self-seizing phenomenon for them.

Using brakes in combination with a planetary gear set is preferable to using clutches. The load on the brake is less and it can be made more compact.

The pressure in the drive system is controlled by an electronic-hydraulic control system [8, 9]. Various pressure control algorithms have been researched and realized - from elementary (static oil supply) to complex (pulse-width modulation of fluid pressure) [9]. The nomenclature of working fluids (oils) [8, 10, 11] has been worked out and this nomenclature is expanding. For transmissions designed for high power, it is advisable to use the materials recommended in research paper [8].

Brake T_0 is designed to stop the output shaft (connected to the row carriers) of the gearbox.

If we complicate the design of the gearbox (add a blocking clutch between the shafts connected to the motors), then the traction electric motor can be used as a starter when this clutch and brake T_0 are turned on. Such solution is possible, but does not seem appropriate, because standard ICE starters are compact and reliable.

An energy source is required to power the traction electric motor. As such, lead-acid storage batteries and a voltage converter with the required characteristics can be used. But at present, domestic manufacturers have mastered the production of more efficient energy storage devices based on

IOP Conf. Series: Materials Science and Engineering 1111 (2021) 012065 doi:10.1088/1757-899X/1111/1/012065

ionistors. The method for calculating the parameters of an ionistor storage as applied to a ground transport vehicle is given in [6]. Charging the storage device is advisable from an external electrical network when the ship is at rest. Recharging is carried out from an internal combustion engine operating in a stationary mode. In this regard, it is necessary to provide for the installation of a generator connected to the internal combustion engine and a rectifier unit.

The reducer in the load branch 5 (Figure 1) allows to adjust the transmission of the m wearing transmission based on the specific situation. In principle, this gearbox may not be available.

Table 1 lists the main operating modes of the power plant and included controls.

Let us consider the possible modes of operation for the circuit implemented in Figure 1 based on the diesel and traction electric motor.

1. The main mode is the use of a traction electric motor (which will provide less noise): the brake of a row connected to the traction electric motor, for example, T_2, is turned on. The internal combustion engine works primarily to recharge the onboard power source and can be turned off. In this mode, the internal combustion engine is advisable to use in the mode of maximum efficiency.

2. Standby mode - using the internal combustion engine (an increase in fuel consumption is expected, it is advisable to use the mode if the traction electric motor fails): the brake of the row associated with the internal combustion engine is on, for example, T_1. The internal combustion engine can work partially to recharge power sources or, including at full power, to drive the propeller (propellers).

3. Maximum power mode (forced mode) - the internal combustion engine and the traction electric motor operate in parallel: brakes T_1 and T_2 are engaged in the transmission. The mode is used to obtain maximum power on the propeller shaft.

4. Scrolling mode - using a traction electric motor for scrolling the internal combustion engine shaft, for example, when starting. Brake T_0 is on, power is not supplied to the output from the transmission.

Table 1. The main operating modes of the power plant

Mode	Included the elements management	Used engines
Main	T_1	1
Spare	T_2	2
Forced	T_1 and T_2	1 and 2
Scrolling	T_0	1 and 2

The scrolling mode, in principle, makes it possible to abandon the electric starter for the internal combustion engine. When a gas turbine and a diesel engine are used in the power plant, the transmission modes are similar.

1. The main mode is the use of a diesel engine as a more economical engine. The GTE is turned off and its shaft is stopped. The mode is used on the march when the maximum speed of the vessel is not required.

2. Standby mode - using a gas turbine. In most cases, this mode is not justified due to the increased fuel consumption, but it can be used when it is necessary to obtain a high speed of the vessel, for example, during combat maneuvers.

3. Maximum power mode (boost mode) - both motors are used in parallel. The mode is designed to obtain maximum power on the propeller shaft.

4. Scrolling mode - if necessary, the diesel engine can be used to scroll the GTE shaft. Brake T0 is on, power is not supplied to the output from the transmission. Nevertheless, it is advisable to make a regular start of a gas turbine engine from an electric starter.

The kinematics of the transmission is described by the following equations (the basic equations of the kinematics of planetary mechanisms are a kind of Willis equations) [8, 12, 13].

Let us introduce the following about the values of angular velocities:

ω_x - for the output link of the gearbox associated with the carriers of both rows;

ω_{01}, ω_{02} - for sun gears indirectly connected with motors 1 and 2 ;

ω_{T1} and ω_{T2} - for epicyclic gears included in the rows, connected, respectively, with engines D_1 and D_2.

Let us designate k_1 and k_2 - kinematic parameters of three-link planetary mechanisms, the sun gears of which are indirectly connected with motors 1 and 2.

Then the gearbox kinematics can be described by the equations:

$\omega_{01} = k_1 \omega_{T1} + (1 - k_1) \omega_x$;

$\omega_{02} = k_2 \omega_{T2} + (1 - k_2) \omega_x$;

If we take constant angular velocity of one of the sun gears $\omega_{01} = 1$; then the angular velocity ω_{02} is an independent parameter and can be expressed in fractions of ω_{01}.

Additionally, you should set the laws of motion of the links associated with the controls: when the brake is turned on, the angular velocity of the links associated with it is zero.

Then, when the traction electric motor and the internal combustion engine work together, equality must be observed:

$\omega_{01} / (1 - k_1) = \omega_{02} / (1 - k_2)$.

Torque in the output unit composed of moments on the carrier n of radiation of a respective planetary rows. Moments in ulcerative s s planetary gear sets are determined based on the traditional equilibrium equations [8, 12, 13].

Let us introduce the notation for the moments:

M_x - gear on the output unit associated with the guide both p i rows;

M_{01} and M_{02} - on sun gears, indirectly associated with motors 1 and 2;

M_{T1} and M_{T2} - on the epicycles of the rows, indirectly associated with engines 1 and 2.

The moments on the brakes associated with the listed epicycles will be equal to the moments on the epicycles, but have the opposite sign.

Then on any operating mode:

$M_x = - [D_1(1 - k_1) M_{01} + D_2(1 - k_2) M_{02}]$;

$M_{T1} = - k_1 M_{01}$;

$M_{T2} = - k_2 M_{02}$.

In this record, $D_{1(2)}$ is a coefficient that takes values 1 if the corresponding control is enabled, or 0 if the control is disabled.

At partial load of the hybrid powertrain: $D_1 = 0$ or $D_2 = 0$.

In forced mode: $D_{11} D_{21} D_{12} D_{22} = 1$.

Ensuring the joint operation of the engines is carried out by selecting the gear ratios of the transmission branches.

Considering modern publications [14–17], we can conclude that the Russian industry today has technologies that allow you to create a traction electric motor and energy storage devices that can provide a sufficient resource and reliability when working in a transmission, operating in severe conditions.

The variants of the proposed transmission can be adapted for use on ships for various purposes, including driverless.

4. Conclusion

Thus, there is reason to expect that the proposed set of modernization measures will lead to an improvement in the following operational properties of the chassis.

1. The use of a parallel-type hybrid power plant on a driverless ship will improve its reliability.
2. In transport engineering (design of high-speed tracked vehicles), there is a set of methodological and technological solutions (materials, manufacturing technologies for individual units) that can be used to create such a power plant.
3. The electrical, mechanical and electronic components necessary for the creation of a hybrid power plant are already being produced by the domestic industry, and the problem of import substitution, in principle, seems to be solved in this case.

References

[1] Akimov P P 1965 *Power plants of sea ships (*Moscow: Transport) p 291
[2] Kotlya I V 1967 *Shipe gas turbine units* (Leningrad: Shipbuilding) p 823
[3] Artemov G A et al 1987 *Marine gas turbine installations* (Leningrad: Shipbuilding) p 477
[4] Usov O A et al 2014 The main options for the hybrid-type electromechanical transmission scheme and the methodology for determining its parameters units for military tracked vehicles *Proc. of the actual problems of protection and security* **3** 111-122
[5] Usov O A et al 2015 Electromechanical transmission with a hybrid power plant for a military tracked vehicle *JEST* **2(219)** 167-174
[6] Dobretsov R U, Porshnev G P and Uvakina D V 2018 Performance improvement of Arctic tracked vehicles *MATEC Web Conf. Volume Int. Sci. Conf. on Energy, Environmental and Construction Engineering (EECE-2018)* **245** 10
[7] Borisiv E G et al, RU Patent No. 2652371, The vehicle drive with infinitely variable double ended mechanism of transmission and rotation (25 April 2018)
[8] Nosov N A, Galyshev V D, Volkov Yu P et al 1972 *Calculation and design of tracked vehicles: Textbook for universities* (Leningrad: Mechanical Engineering) p 559
[9] Dobretsov R Y and Uvakina D V 2020 The Mechatronic Device Impulse Control in Vehicle Powertrains *Lecture Notes in Mechanical Engineering* pp 63-73
[10] Kharitonov S A 2003 *Automatic transmissions* (Moscow: Publishing House and stances Aristel, Izdatelstvo AST) p 335
[11] Fischer R 2015 *The Automotive Transmission Book* (Springer International Publishing Switzerland). DOI: 10.1007/978-3-319-05263-2
[12] Shelomov V B 2013 *The theory of motion of multipurpose tracked and wheeled vehicles. Traction calculation of curvilinear motion: a textbook* (St. Petersburg Polytechnic University) p 90
[13] Zabavnikov N A 1975 *Fundamentals of the theory of transport tracked vehicles* (Moscow: Mechanical engineering) p 448
[14] Belko V, Glivenko D, Emelyanov O, Ivanov I and Plotnikov A 2017 Current Pulse Polarity Effect on Metallized Film Capacitors Failure *IEEE Transactions on Plasma Science* **45(6)** 1020-1025
[15] Belko V O, Petrenya Y K, Andreev A M, Kosteliov A M and Roitgarz M B 2019 Numerical Simulation of Discharge Activity in HV Rotating Machine Insulation *Proc. of the 2019 IEEE Conf. of Russian Young Researchers in Electrical and Electronic Engineering* pp 800-802
[16] Kozlov M A et al 2016 New generation combined power and energy storage devices *Coll. of articles of sci.-pract. conf. "Development and use of electric transmissions for weapons and military equipment"* (JSC "VNIITransmash") pp 140-146
[17] Ilyukevich K V and Sibiryakov R V 2016 Prospects for the use of lithium-ion storage batteries as part of electric transmissions *Coll. of articles of sci.-pract. conf. "Development and use of electric transmissions for weapons and military equipment"* (JSC "VNIITransmash") pp 147-150

IOP Conf. Series: Materials Science and Engineering 1111 (2021) 012066 doi:10.1088/1757-899X/1111/1/012066

Design and technological support for the assembly of long thin-walled buildings

A S Yamnikov[1], E N Rodionova[1,2] and I A Matveev[2]

[1] Tula State University, 92, Lenina Avenue, Tula, 300012, Russia
[2] Scientific-Production Association SPLAV, 33, Shcheglovskaya Zaseka Street, Tula, 300004, Russia

E-mail: info@tsu.tula.ru

Abstract. The article considers the issue of design and technological support for the assembly of extended thin-walled buildings, in particular, the problem of ensuring their entry into the reciprocal parts that ensure the functioning of the assembled products. A new method for controlling the size of the entry of buildings is proposed. The objective of the technical solution is to reduce the complexity of control of the hulls while ensuring their entry into launch tubes. The method consists in the fact that the housing is installed in two prisms according to the extreme centering thickenings. The readings of the measuring device are taken, which judge the suitability of the housing by the diameter of the adjacent cylinder, characterized in that the indicator is installed in a plane perpendicular to one of the faces of the prism, adjusted according to the standard when touching the tip of the middle centering thickening of the standard. As a result of testing in the conditions of the current production, it was found that the verification of the product is reduced several times.

1. Introduction

The conducted studies prove that to ensure the entry of the rocket engine into the starting guide, the designer must take into account the assembly conditions, which allows one to expand the tolerances on the geometric dimensions of the joined sections and the dimensions of the mating surfaces. The designer must determine the accuracy of the initial mutual position of the parts and the sequence of actions (movements of a particular part), as a result of which the required accuracy of the mutual position of the assembled parts can be achieved. If the designer does not give recommendations on the choice of the assembly method, the technologist uses the standard assembly methods applicable for this class of products and is not able to ensure guaranteed product fit, as the conditions for screwing together the parts laid down by the designer, as a rule, have a fairly wide tolerance. Therefore, in order to reduce the percentage of rejects, technologists are forced to introduce technological tolerances on the connection parameters, tightening design tolerances by a factor of 1.5-2.

Thus, the entire technological effect obtained by the designer from expanding tolerances is reduced to zero by the technologist, otherwise the latter is forced to increase labor intensity by introducing into the technological process the operations of sorting or selecting parts before assembly. The latter process, as a rule, is left to the hands of a highly skilled labor collector, who, on the basis of one's intuition, by successive selection by the test method, achieves a suitable assembly. Therefore, the designer must lay the base surfaces in the product with the accuracy necessary to achieve the desired relative position of the parts after assembly. To ensure suitability, it is technologically necessary to withstand the requirements laid down by the designer.

2. Materials and methods

There are a number of products with prefabricated cylindrical bodies, which differ from most engineering structures by the presence of a centering thickening in the middle of the body, the diameter of which is smaller than the diameters of the base extreme centering thickenings. For such cases, the general principle of interchangeability is not applicable based on a comparison of the diameters of adjacent cylinders (GOST 31254-2004, GOST 30893.2-2002); therefore, there is a method for directly monitoring the suitability of an assembled extended cylindrical body, for example, a jet engine body, along adjacent contours [1, 2].

3. The study of the influence of errors in the shape of the base surfaces of the precast housing in the form of ovalities

To ensure the functioning of the studied products, it is required to guarantee the case enters the launch guide. Free entry is determined by the presence of a gap between the adjacent cylinder of the housing and the adjacent cylinder of the starting guide.

As generalized boundaries, covered and covering real elements, adjacent surfaces and element profiles are accepted. From adjacent surfaces and profiles, the deviations of the shape of the surfaces are counted towards the body of the part. In this regard, the condition for the entry of simple geometric bodies having a considerable length, characterized by the ratio of length to diameter: L/D> 10. Figure 1 shows a diagram of the position of the housing in the start guide.

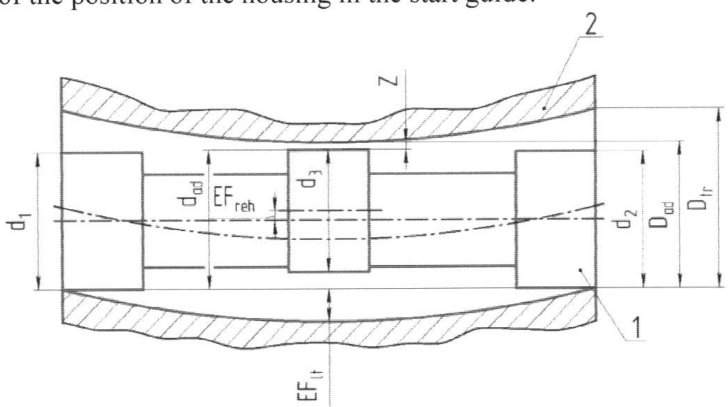

Figure 1. Scheme of the position of the housing in the start guide: 1 - prefabricated extended cylindrical body; 2 - starting guide.

Figure 1 shows that in order to comply with the principle of interchangeability of the housing and the starting guide, the condition:

$$\begin{cases} d_1 < D_{pt\,\min} > d_2, \\ 0,25(d_1+d_2)+0,5d_3+EF_{reh}+EF_{ovreh\Sigma}=d_{0reh}\le D_{0d}=A_0 \end{cases} \quad (1)$$

where: $D_{lt\,\min}$ - minimum diameter of the starting guide, d_1, d_2 - diameters of extreme centering thickenings, d_3 - diameter of the middle centering bulges, A_0 - diameter of the cylinder of zero curvature of the starting guide; d_{0reh} - step housing entry size; EF_{reh} - body axis curvature; $EF_{ovreh\Sigma}$ - the sum of the projections of the ovality of the centering thickenings of the body on the vector EF_{reh}.

If one critical section is laid in the structure in the housing, which ensures guaranteed entry of the housing into the launch guide, then the control method can be easily implemented using the installation used to control radial runout in the current production. In the current production, as a rule, roller prisms are used, since they facilitate the control process and do not damage the surfaces of the product being monitored, since there is no sliding of the product surfaces relative to the prisms. The diameter of the rollers is selected so that they simulate a 90° prism, so that the subsidence of the axes of the base surfaces would be minimal. When controlling the radial runout, the housing is installed in two prisms according to the centering thickenings, and the radial runout of the average centering thickening is determined by the indicator installed in the plane of the bisector of the angle of the prisms, the average thickening [3].

The actual size of the housing entry can be expressed from the second equation of the system (1), replacing the diameters of the centering bulges - d_1, d_2, d_3 twice the size of the minor axes of the ellipses of the corresponding centering thickenings - $b_{1.2.3}$:

$$d_{0rehf} = 0{,}5\left(b_1 + b_2\right) + b_3 + EF_{reh} + EF_{ovreh\Sigma}. \quad (2)$$

In equation (2) a priori it is assumed that the curvature vector of the axis of the housing - EF_{reh} perpendicular to the measuring base, as from the experience of indirectly checking the size of the entry by measuring the radial runout of the average centering thickening, in almost all cases they coincide.

It is more difficult to establish the value and direction of the total vector of projections of the ovality of the centering thickenings of the body on the vector EF_{reh} - $EF_{ovreh\Sigma}$.

According to the rules of mathematics, the projection of the sum of vectors in a given direction (in the case under consideration on the Y axis) in accordance with the gear ratios of the diameters on the size of the occurrence is determined by the dependence:

$$EF_{ovreh\Sigma} = 0{,}5\left(EF_{ovreh1} \cdot \sin\varphi_1 + EF_{ovreh2} \cdot \sin\varphi_2\right) + EF_{ovreh3} \cdot \sin\varphi_3. \quad (3)$$

On the right side of equation (3) are 6 independent quantities. Tolerances for ovality of centering thickenings are given, usually such values are distributed according to the Rayleigh law. Phase angles of inclination of the major semi-axes of elliptical sections, respectively, of the centering bulges 1, 2, 3 relative to the measuring base when controlling the size of the entry - $\varphi_{1.2.3}$, obey the uniform distribution law. Equation (3) shows that the ovality of the centering thickenings have some effect on the size of the entry. It is difficult to predict such an effect due to the combination of the influence of six random variables. In such cases, it is advisable to use a statistical study of the influence of errors or mathematical modeling of this effect [4].

In indirect control of the entry size by measuring the radial run-out of the average centering bulge of the assembled case, only the ovality of the centering bulge 3 is affected by the readings, and the influence of the ovality of the centering bulge 1 and centering bulge 2 is hidden by using the special properties of prisms with an angle of 90°.

If the linear displacement sensor 3 is installed at an angle of 45° to the bisector of the angle of the prism 1, that is, perpendicular to one of the faces of the prism, then the actual case entry size will be measured directly - d_{0rehf}. To configure the indicator, we use standard 2, which is made close to the nominal dimensions of the case (Figure 2).

Figure 2 shows that the method for directly controlling the diameter of the adjacent body contour directly takes into account the influence of all six random components of equation (3), therefore, the shape error of all centering thickenings of the body and their phase angles do not create errors in determining the size of the entry by the studied method.

In equation (2) it is assumed that the curvature vector of the axis of the housing - EF_{reh} directly affects the size of the entry, however, the displacement of the center of the centering thickening 3 relative to the common axis of the body, realized by the axis O_1O_2, passing through the centers of the extreme centering thickening 1 and centering thickening 2, causes the axes of the pipes making up the assembled body to skew (Figure 3).

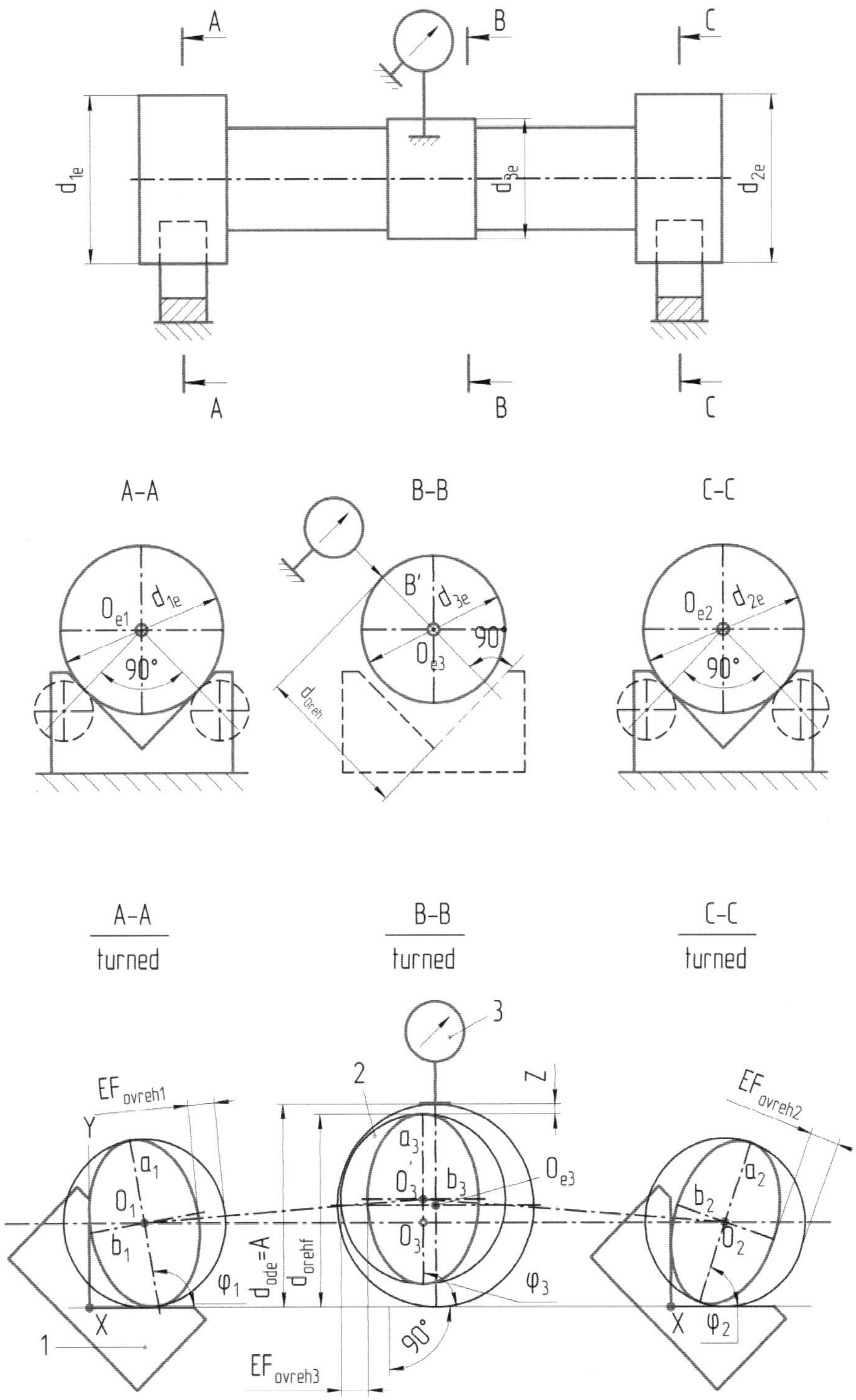

Figure 2. The control scheme for the size of the entry of the case: 1 - prisms; 2 - standard; 3 - indicator.

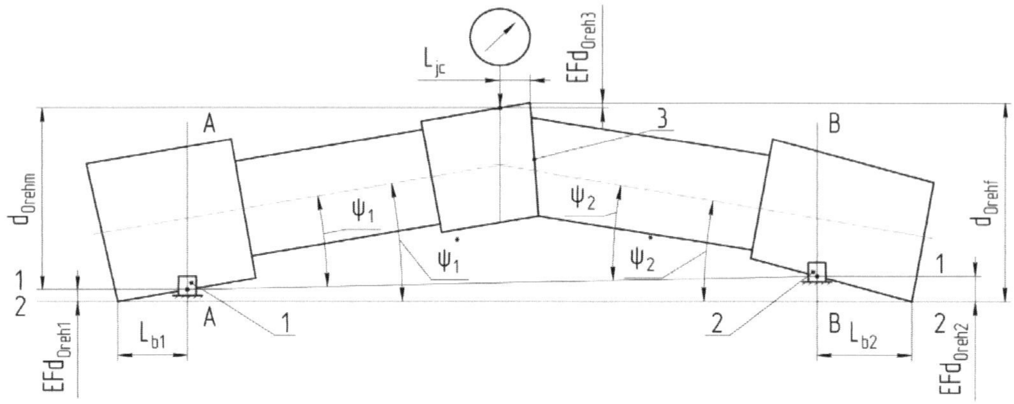

Figure 3. The scheme of the influence of the skewness of the axes of the pipes of the assembled housing on the error in measuring the size of the entry

Housing pipes are connected using threaded half-locks, including threaded threads, short centering bands and ends. In the connection, both parallel displacements of the axes and their skew are possible [5, 6].

Typically, prisms are located approximately in the middle of the length of the centering bulge at a distance L_{b1} and L_{b2} from the edge of the pipes, so there is a shift in the line I-I of the size control (Figure 3) relative to the line that serves as the reference d_{0rehf} on the product [7, 8]. Measurement error $EFd_{0reh1,2}$ is approximately determined by:

$$EFd_{0reh1,2} = 0{,}5\left(EFd_{0reh2} + EFd_{0reh1}\right), \tag{4}$$

where: $EFd_{0reh1} = L_{b1}\psi_1$ $EFd_{0reh2} = L_{b2}\psi_2$ - size measurement errors from "sagging" of the product in prisms; L_{b1} and L_{b2} - distance to the base section from the ends of the pipes 1 and 2.

Equality can only be exact when $EFd_{0reh1} = EFd_{0reh2}$, when lines 1-1 and 2-2 are parallel, and the corners $\psi_1 = \psi_1^*$ and $\psi_2 = \psi_2^*$ tilt axes of pipes with an axis of control (reference of the size d_{0rehf}) are equal.

Offset control plane distance L_{jC} from the junction plane, leads to measurement error:

$$EFd_{0reh3} = L_{jc}\psi_1, \tag{5}$$

where: L_{jC} - distance from the control plane to the joint plane of the joined pipes (to the edge of the surface to be monitored); ψ_1 - the angle of inclination of the axis of the first pipe with the axis of control [9, 10].

To determine the order of angle values ψ, combinations of sizes and tolerances on a typical case are considered: pipe lengths of about 1000 mm, lengths of centering thickening 1 and centering thickening 2 - about 80 mm, maximum allowable radial runout of the average centering thickening 3 - 0.5 mm, from where the permissible curvature of the body axis is 0.25 mm.

Figure 3 defines:

$$\sin\psi_1 \approx \sin\psi_2 = \frac{TEFd_{0reh}}{L}, \tag{6}$$

where: L - pipe length.

Substituting the numerical values of the components (6) into the equation, we obtain the maximum possible value $\sin\psi_1 \approx \sin\psi_2 \approx \frac{0{,}25\cdot}{1000} = 0{,}00025$, given that at small angles the functions of the sine, tangent, and angles in radians are equal, we get $\psi_1 \approx \psi_2 \approx \psi_3 = 0{,}00025\,radian = 0{,}143^\circ$.

$EFd_{0reh1} \approx EFd_{0reh2} \approx 0{,}00025 \cdot 40 = 0{,}01$ mm.

Similarly, we find this by the formula (5) - $EFd_{0reh3} \approx 0{,}01$ mm. In this case, the total error of measurement of the size of the entry caused by the misalignment of the axes of the pipes of the housing is determined from the expression:

$$EFd_{0reh\Sigma} = 0{,}5\left(EFd_{0reh1} + EFd_{0reh2}\right) + EFd_{0reh3}, \quad (7)$$

substituting into which the maximum possible values of the components $EFd_{0reh\,i} \approx 0{,}01$ mm, we get $EFd_{0reh\Sigma} = 0{,}02$ mm.

As a rule, we set the real value of the angles ψ_1 and ψ_2. In practice, it is impossible; therefore, it is possible to take into account the measurement error only using the calculated limiting angles ψ_1 and ψ_2, tightening size tolerance $TEFd_{0reh\Sigma}$ when setting up the device.

To reduce the measurement errors of the case entry size d_{0reh} during control, prisms should be installed as close as possible to the edges of the case. However, they should not go beyond it; therefore, the distance from the edge of the base centering bulges is taken to be at least 2-3 mm, which usually corresponds to the error $EFd_{0reh1,2} \approx 0{,}001 \div 0{,}00075$ mm.

4. Conclusion

A study of the design and technological support for the assembly of extended thin-walled cases allowed us to develop a new way to control the size of the entry of thin-walled buildings, showing that the existing ovality of the centering bulges has a significant effect on the size of the controlled size, but do not affect the accuracy of its measurement.

When using the traditional arrangement of supporting prisms, approximately in the middle of the length of the extreme centering bulges, the skewness of the axes of the bodies significantly affects the error in measuring the size of the entry. With the maximum permissible radial runout of centering thickenings, a measurement error of up to 0.02 mm is obtained.

To reduce the control error, it is necessary to place the elements as close as possible to the edges of the parts to be joined. When testing, it is recommended to use an extended contact, which should overlap the joint to determine the size of the entry on two adjacent surfaces (two parts or one).

References

[1] Yamnikov A S, Yamnikova O A, Matveev I A and Rodionova E N 2019 Russian patent **2682572** 8
[2] Yamnikov A S, Rodionova E N, Yamnikova O A and Matveev I A 2019 *Measurement Techniques* **8(62)** 692-696
[3] Pavlov D A, Erpalov M V, Shimov G V and Pavlova E A 2018 *Ferrous metals* **10** 17–21
[4] Bazhukov I N, Serebryakov A V and Parshakov S I 2018 *Ferrous metals* **8** 51–57
[5] Sosenushkin E N, Yanovskaya E A and Emelyanov V V 2015 *Russian Engineering Res.* **6(35)** 462–465
[6] Grigoriev S N, Teleshevskii V I and Glubokov A V 2012 *Measurement Techniques* **5(55)** 526–529
[7] Madureira L R and Melo F Q 2015 *Mechanics Res. Communications* **70** 79–84
[8] Yakovlev S S, Tregubov V I and Pilipenko O V 2015 *Bull. of mechanical engineering* **1** 73–78
[9] Shinkin V N 2017 *CIS Iron and Steel Review* **14** 28–33
[10] Yakovlev S S, Tregubov V I and Osipova E V 2016 *Russian Engineering Res.* **6(36)** 472–475

IOP Conf. Series: Materials Science and Engineering 1111 (2021) 012067 doi:10.1088/1757-899X/1111/1/012067

Automation security in a machine-building plant

K Yu Zhigalov[1], E R Guzueva[2] and S I Polovneva[3]

[1]V. A. Trapeznikov Institute of Control Sciences of Russian Academy of Sciences
[2]Chechen State University, 17a, Dudaev Boulevard, Groznyy, 364015, Russia
[3]Irkutsk National Research Technical University, 83, Lermontov str., Irkutsk, 664074, Russian Federation

E-mail: kshakalov@mail.ru

Abstract. The operating system is the most important software component of any computer, and therefore, the security of the enterprise information system as a whole depends on the level of implementation of the security policy in a single operating system. In general, Linux operating systems (OS) cannot be considered as a ready-to-use security system. There are many recommendations on improving Linux security, but the authors of these recommendations do not guarantee that their recommendations are sufficient to achieve the required level of system security. This study attempts to systematize available methods for increasing the level of security of application software for Linux-like OSs operating as part of control systems at nuclear power facilities. The required level of security of the entire system can be achieved at the application software level.

1. Introduction

Due to increased security threats and the modification of viruses, Linux cannot be considered as a ready-to-use security system.

The main and crucial problem in all security systems for control systems is unauthorized access (UA). UA threats and attempts are preventable and may be suppressed (thus not creating the possibility of their implementation), as well as their successful implementation may be prevented [1, 2].

This requires the following:

• clearly identify the codes of applicable operations in the system and all entities/objects that have the right to perform them. The operating system software shall monitor the possible UA attempts, notify their launch and block the operations used to launch them in time. For example, in case of repeated application of the wrong password at logon, the activity of the user under whose name the login is attempted is blocked;

• reduce the codes of applicable operations when working with the system to the minimum level. For example, using a single utility for data backup. In other words, specific actions in the system should be performed by the only possible way, and not several;

• use of sets of interoperability rights that entitle only authorized transactions;

• completely exclude the use of private rights to obtain new ones belonging to other users. To do this, it is necessary to develop a private security policy (SP) as an addition to the Linux OS SP to comply with the above requirements;

• all network infrastructure, PCs and operating systems will be considered the security policy objects Everything else, personnel, software and equipment that directly interact with the objects are subjects of the SP.

2. Methods and materials

The modeling of concepts taking into account the possible harm of developed SP is often classified as the "violator". A violator is a subject with the right of access to objects or when trying to obtain it in an accidental or deliberate manner, in the future to use these rights for private purposes. In this case, to access privileges and handle data. The violator by its actions may not harm by making changes that interfere with the correct functioning of the system or change the settings of the system, but nevertheless this ability to carry out such activities remains, thereby posing a potential threat. It may be contemplated herein that any subject may be a potential violator. It should also be noted here that any subject that does not plan any malicious activities may be classified as the "violator". For example, a user may mistakenly be assigned the superuser rights, which he or she can accidentally use [3].

The human factor is one of the most important and pressing security issues. Here, even an experienced user from the support services that can use an identifier and a key with system administrator rights may be considered a "violator". In this case, the attributes of a user with higher access privileges are usually used. Weakly protected systems usually apply a security policy, where a user with low privileges uses their access rights to expand them. In this case, the "violator" may not deliberately compromise the system by leaving the software active after the end of the session, thereby allowing further access to passwords or IDs [4, 5]. The above actions are classified as a dangerous type of attack due to the fact that it is difficult to track. Besides, the threat of unauthorized access may be performed not only by the subject, but also by objects, and may manifest itself with an incorrectly configured operating system or application programs with incorrect algorithms. In this case, the threat comes from assigning erroneous rights to the default user, who should not have such rights according to the SP.

3. Results

The problem of emergency should be considered separately at the preliminary stages of SP development and be independent of the developed SP. In the future, these tasks will be closely interrelated and will have to be solved as a whole. In fact, the development of a safe UA system can become the main problem of eliminating or preventing emergency or critical situations, in particular if the actions are carried out in the manual control mode. Therefore, on the one hand, in case of emergency, the SP shall switch to the reduced safety mode so as not to interfere with the operational actions of the maintenance personnel. However, on the other hand, in an emergency state, the system will be in the most vulnerable state, so an attacker can create or falsify such a situation and take advantage of the moment, which may further lead to disastrous consequences.

Therefore, Remote Access or Remote Administration should only be implemented if absolutely necessary, because they are a major vulnerability to external network attacks.

A simple rule to improve security – the smaller the number of users in the system, the lower the risk of dangerous situations associated with improper assignment of rights. Based on this, it is necessary to take actions to reduce the number of subject, thereby reducing the risks for violator activity in relation to objects (network administrators, technical support staff with access rights to the system). Here, it is important to exclude the possibility of exposure of subjects/objects on behalf of other subjects/objects. The communication of objects via information channels is desirable with authentication of received and transmitted information. It is mandatory to number packets with their further encryption when transmitted to network objects. The receiving subjects shall decrypt packets and analyze counters. In many cases, it can be considered a failure if the values are not equal to the actual ones when transmitting a packet (calculated by the receiving party) or a network segment, a new device may appear, which may mean an intrusion attempt. Modern network devices (switches, gateways) allow automatically identifying unauthorized packets or packets with unknown network addresses.

The most reliable but difficult solution is the development of personal software, which has all the necessary tools and a single operator environment. Here, it is desirable to have a console with the right to access only favorite users. The developed software shall have all the system administration tools.

It is necessary to highlight the recommended stages in the development of such software:

• identifying elementary operations and compilation of their full list, execution of which will be carried out at the request of the personnel (operators);

• identifying and developing a complete list of activities to be performed when working with the system. Work refers to the sequential execution of interrelated operations. If any operation that is part of the required work cannot be performed by an operator, then this operator cannot perform the work;

• developing a complete list of responsibilities of maintenance personnel;

• grouping of work by individual responsibilities;

• assigning responsibilities to individual users (individuals).

The hierarchy is built during all stages, an example of which is shown in Fig. 1, where uniquely mapped operations, work, responsibilities and users themselves are given.

After building such a hierarchy, at the application software level it is possible to control – allow/prohibit the execution of all operations and works by a single operator. Ideally, the correctly designed system itself provides the operator with the means for which he has the rights.

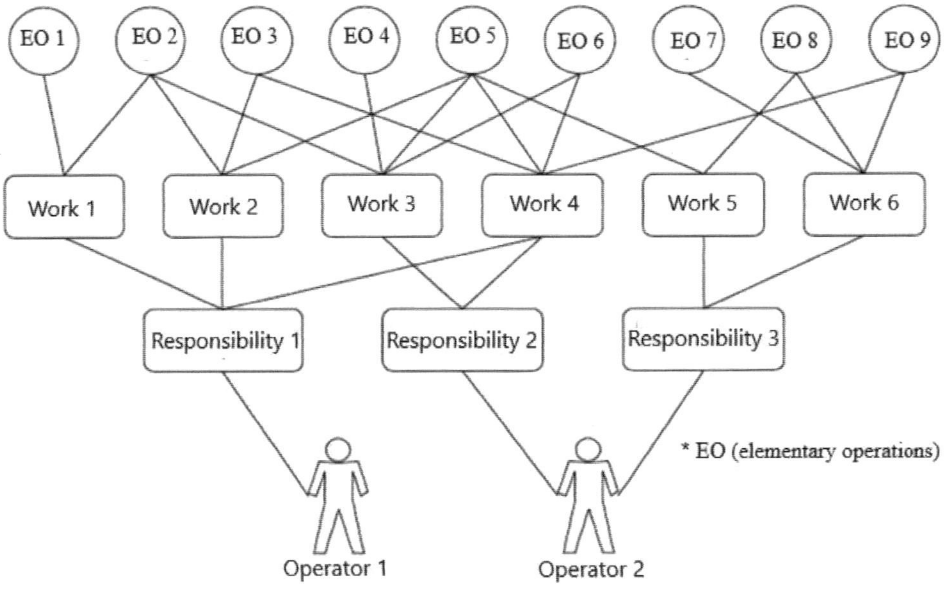

Figure 1. Distribution of rights and responsibilities among users

The proposed approach to the problem is unique since the hierarchy tree is developed once at the stage of software design and development, and is not subsequently modified while using the system. The main thing is that the SP does not require continuous support and the system administrator is not present in the hierarchy at all, thus being limited to personnel who change users' passwords as a password administrator. Such a person will not be able to harm the system, as in the case of using standard Linux SP, in which the system administrator can perform any actions (copying, deleting, editing or formatting).

In any case, the most vulnerable chain in such a system is the password administrator, since no one restricts it to assigning a password to any user and login to the system using his "login". There are many tools for preventing such actions, such as biometric data identification systems, for example, fingerprint scanners. The user ID shall correspond to two confirmations – password and fingerprint.

Login actions are carried out in stages: entering a "login", entering a password, and as a confirmation – a fingerprint. Such a system should store fingerprints of all users registered and having access to the object in its memory.

The main difference in this approach is the inability to act on behalf of someone else. Such applications can guarantee the security of the system. If such software cannot be developed, then the user is allowed to use only the software necessary for his work. This approach is quite common and all its disadvantages are listed above.

Restriction and access control to objects. All attempts to access objects must be stopped at the stage of processing network packets. Only packets with known MAC addresses must be sent over Ethernet networks. The appearance of an "alien" Ehternet packet in the network is a connection to the network of another device and the security system must respond accordingly with a corresponding signal. For full tracking of non-native packets, specialized software can be used, which function is to control all traffic on a certain segment of the network.

To prevent non-native packets from spreading to other segments of the network, the functionality of switches is also used, the configuration of which allows them setting up their internal inter-port forwarding and routing tables with sufficiently secure routing tables.

Prerequisites for creating a unified operator environment. Building a fairly easy-to-operate and UA-protected multi-user information system is quite difficult. This is mainly caused by the fact that in the initial stages of software design it is quite difficult to answer the following questions:

- which users the software is calculated for and what is the range of end users;
- how fully the system should support multi-user mode;
- what risks and where the UA threat may come from.

Linux functionality is designed in such a way that most developers and users who rely on this OS in their projects can solve any of their tasks based on standard SP. In other words, the operating system is universal, but there are many packages that provide a wide range of tasks and require specific conditions for their operation. The OS concept is based on a time-tested security policy.

However, the OS requires users of a rather professional level to work with standard tools included in the distribution, which is also not guaranteed in many cases.

The versatility of the system also poses a threat, with possible incorrect configuration of the system, since with a variety of configurations it is very difficult to take into account many factors. However, Linux followers argue that the simpler the system, and the smaller the use of programs developed by third-party manufacturers, the less likely it is to occur that failures and situations that arose due to software conflicts in the software are not taken into account at all stages of development.

The Linux security policy requires increased attention and support from the system administrator. Like any user, an administrator may make a mistake even having many rights in the system. Therefore, the presence of such a person with system rights is not desirable, but on the other hand it is necessary only when using Linux OS SP.

The problem of having an administrator is supplemented by the already expressed opinion about the presence of professional users at the facility.

The following recommendations can be considered to solve these problems:

1. Persist number of employees, and the employees rarely change. In this situation, there is no need to create new accounts, change access rights and privileges, which excludes the use of standard tools for configuring Linux OS SP and the use of qualified personnel.
2. At the facility, the responsibilities of all staff are strictly limited and cannot change over time, and each employee is assigned a unique set of software tools, which he can use only. In the case of a change of employee, the unique set of rights remains unchanged.
3. For the SP system, it is quite critical to support multi-user mode, i.e. only one person can work with the system at a time. Remote work with the system cannot be allowed.
4. Each technical support employee should use only software created by the system developer, where there is no need for standard Linux tools.

Linux functionality allows combining everything listed in the Unified Operator Environment (UOE). By using UOE, you can simplify and automate all possible activities. In this case, all users will not be able to go beyond the scope of the UOE and do not allow running standard administration utilities. The above features are not available to users.

UOE serves as some kind of intermediary between utilities and the user allowing only a certain user operating and to the extent authorized by the internal security policy of UOE. Each user of the system works only in a specially designed shell and uses only the means that this shell provides. No other tools are available to the user. The shell itself is presented in the form of windows with access to various tools and a strictly defined set of actions is performed working with a certain window. For all users, working in the system is presented as a session. Connecting to the system, the user opens a session, when he leaves the system, closes it. Another user can log in only if the previous user has closed the session, essentially only one user can work with the system. Only one window is available before the session is opened when the user performs the authentication. After confirming his identity, the shell is transferred to the appropriate mode. Self-tuning for the type of work performed by the connecting operator is meant by the mode.

The following are options for shell self-tuning:

• each user has a strictly defined list of rights that specifies the windows he can work with. At the same time, the windows that the user is not allowed to work with are not displayed;

• each user has a specific shell, which is previously created as a separate program and loaded immediately after the session is opened, or in a single program, a separate part of it is responsible for working with a specific user.

4. Conclusion

The study suggests the following recommendations on the design of the user interface:

1. Try to use as few windows as possible.
2. Group control elements in the window by their functionality.
3. The interface and its design shall prevent erroneous operator actions.
4. Correspond to a single interface element style.
5. It is recommended to avoid "highlighting".
6. Reduce the color palette as much as possible and use a strict stylistic character.
7. All application programs must be designed with a strictly defined screen resolution in mind.
8. Activation of any control must always result in unambiguous visual changes expected by the operator.
9. The status of all displayed control elements must correspond to the current state of the control elements.
10. Ensure confirmation to the user that the system has accepted his request and that the request has been completed correctly.
11. All fault or error notifications must be displayed in a short form (1-2 lines of text), and it must be possible to obtain detailed information on the error.
12. Avoid complex and intricate reference systems as much as possible.
13. Other windows should not be available until one window is closed.
14. Avoid the ubiquitous use of control elements with "Apply" response.
15. Remove the ability to drag windows. Upon activation it is desirable that each window should appear strictly at a specific location and have a predetermined, constant size.
16. The availability of the "Freeze" button is everywhere, the function of which is that after clicking it any input will be blocked until the password of the user who clicked the button is entered.
17. Log availability where all operator actions should be displayed and recorded.

References

[1] Klyuev R V, Bosikov I I, Mayer A V, Gavrina O A 2020 *Sustainable Development of Mountain Territories* **V.12 2(44)** 283-290

[2] Klyuev R, Madaeva M, Umarova M 2020 *International Ural Conference on Electrical Power Engineering (UralCon)* INSPEC Accession Number: **20034041** 356-361
[3] Bovkun A, Korodyuk I 2019 Analysis of the development of small innovative enterprises in the construction industry. *IOP Conf. Ser.: Mater. Sci. Eng.* **667**
[4] Skorobogatova Y A, Bovkun A S, Ivanov M Yu, Shilova O S 2020 The role and place of small innovative enterprises in the construction industry in the modern economy of single industry towns. *IOP Conf. Ser.: Mater. Sci. Eng.* **880**
[5] Repinskiy O D, Konyukhov V Yu, Bovkun A S, Schupletsov A F 2021 Improving the competitiveness of Russian industry in the production of measuring and analytical equipment. *Journal of Physics: Conference Series* **1728**

Technology for prediction of salt deposition in oil production

L E Lenchenkova[1], L Z Zainagalina[2], N D Bulchaev[3], Yu A Kotenev[1], F Z Bulyukova[1] and A R Safiullina[2]

[1] Ufa State Petroleum Technological University, 1, Kosmonavtov St., Ufa, Republic of Bashkortostan, 450062, Russian Federation
[2] Ufa State Petroleum Technological University, Branch of the University in the City of Oktyabrsky, 54a, Devonskaya St., Oktyabrsky, Republic of Bashkortostan, 452607, Russian Federation
[3] Grozny State Oil Technical University named after Academician M.D. Millionshchikov, 100, Kh.A. Isaev prospect, Grozny, Chechen Republic, 364051, Russian Federation

E-mail: lenchenkoval@mail.ru, info@of.ugntu.ru

Abstract. The formation of inorganic salts in the bottom-hole area leads to a deterioration of the reservoir filtration properties and a decrease in the productivity of oil wells. The conducted research shows that the decrease in rock permeability due to salt deposition during flooding mainly depends on the number of filtered pore volumes of the injected solution, the rate of injection, temperature, differences in pressure, initial permeability of the rock and the concentration of the solution. The development of an empirical formula for predicting salt deposits in the bottom-hole area and describing the dependence of rock permeability on these factors involves the use of reliable experimental and field data. The authors propose a method for determining the probability of precipitation of some inorganic salts in the well.

1. Introduction

A fundamental part of evaluating an oil field to control salt deposition is to predict the possibility and intensity of salt deposition and determine sedimentation during oil production. Prediction of hard mineral precipitation is one of the ways to control salt deposition. Therefore, it is necessary to forecast the conditions of salt deposition in the oil industry. The first stage of salt deposition control is the analysis of waters to determine their propensity to form salt deposits. Thus, to make a predictive assessment of salt deposition, it is necessary to know the chemical composition of water and the thermobaric conditions of the formation.

2. Materials and methods

There are many models for determining the probability of salt deposition. Available models for predicting salt deposition conditions are based on thermodynamic data on limited solubility [4, 10]. The corresponding thermodynamic model should determine the type of salt formed and the maximum amount of precipitation when mixing the injected and reservoir water. Determining the probability of precipitation of some inorganic salts in the well requires the use of reliable experimental and field data. The method proposed by the authors consists of developing an empirical formula for predicting salt deposits in the bottom-hole area and describing the dependence of rock permeability on the number of filtered pore volumes of the injected solution, the injection rate, temperature, pressure drop, initial rock permeability and solution concentration. The authors have developed a model based on multi-factor correlation and regression analysis for predicting a decrease in the permeability of BHA as a result of the deposition of mineral deposits (barium sulfate).

3. Results and Discussion

In [2], the ability of water to precipitate in oilfield equipment was determined by calculating the stability of injected and reservoir water based on the Stief-Davis approach. The saturation index (SI) of water with salts (calcium carbonate, barium sulfate, calcium sulfate, and strontium sulfate) was first proposed by Oddo and Thomson [6]. The proposed water salt saturation index was a polynomial equation that was a function of temperature, pressure, and ionic force. When determining the water saturation index with calcium carbonate, the concentration of carbon dioxide (CO2) in the solution was included in the formula. In [8], we present models and semi-empirical relationships based on experimental data that could determine the statistical probability of the formation of insoluble salts when water is injected into the reservoir using the saturation index. In [3], we have successfully determined the probability of precipitation of calcium carbonate in the system of reservoir pressure maintenance by applying a model based on the saturation index when mixing incompatible waters in the aquifer. Models often present the effect of temperature, pressure, and pH of the solution on the intensity of salt deposition. Thus, the possibility and prediction of the formation of salt deposits in static conditions are calculated by the saturation index specified in the following formula [6, 9]:

$$SI = \lg(\frac{[\mathrm{Ca}][An]}{K_{sp}}), \quad (1)$$

where SI is the saturation index, a dimensionless value; [Ca] and [An] are square brackets representing the molar concentration of cations and anions, mol/l; Ksp is the product of solubility at equilibrium, mol/l.

In formula (1), the numerator (the product of two concentrations) is an indicator of ion activity. Ksp depends on the thermobaric conditions and the ionic strength of the solution (I), expressed as follows [8]:

$$pK_{sp} = -lg(K_{sp}), \quad (2)$$

$$SI = -\lg([Ca][An]) + pK_{sp}, \quad (3)$$

$$pK_{sp} = a_1 + a_2T + a_3T^2 + a_4P + a_5I^{0,5} + a_6I^{0,5}T + a_7I, \quad (4)$$

$$SI = \lg([Ca][An]) + a_1 + a_2T + a_3T^2 + a_4P + a_5I^{0,5} + a_6I^{0,5}T + a_7I, \quad (5)$$

where T is the temperature, 0F; P is the pressure, per square inch (psi); a1, a2, a3, a4, a5, a6, and a7 are the empirical coefficients (Table 1), I is the ionic force, mol/l.

If SI>0, salt deposition occurs, and if SI<0, there is no risk of salt deposition [1].

Table 1 shows the values of the empirical coefficients for sulfate salts.

Table 1. Empirical coefficients

Salt	a1	a2	a3	a4	a5	a6	a7
$CaSO_4 \cdot 2H_2O$	3.47	$1.8 \cdot 10^{-3}$	$2.5 \cdot 10^{-6}$	$-5.9 \cdot 10^{-5}$	-1.13	0.37	$-2.0 \cdot 10^{-3}$
$CaSO_4 \cdot H_2O$	4.04	$-1.9 \cdot 10^{-3}$	$11.9 \cdot 10^{-6}$	$-6.9 \cdot 10^{-5}$	-1.66	0.49	$-0.66 \cdot 10^{-3}$
$CaSO_4$	2.52	$9.98 \cdot 10^{-3}$	$-0.97 \cdot 10^{-6}$	$-3.07 \cdot 10^{-5}$	-1.09	0.50	$-3.3 \cdot 10^{-3}$
$SrSO_4$	6.11	$2.0 \cdot 10^{-3}$	$6.4 \cdot 10^{-6}$	$-4.6 \cdot 10^{-5}$	-1.89	0.60	$-1.9 \cdot 10^{-3}$
$BaSO_4$	10.03	$-4.8 \cdot 10^{-3}$	$11.4 \cdot 10^{-6}$	$-4.8 \cdot 10^{-5}$	-2.62	0.89	$-2.0 \cdot 10^{-3}$
$MgSO_4$	2.30	$1.74 \cdot 10^{-3}$	$4.55 \cdot 10^{-6}$	$-7.8 \cdot 10^{-6}$	2.28	-0.46	$-0.60 \cdot 10^{-3}$

For calcium carbonate the saturation index is defined as follows:

$$SI = \lg([Ca^{2+}][CO_3{}^{2-}]) + pH - 2{,}76 + 9{,}88 * 10^{-3}T + 0{,}61 * 10^{-6}T^2 - 3{,}03 * 10^{-5}P - 2{,}348\, I^{05} + 0{,}77\, I \quad (6)$$

When determining the water saturation index of calcium carbonate by formula 6, the value of the hydrogen index is determined depending on the presence or absence of a gas phase in the system. However, it is difficult to determine the exact value of the hydrogen index due to theoretical and practical limitations [5].

There are various software applications for predicting hard mineral precipitation and getting an idea of the chemical equilibrium that characterizes water systems. The computer program provides the prediction of salt deposition based on thermodynamics for determining physical and chemical properties in multiphase water systems. This basis applies to most multicomponent mixtures and allows predicting the likely possibility of salt deposition at any values of temperature, pressure, and ion concentration. Some authors have developed a computer program for predicting the probable possibility of salt deposition and determining the water saturation index by applying reservoir temperature and pressure.

To determine the effect of the liquid-vapour equilibrium on the intensity of salt deposition, the PHREEQC program was used, which simulates the conditions of salt deposition during degassing with different initial concentrations of dissolved gas. They concluded that it is necessary to choose the best liquid-vapour equilibrium model for simulating the prediction of salt deposition, especially in the presence of gas mixtures. Some programs give values for the tendency to form salt deposits (the ability of water to form deposits). However, the tendency to form salt deposits (ST) and the saturation index (SI) of water are strongly related to each other. There is the following relationship between them:

$$SI = \lg(ST)\,, \quad (7)$$

To predict changes in rock permeability due to salt deposition, it is necessary to develop a complex correlation, which is the parameters that affect the degree of deterioration of the reservoir properties of the BHA. Many scientists have studied the deposition of inorganic salts in porous media to develop mathematical models that can predict a decrease in the permeability of rocks in reservoir conditions. The development of models to describe the process of salt deposition in a porous medium when mixing reservoir and injected water, including dispersion/diffusion modelling, leads to the prediction of salt deposition in the BHA.

Since the process of salt deposition is caused by the mixing of two liquids, it is necessary to determine the main parameters in the "reaction-diffusion-advection" control to develop an asymptotic model to determine the deposition rate in porous media. The resulting model can be compared with a numerical model to verify the accuracy of the prediction of a decrease in the permeability of the BHA due to salt deposition. Based on the laboratory data, some empirical formulas are proposed for predicting the decrease in permeability due to salt deposition during flooding. Most of these studies focused on only one type of salt during the water injection process. Some authors have developed a mathematical model [11, 12] for predicting a decrease in permeability due to simultaneous deposition of several types of inorganic salts based on thermodynamics, kinetics and hydrodynamics when a liquid moves in a porous medium. It was also noted that obtaining laboratory data on the formation of salt deposits makes it possible to increase the accuracy of predicting the precipitation of solid mineral precipitation in porous media.

Modelling the deposition of one type of salt is not a difficult task, but the simultaneous deposition of several types of salts and, respectively, the modelling of these processes is a difficult task due to the complexity of the water system and the interaction of different ions during salt deposition. The complexity of the crystallization process limits most of the developed models for porous media since the deposition of one salt does not pay enough attention to the co-deposition of other salts, especially those that have common ions. In the presence of several types of salts and their co-deposition in porous media, each salt is deposited according to its scenario, which differs from the others. Therefore, the developed models for predicting the risk of deterioration of the permeability of the BHA as a result of precipitation of a single salt can not give correct results when several salts fall out simultaneously. This happens mainly because the co-deposition of inorganic salts and the formation of their mixture affect the thermodynamic and kinetic behaviour of each salt in the mixture. Deposits in oil fields are rarely pure and often occur as mixtures of two or more inorganic compounds with the presence of paraffin, silicon dioxide, or some other impurities. They further complicate the process of salt deposition and, respectively, the development of technological schemes to control them.

The conducted research shows that the decrease in rock permeability due to salt deposition during flooding mainly depends on the number of filtered pore volumes of the injected solution, the rate of injection, temperature, pressure drop, initial permeability of the rock and the concentration of the solution. The development of an empirical formula for predicting salt deposits in BHA and describing the dependence of rock permeability on these factors involves the use of reliable experimental and field data. However, the accurate measurement of experimental data is almost impossible, and some incorrect and undesirable measurements can still lead to significant distortions of the results.

Recently, a new model has been proposed using the MATLAB program based on an annealing simulation algorithm to predict the decrease in the permeability coefficient caused by the deposition of inorganic salts in porous media. In [7], the authors conducted some experimental and theoretical studies to examine the decrease in permeability due to the deposition of calcium sulfate in porous media. It is emphasized that when studying the decrease in permeability as a result of salt deposition, it is necessary to systematically take into account various important parameters that affect this complex process to one degree or another. A new formula with high accuracy for determining the permeability coefficient as a result of the formation of salt deposits is proposed. The formula developed by the authors is based on data obtained in models of carbonate-type cores at low pressures. A slight difference between the standard deviations of the formula and the experimental data confirms the validity of the formula. The developed formula for estimating the decrease in permeability of a porous medium is expressed by the following equation [7]:

$$\frac{K_C}{K_I} = \exp(-PV_{inj}\frac{f_0 A\left[\left(\frac{T}{40}\right)(C_{sat}-1)\right]^{\frac{1}{2}}}{bq}), \qquad (8)$$

where K_C – the current permeability (after salt deposition), microns; K_I – the initial sample permeability, microns; PV_{inj} – number of filtered pore volumes, m/m; f_0–the initial core porosity, unit fraction; A – the cross-sectional area, cm^2; T – temperature, °C; C_{sat} – the degree of saturation of water with calcium sulfate, unit fraction; b – the empirical coefficient, min/cm; q – the rate of injection, cm^3/min.

4. Conclusion

The authors have developed a model based on multi-factor correlation and regression analysis for predicting a decrease in the permeability of BHA as a result of the deposition of mineral deposits (barium sulfate). The model is developed after analyzing experimental data and statistical analysis, and is interpreted as follows:

$$\exp\left(\frac{K_C}{K_I}\right) = b_1 + b_2 T^{0,17} + b_3 (qt)^{0,25} + b_4 \ln(V_{pore}) + b_5 \ln(C_K) + b_6 \ln(\Delta P), \qquad (9)$$

where K_C is the current permeability of the sample (after salt precipitation), microns; K_I is the initial permeability of the sample, microns; b_1, b_2, b_3, b_4, b_5, *and* b_6 *are* the empirical coefficients the values of which depend on the concentration of barium; T – the reservoir temperature, °F; q – injection rate, ft/min; t – the time of injection, min; V_{pore} – the volume of pore space, cubic ft; C_K is the concentration of barium sulfate, mg/l; ΔP – the differential pressure, pound per square inch (Pa).

The statistical analysis of the results shows that at medium and high concentrations of barium, the model accurately estimates the decrease in rock permeability with determination coefficients of 0.941 and 0.948, respectively. The analysis of graphical errors also shows that the model corresponds to experimental data.

References

[1] Garda A V, Thomsen K and Stenby E H 2006 Prediction of Mineral Scale Formation in

Geothermal and Oilfield Operations Using the Extended UNIQUAC Model: Part II. Carbonate-Scaling Minerals *Geothermics* **35(3)** 239-284. DOI: http://dx.doi.org/10.1016/fgeothermics.2006.03.001

[2] Rogachev M K and Mukhametshin V V 2018 Control and Regulation of the Hydrochloric Acid Treatment of the Bottom-Hole Zone Based on Field-Geological Data *J. of Mining Institute* **231** 275-280. DOI: 10.25515/PMI.2018.3.275.

[3] Zainagalina L Z, Suleimanov R I, Gabdrakhimov M S and Khabibullin M Ya 2018 Determining Oscillating System Dynamic Parameters of a Near-Bit Junk Pulper *Advances in Engineering Research* **157** 642-645

[4] Shaidullina R M, Amirov A F, Muhametshin V Sh and Tyncherov K T 2017 Designing Economic Socialization System in the Educational Process of Technological University *European J. of Contemporary Education* **6(1)** 149–158. DOI: 10.13187/ejced.2017.1.149

[5] Tahmasebi H A, Kharrat R and Soltanieh M 2010 Dimensionless Correlation for the Prediction of Permeability Reduction Rate due to Calcium Sulphate Scale Deposition in Carbonate Grain Packed Column *J. of the Taiwan Institute of Chemical Engineers* **41(3)** 268-278. Retrieved from: http://dx.doi.Org/10.1016/i.itice.2009.11.006

[6] Khabibullin M Ya, Samigullina L Z and Zainagalina L Z 2018 Laboratory Studies for the Momentum of Liquid Optimal Impact on Hydro-Carbon Formation *Advances in Engineering Research* **157** 276-279

[7] Jordan M M, Kemp S, Sorhaug E et al 2003 Effective Management of Scaling From and Within Carbonate Oil Reservoirs, North Sea Basin *Chemical Engineering Research and Design* **81(3)** 359-372. Retrieved from: http://dx.doi.org/10.1205/02638760360596919

[8] Stenkin A V, Kotenev Yu A, Mukhametshin V Sh and Sultanov Sh Kh 2019 Use of Low-Mineralized Water for Displacing Oil From Clay Productive Field Formations *IOP Conf. Ser.: Mater. Sci. Eng.* **560(1)** 012202. DOI: 10.1088/1757-899X/560/1/012202

[9] Almukhametova E M, Shamsutdinova G F, Sadvakasov A A, Tyncherov K T, Petrova L V and Stepanova R R 2018 Modelling development of Fyodorovsky deposit *IOP Conf. Ser.: Mater. Sci. Eng.* **327(4)** 042100. DOI: 10.1088/1757-899X/327/4/042100.

[10] Kan A T and Thomson M B 2012 Scale Prediction For Oil and Gas Production *SPE J.* **17(2)** 362-378. Retrieved from: http://dx.doi.org/10.2118/132237-PA

[11] Kuleshova L S, Kadyrov R R, Mukhametshin V V and Safiullina A R 2019 Design Changes of Injection and Supply Wellhead Fittings Operating in Winter Conditions *IOP Conf. Ser.: Mater. Sci. Eng.* **560(1)** 012072. DOI: 10.1088/1757-899X/560/1/012072

[12] Kuleshova L S, Kadyrov R R, Mukhametshin V V and Akhmetov R T 2019 Auxiliary Equipment for Downhole Fittings of Injection Wells and Water Supply Lines Used to Improve Their Performance in Winter *IOP Conf. Ser.: Mater. Sci. Eng.* **560(1)** 012071. DOI: 10.1088/1757-899X/560/1/012071

Selection of the drilling fluid density as specified with the expected deformation of the cross-section of the well

L Z Zainagalina[1] and F Z Bulyukova[2]

[1] Ufa State Petroleum Technological University, Branch of the University in the City of Oktyabrsky, 54a, Devonskaya St., Oktyabrsky, Republic of Bashkortostan, 452607, Russian Federation

[2] Ufa State Petroleum Technological University, 1, Kosmonavtov St., Ufa, Republic of Bashkortostan, 450062, Russian Federation

E-mail: flura2003@mail.ru, info@of.ugntu.ru

Abstract. During inclined and horizontal drilling, the cross-section of the well is deformed in the form of uneven displacement of its walls, which is one of the reasons for tightening and tool sticking. To limit the displacement of the walls, it is proposed to adjust the density of the drilling fluid taking into account the elastic properties of the rock and the angle of curvature of the well.

1. Introduction

Nowadays, the choice of drilling fluid density is carried out under the conditions for preventing three types of complications: the inflow of reservoir fluids; the opening of the absorption of drilling fluid and the instability of the well walls [1, 5, 6]. The condition for stability of the well walls is based on strength calculations of rocks.

In [2] it is shown that ensuring the stability of the well walls is a necessary condition for preventing complications caused by deformation and destruction of the well walls, but it is not sufficient. When drilling, the bit forms a round borehole with a radius $R_n \approx R_b$ (R_b - the radius of the bit), but as the bit moves away from the fixed cross-section of the wellbore, its deformation occurs, changing not only the size but also the shape of the section, as shown in figure 1.

At the same time, the more the angle α of curvature of the well, the more the shape of the section changes (in figure 1, section *A-A*), becoming more elliptical. This statement is confirmed by viewing the cavernograms of inclined wells recorded by the cavernometer in two mutually perpendicular directions. Examples of such recording in wells in Western Siberia are shown in figure 2. The interval is composed mainly of heavy clays, for which the Poisson's ratio μ is assumed to be 0.35 [3].

Figure 2 shows that the dimensions of the well in two mutually perpendicular directions differ significantly. In this case, the radius of the well can be either more or less than the nominal one.

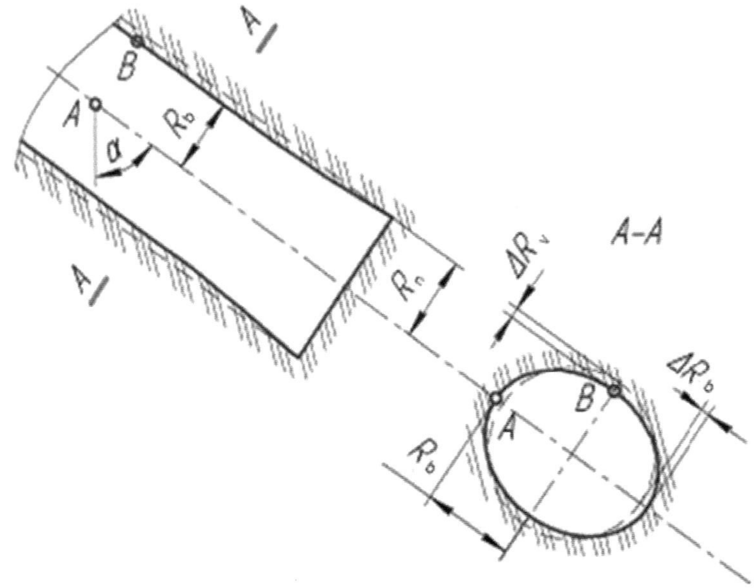

Figure 1. The bottom-hole part of an inclined well

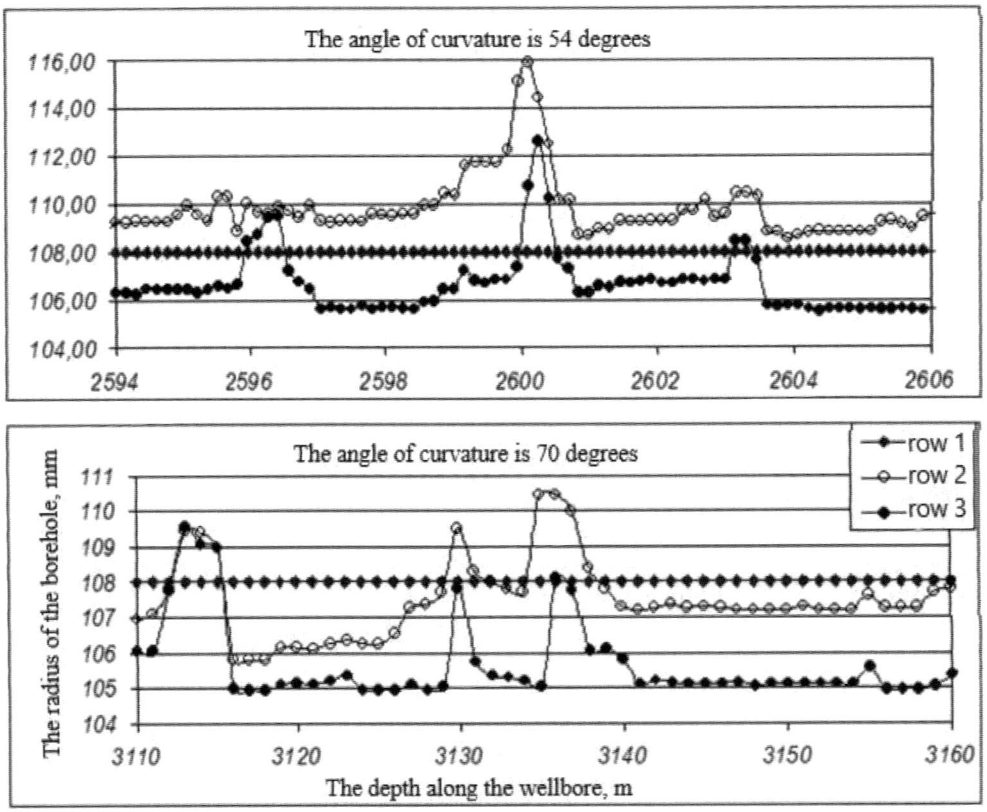

Figure 2. The cavernograms of inclined wells recorded by a cavernometer in two mutually perpendicular directions: row 1 – nominal radius; rows 2 and 3 – measured radii of the well

As the angle of curvature increases, the displacement of the lateral wall ΔR_b of the well decreases and changes sign (rows 2), and the displacement ΔR_v of the upper wall monotonously increases, reducing the size of the well (rows 3). This displacement of the walls above the bit can cause tightening and even jamming of the bit when lifting from the well. Therefore, the measures should be taken to prevent this phenomenon.

2. Materials and methods

The solution of the problems was performed analytically with the formulation and conduct of special experiments and the involvement of published results of testing of rocks on well models, as well as the results of cavernometry of inclined wells in Western Siberia.

3. Results and Discussion

The main means of controlling the stress-strain state of the rock composing the well wall is the selection of the required density $\rho_{d.f.}$of the drilling fluid. In vertical wells, there was almost no need to take into account the displacement of the well walls. But as the angle of curvature increases and, especially in cases of the horizontal borehole, neglect of this phenomenon can lead to serious complications. In [4], a program for calculating the displacement of the walls of an inclined well is presented. Using this program, the values of the actual lateral expansion coefficient (λ = 0.69) and the modulus of deformation (C = 1,870 MPa) for the range of 3,140-3,160 m (see figure 2) were estimated using the cavernogram as one of the dangerous ones from bit jamming.

The obtained rock characteristics were used to calculate the dependence of the displacement $\Delta R$of the well wall on the ratio r_s/r_g (r_s– static pressure of the drilling fluid in the well; r_g– geostatic pressure) at different curvature angles of the well and to assess the influence of the well curvature angle and geostatic pressure on the choice of the drilling fluid density. The calculations were performed at three levels of geostatic pressure: 57; 70 and 100 MPa. The r_g value = 57 MPa corresponded to an interval of 3,140-3,160 m (see figure 2), as well as at three levels of well curvature: 0 (vertical well); 45 (inclined well) and 90 degrees (horizontal well).

Figure 3 shows the obtained dependencies of ΔR on r_s/r_g.

Figure 3 shows that when $r_s/r_g = 0$, the displacement is negative and the largest in absolute value. As the r_s/r_g ratio increases, the displacement changes linearly, reaches zero and then becomes positive. As the angle of curvature increases, the difference between the upper and lateral wall displacements increases.

From complications, the main role is played by the displacement of the upper wall. Ideally, the displacement of the upper wall should be zero, and the displacement of the lateral wall should be positive [7, 8]. The second condition can be assumed to be equal to the module of displacements of the upper and lateral walls. Finally, the third condition is to accept the maximum allowable amount of displacement of the upper wall of the well. For roller-cone bits, the value of the roller plugs can be taken as an acceptable value [9-12]. For example, for bits with a diameter of 215.9 mm, this value is 2.5 mm.

The equations in figure 3, describing the dependence of ΔR on r_s/r_g, allow calculating the values of r_s/r_g and the corresponding density $\rho_{d.f.}$of drilling fluid for the accepted conditions at a known average density ρ of the overlying rocks, which was about 2.4.

It is known that

$$\frac{p_s}{p_g} = \frac{\rho_{d.f.} g h_d}{\rho g h_d} = \frac{\rho_{d.f.}}{\rho}, \tag{1}$$

g – acceleration of gravity; h_d – the depth of occurrence of the studied rock. So

$$\rho_{d.f.} = \frac{p_s}{p_g}\rho. \tag{2}$$

IOP Conf. Series: Materials Science and Engineering 1111 (2021) 012069 doi:10.1088/1757-899X/1111/1/012069

Figure 3 shows that the first condition for this example can only be implemented for a vertical well (r_s/r_g = 0.69; $\rho_{d.f.}$ = 1.66). With increasing α, these values grow rapidly.

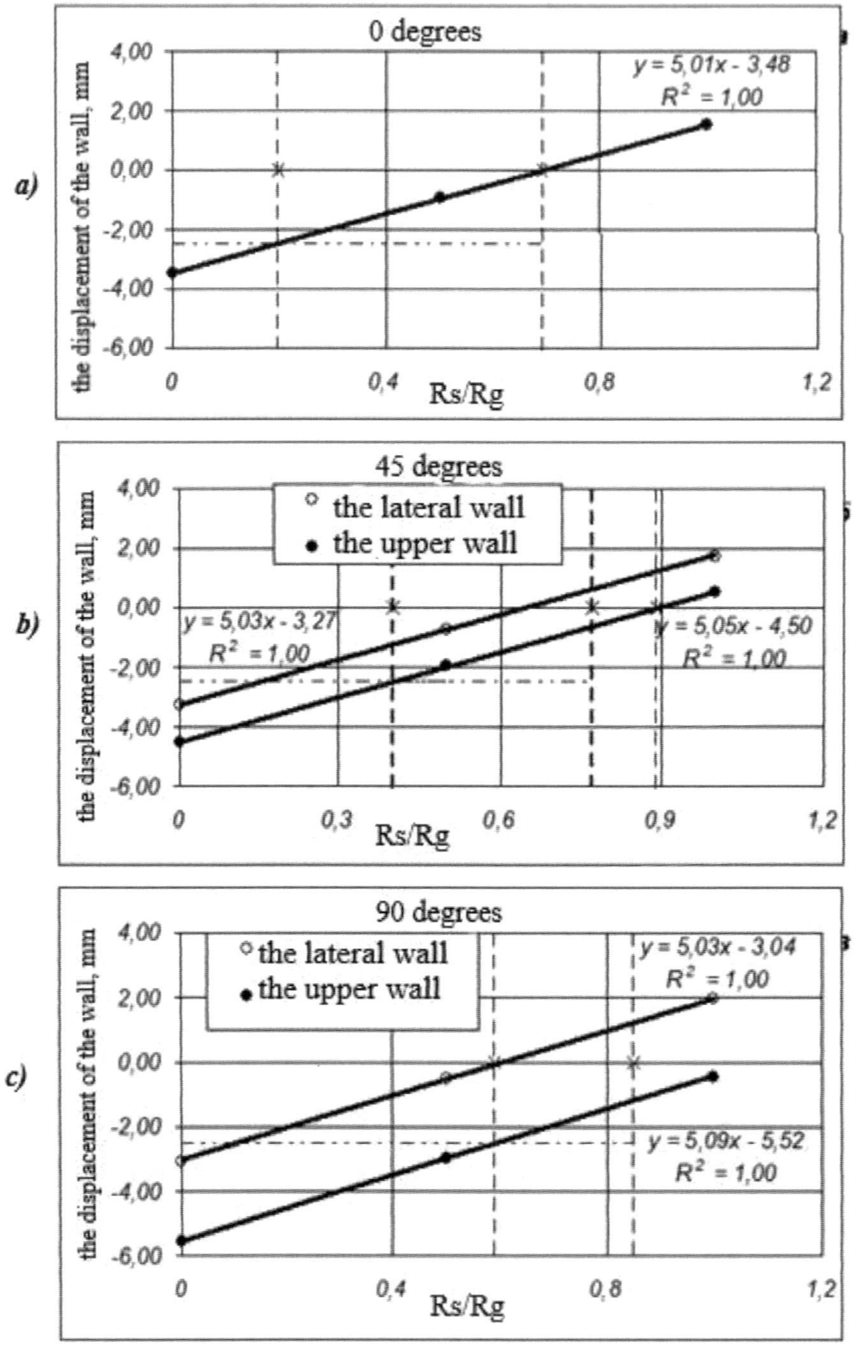

Figure 3. The dependencies of ΔR on r_s/r_g and their corresponding equations

Table 1 shows the results of calculating the density of drilling fluid for two other conditions.

Table 1. The recommendations for the density of the drilling fluid

α, °	The drilling fluid density is not less than for the conditions					
	r_g = 57 MPa		r_g = 70 MPa		r_g = 100 MPa	
	ΔR_{uw} = = -2.5 mm	ΔR_{uw}= =-ΔR_{lw}	ΔR_{uw} = = -2.5 mm	ΔR_{uw}= =-ΔR_{lw}	ΔR_{uw} = = -2.5 mm	ΔR_{uw}= =-ΔR_{lw}
0	0.48	-	0.70	-	1.00	-
45	0.96	1.85	1.18	1.85	1.47	1.85
90	1.42	2.04	1.66	2.04	1.94	1.99

4. Conclusion

The table and figure 3 demonstrate that the lower limit of the drilling fluid density for all three conditions depends significantly on the angle of curvature of the well. The drilling fluid densities for the first and second conditions do not depend on the geostatic pressure value, while for the third condition the drilling fluid density depends on all the factors considered. The third condition gives the smallest restrictions on the density of the drilling fluid.

Consider the provision of the third condition depending on the angle of curvature of the well. These data of this table allow compiling the dependency ρd.f. on α. This dependence is linear and for Rg = 57 MPa has the form

$$\rho d.f.= 0{,}0104\alpha + 0{,}483, \quad (3)$$

where αis the angle of curvature in degrees. For example, for the drilling interval 3,140-3,160 m considered in the example (see figure 2), the value of the drilling fluid density from the third condition is

$$\rho d.f.= 0{,}0104\cdot 70 + 0{,}483 > 1{,}21.$$

This interval was drilled with a fluid density of 1.14. At the same time, the displacement of the upper wall of the well was about 3 mm. The well report shows the tool tightening in this drilling interval.

References

[1] Khisamov R S, Abdrakhmanov G S, Kadyrov R R and Mukhametshin V V 2017 New Technology of Bottom Water Shut-off *Oil Industry* **11** 126–128. DOI: 10.24887/0028-2448-2017-11-126-128

[2] Malyarenko A M, Bogdan V A, Kotenev Yu A, Mukhametshin V Sh and Umetbaev V G 2019 Wettability and Formation Conditions of Reservoirs *IOP Conf. Ser.: Earth Env.* **378(1)** 012040**.** DOI: 10.1088/1755-1315/378/1/012040

[3] Suleimanov R I, Zainagalina L Z, Khabibullin M Ya, Zaripova L M and Kovalev N O 2018 Studying Heat-Affected Zone Deformations of Electric Arc Welding *IOP Conf. Ser.: Mater. Sci. Eng.* **327(3)** 032053. DOI: 10.1088/1757-899X/327/3/032053.

[4] Petrova L V and Yarullin D R 2019 Evaluation of the Effect of Asphalt Resin Paraffin Deposits on Oil Well Performance *IOP Conf. Ser.: Mater. Sci. Eng.* **560(1)** 012084. DOI: 10.1088/1757-899X/560/1/012084.

[5] Polyakov V N, Chizhov A P, Kotenev Yu A and Mukhametshin V Sh 2019 Results of System Drilling Techniques and Completion of Oil and Gas Wells *IOP Conf. Ser.: Earth Env.* **378(1)** 012119. DOI: 10.1088/1755-1315/378/1/012119

[6] Zainagalina L Z, Petrova L V and Petrov V A 2019 Analysis of the Efficiency of Telemetric Systems for Drilling Wells *IOP Conf. Ser.: Mater. Sci. Eng.* **560(1)** 012096. DOI: 10.1088/1757-899X/560/1/012096.

[7] Nie Z, Liu Z and Sun X A 2018 Multirelay Cooperation Method for Wireless Transmission of MWD and LWD Signals *IEEE Transactions on Geoscience and Remote Sensing* **56(3)** 1229-1237

[8] Saikia K 2018 A proposed methodology of 3D geomodeling-while-geosteering for optimum

horizontal well placement and enhanced geological risk management *Indian J. of Geo-Marine Sciences* **47(4)** 826-830

[9] Gillen M E, Moody B and Dymmock S 2018 New LWD technology provides high-resolution images in oil- and water-based muds for improved decision making in real-time *Proc. of the Annual Offshore Technology Conf.* **4** 2617-2626

[10] Liu Z, Nie Z, Sun X, Wen D and Tan J 2017 A low-frequency forward-looking antenna array for LWD and MWD *Progress in Electromagnetics Research Symp.* pp 151-154

[11] Rogachev M K and Mukhametshin V V 2018 Control and Regulation of the Hydrochloric Acid Treatment of the Bottom-Hole Zone Based on Field-Geological Data *J. of Mining Institute* **231** 275-280. DOI: 10.25515/PMI.2018.3.275.

[12] Kuleshova L S, Kadyrov R R, Mukhametshin V V and Akhmetov R T 2019 Auxiliary Equipment for Downhole Fittings of Injection Wells and Water Supply Lines Used to Improve Their Performance in Winter *IOP Conf. Ser.: Mater. Sci. Eng.* **560(1)** 012071. DOI: 10.1088/1757-899X/560/1/012071

AUTHOR INDEX

Akberov, K Ch ... 237
Akhmetshin, A R ... 1, 27
Akhmetshin, D A ... 307
Aksenova, N A ... 9
Ali, M S ... 15
Andreev, A M ... 148
Andryashina, N S ... 199
Antipin, D Ya ... 69, 75
Astafyeva, N A ... 20
Astashkov, N P ... 1, 27
Balanovskiy, A E ... 288
Balli, M Kara ... 143
Bandurin, A P ... 39
Bandurin, M A ... 39
Bandurina, I P ... 39
Baranov, A Y ... 148
Bataev, D K-S ... 135, 219
Bazhanov, A P ... 45
Beglyarov, D S ... 15
Belyaev, A N ... 53
Beregova, G M ... 397
Blinov, A S ... 61
Bolkov, Ya S ... 307
Bondarenko, D A ... 69, 75
Boyarshinov, M G ... 90
Bozhko, A N ... 99
Bulchaev, N D ... 436
Bulyukova, F Z ... 436, 441
Cherney, O T ... 107
Chernyakov, M K ... 237
Daudov, I M ... 163, 192, 282
Daudov, Kh A ... 113, 153, 330
Davydov, A Yu ... 121, 374, 405
Deren, E A ... 127
Devyanin, S N ... 82
Dobretsov, R Yu ... 418
Dobretsova, S B ... 418
Druzhinina, T Ya ... 135, 393
Dudareva, O V ... 265
Filippov, A A ... 270
Galimullin, M L ... 168
Galkin, V V ... 270
Garina, E P ... 199
Gavrilova, Zh L ... 153, 163, 282
Gorban, A V ... 229, 330
Gorelenkov, A I ... 75
Gozbenko, V E ... 20, 288
Grinkrug, M S ... 143
Guseva, E A ... 179, 184
Guzueva, E R ... 430
Iakovkina, T N ... 163
Ignatev, I V ... 192, 229
Ivanov, L V ... 148
Ivanov, M Yu ... 397
Ivashchenko, G A ... 219
Kamaltdinov, V G ... 82
Kargapol'tsev, S K ... 20, 288
Karlina, A I ... 1, 20, 27, 288
Karlina, Yu I ... 20, 288
Khabibullin, M Ya ... 168, 374, 381, 405
Kharlamova, I Yu ... 237
Kirillov, N ... 345
Klochkova, E ... 135, 219
Kolosov, A D ... 179, 184
Komarnicki, P ... 339
Komarova, T K ... 174
Kondrat'ev, V V ... 179, 184
Konoval, D V ... 127
Konstantinova, M V ... 179, 184
Korabelnikov, M I ... 9
Korchevskaya, Yu V ... 15
Kostin, S V ... 174
Kotenev, Yu A ... 436
Kovalevich, M V ... 352
Kozlova, E P ... 199
Kudashkin, V A ... 282
Kuleshova, L S ... 243, 260
Kurasov, V S ... 314
Kurochkin, V N ... 206
Kutepova, L I ... 107
Kuvshinov, A A ... 211
Kuznetsova, S N ... 199
Lenchenkova, L E ... 436
Listkov, V Yu ... 237
Maksimov, N A ... 229
Malastowski, N S ... 61
Malozyomov, B V ... 237
Markov, V A ... 82
Martyakhin, D S ... 174
Matveev, I A ... 424
Mihkailova, O ... 345
Mikhaylova, S V ... 301
Mikheyev, N V ... 53
Mingulov, I Sh ... 243, 249
Mingulov, Sh G ... 243, 249
Minnivaleev, A N ... 254

Minnivaleev, T N 254
Mukhametshin, V Sh 326
Mukhametshin, V V 243, 260
Mukhina, M V 270
Myagkov, L L 61
Nadrshin, V V 113
Nazaretov, A A 275
Nazarkin, E E 15
Nguyen, Van Chie 288
Nikitchenko, S L 206
Novgorodov, N A 143
Novikova, G 345
Olentsevich, V A 1, 27
Orobey, M N 192
Orobinsky, V I 53
Pachurin, G V 270
Pachurin, V G 270
Patrusova, A M 113
Permovskiy, A A 107
Petrov, V A 121
Pitukhin, E A 295
Pogrebnaya, I A 301
Pokhomchikova, E O 157
Polovneva, S I 430
Pospelov, P I 174
Prikhodko, Iu S 307
Priporov, I E 314, 320, 386
Prosviryakova, M 345
Rakhimov, N 326
Rodionova, E N 424
Romanovskaya, E V 199
Sa, B 82
Safiullina, A R 436
Sakharov, V A 211
Samurganov, E E 314
Saricheva, E N 237
Saveleva, N N 334
Shamanov, R S 45
Shamarova, N 339
Shamin, E 345
Shapshal, A S 359
Shapshal, S A 359
Shatsky, V P 53
Shcherbak, N A 206
Shed, S N 334
Shemonaeva, E S 352
Shepelev, A B 314, 320, 386
Sheredekin, V V 53
Shevkunova, A V 359
Shishkina, O A 69
Shishkov, A N 307
Shtayger, M G 1, 20, 27, 288
Shuraev, I A 237
Skachkova, E G 107
Skrypnik, V I 295
Smirnova, Zh V 107, 270
Starokon, I V 367
Storchevoj, V 345
Suleimanov, R I 168, 374, 381, 405
Suslov, K V 1, 27
Svirbutovich, O A 219
Sygotina, M V 157
Talakhadze, T Z 359
Tkacheva, J I 143
Trishina, T V 53
Trotsenko, I A 15
Tsvetkov, V A 148
Turluyev, R R 265, 393, 397
Tyncherov, K T 326
Usmanova, F K 412
Usmanova, L Z 412
Vakhrusheva, M Yu 135
Vasilenok, A V 148
Vasiliev, A I 418
Vavilin, A S 90
Volkhonsky, A E 127
Vorobiev, V I 69, 75
Vysotskaya, I A 53
Wenge, C 339
Yaitskov, I A 275
Yaitskova, N M 275
Yamnikov, A S 424
Yurchenko, I F 39
Zainagalina, L Z 381, 436, 441
Zaitsev, S 345
Zaripova, L M 121, 374, 381, 405
Zavrazhnov, A I 53
Zhigalov, K Yu 265, 330, 393, 430
Zhigalov, K 135, 157, 219, 229
Zhovtikhanov, D S 9
Zinoviev, A A 153